Visio® 2007
Bible

Visio® 2007 Bible

Bible

Bonnie Biafore

Wiley Publishing, Inc.

Visio® 2007 Bible

Published by
Wiley Publishing, Inc.
10475 Crosspoint Boulevard
Indianapolis, IN 46256
www.wiley.com

Copyright © 2007 by Wiley Publishing, Inc., Indianapolis, Indiana

Published simultaneously in Canada

ISBN: 978-0-470-10996-0

Manufactured in the United States of America

10 9 8 7 6 5 4 3 2 1

For general information on our other products and services or to obtain technical support, please contact our Customer Care Department within the U.S. at (800) 762-2974, outside the U.S. at (317) 572-3993 or fax (317) 572-4002.

Library of Congress Cataloging-in-Publication Data
Biafore, Bonnie.
  Visio 2007 bible / Bonnie Biafore.
    p. cm.
  Includes index.
  ISBN-13: 978-0-470-10996-0 (paper/website)
  ISBN-10: 0-470-10996-3 (paper/website)
  1. Computer graphics. 2. Microsoft Visio. I. Title.
  T385.B523 2007
  006.6—dc22
                        2006101153

About the Author

Bonnie Biafore is a project management consultant, although her career as an author often assigns her one of her toughest tasks—managing herself. Bonnie writes about personal finance, investing, project management, and the technology that makes these endeavors easier to deal with. As an engineer, she's steadfastly—sometimes excruciatingly—attentive to detail. Fortunately, for her readers, she demystifies complex topics and occasionally transforms sleep-inducing material into educational entertainment. Her books, including the *NAIC Stock Selection Handbook, QuickBooks 2006: The Missing Manual,* and *On Time! On Track! On Target!* have won awards from organizations such as the Society of Technical Communication and APEX Awards for Publication Excellence.

Her education and work experience make her the ideal author for this Visio book. With a Bachelor of Science in Architecture and a Master of Science in Structural Engineering, she is well versed in using Visio for architecture and engineering and integrating it with CAD applications. As a project manager and consultant, she constantly applies Visio to office productivity problems. As a software project manager and application developer, she has also used Visio to document databases, software systems, and networks. The engineer in her is fascinated with both the simplicity and power of Visio and she enjoys experimenting with its customization and automation features.

When not chained to her computer, she hikes in the mountains with her dog, cooks gourmet meals, and practices saying no to additional work assignments. You can learn more at her web site, www.bonniebiafore.com. If you have questions or want to relate one of your favorite Visio stories, e-mail Bonnie at bonnie.biafore@gmail.com.

Credits

Contents at a Glance

Contents

Contents

Contents

Contents

Part III: Using Visio for Office Productivity   237

Chapter 11: Collaborating with Others 239

Chapter 12: Building Block Diagrams 255

Contents

Contents

Contents

Part VI: Customizing Stencils, Templates, and Shapes   615

Chapter 31: Creating and Customizing Templates 617

Chapter 32: Creating and Customizing Stencils 623

Chapter 33: Creating and Customizing Shapes 633

Contents

Contents

Acknowledgments

My thanks go to my editors, Jim Minatel and Kenyon Brown, for smoothing what can often be a bumpy road to book publication. I am indebted to John Marshall, my technical reviewer, for graciously sharing his abundant knowledge of Visio with me. As a Visio MVP (Microsoft Valued Professional), John knows a lot more about Visio than I do and helped make this book more accurate and practical.

Introduction

Visio® *2007 Bible* is as comprehensive a guide to the popular Microsoft diagramming software as a book of less than 1000 pages can be. Covering both Visio Standard and Visio Professional, this book explains Visio fundamentals as well as more advanced techniques that help you master any type of diagram. It also describes in detail how to use each of the specialized templates that Visio Standard and Visio Professional offer. Although the book provides an introduction to ShapeSheets and automation tools, one thing it doesn't do is delve deeply into the development tools that Visio offers.

Visio 2007 includes some significant changes and enhancements as well as a few new templates, such as PivotDiagrams and ITIL. Visio 2007 offers new tools to help you visualize information—from Themes, which apply professional-looking formatting with a few clicks, to Data Graphics, which transform textual data into eye-catching graphics. Speaking of data, the new Data Link feature greatly simplifies integrating diagrams with external data sources. As always, a few features have been discontinued, including the Novel Directory Services template, and a few have earned new names, such as custom properties changing to shape data. *Visio 2007 Bible* identifies these new features, enhancements, and changes and differentiates the capabilities available in both the Standard and Professional versions versus those available only in Visio Professional.

Visio 2007 is a powerful combination of simple concepts and straightforward tools with far-reaching application. Whether you want to communicate basic business processes or highly specialized technical topics, Visio offers tools to simplify your work. This book strives to follow the same model. It explains Visio's concepts and basic tools in a way that helps beginners get started and more advanced users get better. In addition, the book includes dozens of chapters on specialized templates that describe how the template, tools, and shapes support the work required and simplify typical tasks.

Is This Book for You?

Visio 2007 covers a lot of ground, and this book is right there with it. Beginners can learn the basic concepts and techniques that are the foundation of Visio's power and then apply those techniques to create the types of diagrams they need. Readers with some Visio experience can learn how to increase their productivity, use specialized templates and employ advanced techniques to draw more effectively or customize solutions. Advanced users can learn about new features, changes, and how to replace the features that have been discontinued in Visio 2007.

Although the book is fast paced, beginners can learn to use Visio, whereas more advanced users can notch up their productivity by applying tips and advanced techniques. Readers in a hurry will appreciate the topic organization that makes it easy to find a solution as well as Tips and Cautions that help solve problems quickly.

This book includes many tips for applying the methodologies and knowledge behind many of the specialized templates. In addition, you'll learn about web sites and books that you can explore to increase your knowledge of the notations that the specialized templates are designed to support. However, you won't find a complete education on all the methodologies and industries covered by Visio's templates.

Conventions Used in This Book

To help you get the most from the text and keep track of what's happening, a number of conventions are used throughout the book:

- When important terms are first introduced, they are highlighted in *italic*.
- Characters that need to be typed in are in **bold**.
- Keyboard strokes appear as follows: Ctrl+A.
- URLs, file names, directory names, and other program elements are contrasted from regular text in a monospaced font like this.

Icons Used in This Book

Following is a brief description of the icons used to highlight certain types of material in this book:

 This icon highlights helpful hints, time-saving techniques, or alternative methods for accomplishing tasks.

 This icon identifies additional information about the topic being discussed.

 This icon alerts you to potential problems or methods that can impede your work if not used properly.

 This icon points you to other chapters or books that contain additional information about a topic.

 This icon emphasizes new or significantly enhanced features in Visio 2007.

How This Book Is Organized

Visio 2007 Bible contains 42 chapters, divided into seven parts. In addition, the book is accompanied by a web site (www.wiley.com/go/visio2007bible) that provides downloadable sample Visio files you can use to practice what you've learned. The following sections provide an overview of each part of the book.

Part I: Understanding Visio Fundamentals

Part I highlights the features that distinguish Visio 2007 from the 2003 version and introduces Visio's basic concepts and techniques. The first chapter provides an overview of the new Visio 2007 features and explains concepts such as templates and stencils, drag and drop drawing, and the components of the Visio interface. Chapters 2 and 3 explain how to work with Visio files, drawing tools, drawings, and drawing pages. Chapters 4 and 5 show you how to produce diagrams by creating and editing shapes and connectors. Chapters 6 and 7 introduce techniques to improve the appearance and readability of diagrams using text and formatting.

Part II: Integrating Visio Drawings

Visio 2007 includes many new and improved integration features, as explained in this part. Chapter 8 discusses methods for linking and embedding objects into Visio or linking and embedding Visio objects into other applications. Chapter 9 describes techniques for publishing Visio diagrams to the Web. Chapter 10 covers the data-oriented features that Visio 2007 offers, including basic concepts, such as shape data, as well as significant new features, such as Data Graphics and Data Link.

Part III: Using Visio for Office Productivity

Part III is the first of three parts in this book that cover specialized templates. It begins with Chapter 11, which describes new and existing tools for collaborating with others, a critical element to office productivity. Chapters 12 and 13 cover templates for building block diagrams and charts and graphs. Chapter 14 explains the many productivity tools, shapes, wizards, and data-sharing features for documenting organizations in the Organization Chart template. Chapters 15 and 16 explain tools and techniques for documenting flowcharts and business processes. Chapter 17 discusses the Visio tools for documenting and scheduling projects. Chapter 18 describes the Visio Brainstorming template. Finally, Chapter 19 introduces the new PivotDiagram template and explains how it helps you analyze business data.

Part IV: Using Visio in Information Technology

Part IV describes the tools, wizards, and shapes that make Visio the most popular tool for documenting software systems and networks. Chapter 20 provides detailed instructions for modeling and documenting databases and database systems using a variety of notations. Chapter 21 describes how to document software systems with the Unified Modeling Language using the modeling tools available with the UML Model template and how to create different types of UML diagrams. Chapter 22 introduces several additional templates for documenting software systems. Chapter 23 describes the Visio template for mapping web sites. Chapter 24 describes techniques for creating effective network diagrams and identifies the network features no longer available in Visio 2007.

Part V: Using Visio for Architecture and Engineering

Visio 2007 works for scaled drawings as well as it does for diagrams. Part V covers the Visio tools for scaled drawings and discusses what Visio can and can't do for architectural and engineering drawings. Chapter 25 is an introduction to the concepts that underlie scaled drawings, such as scale, units, and dimensions. Chapter 26 describes different methods for creating scaled drawings and how to use layers to manage information. Chapter 27 delineates procedures for adding basic plan components, such as walls, windows, doors, and furniture, as well as how to create other types of architectural and engineering plans. Chapter 28 discusses how to use the Visio Space Plan template to plan space and manage facilities. Chapter 29 describes the Visio tools for integrating Visio and CAD drawings, which are all based on AutoCAD file formats. Chapter 30 covers the Electrical Engineering, Mechanical Engineering, and Process Engineering templates.

Part VI: Customizing Templates, Stencils, and Shapes

Part VI returns to Visio concepts and techniques with a focus on customization. Chapter 31 discusses how to create and customize templates so you can start new drawings with the settings you want. Chapter 32 describes techniques for creating and customizing stencils to create custom collections of built-in shapes, shapes you've modified, or custom shapes you've developed. Chapter 33 provides techniques for customizing shapes or creating your own and explains how to use custom properties to store data. Chapter 34 digs deeper into customizing shapes by introducing techniques for modifying fields in Visio ShapeSheets or writing custom formulas to control shape appearance and behavior. Chapter 35 explains the benefits and techniques for formatting with styles and describes how to create custom line patterns, fill patterns, and line ends. Chapter 36 describes techniques for customizing or creating your own toolbars and menus. Chapter 37 introduces the techniques available for automating Visio, including macros and writing add-ins.

Part VII: Quick Reference

Part VII includes helpful information and reference lists. Chapter 38 describes the process for installing Visio 2007. Chapter 39 provides different sources of help available for Visio 2007, both within the product and online. Chapter 40 identifies additional sources for customized and specialized templates, stencils, and Visio-based solutions. Chapter 41 is a reference to the most helpful keyboard shortcuts. Chapter 42 identifies the templates that Visio Standard and Visio Professional provide and the stencils each one opens.

Visio® 2007 Bible

Part I

Understanding Visio Fundamentals

Chapter 1

Getting Started with Visio

Humans are visual creatures, so it isn't surprising that we want to visualize our ideas, designs, and final products and communicate them graphically. In the past, high-quality presentations were the work of professional graphic artists and illustrators, but with Visio 2007, anyone can produce informative and attractive diagrams, drawings, and models. Visio is so straightforward that you can use it to capture the fast-paced output of brainstorming sessions or the frequent changes made to initial designs. At the same time, Visio is powerful enough to develop sophisticated models, and precise enough to document the details of existing systems.

Visio 2007 jumpstarts your efforts with solutions designed specifically to produce different types of diagrams. Visio templates set up your work environment with menus of specialized tools, sets of predefined shapes, and drawing settings such as page size and orientation typical for the type of drawing you want to create. Visio stencils categorize thousands of predefined symbols by industry, drawing type, and application. These Visio *SmartShapes* have built-in behaviors and properties to help you quickly assemble drawings and collect information.

Simplicity and convenience are key to the power of Visio. To construct a drawing, you drag and drop predrawn shapes from stencils onto drawing pages. Defining relationships between shapes has gotten even easier in Visio 2007 with the introduction of AutoConnect, which helps you place and connect new shapes on drawing pages. Specialized tools help lay out drawings and perform typical tasks. The simplicity of integrating Visio with tools such as Microsoft Office, AutoCAD, and database management systems makes it easy to maintain drawings and documentation of systems.

IN THIS CHAPTER

Discovering features new to Visio 2007

Learning the basic concepts behind Visio's power

Exploring the Visio interface

What's New in Visio 2007?

Visio 2007 has dramatically simplified connecting shapes, formatting diagrams, and visualizing data. As always, Visio 2007 includes several new templates and stencils as well as improvements and enhancements to many existing ones. This section highlights some of the new features and tells you which chapter to read to learn more.

NEW FEATURE   Look for the New Feature icon throughout this book to discover what's new and improved.

Productivity Enhancements

Visio enhances its reputation for being quick and easy with the following new or improved features:

- **AutoConnect** — This brand-new method for connecting shapes helps you add shapes to drawing pages, connect them to existing shapes, and align them in one step. However, not all shapes work with AutoConnect, which is the case if you don't see the distinctive blue AutoConnect arrows when you hover the pointer over existing shapes. See Chapter 5 for more on connecting shapes with AutoConnect.

- **Themes** — Whether you don't have time for fancy formatting or tend to wear stripes with checks, the new Themes feature makes attractive, color-coordinated diagrams incredibly easy. In Visio 2007, the Themes feature offers sets of designer-coordinated colors that look good together as well as Theme effects for applying predefined consistent fills and borders to the shapes on your diagrams. See Chapter 7 for the details on Themes.

- **Data Link** — The new Data Link feature simplifies connecting Visio diagrams to data sources, connecting data source fields to Visio shape data fields, and managing the data connections you've made. In addition, for two-way data links, the Database Wizard is still available. See Chapter 10 for more information on integrating Visio drawings with data sources.

- **Data Graphics** — Data Graphics can visualize data associated with shapes in several ways, including text callouts, status bars, and icon sets (for example, to represent status stoplight fashion with red, green, and yellow circles). Unlike the extensive shape customizations required in earlier versions of Visio, Data Graphics provide several pre-defined methods for showing data associated with shapes. However, if you want to build your own Data Graphics, that's easy, too. See Chapter 10 for more on working with Data Graphics.

- **Getting Started** — When you choose File ➪ New ➪ Getting Started, the reorganized Getting Started window provides easy access to templates and recent documents. The categories for templates have been simplified so you don't have to scan the list for the category you want. In the main area of the window, you can choose thumbnails of templates you've used recently or Visio drawings you've opened recently. The Recent Documents pane even shows thumbnails of the contents of your drawings. In addition, templates appear in multiple categories if they support different diagramming tasks. See Chapter 2 for more information on the Getting Started window.

- **CAD Integration** — The overall design of CAD integration hasn't changed, but the Visio team has enhanced the integration in many ways, as discussed in Chapter 29.

New and Improved Templates and Stencils

See Chapter 42 for a guide to all the Visio 2007 templates and the categories in which you can find them. The following templates and stencils are new in Visio 2007:

- **PivotDiagram template** — This new template available in Visio Professional 2007 emulates Excel pivot tables and helps visualize business information from multiple viewpoints.

- **ITIL Diagram template** — This new template available in Visio Professional 2007 includes a new ITIL stencil, but also opens stencils from other categories that support the ITIL approach to documenting IT services.

- **Value Stream Map** — This new template available in Visio Professional 2007 helps illustrate process performance, which you can use to eliminate delays and waste.

- **Rack Mounted Servers stencil** — The Detailed Network Diagram template opens this new stencil, which contains 16 rack servers drawn in the isometric perspective.

- **XML Web Service shape** — This one new shape on the Detailed Network Diagram stencil represents an XML web service.

- **Department stencil** — This new stencil for the Work Flow Diagram template shows departments and people in three-dimensional shapes that work with the new Themes feature.

- **Work Flow Objects and Work Flow Steps stencils** — These are additional new stencils for the Work Flow Diagram template.

 **To find other sources for Visio stencils and templates, refer to Chapter 40.**

What's Missing in Visio 2007

As much as Visio gives, it also takes away a few things in each new release. Here are some of the items that are no longer available in Visio 2007:

- The Novell Directory Services template and its stencils are no longer available.

- File Search is no longer available on the File menu.

- The Insert Picture from Scanner or Camera feature is gone.

- The Getting Started Tutorial and Diagram Gallery are no longer available on the Help menu.

- The New Drawing and Template Help task panes are no longer available.

- The Org chart theme feature has been replaced by the Themes feature.

New Homes for Features in Visio 2007

Microsoft loves to move things around. Here's how to find some of your old but relocated favorites:

- The Color Schemes add-on appears only when you right-click the page in an old drawing created in an earlier version of Visio, because Color Schemes have been replaced by Themes.

- Custom Properties Sets, now called Shape Data Sets, are available only when you right-click the Shape Data window.

- The Shared Workspace task pane has been renamed to Document Management.

- Export to Database is available only by choosing Tools ⇨ Add-ons ⇨ Visio Extras.

- Rotate Left, Rotate Right, Flip Vertical, and Flip Horizontal are available only on the main Shape menu.

- Format ⇨ Special appears only if you run Visio in developer mode.

- The option to save the drawing workspace with a Visio document is available by choosing File ⇨ Properties.

What Visio Is and Isn't

Visio can be many things to many people. Applied properly, Visio can help you produce simple diagrams as well as complex models. However, Visio's far-reaching capabilities can be confusing if you don't understand how they differ. Even worse, Visio can be downright diabolical if you try to use it for tasks too far afield from what it's designed for.

Many drawings are simple diagrams with some basic connections and little or no associated data. For these drawings, dragging and dropping shapes and gluing connectors are about all you need to know. If you use the new AutoConnect feature, you don't even have to align shapes after you connect them. The remaining chapters in Part I, Understanding Visio Fundamentals, describe the basic tools you need to diagram with Visio.

However, Visio Professional can also produce intelligent models and specialized documentation for numerous fields, including software engineering, architecture, mechanical and electrical engineering, and business process modeling. Templates for these purposes often contain specialized tools. However, the real stars are usually the shapes in the stencils for these templates, which have smart features — built-in behaviors and attributes that fit the shapes to their roles. For example, intersecting walls in building plans are smart enough to clean up their overlapping lines. Cubicle shapes might contain shape data for identifying the people who occupy the space for occupancy reports. These features are time-savers when you know how to use them but can make Visio seem to have a mind of its own when you don't. Parts III, IV, and V of this book teach you the ins and outs of more sophisticated Visio solutions.

You can draw precise plans to scale with Visio. Visio Standard supports only basic building plans, whereas Visio Professional supports a variety of architectural and engineering plans. Nonetheless, you'll probably want the extra power of a CAD application, such as AutoCAD, to design and document large or complex plans. Even so, Visio can complement your CAD application. You can create shapes faster and more easily in Visio and then import them for use in AutoCAD or other CAD applications. Team members who don't have access to AutoCAD can create their drawings in Visio using CAD drawings as a backdrop and import their work into AutoCAD if necessary. Visio also simplifies preparing presentations for large projects.

Understanding Visio Concepts

Visio enhances your drawing and modeling productivity because so many of its elements include features that incorporate industry expertise. Most of the time, you don't even think about how much Visio does for you because the templates, stencils, and shapes do just what you would expect. However, some of Visio's specialized capabilities might surprise or even confuse you at first. By understanding the concepts that make Visio so powerful, you can prevent problems and keep your productivity at a blistering pace.

Using Templates and Stencils

In the real world, templates are patterns you use to build something. For example, you could use a standard design for a log house to simplify the construction of your home. In Visio, a template is a special type of Visio drawing file that contains the tools you need to produce a specific type of drawing. Each template comprises settings, stencils, styles, and sometimes special commands to make your work on a drawing as easy as possible.

When you create a drawing based on a template, Visio does the following things:

- **Opens stencils with shapes** — Visio opens stencils that contain the shapes you need for the type of drawing you are creating.

- **Includes styles** — Visio provides special formatting styles typical for the current drawing type. For example, a construction project created from a floor plan template includes line styles typically used to dimension architectural plans.

- **Automatically displays menus and toolbars** — If the template contains a special menu, Visio adds an entry for the menu to the menu bar. If the template contains a special toolbar, Visio floats the toolbar in the drawing area.

- **Specifies settings** — Visio specifies settings typical for the type of drawing. For basic block diagrams, Visio uses letter-size paper, portrait orientation, one-to-one scale, and inches for measurement units. For site plans, it specifies a 36" × 42" architectural drawing size in landscape orientation, a scale of 1 inch to 10 feet, and measurement units of feet and inches.

- **Displays rulers and grid** — To make positioning shapes easy, the rulers and grid take into account the scale and units for the drawing. For example, a block diagram shows inches on the rulers with each grid cell equal to one-quarter inch. Conversely, rulers for a site plan display feet in the rulers with each grid cell equal to 10 feet.

Visio stencils are categorized collections of shapes. To continue the house analogy, a Visio stencil is like a catalog of cedar logs and connecting brackets that are available from your local building supply store. To build your home, you order the components you need from the store and assemble them according to your house design. In Visio, you assemble your drawings by dragging and dropping shapes from stencils onto your drawing page.

What Makes Shapes Smart

The Visio philosophy is elegantly simple — you construct drawings primarily by dragging and dropping predefined shapes onto drawing pages. Although working with Visio can seem like copying clip art into a document, Visio shapes are much more powerful. In fact, Visio calls them *SmartShapes* because their built-in properties and behaviors make them seem, well, smart.

Shape behaviors help you position shapes and connect them appropriately to other shapes. For example, when you place a door shape in a wall, the door lines up with the wall and creates an opening into a room, as shown in Figure 1-1. That same door might contain shape data to configure the shape or identify it, also shown in Figure 1-1. For example, one door property specifies whether the door is centered in the wall. Other door properties can define a door's dimensions, its catalog number, or its associated room number, so you can produce a schedule of the doors you need and where they belong in your building.

FIGURE 1-1

Shapes include behaviors and properties that make them seem smart.

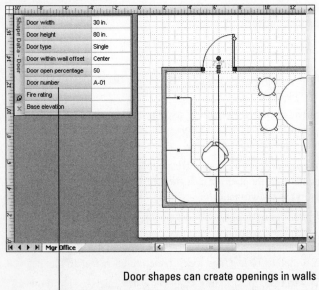

Door shapes can create openings in walls

Shape data configures and identifies shapes

CROSS-REF To learn about how to define properties and behaviors for shapes, see Chapter 33.

In Visio, predrawn shapes are called *masters*, which are stored and categorized in stencils. When you drag and drop a master from a stencil onto your drawing page, you create a copy, called an *instance* of that master. Each instance inherits its master's behaviors, so it knows how to act when you add it to your drawing. It also inherits its master's properties, so you can assign unique values to an instance.

As you'll learn in detail in Chapter 4, shapes have handles to help you position, resize, and connect them to one another. When you select a shape, Visio marks these features with colored graphics, so you can drag them to make the changes you want. Shapes include the following types of handles:

- **Selection handles** — Red or green boxes appear when you select a shape. You can drag these selection handles to resize a shape or attach connectors to them.

- **Connection points** — Blue Xs mark locations where you can glue connectors or lines.

- **Rotation handle** — This is a red or green circle that you can drag to rotate a shape.

- **Control handles** — Yellow diamonds appear on some shapes. You can drag control handles to modify a shape's appearance — for example, to change the swing on a door.

- **Eccentricity handles** — You can drag these green circles to change the shape of an arc.

Connecting Shapes

Relationships often convey as much information as the elements they connect. Whether you are showing who reports to a manager in an organization or defining the relationship between two database tables, connections between Visio shapes not only provide information about a relationship, but they also help you lay out and rearrange the shapes on your drawing.

CROSS-REF To learn more about connecting shapes, see Chapter 5.

What Connectors Do

At their simplest, connectors are linear shapes with other shapes attached to each end. However, connectors have smarts of their own. When you move two connected shapes, the connector between them adjusts to maintain that connection. Likewise, connectors reconfigure to keep shapes connected when you use automatic layout tools in Visio. For example, you can change the layout of an organization chart from horizontal to vertical and the connectors alter their paths as the employee shapes take up their new locations.

Connectors have start and end points that define the direction of the connection between shapes. Which end you connect to a shape can make a big difference in behavior. For example, in a database model, the table shape at the start of a connector is the *parent*, whereas the table at the end of a connector is the *child*. When you define a one-to-many relationship between those connected tables, the one is associated with the table at the connector's start point and the many belongs to the table at the connector's end point.

TIP When you want to differentiate the predecessor and successor for two connected shapes, such as in a data flow diagram or project schedule, make sure you glue the start point of the connector to the shape you are connecting from and the end point to the shape you are connecting to.

Straight versus Dynamic Connectors

Straight connectors are straight lines that connect shapes. They lengthen, shorten, and change their angle to maintain shape connectivity, but they draw straight over shapes that are in their path, as shown in Figure 1-2. Dynamic connectors are smarter. They automatically bend, stretch, and detour around shapes instead of overlapping them, as shown on the right side of Figure 1-2. They can also jump over other connectors to make connections easier to follow on a drawing. By default, dynamic connectors use right angles to bend around shapes. You can change the path of a right-angled connector by moving any of its vertices, or add or move segments of a right-angled connector by dragging a midpoint of a segment. Curved connectors are dynamic as well. You can drag their control points and eccentricity handles, to modify the shape of the curve.

FIGURE 1-2

You can connect shapes with straight or dynamic connectors.

Dynamic connectors route around shapes

Straight connectors draw over shapes

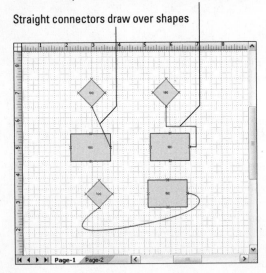

Using Glue

Similar to glue in real life, Visio glue makes things stick together. Visio glue comes in two varieties: shape-to-shape and point-to-point. *Shape-to-shape glue,* also known as *dynamic glue,* builds dynamic connections between shapes. When you reposition shapes connected with shape-to-shape glue, the end points of the connector move to the closest available connection points, as shown in Figure 1-3. *Point-to-point glue,* also known as *static glue,* keeps the connector end points glued to the specific points you selected on the shapes, also illustrated in Figure 1-3. In addition, you can combine dynamic and static glue, gluing a connector to a shape at one end and a specific point at the other.

FIGURE 1-3

Dynamic glue draws the shortest connectors between two shapes.

Shape-to-shape glue draws the shortest connections

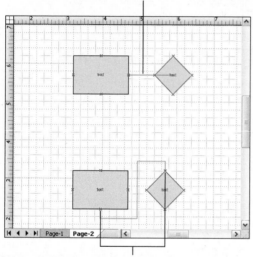

Point-to-point glue always connects the points you select

By default, you can glue to entire shapes, connection points, or guides. You can change glue settings to also glue to shape handles, shape vertices, or any point on a shape's geometry. As you draw a connector, a red box appears around a shape when you are connecting to that shape. If you are connecting to a point, the connection point turns red.

Exploring the Visio 2007 Interface

The Visio interface makes it easy to access the features you use. By default, the Visio environment positions menus and toolbars across the top, the Shapes window with stencils and shapes to the left, the task pane to the right, a status bar along the bottom, and the drawing window in the center, as shown in Figure 1-4.

FIGURE 1-4

The Visio environment provides convenient access to tools.

Menu bar

Shapes window   Toolbars   Task pane

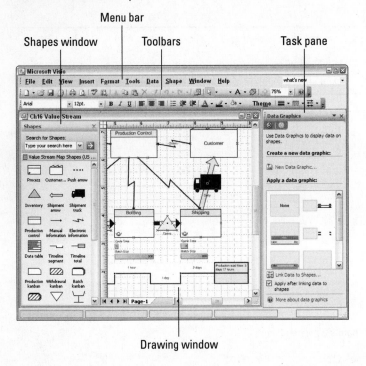

Drawing window

Menus and Toolbars

Most Visio features appear on one or more of the plentiful built-in menus or toolbars. In the Visio menu bar, in addition to the menus so familiar to Microsoft Office users, specialized menu entries sometimes appear, which include commands specifically for the template you're using. For example, the Organization Chart template adds an Organization Chart entry to the menu bar, which opens a drop-down menu with commands, such as Arrange Subordinates or Import Organization Data.

However, the fastest route to many tasks is right-clicking something on the drawing page or within the Visio windows to display a shortcut menu. In addition, shortcuts for the most popular commands are available on the Standard or Formatting toolbars, which appear by default. Some templates include specialized toolbars, which float in the drawing window. You can easily show or hide a toolbar:

- To display a toolbar, choose View ➪ Toolbars and choose the toolbar you want to use. A check mark appears when the toolbar is displayed. A specialized toolbar appears in the toolbar list when a drawing of its type is active.

- To hide a toolbar, choose View ➪ Toolbars and click the checked toolbar that you want to hide.

You can dock a toolbar along the top, bottom, or sides of the Visio window. When you dock a toolbar to the left or right, the toolbar hangs vertically along the side. Toolbars are easily manipulated:

- To reposition a docked toolbar, drag its move handle to a new location. The move handle is a series of dots to the left of a horizontal docked toolbar and along the top of a vertical docked toolbar.

- To float a toolbar in the middle of the window, drag its move handle to a new position.

- To reposition a floating toolbar, drag its title bar to a new location.

Task Panes

The more sophisticated features that you might use for long stretches during a session typically have their own task panes, such as the new Themes and Data Graphics features. Task panes dock on the right side of the screen by default. To show or hide a task pane, choose View ➪ Task Pane. You can also display the task pane by pressing Ctrl+F1.

The Visio 2007 Drawing Area

The drawing window, which contains your active drawing, takes center stage in the Visio drawing area. However, you can display several other windows to facilitate your work. To display one of these other windows, choose View and then the window name.

The Drawing Window

Drawing pages appear in the drawing window, where you can add shapes or modify and format the contents of your drawing. You can view different areas of a page using the horizontal and vertical scrollbars. To view another page, at the bottom of the drawing window select the tab for that page.

A drawing grid and rulers make it easy to position and align shapes on a page. To display a grid in the drawing window, choose View ➪ Grid. To display rulers, choose View ➪ Rulers. The units that rulers display vary depending on the type of drawing and scale you are using. For example, the rulers for a block diagram use inches, whereas rulers for a site plan use feet.

To change the ruler units, choose Tools ➪ Options and select the Units tab. Click the Change button and, from the Measurement Units drop-down list, choose the units you want.

The Shapes Window

You drag and drop shapes from the stencils in the Shapes window onto a drawing page. The Shapes window docks by default on the left, as previously shown in Figure 1-4. However, you can reposition the Shapes window or even individual stencils to suit your needs. For example, you can dock the Shapes window at the top or the bottom of the drawing area to provide more room for pages set to landscape orientation.

Here are some methods for configuring the Shapes window:

- To add another stencil to the Shapes window, choose File ➪ Shapes, and choose the category and then the stencil you want.
- To display the shapes for an open stencil in the Shapes window, click the stencil's title bar.
- To resize the Shapes window, drag the vertical divider between the Shapes window and the drawing window to the left or right.
- To change the information displayed in the Shapes window, right-click the Shapes window title bar and choose one of the options, such as Icons Only, from the shortcut menu.

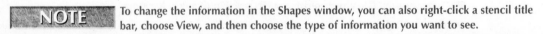 **NOTE**   To change the information in the Shapes window, you can also right-click a stencil title bar, choose View, and then choose the type of information you want to see.

By default, in the Shapes window you see the title bars for all open stencils, but only the shapes for the active stencil. To view multiple stencils at the same time, you can

- Drag a stencil from the Shapes window and float it on the screen, as shown in Figure 1-5.
- Drag a stencil to the top or bottom of the Shapes window to create a second stencil pane, also shown in Figure 1-5.

The Drawing Explorer

The Drawing Explorer, shown in Figure 1-6, offers a hierarchical view of your drawing. You can use the Drawing Explorer to find, add, delete, or edit the components of your drawing, including pages, layers, shapes, masters, styles, and patterns — sometimes, more easily than performing the same task on the drawing page. For example, you can select and highlight a shape on a drawing by double-clicking its name in the Drawing Explorer window. To display the Drawing Explorer, choose View ➪ Drawing Explorer Window.

TIP   You can conserve screen real estate by docking and merging view windows. You can dock other view windows within the Shapes window or merge several windows into one. To dock a view window, such as Pan & Zoom, drag it into the Shapes window. To merge view windows, drag one window by its title bar into the center of another window. To switch between merged windows, select the tab for the view you want.

FIGURE 1-5

You can dock stencils in the Shapes window or float them on the screen.

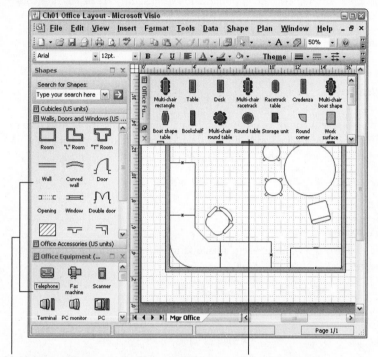

Multiple stencil panes in the Shapes window   A floating stencil

FIGURE 1-6

You can exploit the hierarchy of drawing components in the Drawing Explorer window.

The Size & Position Window

The Size & Position window is particularly useful when you work on scaled drawings such as building plans, where precise measurements are important. You can use the Size & Position window to view and edit a shape's dimensions, position, or rotation.

 To learn more about how to use the Size & Position window, see Chapter 4.

The Shape Data Window

The Shape Data window is the best place to modify the shape data for a number of shapes. The window stays open until you close it and displays the shape data values for the currently selected shape. To edit a property in the Shape Data window, click the property box and enter or edit a value.

 To learn more about shape data, see Chapters 10 and 34.

The ShapeSheet Window

Every aspect of a shape is defined in its ShapeSheet. You can display the ShapeSheet by choosing Window ➪ Show ShapeSheet.

 To learn more about ShapeSheets, see Chapter 34.

Summary

Visio is a great tool for communicating ideas, documenting business results, and modeling systems. Using drag-and-drop drawing techniques, anyone can produce great-looking diagrams, drawings, and models. This chapter introduced the new and improved features in Visio 2007 and mentioned features that have been discontinued. You also learned a little about the concepts that make Visio so powerful. Finally, you learned about the Visio interface and the windows you can use to make your drawing work as simple as possible.

Chapter 2

Getting Started with Drawings

With Visio, there's no reason to consume your precious time or strain your creativity by starting diagrams from scratch. In fact, creating diagrams from templates or existing drawings is a good idea even with the most luxurious deadlines or robust aesthetic sense. A template automatically opens a set of Visio stencils with the shapes you're most likely to use for that type of diagram; applies the typical drawing options, such as paper size and scale; and presents any additional tools, such as menus, toolbars, and wizards, if they exist for that template.

New drawings can originate from a variety of sources, including existing Visio drawings, Visio templates on your computer, and Visio templates from numerous online sources. Furthermore, Visio provides several methods for creating drawings, none of which requires more than a few clicks of your mouse.

Saving your work is a task you should learn to perform often, and particularly before any experimentation with your diagrams or Visio commands. You also call the Save command into play to create your own stencils and templates, as well as other file formats, for example, HTML for web pages or graphics formats to create a picture of a diagram. Visio offers several easy methods for saving files, so there's no excuse for not doing so.

After you perfect a diagram, you want it to stay that way until you decide to change it. You can protect diagrams to prevent unwanted changes, whether they are edits you make inadvertently or modifications from well-meaning yet misguided colleagues.

This chapter begins by describing the different methods for creating drawings and the advantages of each. You'll also learn how to open drawings and save them in different formats. Finally, this chapter explains how to protect drawings and the content within them.

IN THIS CHAPTER

Creating drawings from templates or existing drawings

Opening Visio drawings and other types of files in Visio

Opening recently used files quickly

Specifying save options

Saving different types of files

Protecting files

Creating Drawings

Most of the time, the easiest way to create a new drawing is by starting with a Visio template. Visio itself offers dozens of templates covering all kinds of diagrams, and hundreds more are available online, if you're looking for something that the program doesn't offer. If you have an existing Visio drawing that is similar to your current diagramming assignment, you can save that drawing as a new file to take advantage of its content *and* template. This section describes several methods for creating drawings using templates or existing drawings.

The Advantages of Templates

A Visio template takes care of a lot of diagram housekeeping, so you can start building your diagram as quickly as possible. When you use a template to create a new drawing, it handles the following tasks:

- Sets drawing options, such as drawing scale, page size, and paper orientation. For example, a flowchart template might set the drawing scale to 1-to-1, and the paper to an 8 1/2" × 11" page with portrait orientation. A template for an organization chart might use a larger page size with landscape orientation.

- Opens an appropriate set of shape stencils, for example, specific diagram shapes, annotation shapes, and dimensioning shapes.

- Adds one or more blank drawing pages.

- In some cases, adds a specialized menu to the Visio menu bar or a floating toolbar to the drawing area.

 Many templates that are available online, from Microsoft Office Online or other web sites, also include shapes on the drawing page to jump-start your diagram.

Methods for Choosing a Template

Visio provides three different methods for choosing a template for a new drawing. Pick the method that you prefer or use different methods, depending on the situation:

- **Getting Started screen** — Out of the box, Visio displays the Getting Started screen every time you launch the program. The Getting Started screen, which is an improvement over the Choose Drawing Type screen in Visio 2003, provides easy access to templates and recent Visio documents alike.

  - To use a template that you applied recently, in the Recent Templates area, click the template image.

  - To open a Visio drawing you worked on recently, in the Recent Documents area, click the image for the drawing file. The images actually show thumbnails of the drawing contents, so you can readily identify the file you want.

  - The Template Categories area, on the left side of the Getting Started screen, lists folders for each category of template, such as Business, Engineering, and Network. Click a category and then click a template to preview or use it.

NEW FEATURE Visio 2007 has reduced the template categories to only eight: Business, Engineering, Flowchart, General, Maps and Floor Plans, Network, Schedule, and Software and Database. However, Visio makes templates easier to find by listing them in every category to which they apply. For example, the Work Flow Diagram template appears in the Business and Flowchart categories.

TIP Although the Getting Started screen provides access to new and existing drawings, minimalists can start Visio with nothing more than the standard menus. To do this, click Tools ➪ Options. Select the View tab and clear the Getting Started screen check box.

- **File ➪ New** — Regardless where you are in Visio, choosing File ➪ New is the quickest way to any template. Choose File ➪ New ➪ Getting Started to preview and choose from available templates. Or, choose File ➪ New ➪ and then click the category you want.

- **New Drawing from Template** — On the File ➪ New menu, the New Drawing from Template command opens a dialog box, which navigates folders looking for Visio template files (the Files Of Type box is set to Template). If you use templates you've downloaded, this dialog box is a sure-fire way to open them.

Creating Drawings from Templates

To create a drawing from a template, follow these steps:

1. Display the Getting Started screen by choosing File ➪ New ➪ Getting Started.

2. On the left, click the category for the type of drawing you want to create. Visio displays thumbnails of each type of drawing in that category, as shown in Figure 2-1.

NOTE If you don't need handholding to find the right template, choose File ➪ New and then position the pointer over the category for the template. When the submenu appears, choose the template you want. Take care to choose the template that uses the measurement units you want, indicated by (US Units) or (Metric) to the right of the template name.

3. Click the image representing the type of drawing you want. Visio displays a larger thumbnail of the template you chose as well as a description of common applications for it.

4. On the right, choose the US Units option or Metric option to specify the measurement units you want, and then click Create. Visio creates a new blank Visio drawing file based on the template.

TIP Suppose you want to use Visio drawing tools to mark up an image and you don't want any assistance from a template. To create a new blank drawing without stencils or predefined settings, choose File ➪ New ➪ New Drawing (US Units) or simply press Ctrl+N. To create a metric drawing, choose File ➪ New ➪ New Drawing (Metric).

FIGURE 2-1

The Getting Started screen provides access to templates and existing drawings.

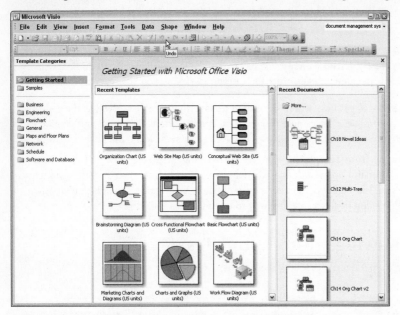

Creating Drawings from Existing Drawings

Sometimes, an existing Visio drawing possesses exactly the settings you want for a new drawing, beyond what a template initially applies. It might even have shapes you want already in place, such as your favorite title block. To save an existing drawing as a new one, follow these steps:

1. Open the existing drawing.

CAUTION   To prevent unwanted modifications to the original drawing file, don't make any changes to the drawing until you save it as a new file.

2. To save the drawing as a new drawing, choose File ➪ Save As.

3. In the Save As dialog box that appears, navigate to the folder you want. In the File Name box, type the new name for the file and click Save.

CROSS-REF   If an existing drawing contains settings and content you use frequently, you can create a template of that drawing, as described in Chapter 31.

Opening Drawings

Visio provides several methods for opening drawing files you want to work on. The Open dialog box in Visio 2007 has all the familiar features for finding and opening Visio drawing files, stencils, templates, and other file formats. More recently opened drawing files appear on the File menu and the Getting Started screen. This section describes the methods for opening drawings as well as what Visio does behind the scenes.

NEW FEATURE Visio 2007 can open drawings created in earlier versions of Visio. However, Visio 2002 can't open files created with Visio 2003 or later because of changes to XML and binary file formats. However, you can save Visio 2007 files as Visio 2002 drawings, stencils, or templates, so you can open them in Visio 2002.

When you open a Visio file, the program takes several steps beyond the obvious display of the drawing page and stencils:

- The workspace list saves the files, windows, and window positions from the last work session so that Visio can reset the same work environment the next time you open the drawing.

- Visio includes styles and color palettes associated with the drawing.

- Visio opens a VBA project, which initially contains an empty class module: ThisDocument.

- Visio opens a ShapeSheet for the file, which you can use to store information about the file and the shapes within it.

CAUTION Macros can increase your productivity, but they can also jeopardize your computer security. To limit the macro features that Visio uses, choose Tools ⇨ Trust Center. Visio selects the Macro Settings section by default, so you can choose an option to specify whether to allow macros to run with or without notification. See the section "Protecting Your Drawings" later in this chapter for more information on the Trust Center.

Opening Visio Drawings

The Open dialog box hasn't changed over the last several versions of Visio, but it provides handy features for opening files that you might not have stumbled across yet. From this dialog box you can browse, access recently opened files, open one or more files, and specify how to open a file.

To open a file or files using the Open dialog box, do one of the following:

- On the Standard toolbar, click Open.

- Press Ctrl+O.

- Choose File ⇨ Open.

Working with Document Management Systems

Visio can retrieve from and save to a document management system (DMS) as long as it supports the Open Document Management Architecture (ODMA) 1.5 standard. When you choose File ⇨ Open and Visio detects an ODMA 1.5–compliant management system on your computer, it opens the DMS Open dialog box instead of the regular Open dialog box. Similarly, when you save a file in Visio, the DMS Save dialog box appears so that you can save and store your Visio drawing within the DMS. If the DMS dialog boxes don't appear, you might need to register Visio with your DMS application. Refer to your DMS documentation for instructions.

You can't use a keyboard shortcut or the Open icon on the Standard toolbar to access the DMS dialog boxes. To open or save files with a DMS, you must choose File ⇨ Open and File ⇨ Save.

After the Open dialog box appears, use any of the following methods to open the file or files you want:

- **Open any file** — Navigate to the folder that contains the file you want to open and double-click the file.

- **Open more than one file** — Press Ctrl and click each file to open. Click Open to open all the selected files. To select several files listed in sequence, click the first file and Shift+click the last file before clicking Open.

- **Open recently opened files** — In the navigation bar on the left, click My Recent Documents and then double-click the file you want.

- **Specify how to open the file** — If you click the down arrow to the right of the Open button, a drop-down menu appears. Choose Open Original to open the file as you would normally. If you want to create a new drawing based on the original, choose Open as Copy. To prevent making inadvertent changes, choose Open Read-Only.

CROSS-REF   Visio opens drawings using the units defined in the template on which the drawing is based and any settings applied in earlier sessions. To learn how to change measurement units or settings for the drawing, see the section "Setting Up Pages" in Chapter 3.

Accessing Recently Used Files

Visio might be easy to use, but, more often than not, more than one work session is required to complete a drawing. Locating and opening files you've worked on recently is easy with one of the following shortcuts:

- **Getting Started screen** — In the Recent Documents window, click the thumbnail image of the file you want to open.

- **My Recent Documents folder** — In the Open dialog box, click My Recent Documents to view Visio files you accessed recently. To open a file, select the file in the file list and click Open.

- **File menu** — Depending on the options you have chosen, a number of your recently opened files appear at the bottom of the File menu. Click a file name to open that file.

TIP You can modify the number of recently used files that appear on the File menu and in the Getting Started screen. To do this, choose Tools ➪ Options, choose the General tab, and, in the Recently Used File List box, click the arrows to change the number of entries that you want to appear. By default, Visio lists four documents; but the Recently Used File List can range from zero to nine.

Opening Other Types of Files in Visio

There are plenty of reasons to open other types of files in Visio, including stencils, templates, workspaces, and several types of graphics files. For example, you must open stencils or templates in order to customize them. Opening a workspace is a shortcut to recreating a complex Visio work environment — it opens all the files and windows for that workspace and positions the windows as they were when the workspace was saved. Visio opens files saved in other graphic formats as a single shape that you can move, resize, and rotate.

Choose File ➪ Open to display the Open dialog box. In the Files of Type drop-down list, choose the type of file you want to open. As you navigate folders, Visio displays only files of the type you specified in the file list. Select the file you want and click Open.

TIP If you want to insert content from a type of file that Visio doesn't support, use the clipboard. Open the file using another drawing application, cut or copy a section of the image, and then paste it into a Visio drawing.

CROSS-REF To learn more about working with other types of files, refer to Chapters 8, 9, and 29.

Saving Visio Files

Nothing reduces productivity like recreating work that was lost. Saving your work in Visio is easy, although developing the habit to do so can be difficult — until you lose something important. Saving also comes into play when you want to use a Visio diagram in another way, such as publishing it to a web site or saving it in graphic format for inclusion in a report.

The next section describes how to save files to other formats, but to simply save the work you've completed in a Visio file, use one of the following methods:

- Press Ctrl+S.
- On the Standard toolbar, click Save.
- Choose File ➪ Save.

When you save a drawing for the first time, Visio opens the Save As dialog box automatically. You specify a file name and location for the drawing and click Save to store the file. After the file is saved, pressing Ctrl+S is the fastest way to save changes.

NOTE When you save a file, Visio not only saves any changes you've made to pages, shapes, and properties; it also saves the position of all open windows to the drawing workspace so that your Visio environment will look the same the next time you open the file. If you don't want to save the workspace for some reason, choose File ➪ Properties. On the Summary tab, uncheck the Save Workspace check box and click OK. Then, save the file.

Saving as Other Formats

Depending on what you have in mind for your diagram, you can save drawings as several other types of files. To save to other formats, choose File ➪ Save As and, in the Save As dialog box, in the Save As Type drop-down list, choose the format you want. The following list identifies the types of Visio files you can save and what each one represents:

- **Drawing** — Use this file type to save a drawing as a copy of the original in the current Visio file format.

- **Stencil** — Save a drawing file as a specialized Visio drawing file, which can act as a stencil and contains master shapes that you drag onto drawings. (See Chapter 32 for detailed instructions.)

- **Template** — Save the settings, stencils, windows, styles, macros, and content of a drawing file as a template for new drawings. (See Chapter 31 for detailed instructions.)

- **XML Drawing, XML Stencil, XML Template** — Save a drawing file in the Visio XML format for drawings, stencils, and templates, respectively.

NOTE Visio can produce both binary and XML versions of files. The files extensions VSD, VSS, and VST represent the binary format for drawings, stencils, and templates, respectively. For XML files, the extensions are VDX, VSX, and VTX. These files act like regular Visio files, but you can also open them in a text or XML editor and access your Visio data with XML tools. For example, developers can write programs to construct or read drawings without launching Visio.

- **Visio 2002 Drawing, Visio 2002 Stencil, Visio 2002 Template** — Share drawings with people who use Visio 2000 or 2002 by saving to one of these formats.

- **Scalable Vector Graphics, Scalable Vector Graphics - Compressed** — Save Visio drawing files to either of these formats if you want to transfer your diagram to another type of graphics program that supports this W3C 2-D graphic standard.

- **AutoCAD Drawing, AutoCAD Interchange** — Choose one of these types to save your drawing file as an AutoCAD drawing. For example, you can start a plan drawing in Visio and then transfer it to AutoCAD for final engineering drawings. (See chapter 29 for detailed instructions.)

- **Web Page** — Save a Visio drawing to HTML format for publication on the Web, as described in Chapter 9.

- **Graphics formats** — The rest of the file types in the Save As Type drop-down list are different types of graphics files, such as metafiles, GIF, JPEG, PNG, TIFF, and BMP formats. To save a Visio drawing as a picture, choose the graphic format appropriate for its destination. For example, GIF files are ideal for pictures on the Web, whereas TIFF files are more suitable for print publications.

Working Around Visio File Format Incompatibilities

Visio 2007 uses the same file format as Visio 2003. Typically, each version of Visio can save files to both the current and previous format, which means that Visio 2007 can save to the format used by Visio 2003 as well as the one for Visio 2002. When you save to an earlier format, Visio omits information or formatting that the earlier version doesn't support. Although Visio 2007 can read Visio 2002 XML files, it can't save to Visio 2002 XML format. To do that, you must save a file as a Visio 2002 drawing and then convert it to XML in Visio 2002.

Every version of Visio can read any Visio file from previous versions. For example, Visio 2007 can read a Visio 1.0 document, the very first Visio file format. However, opening a drawing file created in a newer file format isn't so accommodating, because earlier versions of Visio have no idea what new Visio features might be introduced in future versions.

If you can't work with a Visio drawing file due to file format incompatibility, the most common solution is to ask the author of the Visio drawing to save the drawing as an older file version. Although you're able to work on the drawing file, the file might lose newer features during the conversion. If your version of Visio is embarrassingly old, another option is to ask the author to publish the drawing as a web page (see Chapter 9). The Visio Viewer is a free download that displays and prints Visio drawings from other versions (see Chapter 11) and is also built into Internet Explorer 7.

The Visio File menu includes a Publish as PDF or XPS command for producing PDF (Portable Document Format) or XPS (XML Paper Specification) files that you can publish to the Web. To specify what you want to publish, in the Publish as PDF or XPS dialog box, click Options to specify the page range, whether to publish in color or black and white, and whether to include non-printing data, such as accessibility tags.

TIP You can save a Visio drawing as a print file, which you can then print from any computer connected to the type of printer specified for that file, even if Visio is not installed on that computer. To do this, choose File ➪ Print and select the Print to File check box. Click OK and specify the name of the print file and where you want to save it. Then, in the Windows Command Prompt window, enter the `lpr` command to redirect the print file to the printer you want to use. In addition, although Visio no longer supports PostScript formats, you can save Visio to a PostScript file by choosing a PostScript printer in the Visio Print dialog box.

Configuring Visio Open and Save Options

Every person who uses Visio has his or her own preferences, such as where to save files. Visio has options for several aspects of the program's behavior, including what Visio does when you open and save files. To specify open and save options, open the Options dialog box by choosing Tools ➪ Options. Choose the Save/Open tab and then configure one or more of the following settings:

- **Prompt for Document Properties on First Save** — If you store information in document properties to make files easier to find, check this check box. Visio asks you whether you want to fill in document properties when you save a file for the first time.

- **Save AutoRecover Info Every__Minutes** — For those who can't seem to build a habit of saving their work, this option can be a lifesaver. Select this check box and specify the number of minutes between AutoRecover saves. Visio automatically saves a copy of your file, which you can restore in case a problem arises.

- **Default File Type** — In the Save Visio Files As drop-down list, choose the file format that you want Visio to use by default: Visio Document for the current file format, Visio XML Document for the current XML format, or Visio 2002 Document for the previous file format. You can override this choice in the Save As dialog box.

- **Language options** — The language options apply to files that you convert from different versions of Visio. For example, the Let Visio Decide Language option tells Visio to determine the language based on the contents of the converted file. If you want to choose the language when you convert, choose Prompt For Language. To always use the same language, choose Use The Following Language and select the language in the drop-down list.

NOTE If you organize your files in specific folders, choose the Advanced tab and click the File Paths button to specify the location for drawings, templates, stencils, help files, add-ons, and startup paths.

Protecting Files

In Visio, file protection comes in several forms. Read-only files prevent other users from modifying your Visio masterpieces. For more pinpoint control of the protection you apply, the Protect Document feature enables you to specify which items to protect. And new to Visio 2007 is the Trust Center, a window that presents all types of security and privacy settings in one place.

CROSS-REF With layers on a drawing (described in Chapter 26) you can lock the contents of one or more layers against changes.

Saving a Read-Only Copy

A read-only copy is not a surefire way to protect the contents of a drawing. When people open read-only drawings, they can modify the contents all they want. The catch is that they must save modified files with different names. Moreover, anyone can remove the read-only protection in Windows Explorer. The protection comes mainly from the warning message that appears when someone tries to save a read-only file.

To save a read-only copy of a file, choose File ➪ Save As. After specifying a file name and location for the file, click the Save drop-down arrow to display Save options. Select Read Only and then click the Save button.

NOTE As an extra precaution, create a backup copy of the file before saving.

The Protect Document Command

Protecting specific elements on a drawing is possible when you use the Protect Document command, as described in the following steps:

1. Choose View ⇨ Drawing Explorer Window.

2. Right-click the name of the drawing file (not the pages within the file) you want to protect and, on the shortcut menu, choose Protect Document.

NOTE If Protect Document doesn't appear on the shortcut menu, you can add the command to a Visio menu or toolbar. To do this, click the Toolbar Options arrow at the end of a toolbar, and choose Add or Remove Buttons ⇨ Customize. In the Customize dialog box, select the Commands tab, select Tools in the Category list, and then drag Protect Document to a position on the toolbar.

3. Check the items you want to protect from unauthorized changes and click OK. You can protect the following elements of a Visio drawing:

   - **Styles** — Although you can still apply styles when this check box is selected, you can't create new styles or edit existing ones.

   - **Shapes** — This setting combined with the From Selection setting in the shape Protection dialog box prevents you from selecting shapes.

   - **Preview** — This setting prevents changes to a Visio file's preview image when you change the contents of a drawing page. For example, you can set the preview image when the drawing file is simple and then prevent it from updating when the drawing contents grow dense. For templates, a preview shows the purpose of the template. You save a sample drawing with the template and lock the preview image. Then, you can delete the sample drawing to clean out the template, but the preview image remains.

   - **Backgrounds** — Use this to prevent the deletion or editing of background pages.

   - **Master shapes** — This setting prevents the creation, editing, or deletion of masters. However, you can still create instances of masters on drawing pages.

To remove protection from a drawing, clear the check boxes in the Protect Document dialog box and click OK.

Security and Privacy Settings

Unfortunately, life online isn't completely safe. Macros can be great productivity tools, but miscreants write macros that hide in files and spread viruses to your computer when the host files are opened. Add-ins and ActiveX controls are also susceptible to tampering for malicious purposes. The Trust Center is a new Office 2007 feature, shown in Figure 2-2, which checks for potential security breaches in your Visio and other Office 2007 files. You can specify the level of security you want. This section describes the Trust Center settings that are available for add-ins, ActiveX controls, and macros. To open the Trust Center, choose Tools ⇨ Trust Center.

FIGURE 2-2

The Trust Center provides settings that control whether macros, add-ins, and ActiveX controls can run and what they can do.

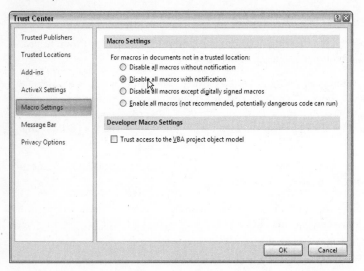

Macro Settings

The Trust Center window opens to Macro Settings automatically. The settings in this section apply only to macros in documents that come from an untrusted location, for example, a web site that is unfamiliar to you. If a document resides in a trusted location, Visio assumes that the macros within that document are safe. The Trust Center checks whether macros have valid and current digital signatures either from trusted publishers or documented by a certificate from a certificate authority.

Choose one of the following options to specify how Visio should respond to macros:

- **Disable All Macros Without Notification** — For untrusting types, choose this option if you want Visio to disable untrusted macros without telling you that it has done so.

- **Disable All Macros With Notification** — This option tells Visio to disable untrusted macros, but to inform you that the macros are disabled in case you want to enable them.

- **Disable All Macros Except Digitally Signed Macros** — Choosing this option says that you trust macros with digital signatures. Visio digitally signed macros even if they come from untrusted locations.

- **Enable All Macros** — This option is for gamblers or the perpetually optimistic folks who don't worry about malicious code running on their computers. The only time you might choose this option is if you receive a drawing file that you know is secure that also contains unsigned macros needed to edit the drawing.

If you are a developer working on customizing Visio with VBA, check the Trust Access To The VBA Project Object Model, so you can run code you develop with VBA.

TIP To tell Visio that a location is trusted, in the Trust Center navigation bar click Trusted Locations. In the Trusted Locations screen, click Add New Location, specify the path to the location, and click OK.

Add-ins

Add-ins are code that expands what your Office programs do. The Trust Center checks add-ins for their trustworthiness as it does for macros: valid and current digital signatures from trusted publishers or backed by certificates from reliable authorities. In the Trust Center window, click Add-ins to open the Add-ins screen. Use one of the following methods to specify how Visio should handle add-ins:

- **Require Application Add-ins To Be Signed By Trusted Publisher** — Select this check box if you want Visio to disable add-ins that are not signed by trusted publishers. This is a common compromise between security and convenience.

- **Disable All Application Add-ins** — This option tells Visio to disable add-ins, which is highly secure but can disable features you depend on, such as global templates in Office.

- **Manage Add-ins** — To specify the add-ins that you want to run, in the Manage drop-down list, choose COM Add-ins and click Go. In the COM Add-ins dialog box, click Add to specify an add-in that you want Visio to run.

ActiveX Settings

The settings in this section apply only to ActiveX controls in documents that come from an untrusted location.

Choose one of the following options to specify how Visio should respond to ActiveX controls:

- **Disable All Controls Without Notification** — Choose this option if you want Visio to disable untrusted ActiveX controls without telling you that it has done so.

- **Prompt Me Before Enabling Unsafe for Initialization Controls** — This option tells Visio to ask you how you want to work with ActiveX controls designated as Unsafe For Initialization or Safe for Initialization (certified as safe by the developer).

- **Prompt Me Before Enabling All Controls With Minimal Restrictions** — This is the default option, which sets behavior based on whether the document contains a VBA project and the designation of the ActiveX controls.

- **Enable All Controls Without Restrictions and Without Prompting** — It should be obvious that this option allows ActiveX controls to run regardless whether they are designated as safe.

Select the Safe Mode check box to restrict the behavior of ActiveX controls that are designated as Safe for Initialization. For example, Safe Mode could restrict access to read only, whereas in Unsafe Mode, controls could read and write to files. ActiveX controls that are designated as Unsafe for Initialization always run in Unsafe Mode, regardless of this setting.

Summary

By creating drawings using templates, you can get to work without worrying about settings such as page size or scale. You can also use existing drawings as a foundation for new drawings. Whether you create drawings from templates or existing drawings, Visio provides several methods for creating new drawing files, opening existing files, and saving files.

Keeping the work you've completed safe is important. Visio includes methods for protecting your work from inadvertent changes, whether you modify a drawing by mistake or someone else tries to enhance your diagram without your permission. In addition, the new Trust Center presents all sorts of security and privacy settings for specifying whether Visio should run macros and other code and under what conditions.

Chapter 3

Working with Visio Files

As you work on Visio drawings, you'll find yourself zipping around the drawing area, editing here, adding there, and generally tweaking everywhere. Visio includes the usual panning and zooming commands, but you can also look at different aspects of your Visio drawings through several special Visio windows, such as the Shape Data Window, which shows shape data fields and their values.

As your drawings grow in size and complexity, you can add pages to hold more content. Background pages work like watermarks, displaying company logos or background graphics, such as title blocks, that appear on each page of a drawing. Working with multiple drawing pages is easy — a page-level shortcut menu includes commands for adding, deleting, renaming, and reordering pages. For extra credit, you can rotate pages temporarily to add objects at an angle — for example, to draw offices positioned at different angles in a modernistic floor plan.

Page settings further fine-tune how your drawings look and behave, for example, whether you want a 2-inch thumbnail with shapes at their actual size or a 22-inch by 34-inch architectural plan with shapes scaled to $\frac{1}{4}$ inch equal to 1 foot. Layout and Routing settings influence the appearance of connections and the readability of your drawings, whereas Shadow settings add visual impact.

After you complete a diagram, printing is often the next task on the agenda. In Visio, you can preview your diagrams to make sure they are what you want before committing them to paper. And, for the printing itself, you can specify what parts of the Visio drawing you want to print.

31

Viewing Drawings

Moving in, out, and around the drawing page is a continuous part of drawing construction and fine-tuning your drawings. Panning and zooming is the most common way to see what you want, but Visio offers other tools for viewing drawings. Suppose you want to see close-ups of two shapes that reside on opposite sides of the drawing. By creating additional windows for your Visio drawing, you can zoom in on both shapes at the same time.

CROSS-REF People who don't have Visio installed on their computers can still view Visio drawings. To learn about using the Microsoft Visio Viewer or viewing Visio drawings on web pages, see Chapter 11.

Panning and Zooming

Sometimes, you want to see the big picture of a diagram, and at other times close-ups are required. For example, taking in the entire drawing page at once helps keep the drawing within the page borders or highlights areas that are too dense with shapes. Close-ups are ideal for selecting shapes that you want to edit, reading and editing text in shapes, or performing shape operations with surgical precision. Whenever you zoom into smaller areas, you typically start to pan around the drawing to see other areas.

Fast Pan and Zoom with Keyboard Shortcuts

Keyboard shortcuts provide the fastest way to pan and zoom and are the hands-down choice for Visio mavens:

- **Zoom in** — Use Ctrl+Shift+left-click.
- **Zoom out** — Use Ctrl+Shift+right-click.
- **Zoom into an area** — Use Ctrl+Shift+drag across the area with left mouse button.
- **Change the center of the zoom area** — Press Ctrl+Shift and drag the magnifying glass to the new center.
- **Pan** — Use Ctrl+Shift+drag with right mouse button.

TIP Visio can center the zoom area on the selected shape when you zoom in or out. To set this behavior, choose Tools ➪ Options, select the General tab, and check the Center Selection On Zoom check box.

To pan up or down, roll the mouse wheel. Press the Shift key while rolling the wheel to pan from side to side. You can zoom in and out by pressing the Control key while rolling the mouse.

Visual Cues with the Pan & Zoom Window

The Pan & Zoom window gives you a bird's-eye view and a close-up at the same time. If you want to zoom into a specific area while keeping tabs on where you are in the drawing page, this window can't be beat. As illustrated in Figure 3-1, the Pan & Zoom window shows the entire drawing page with the zoom area outlined in red (gray in the figure). To open the Pan & Zoom Window, choose

View ⇨ Pan & Zoom Window. Then, use one of the following methods to specify the area you want to see:

- **Zoom in or out** — On the right side of the Pan & Zoom window, drag the zoom slider up or down to zoom in or out, respectively.

- **Zoom into an area** — Drag a side or corner of the red outline to resize it, thereby changing the part of the drawing visible in the drawing window.

- **Zoom into an area** — Click and drag to define a new zoom area box in the Pan & Zoom window.

- **Reposition the center of the zoom area** — Click a point in the Pan & Zoom window to relocate the center of the zoom area box.

FIGURE 3-1

The Pan & Zoom Window indicates the zoom area with a red outline, which you can drag or click to change the zoom area.

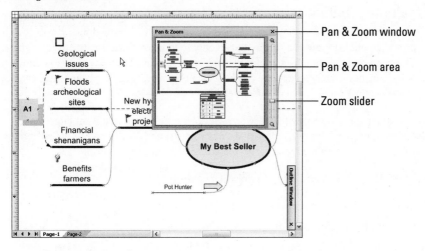

TIP Visio docks the Pan & Zoom window within the drawing window, but if space in the drawing window is at a premium, you can drag the Pan & Zoom Window by its title bar and dock it within the Shapes window.

Panning and Zooming from Menus and Toolbars

The View menu and the Standard toolbar both contain zoom options, but the zoom list on the Standard toolbar is the faster of the two. Click the down arrow to the right of the zoom percentage to display a menu with several predefined zoom percentages as well as Page, Width, and Last to display the entire page, the entire width of the page, and the last zoom used, respectively.

Working with Drawing Windows

Sometimes, one drawing window just isn't enough, for example, when you want to copy shapes from one drawing to another or view details in two widely separated areas of the same drawing. Just as Visio automatically creates a new drawing window when you open a second Visio drawing file, you can create additional windows to look at different drawing pages or sections within the same page. Of course, with a gaggle of drawing windows to look at, the next task is to arrange them, so you can see everything you need at once. Visio provides several methods for arranging drawing windows, as described in this section.

Creating New Windows

To create a new drawing window, choose Window ⇨ New Window. Visio creates the new drawing window and displays the same drawing contained in the previous window, identified in the Visio title bar by the :2 that follows the file name. The Shapes window for this new drawing window doesn't contain any open stencils. However, windows docked in the Shapes window are docked in the new Shapes window as well.

Managing Multiple Windows

Initially, Visio places a new drawing window over any other windows, filling the drawing area completely. To see more than one window at a time, you can choose tiling or cascading to create a separate pane for each open window. Tiling is great for viewing several areas of detail at the same time, because it arranges the panes side by side in the drawing area. All the windows appear at the same time, but each one takes up a smaller area of the screen.

Cascading is better when you want to switch between windows quickly but also want larger panes for each window. Cascaded windows overlap, with each window slightly lower and to the right of the previous one. When you cascade windows, the current window appears in front.

Here are several methods for switching between drawing windows:

- **Tile windows** — Choose Window ⇨ Tile.
- **Cascade windows** — Choose Window ⇨ Cascade.
- **Bring a window to the forefront** — Click any visible part of that window. When a window is hopelessly buried under other windows, choose Window and then, at the bottom of the menu, choose the name of the window.
- **Fill the drawing area with one window** — Click that window's Maximize button (the button in the upper-right corner with an icon that looks like a window).

Working with Drawing Pages

A Visio drawing can require multiple pages for several reasons. For example, the database model for an airline reservation system could require hundreds of pages to show every table. A construction project could show the building plan on one drawing page, with elevations and construction details on others. Moreover, background pages act like watermarks — for example, displaying a company logo and standard title block for every foreground drawing. Every page in a drawing file can have its own settings for drawing size, orientation, margins, units and scale for architectural and engineering drawings, and more.

Background and Foreground Pages

When you create a page in Visio, you see two options for page type: foreground and background. By default, Visio selects Foreground, but Visio experts know that background pages can be incredibly useful. Before you create pages in drawings, read on to learn what each type of page is for and how to make the most of them.

- **Foreground pages** are where the meat of diagrams lies. These pages contain the flow-chart shapes, organization chart shapes, database shapes, and so on. When you print in Visio and specify the All option for the page range, Visio prints all the foreground pages. The only time background pages print is when they are associated with one or more foreground pages, and then they print whenever their associated foreground pages print.

- **Background pages** contain supporting material, which might apply to one or more foreground pages — for example, a title block, company logo, legend, and drawing border. Visio background pages are a lot like Headers and Footers in other programs such as Word and Excel — they can hold common information that you want to appear in the same location on each page. You might then wonder why Visio also offers headers and footers. In Visio, headers and footers have a very specific role — enabling you to add page numbers and dates on individual pages when a single Visio drawing page prints onto several pieces of paper. If the Visio drawing page and the printed page size are identical, background pages are the best way to show information on every page.

TIP Each foreground page can have only one associated background page. However, you can build a hierarchy of supporting pages by assigning one background page to another background page. For example, you could create a background page for the drawing title block, which applies to every foreground page in the drawing file. Then, you might create two other background pages: one for a floor plan legend and the second for the legend for detail pages. Then, associate the title block background page to each legend background page. Now, you can associate the floor plan legend page to every floor plan foreground page and the detail legend page to every detail foreground page. The final result is floor plans and detail pages that show the appropriate legend and the title block.

Creating Pages in a Drawing

A Visio drawing file initially contains one page, but you can create as many additional pages as you want. Although new pages inherit the page settings of the active page in the drawing window, Visio displays the Page Setup dialog box, so you can modify page settings at the time of creation.

Creating New Pages

Use any of the following methods to create foreground or background pages:

- Right-click any page tab and, on the shortcut menu, choose Insert Page.
- Choose Insert ⇨ New Page.
- In the Drawing Explorer window, right-click the Foreground or Background folder and then click Insert Page.

Regardless of which method you use, Visio creates a new page and gives it a default name with a name such as Page-1. You can rename the page to whatever name you want, as described in the section "Renaming Pages" later in this chapter.

Creating Background Pages

Background pages require two steps — you must create a background page and then assign it to a foreground or background page. To set up a background page, follow these steps:

1. Choose Insert ⇨ New Page.

2. Select the Background option. Visio fills in the Name text box with a default name, such as Background-1.

3. Optionally, to assign a different name, in the Name text box, type a name for the new background page.

4. Make any other page setting changes you want and click OK. Visio creates the new page and a new page tab. The page tabs for background pages appear to the right of the page tabs for all foreground pages.

5. To assign the background page to a foreground or background page, select the page to which you want to assign the background page by clicking its page tab. Choose File ⇨ Page Setup.

6. Select the Page Properties tab; in the Background drop-down list, select the background page name; and click OK.

Editing Pages

When a drawing contains several pages, there's bound to be some changes to those pages — deleting pages as you move shapes around, renaming pages to make them easier to identify, or reordering the pages.

The quickest way to access page commands is by right-clicking a page tab at the bottom of the drawing window. From this shortcut menu, you can do all of the following:

- Insert new pages
- Delete the selected page
- Rename the selected page
- Modify the order of the pages

Protecting Background Pages from Modification

Background pages typically contain content, such as title blocks, graphic backgrounds, and logos, which apply to several foreground pages. When you have a background page set up the way you want, you want it to stay that way. But the page tab at the bottom of the drawing area displays a tab for background pages, just begging to be selected — and from there, it's a small step to editing the background page by mistake. Visio is one step ahead of you, because the Protect Document command can hide background pages, so their tabs don't appear in the drawing window, although the names still appear on the Background drop-down list in Page Setup. With Protect Document in place, you must unprotect background pages in order to change background pages or create new ones.

The Drawing Explorer Window is the only place to issue the Protect Document command. To hide background page tabs, do the following:

1. Choose View ➪ Drawing Explorer Window.

2. At the top of the Drawing Explorer Window, right-click the name of the drawing file and, on the shortcut menu, choose Protect Document.

3. In the Protect Document dialog box, select the Backgrounds check box.

4. Click OK. In the drawing window, Visio hides the tabs for all background pages.

5. To redisplay the tabs for background pages, repeat the previous steps, but clear the Backgrounds check box. In the Drawing Explorer Window, select a background page and the page tab reappears.

Renaming Pages

Visio assigns names, such as Page-1, to new pages you create. When a drawing has more than one page, meaningful page names make it so much easier to find the data you want. To rename a page, use one of the following methods:

- Double-click the page tab and type the page's new name.

- Right-click the page tab, click Rename Page, and type the page's new name.

 **If the Page Setup dialog box is open, you can rename the current page by clicking the Page Properties tab and typing a new name in the Name text box.**

Deleting Pages

To delete only one page, right-click its page tab and click Delete Page. To delete several pages at once, follow these steps:

1. Choose Edit ➪ Delete Pages.

2. In the Delete Pages dialog box, Ctrl-click each page you want to delete. If the pages still have the default names that Visio assigned and you want Visio to renumber the page names, select the Update Names check box.

3. Click OK.

 A background page remains immune to deletion as long as at least one foreground page uses it. To remove a background page assignment, select the foreground page to which it is assigned, open the Page Setup dialog box, click None in the Background drop-down list, and then click OK. After you have removed all the background page assignments, you can delete the background page.

Reordering Pages

As you insert and delete pages, the remaining pages might end up out of order. You can reorder foreground pages. If most of the page tabs are visible, the easiest reordering method is to drag and drop pages into the positions you want. To do this, select the tab of the page you want to move and drag and drop the tab into a new position.

For drawings teeming with pages, it's easier to reorder them in the Drawing Explorer window. To do this, follow these steps:

1. Choose View ➪ Drawing Explorer Window.

2. Right-click either the Foreground folder or the Background folder and, on the shortcut menu, click Reorder Pages.

3. In the Reorder Pages dialog box, select the name of a page that you want to move and use the Move Up or Move Down buttons to reposition it in the order. To instruct Visio to renumber default page names, select the Update Names check box.

4. Click OK when you have finished reordering all the pages.

Rotating Pages

If a drawing has areas in which all the shapes appear at angles, such as an office layout that angles some of the furniture, rotating the drawing is much easier than rotating each shape into position. By rotating the drawing page, you can construct each section orthogonally — that is, placing each shape at right angles to the drawing grid.

Although existing shapes and guides rotate with the page, the rulers and drawing grid remain fixed. While the page is rotated, you add new shapes at right angles to the rulers and grid. When you rotate the page back to its original orientation, the shapes you added rotate with it, as illustrated in Figure 3-2.

NOTE Rotating pages does not affect print and page orientation settings. A rotated page prints as if it were not rotated.

To rotate a page, follow these steps:

1. Display the page you want to rotate in the drawing window.

2. Press Ctrl and position the pointer over a corner of the page. The pointer changes to a rotation pointer.

3. Drag the corner until the page is rotated to the angle you want. If you want to rotate to a specific angle, release the mouse button when the rotation angle value in the status bar equals the angle you want.

FIGURE 3-2

Rotating a page enables you to easily add shapes that you want angled on the page.

Add a shape while the page is rotated.   Shape rotates with page.

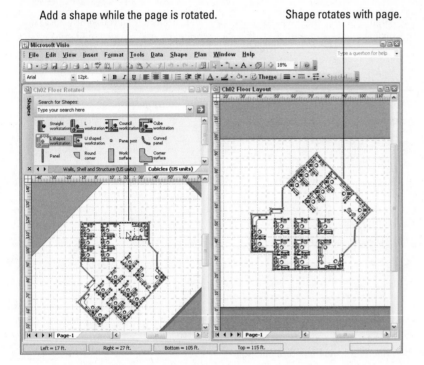

| Left = 17 ft. | Right = 27 ft. | Bottom = 105 ft. | Top = 115 ft. |

Setting Up Pages

Each page in a drawing can have its own unique settings. For example, a floor plan page might use a D-size sheet of paper with landscape orientation and an architectural scale, whereas the door and window schedule for the floor plan might apply a standard letter-size sheet of paper with portrait orientation and no scale. The Page Setup dialog box is the place to specify the settings for a page, which you open by clicking a page's page tab and then choosing File ➪ Page Setup. This section describes the settings you can apply and when they come in handy.

Specifying the Printing Settings

The first step in setting up a drawing page is to specify the printer paper you want to use and how Visio should fit the drawing onto the paper. For that reason, the Print Setup tab is the very first tab in the Page Setup dialog box. Each drawing page within a Visio drawing file can have its own page size and orientation.

To prevent incompatibility between drawing pages and printer paper, Visio displays a preview window that shows the current settings for both, as shown in Figure 3-3. If a drawing page and the printer paper don't fit like a hand in a glove, you can resolve the discrepancy by modifying settings on either of the Print Setup or Page Size tabs. For example, in Figure 3-3 the printer paper is set to portrait orientation, whereas the drawing page is set to landscape. If you print the drawing with these settings, Visio prints the drawing on multiple pieces of paper.

 **If you try to print a drawing when the paper and page are oriented differently, Visio displays a warning message and suggests that you change the orientations to match.**

FIGURE 3-3

The Print Setup and Page Size tabs preview your print settings to highlight incompatibilities.

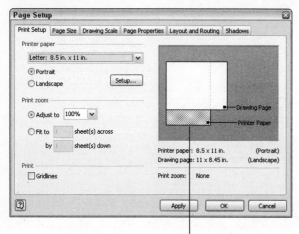

Discrepency between Print Setup and Page Size settings

Here are the Print Setup tab settings and what you can do with them:

- **Printer Paper** — From the drop-down list of standard paper sizes, choose the size of the paper in the printer.

- **Printer Page Orientation** — Choose the Portrait or Landscape option based on how you position shapes on the page. For example, for a report, choose Portrait. For a wide diagram, choose Landscape.

- **Setup Options** — To specify more characteristics for printer paper, click the Setup button and then use one or more of the following settings:

  - **Page Margins** — Click the Setup button to set page margins.

  - **Small Drawing Centering** — If a drawing is smaller than the size of the paper, you can specify how you want Visio to position the drawing on the paper. Click the Setup button and select the Center Horizontally and Center Vertically check boxes. If you leave these check boxes unchecked, the drawing prints at the top left section of the paper.

- ▪ **Print Zoom** — Print Zoom enlarges or shrinks a drawing only for printing. For example, you can use Print Zoom to print a larger drawing on a letter-size sheet. Select the Adjust To option to specify a percentage for the print zoom or select the Fit To option to specify the number of sheets across and down.

- ■ **Print Gridlines** — Select the Gridlines check box to print the drawing grid along with the contents of your drawing.

CAUTION The Print Setup tab in the Page Setup dialog box controls page size and orientation for a page. Clicking the Properties button in the Print dialog box sets page size and orientation settings for a printer. If you choose different settings in these two locations, Visio will display a warning before printing, indicating that drawing pages will print across multiple pages because they are oriented differently than the printed page. To prevent this behavior, make sure that your drawing page dimensions are compatible with printer page size and orientation.

Specifying Page Size

By default, Visio sets the drawing page size to the printer paper size. To specify a page size that is different from that of the printer paper, in the Page Setup dialog box, select the Page Size tab. For example, you can use a Ledger page (17 inches × 11 inches) for a large diagram, but print it on letter-size paper. The Page Size tab displays a preview window that shows the current settings for both Print Setup and Page Size.

On the Page Size tab, you can specify the following:

- ■ **Page Size Same As Printer Paper Size** — Visio selects this option by default.

- ■ **Pre-defined Size** — For pre-defined paper sizes, select a category of sizes, such as Standard or ANSI Architectural, and then select one of the standard sizes for that category.

- ■ **Custom Page Size** — If you use specially sized paper, specify the dimensions for the height and width of the custom page.

- ■ **Size To Fit Drawing Contents** — Selecting this option defines a page size just large enough to hold the contents of the current page.

- ■ **Page Orientation** — Choose Portrait or Landscape for the page. Make sure that the page orientation and printer paper orientation match.

TIP You don't need the Page Setup dialog box open to change the size of a page. To resize a page in the drawing area, on the Standard toolbar, click the Pointer tool. Press and hold the Ctrl key and position the pointer along the edge of the page, until the pointer changes to a double-headed arrow. Drag the edge to change the dimension of the page. As you drag, the current dimensions for the page appear in the status bar.

Defining Drawing Scale

When you work on scaled drawings such as construction plans, the drawing scale is what makes real-world objects fit onto sheets of paper. The drawing scale in Visio is the ratio that transforms real-world measurements to smaller distances on a drawing page. For example, the architectural scale, 1/4" = 1' 0" means that one foot in the real world shows up as a quarter of an inch on paper. In Visio, the No Scale option shows objects at their actual size — one-to-one scale. On the Drawing Scale tab, you can specify the following options:

- **No Scale** — Objects appear at their actual size.
- **Pre-defined Scale** — Select a category of scales, such as Architectural or Metric, and then select one of the standard scales for that category.

NOTE The page size at the bottom of the Drawing Scale tab indicates the real-world size of your scaled drawing page (a letter-sized page represents 96 feet by 144 feet at 1/4" = 1' 0" scale). You can modify these values to change the size of the page, at which point the preview window shows the new relationship between your page and the printer paper.

- **Custom Scale** — Specify the measurement unit and the associated size in the real world to define a custom scale.

CROSS-REF To learn more about using drawing scales, see Chapter 25.

TIP The Title Blocks stencil includes masters that display the current drawing scale. These shapes reference a page field, so the scale in the title block updates automatically if you modify the scale for the drawing page.

Specifying Measurement Units

On the Page Properties tab, the Measurement Units drop-down list provides choices for how Visio measures objects on a page and the units that appear on the drawing rulers, such as Meters or Feet and Inches. The Measurement Units list includes many familiar units of distance, units of time, and a few options that you might not know. For example, with drawings that contain graphs showing performance over time, you can specify the time units that appear on the drawing rulers. The remaining units in the list are typesetting measurements; didots and ciceros represent metric measures, whereas points and picas represent U.S. typesetting measures.

Setting Up Layout and Routing

The Layout and Routing tab is where you set the defaults for how shapes and connectors interact on a page, such as where connectors attach to shapes, whether connectors that share a shape overlap or not, and whether connectors use straight or curved lines. Visio uses these settings, unless you specify different settings with the Configure Layout command on the Shape menu. As illustrated in Figure 3-4, the Layout and Routing tab previews the settings you choose, so you can decide whether to apply them. On the Layout and Routing tab, you can specify the following options:

- **Style** — This setting specifies the layout of shapes, for example, Tree or Flowchart.
- **Direction** — Only applicable to some styles, such as Tree and Organization Chart, the direction specifies whether layout proceeds from top to bottom, bottom to top, left to right, or right to left.

- **Separate** — This setting specifies which type of connections should be separated. For example, when you choose Unrelated Lines, connectors from the same manager to multiple subordinates are related, so overlapping paths are acceptable. However, Visio would separate connectors between different managers and subordinates. This setting has no effect on connectors that are already separated.

- **Overlap** — As a companion to the Separate setting, Overlap specifies which connections should be overlapped. For example, a common choice for Overlap is Related Lines, which would overlap all the connectors between one manager and the manager's subordinates. Overlapping all lines makes drawings less cluttered, but relationships between shapes are more difficult to identify.

NOTE The Separate and Overlap settings can contradict each other, and, in that case, one setting overrules the other. For example, if you set Separate to All Lines, but set Overlap to Related Lines, Visio overlaps related lines.

FIGURE 3-4

In the Page Setup dialog box, the Layout and Routing tab defines the default settings for how Visio lays out shapes and connectors.

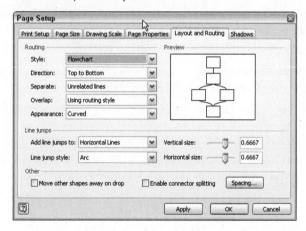

- **Appearance** — This specifies whether the connectors are straight or curved.

In the Page Setup menu, you can also configure the appearance of line jumps, which visually indicate that lines do not connect:

- **Add Line Jumps To** — Choose which connectors use line jumps in the Add Line Jumps To drop-down list. For example, you can display line jumps in horizontal or vertical lines, with the first or last line displayed.

- **Line Jump Style** — This setting specifies the appearance of line jumps, such as arcs or gaps.

- **Line Jump Size** — Set the Horizontal Size to specify the width of line jumps on horizontal lines. Set the Vertical Size to specify the height of line jumps on vertical lines.

The Layout and Routing tab includes a few other settings, including the following:

- **Move Other Shapes Away on Drop** — When this check box is selected, shapes automatically move on a page to make room for a shape that is dragged, moved, or resized on a page.

- **Enable Connector Splitting** — When this check box is selected, a shape dropped onto a connector splits the connector in two. The new shape automatically has connections to the two shapes that were previously connected.

- **Spacing** — Click this button to open the Layout and Routing Spacing dialog box. You can specify spacing between shapes, spacing between connectors, spacing between connectors and shapes, and the average shape size.

Setting Up Shadows

The Shadows tab controls the appearance of shadows associated with shapes on a page, such as the following options:

- **Style** — This determines the angle and connection between the shadow and the shape, much like the position of a light source determines the shape of a shadow. The entries in the drop-down list provide thumbnail samples of the shadow effect of each style.

 If you choose an oblique style, in which the shadow is oriented at an angle to the shape, you can specify the angle of rotation for the shadow.

- **Size & Position** — These settings specify the distance between shapes and their shadows. You can specify distances for horizontal and vertical offset or use the direction buttons.

- **Magnification** — Drag this slider to specify the size of the shadow relative to the original shape. For example, 100% makes the shadow the same size as the shape.

Previewing and Printing Drawings

Visio tries hard to ensure that your drawings print as you would expect. In most Visio templates, the drawing page and printed page settings are the same, so you don't have to adjust page settings before printing. Visio also adjusts colors in your drawing to your printer's resources, for example, swapping out colors for shades of gray, if you don't have a color printer. To get the best results the first time, it's always a good idea to preview a drawing before printing.

NOTE Shapes that lie outside of the drawing page do not print. To include these shapes when you print your drawing, first move them within the borders of the drawing page or resize the drawing page.

Previewing Drawings

Previewing drawings isn't just for admiring your handiwork — it gives you a chance to correct print settings before ruining stacks of paper with potentially flawed diagrams. Visio Print Preview features are similar to those in other Microsoft Windows applications.

To preview a drawing, choose File ➪ Print Preview or, on the Standard toolbar, click Print Preview, which looks like a piece of paper with a magnifying glass. Visio shades the margins for the printer paper, so you can see how your drawing fits on it. For example, if shapes are too close to the margins, you might reposition the shapes or move to a larger size paper. For drawings that require more than one page to print, Visio shades the page breaks as well.

The Page Setup dialog box also previews the drawing page as it fits on the paper — without showing the drawing contents. This is handy for checking the compatibility between drawing page and printer paper, for example, spotting a portrait drawing that you are trying to print to landscape printer paper. Choose File ➪ Page Setup. The Print Setup, Page Size, and Drawing Scale tabs all include a preview window. If the preview indicates a discrepancy, modify settings on those tabs to correct the problem.

TIP The easiest way to ensure that the drawing and printer paper match is to use the Same As Printer Paper Size option. Choose File ➪ Page Setup, select the Page Size tab, and select the Same As Printer Paper Size option.

Printing Drawings

Sure, you can print entire drawings, but, many times, you want to print only specific elements or areas of drawings. For example, for a staff meeting, you might print the organization chart shapes for only your department, not the hundreds of other employees in the company. Before you print, take the time to set up your print options, which are described in the rest of this section. However, with print settings in place, the methods for telling Visio to print are easy:

- Press Ctrl+P.
- On the Standard toolbar, click Print.
- Choose File ➪ Print.

Printing Parts of a Drawing

The Print dialog box contains options for specifying pages or portions of your drawing that you want to print. To print only part of a drawing, choose one of the following methods:

- **Selected Pages** — To choose pages to print, choose File ➪ Print and, in the Print dialog box, choose one of the following options:
  - **All** — This option prints all the foreground pages in the drawing file.
  - **Current Page** — This option prints the active page and works whether the page you select is a foreground or background page.
  - **Pages From and To** — Type the numbers for the first and last page you want to print.

When Shapes Don't Print

If shapes are missing on the printed drawing, they might be configured as nonprinting shapes. To reset a nonprinting shape, right-click it and choose Format ➪ Behavior. Clear the Non-Printing Shape check box and click OK.

Layers that are set not to print are also potential culprits for unprinted shapes. To check the shape layer, right-click the shape and choose Format ➪ Layer. Choose View ➪ Layer Properties and make sure the Print column for the layer is selected.

If a shape *still* doesn't appear, the printer driver might have misinterpreted the shape's colors. To verify the presence of shapes, choose File ➪ Print and then select the Color As Black check box to print all lines and fills with black.

- **Printing a Portion of a Drawing** — To specify an area of the drawing you want to print, choose File ➪ Print and choose one of the following two options:

  - **Selection** — If you have selected shapes on your drawing, click this option to print only the selected shapes.

  - **Current View** — Click this option to print the portion of the drawing that appears in the Visio drawing window.

- **Printing a Background Page** — Display the background page you want to print and then choose File ➪ Print. Select the Current Page option and click OK to print the background page.

- **Printing Only a Foreground Page** — You must remove the background page associated with a foreground page if you want to print only the foreground page. To do this, display the foreground page and choose File ➪ Page Setup. Select the Page Properties tab, click None in the Background box, and click OK. Use the Current Page option in the Print dialog box to print the page.

- **Printing Drawing Markup** — Display the drawing markup and then print the drawing.

TIP By default, guides are nonprintable objects, but you can print a guide by modifying the guide's ShapeSheet. To do this, select the guide you want to print and choose Window ➪ Show ShapeSheet. Scroll to the Miscellaneous section and type False in the NonPrinting cell.

TIP In addition, you can use other Visio features, such as layers and markup, to control what you print. You can specify whether layers — and the shapes on them — print. For example, nonprinting layers are perfect for shapes you use as reference points, guides, or feedback. To prevent a layer from printing, choose View ➪ Layer Properties and clear the check mark in the Print column for the layer.

Even without layers, Visio can configure shapes that appear only while you are working on a drawing, but not when you print. To do this, select the shape or shapes and choose Format ➪ Behavior. Select the Non-Printing Shape check box and click OK.

Printing Large Drawings

When a drawing is larger than the largest paper size that your printer can accommodate, you can choose from several solutions, depending on your needs. If you're fortunate enough to have a larger format printer available, you can add access to that printer to your computer and then print to a larger sheet of paper. But for most folks, one of the following solutions usually does the trick:

■ **Squeezing a drawing onto a page** — If your drawing **almost** fits on one sheet of paper, the easiest solution is to shrink the drawing to fit on one sheet. To do this, choose File ➪ Page Setup and select the Print Setup tab. In the Print Zoom area, click the Fit To option and type **1** in both the Across and Down boxes.

NOTE The Fit To option is ideal any time you want to print a drawing to a specific number of pages. Type the number of pages you want in the horizontal direction in the Across box, and then do the same for the number of pages wanted vertically, in the Down box.

■ **Tiling a drawing across several pages** — Print Zoom doesn't help if your drawing is much larger than your printer paper, or when you want to print your drawing to scale. For these situations, you can tile your drawing across several sheets of paper. The preview in the Page Setup dialog box shows the relationship between your drawing size and printer paper, as demonstrated in Figure 3-5. With tiling, shapes that overlap page breaks might print twice — once on each page on either side of the page break. To prevent shape duplication, display the page breaks on your drawing and relocate any shapes that overlap them. Overlaps eliminated, go ahead and print the drawing to multiple sheets. To eliminate shapes overlapping page breaks, follow these steps:

1. To view the page breaks on your drawing, choose View ➪ Page Breaks. Visio indicates page breaks with gray shading. The thickness of the shaded lines represents the margins set for the printed page.

2. To reduce the thickness of the page breaks, choose File ➪ Page Setup and select the Print Setup tab. Click the Setup button and specify narrower margins. Click OK in the Print Setup dialog box and then click OK in the Page Setup dialog box.

NOTE Printers have minimum margins that you can't reduce. If you specify margins smaller than the minimum for the current printer, Visio sets the margins to the smallest margin that the printer can handle.

Correcting Orientation Mismatches

If you happen to assign one orientation on the Print Setup tab and a different orientation on the Page Size tab (which is usually an oversight), trying to print a Visio drawing page generates an error message. If you click OK in response to this error, Visio prints the drawing on multiple pages. To correct the problem before printing and print the entire drawing on one page, click Cancel and then reset either the drawing page or printer paper orientation so they match.

3. On the drawing page, drag any shapes that overlap the page break to one side or the other.

FIGURE 3-5
_____

The Page Setup preview shows page breaks for tiled drawings.

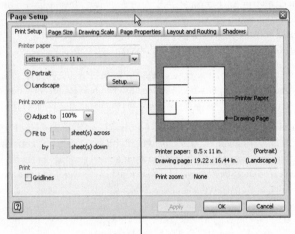

Page break indicators

Printing Drawings in the Center of the Paper

Visio offers several methods for centering drawings on paper. Moving the contents of a drawing to the center of the page or resizing the drawing page to match the contents both work:

- **Centering shapes on a page** — If your drawing page and printer paper are the same size, simply center your drawing by pressing Ctrl+A to select the contents of the page and then choosing Shape ➪ Center Drawing.

- **Resize the page to its contents** — To resize the page to exactly contain the shapes it holds, follow these steps:

  1. Choose File ➪ Page Setup and select the Page Size tab.

  2. Select the Size to Fit Drawing Contents option. The new drawing size appears in the preview area.

  3. Select the Print Setup tab and click the Setup button.

  4. Select the Center Horizontally and Center Vertically check boxes to center the drawing on the printer paper and click OK.

  5. Click OK in the Page Setup dialog box.

Summary

Visio provides tools for viewing drawings as you work and when you're ready to print. While you construct drawings, panning and zooming are constant companions. Then, when you want to print, you can switch to using Print Preview and Print. The Visio Print Setup options give you precise control over what ends up on paper.

Drawing pages hold the content of your diagrams. Foreground pages are for the diagram shapes themselves. Background pages display shapes that apply to several drawing pages, such as company logos or title blocks. A page-level shortcut menu includes a few commands for managing pages, but the Drawing Explorer Window helps you keep pages in line.

Page Settings fine-tune the appearance and behavior of drawings, whether you want to scale shapes to fit on paper, change the size of pages, or squeeze a page onto a slightly too small piece of paper. Other settings influence the appearance of connections and the readability of your drawings, such as those found on the Layout and Routing tab and Shadows tab in the Page Setup dialog box.

Chapter 4

Working with Shapes

Every drawing produced in Visio is comprised almost entirely of shapes, albeit with occasional off-the-cuff lines, curves, and text. No matter what type of diagram you want to develop, creating content is mostly a matter of dragging and dropping shapes onto drawing pages. Known as *SmartShapes,* these predefined shapes have built-in properties and behaviors that help you every step of the way.

As you work, you can select shapes to work on in several ways. With Visio tools, shape handles, and behaviors, you can position shapes easily — and as precisely or as casually as you want. Then, with shapes on the drawing page, other Visio tools and add-ons help you move, align, duplicate, and fine-tune those shapes. To simplify working with many related shapes, groups of shapes act as one monolithic shape, for instance, a title block that includes shapes for each type of title block data.

Visio provides hundreds of built-in shapes for dozens of different types of drawings. With so many shapes to choose from, you might wonder how you would ever find the shapes you want. In addition to categorizing shapes by placing them on stencils, the Visio Search for Shapes tool helps you find shapes on your computer or online.

When all is said and drawn, every so often you need some graphics that you can't find anywhere in the Visio stencils. When that happens, the Visio drawing tools enable you to construct what you need from basic lines, arcs, and curves.

If you're anxious to get started, read the section "Shapes 101" to learn what you need to know to jump-start working with shapes. For everything you ever wanted to know about shapes, continue reading the remainder of this chapter.

Shapes 101

The whole premise behind Visio is that you build diagrams by dragging and dropping predefined graphics onto drawings — without the hassle of remembering geometry class from high school, the worry of whether you have any aesthetic sense at all, or the tedium of drawing each element yourself. In Visio, these graphic (and text) elements are called *shapes* and everything on a drawing is part of a shape, from the tables, chairs, and computers you add to an office plan to the text you add to a bill of materials. Even text boxes, clip art, pictures, and objects, such as AutoCAD drawings, become shapes when you add them to a Visio drawing.

Speaking of clip art, the shapes that Visio provides look a lot like clip art, but they are much more powerful. They're called SmartShapes for a reason. Unlike a piece of clip art, which is essentially an image you can drop into a document, Visio shapes are built from vectors that you can resize, scale, edit, and format.

More importantly, Visio shapes contain behaviors that perform different actions depending on where you place them and how you use them. For example, when you drop a Manager shape onto an Executive shape in a Visio organization chart, the shape automatically connects itself to the Executive shape, positions itself according to the organization chart layout you specify, and builds a reporting relationship between the executive and the manager. Shape behaviors can change shape appearance, such as arrows whose tails can open or close to look like a part of a flowchart box or separate.

Types of Shapes

Shapes come in two flavors: one-dimensional (1-D) and two-dimensional (2-D). Understanding the difference between the two is important, because each type of shape behaves differently and you can't coax 1-D shapes to do things that belong to a 2-D shape's bag of tricks.

One-Dimensional Shapes

From the true sense of the term, one-dimensional, you might expect 1-D shapes to be limited to single straight line segments, that is, a line that follows either the x axis or the y axis. But that's not how Visio looks at one-dimensional. A Visio 1-D shape has a beginning point and an ending point, but the shape often extends over two dimensions, such as the lines, curves, and paths, shown in Figure 4-1.

FIGURE 4-1

Dragging a beginning point or ending point of a 1-D shape changes its length.

Dragging beginning and ending points changes the length of a 1-D shape.

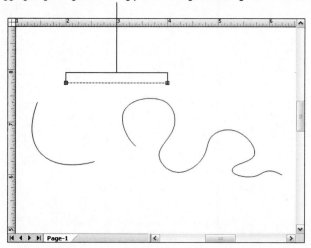

You can identify 1-D shapes by the following two behaviors:

- Green selection points appear at either end of a 1-D shape when you select it (see Figure 4-1). The green selection point at the beginning of the line is an *x* in a green square, and the selection point at the other end contains a plus in a green square. You can drag these points to connect the 1-D shape to other shapes, making them perfect as connectors between 2-D shapes, such as the arrows between boxes in flowcharts.

- When you drag the beginning point or ending point of a 1-D shape, the only thing that happens is that the length of the shape or the position of the point changes.

Two-Dimensional Shapes

2-D shapes are what you typically connect with 1-D connectors — for example, the boxes in a flowchart, the computers in a network, or the positions on an organization chart. Instead of a beginning point and ending point, a 2-D shape has eight selection handles; one at each corner of the shape and one at the midpoint of each side, as shown in Figure 4-2. Dragging corner selection handles changes both the length and width of the shape. (You can drag the selection handles at midpoints to change only one dimension.)

FIGURE 4-2

Dragging 2-D shape selection handles can change the length, width, or both dimensions.

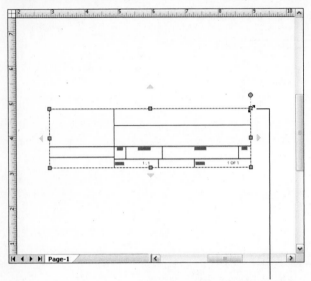

Dragging the two-headed arrow at a corner selection
handle changes both the height and width of a 2-D shape.

Stencils, Masters, and Instances

In Visio, you create drawings by dragging and dropping shapes onto a drawing page. To understand how you can add shapes to drawings, edit them, or develop new ones, some terminology and a few concepts are in order.

- **Stencil** is a file that contains shapes you can drag onto a drawing, for example, the Basic Flowchart Shapes stencil, which includes boxes, annotation shapes, and connectors. The Shapes window contains the stencils associated with a template or additional stencils that you open.

- **Master** is the name of a shape stored on a stencil, that is, the original of a shape. When you drag a master onto the drawing page, the copy or *instance* inherits its master's components, properties, and behaviors. If you open a stencil for editing and edit a master, all future instances that you drag onto a drawing look like the edited version of the master.

- **Instance** is the official name of a copy of a master on a Visio drawing page. Although a shape instance is linked to its master, you can modify an instance on a drawing page and only that instance changes. To keep track of all this, a Visio drawing contains a special stencil called the document stencil that contains a copy of each master dropped on a document. Visio stores only a single copy of the master, called the local master. As you'll learn in Chapter 32, this document stencil is an easy way to create a stencil of customized shapes.

Positioning Shapes

Assembling drawings involves more than dragging and dropping. Whether you're drawing a scaled plan in which dimensional accuracy is critical or aligning shapes to neaten the appearance of a business diagram, you can use Visio tools to position shapes as accurately or approximately as you want.

Rulers, grids, and guides act as reference points for alignment and accurate placement. However, you can snap to many other elements in Visio, including different parts of shapes themselves. When precision is important, you can also position shapes by specifying x and y coordinates in the Size & Position window.

After the initial placement of shapes, several additional tools help you adjust the position of shapes on your drawings. From simple techniques for rotating and flipping shapes to commands that align, distribute, and layout shapes, you can move shapes to make your drawings easy to read and understand.

Modifying Shapes

The shapes you add to a drawing aren't always exactly what you want. After you add shapes, you can modify them in the following ways:

- Change their size by dragging selection handles
- Change their shape by dragging vertices or eccentricity handles (see Chapter 1 for an introduction to Visio elements)
- Change their position in the stacking order
- Duplicate shapes on a page by stamping multiple copies of a master or by copying and pasting one or more shapes

Groups of Shapes

Often, several shapes work together, such as a set of furniture that makes up the furnishing of an office cubicle. By creating groups of shapes, you can move and operate on several shapes as one. Groups include their own behaviors and features to speed up the creation of specialized graphics, such as a bar graph or title block.

In addition, individual shapes can have only one foreground color, background color, line pattern, and text block, as well as several other properties. To apply more than one of these elements, you can assemble individual shapes, each with their own formats and text blocks into a group.

Depending on a group's properties, you can work with the group as a whole or with the individual shapes within the group. For example, when you add a title block to a drawing page, you can move the group into position, but you can also add text to the individual cells to annotate the drawing.

Finding the Right Shapes

When you start a drawing from a template, Visio opens stencils that contain shapes typical for that type of drawing. But why make do with shapes that aren't quite right? Visio includes dozens of built-in stencils with thousands of specialized shapes, but you can also find shapes at Microsoft Office Online, at many vendor web sites, and many other places on the Web. Whatever your diagram requires, you can find the perfect shape from a stencil that isn't open, on another drawing, from someone's custom stencil, or from stencils on the Web.

CROSS-REF For more information about finding templates and stencils on the Web, including URLs for online sources, see Chapter 40.

The search for shapes comes in many forms and travels to many places on its quest. The most powerful approach, the Visio Search for Shapes feature, can find shapes almost anywhere. If the shapes you find will be used over and over, you can save your search results to a custom stencil for easy access the next time you need them. This section describes several ways to find the shapes you want and when to use each one.

Opening Stencils

It's easy to drag shapes onto a drawing when the stencils containing those shapes are already open. Most of the time, the template you use opens one or more stencils with appropriate shapes. Template with associated stencils opened or not, you can open and close stencils to your heart's content to get the shapes you want. For example, feel free to open the Charting Shapes stencil to add a table of information to an organization chart drawing.

First, open the drawing you want to work on or create a new one. Then, to open a stencil, choose File ⇨ Shapes and then use one of the following methods:

- **Open a built-in stencil** — If you know which stencil contains the shapes you want, use this technique. On the Shapes menu, point to a category, then point to a template, and finally, on the lowest-level menu, choose a stencil. For example, if a symbol you used recently to annotate a drawing is perfect for your current drawing, browse to the stencil with that shape and open it.

- **Open a custom stencil** — If you save your favorite shapes to custom stencils, on the Shapes menu, point to My Shapes and choose a stencil.

- **Create a blank stencil** — To add shapes to your own custom stencil, on the Shapes menu, choose New Stencil.

- **Display the document stencil** — To display a stencil that contains all the shapes on the current drawing, on the Shapes menu, choose Open Document Stencil.

CAUTION Some shapes are meant to work within specific templates, so don't be surprised if shapes behave differently when you use them on different types of drawings.

Adding Stencils to the Shapes Menu

If you want the stencils to nestle in the Shapes submenus, along with the stencils built into Visio, the trick is to define a path to your stencils in Visio. Here are the steps:

1. Create a folder for your custom stencils, such as `. . /My Documents/Custom Stencils`.

2. In that folder, create additional folders for each category of stencils you want to see on the My Shapes menu, for example, Water Distribution or Sewer.

3. Choose Tools ⇨ Options and, in the Options dialog box, select the Advanced tab.

4. Click the File Paths button.

5. In the File Paths dialog box, in the Stencils text box, add the path name for your stencils. If the text box already contains a path, append the new path to the end separating it from the existing path names with a semicolon. To browse to the path, click the Ellipsis button (...).

6. Click OK. Visio includes the names of the subfolders within that path as categories on the Shapes menu. For example, you can choose File ⇨ Shapes ⇨ Sewer and then select the custom sewer stencil from the submenu. If the main folder contains stencils, these appear in a folder named Other(s).

NOTE If you choose File ⇨ Shapes and choose a stencil to open when no drawing is open in Visio, the program displays stencil-related toolbars and opens a stencil drawing window with limited functionality. For example, when you right-click a master in the stencil window, you can only copy the master or add it to one of your custom stencils.

Getting Shapes from Others

If you work with a lot of energetic folks, stencils with just the features you want could reside on their custom stencils. In many industries, companies develop stencils with shapes specific to their industry and sell them or make them available for download.

To use a stencil file (files with .vss or .vsx file extensions) you get from someone else, simply copy the file to `C:\Documents and Settings\<your username>\My Documents\My Shapes`. When a stencil file is in that directory, choose File ⇨ Shapes ⇨ My Shapes and the stencil appears on the submenu.

Searching for Shapes

The Search for Shapes feature scans for keywords you specify in shapes within built-in and custom stencils on your computer as well as stencils it finds on the Web. When Visio completes its search, it creates a search results stencil containing the shapes it found. You can drag a shape from the search results stencil onto your drawing or save a shape to another stencil to use it in the future.

CROSS-REF To search custom shapes that you've created, you must add search keywords to the shape masters on your custom stencils. To learn how to do this, see Chapter 33.

To search for shapes, in the Shapes window, in the Search for Shapes text box, type one or more keywords, click the green arrow, and wait for Visio to display its results. (If the Shapes window isn't visible, choose View ➪ Shapes Window to display it.) Visio keeps track of the keywords you used recently so you can repeat a search by clicking the Search for Shapes down arrow, choosing the keyword entry to repeat, and clicking the green arrow.

Several shapes can have the same name, but they come from different stencils. For example, one shape could be metric and the other in U.S. units. If the search returns the shapes in alphabetical order, you change the resulting sort to show shapes by group. Choose Tools ➪ Options. In the Options dialog box, select the Shape Search tab and then, under Results, select the By Group option.

TIP If you want to find more shapes similar to one that Visio retrieved, drag the shape to your drawing, right-click it, choose Shapes ➪ Find Similar Shapes. Visio uses the keywords associated with the selected shape to search for other shapes, and adds them to a search results stencil.

The Search for Shapes text box isn't particularly fussy. It accepts one or more words, separated by spaces, commas, or semicolons. Search for Shapes scans for keywords not shape names, so you'll get better results by typing descriptive words for your search text, such as **round**, **table**, or **furniture**. Keywords can be singular or plural; Visio searches for both forms. For example, typing **buttons** returns shapes with either **button** or **buttons** as a keyword, as illustrated in Figure 4-3. Sadly, wildcards aren't available.

Here are a few techniques that can improve your search success:

- When Visio doesn't find any matching shapes or the results aren't what you expect, try other keywords in your search criteria. For example, instead of **file**, try **cabinet**.

- To rein in an overwhelming number of results, add keywords or phrases to further screen the results. For example, using **table** returns a variety of database shapes in addition to furniture. Use **conference table** to locate shapes for large office tables.

Setting Options for Shape Searches

When you execute a search, the current Shape Search options determine where and how Visio searches for shapes. If the search results don't contain the shapes you want and different keywords don't seem to help, try modifying the Shape Search options.

Open the Options dialog box to Shape Search options by right-clicking the Search for Shapes area in the Shapes window and, from the shortcut menu, choose Search Options. Alternatively, choose Tools ➪ Options and then select the Shape Search tab.

FIGURE 4-3

Search for Shapes finds shapes based on keywords and adds them to a new search results stencil.

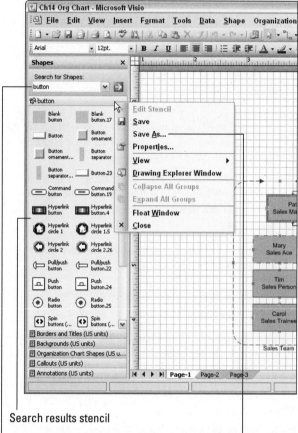

Search results stencil

Search keyword   Click Save As to save search results stencil as a custom stencil.

Here are options you can choose and what they do:

- **Search Locations** — To specify where Visio should search for stencils, check one or more of the following check boxes:
  - Check the My Computer check box to search for shapes in stencils on your computer.
  - Check the Internet check box to search for shapes on the Web.

NOTE Visio Web searches are smart enough to retrieve only the shapes associated with your Visio product. For example, you won't see shapes from Visio Professional in Visio Standard.

- **Keywords to Search For** — To tell Visio you're serious about the keywords you entered, select the All of the Words (AND) option to require that shapes include every keyword you typed. If shapes can contain at least one keyword, select the Any of the Words (OR) option.

- **Viewing Results** — Use one or more of the following options to fine-tune the appearance of search results:

  - **View grouped by stencil** — Viewing the shapes Visio finds grouped by the stencil to which they belong is great for identifying the stencils that you want to open. To do this, select the By Group option.

  - **Create new stencils for each search** — When you search for several types of shapes, you don't want Visio to overwrite your search results. Check the Open Results in New Window check box to tell Visio to create a new search results stencil for every search. You can then save the best stencils to your My Shapes category.

  - **Limit the number of results returned** — Visio is enthusiastic about its hunt for shapes and can retrieve an overwhelming number of results. To keep the results to a manageable number, check the Warn When Results Are Greater Than check box and type a cutoff number in the text box.

 **To display Icons, Names, or Details in the search results stencil, right-click the Shapes Window title bar and choose one of the following options:**

Icons and Names displays icons and shape names just as other stencils appear by default.

Icons Only displays only icons, if the appearance is all you care about.

Names Only displays only shape names, which use less space in the window but require more familiarity with the shapes.

Icons and Details displays icons, names, and a brief description of the shape.

Troubleshooting Shape Searches

If Visio doesn't find shapes you want and you know they exist, check for the following problems:

- **Stencil Path Is Incomplete** — In addition to searching the folders that contain the Visio built-in stencils, the Search for Shapes feature searches for stencils in your stencil path. If you store stencils in several locations, make sure that your stencil path includes those locations. To modify the stencil path, choose Tools ➪ Options, select the Advanced tab, and then click File Paths. To browse folders, click the Ellipsis button to the right of the Stencils box and navigate to the folder you want. To specify more than one stencil path, type a semicolon between each path.

- **Index Being Updated** — If shapes contain keywords and the Indexing Service is enabled, Visio might be trying to search while the index is being updated. Wait a minute and try the search again.

- **No Keywords Associated with Shape** — Open the custom stencil for editing and add keywords to the master.

- **Custom Stencil Open for Edit** — Visio can't search stencils when they are open for edit-ing. If the stencil icon in the stencil title bar includes a red asterisk, the stencil is open for editing. To save it so it can be searched, right-click the stencil title bar and then click Edit Stencil. When the red asterisk in the stencil title bar disappears, you can search the stencil.

- **Stencil Keywords Don't Match** — If you add keywords to a master shape and the Indexing Service is not enabled, the custom stencil in which the master is located might not have the same keywords as the master. Open the custom stencil for editing and check the keywords for the shape you created by right-clicking the master and choosing Edit Master ➪ Master Properties on the shortcut menu.

Saving Shape Search Results to a Stencil

Visio displays the shapes it finds in a search results stencil. If you plan to use these shapes in the future, you can save individual shapes or the entire search results stencil to a custom stencil. Saving shapes to stencils on your hard drive is especially helpful when you find shapes on the Web and don't want to navigate online every time you want to drop those shapes onto a drawing.

NOTE Visio automatically creates a My Shapes folder in your My Documents folder to keep all your custom stencils in one place. It also creates a Favorites stencil in your My Shapes folder for fast access to the shapes you use most frequently.

To save the search results stencil as a custom stencil, follow these steps:

1. Right-click the search results stencil title bar and, from the shortcut menu, choose Save As.

2. In the File Name box, type a name for the custom stencil and click Save. By default, Visio saves stencils in your My Shapes folder.

NOTE To access custom stencils in your My Shapes folder, choose File ➪ Shapes ➪ My Shapes and then click the stencil you want to open.

To save one shape in the search results stencil to a custom stencil, right-click the shape and choose Add to My Shapes ➪ Add to Existing Stencil. You can choose the Favorites stencil, other custom stencils, or Add to New Stencil to create a new custom stencil for the shape.

Speeding Up Shape Searches with the Indexing Service

Shape searches can provide you with plenty of down time to catch up on paperwork, get more cof-fee, or walk the dog. But, when deadlines loom, you want Visio to find shapes fast. The Indexing Service on your computer speeds up searches by maintaining an index of words associated with your shapes so that it can search a database instead of shapes or drawings.

TIP If you don't want shapes from the Web, searches finish faster if you uncheck the Internet check box on the Shape Search tab of the Options dialog box.

To enable indexing on your computer, make sure you have administrator privileges on your computer and then follow these steps:

1. Choose Tools ⇨ Options and select the Shape Search tab.

2. In the Search Locations list, select Visio Local Shapes, and click Properties.

3. Choose Yes, Enable Indexing Service and then click OK. Indexing might take a few minutes; the Shape Search Local Shape Properties dialog box closes when Visio finishes indexing your shapes.

Finding Shapes on Drawings

The Search for Shapes feature is great, but it can't find shapes on your drawings — it searches only for keywords associated with shapes in stencils. For example, as a new manager of a department, you might want to locate your position on a large organization chart by searching for the previous manager's name. In this situation, the Find command does the trick by searching for text in shape text blocks, shape names, shape data, and user-defined cell values in ShapeSheets. See the section "Finding, Replacing, and Correcting Text" in Chapter 6 for instructions on using the Find command.

Selecting Shapes

Before you can do anything with shapes — edit, position, or manipulate them — you have to tell Visio which ones you want to work on. Visio has methods for selecting one shape, several shapes picked one at a time, and groups of shapes.

Selecting Individual Shapes

You can select individual shapes whether they stand on their own or belong to a group. To select one shape, use one of the following methods:

- **One Shape** — Click a shape to select it and display its selection handles.

- **One Shape in a Group** — To subselect a shape in a group, click the shape once to select the group and then click a second time to select the shape and display its selection handles, as shown in Figure 4-4. Double-clicking a shape within a group also selects that shape.

> **NOTE** With any drawing tool active, you can select a line or closed shape by clicking any line. If you pause the pointer, the selection handles appear, followed by control handles at line midpoints, and finally corner control handles.

FIGURE 4-4

Click once to select a group and a second time to select a shape in the group.

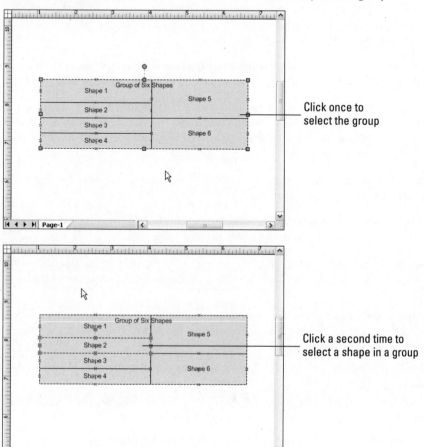

Click once to
select the group

Click a second time to
select a shape in a group

NOTE If a shape remains resolutely unselected when you click it, the shape might be protected against selection, or it could belong to a group. To determine whether a shape belongs to a group, click the shape and, from the shortcut menu, choose Format ⇨ Special. If the Type field value is Group, the shape belongs to a group, and you can double-click the shape to select it. If the shape doesn't belong to a group, check for protection by clicking the shape, choosing Format ⇨ Protection, and seeing whether the From Selection check box is selected. If it is, you can clear the check box and click OK to remove this protection.

To access Format ⇨ Special, you must run Visio in developer mode. Choose Tools ⇨ Options and select the Advanced tab. Check the Run In Developer Mode check box.

Selecting Multiple Shapes

Whether shapes are spread across your drawing or sit side by side, you can select multiple shapes quickly. As you select additional shapes, Visio highlights each selected shape with a magenta box and adds handles for the collection of selected shapes so that you can rotate and resize them all. Select multiple shapes using one of the following methods:

- **Select Box** — To select the shapes within an area, on the Standard toolbar, click the Pointer tool and drag a rectangle that completely encloses the shapes you want to select.

TIP   By default, Visio doesn't select a shape if a portion lies outside the selection rectangle. To include shapes only partially contained within the selection rectangle, choose Tools ⇨ Options and select the General tab. Check the Select Shapes Partially within the Area check box and click OK.

- **Shift-click** — To select shapes that are scattered across your drawing, hold the Shift key and click each shape you want to select.

TIP   You can select one or more groups using the same methods you use to select shapes. To select groups, make sure you click within the group only once.

- **Other Selection Tools** — To access other multiple selection tools, on the Standard toolbar, click the Pointer tool arrow and, from the drop-down menu, choose one of the following tools:
    - **Area** — Select this tool and drag a rectangle to select the shapes within an area. If you drag another rectangle while shapes are selected, the Area tool adds shapes within the new rectangle to the selection.
    - **Lasso** — Select this tool and drag the pointer around an irregular path to enclose the shapes to select, as illustrated in Figure 4-5.
    - **Multiple** — Select this tool and click shapes to add them to the selection. Click a selected shape once more to remove it from the selection.
- **Select All** — Press Ctrl+A to select all shapes on the drawing page.
- **Select By Type** — Choose Edit ⇨ Select By Type and then select the Shape Type or Layer option. If you select by shape type, select the check boxes for each type of shape you want to select. To select layers, check the layer check boxes.

FIGURE 4-5

You can create an irregular selection boundary with the Lasso tool.

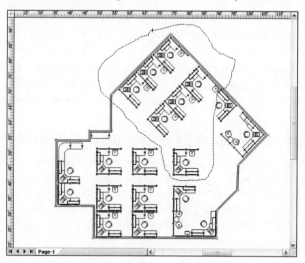

Positioning Shapes

With most business diagrams, you want shapes to line up so the drawing is uncluttered and well-organized. The *exact* position of the shapes doesn't matter at all. On the other hand, precise positions are essential on scaled drawings, such as building plans. For example, a few inches in the wrong direction and a door might end up smack in the middle of a structural column. Visio offers both types of positioning — basic alignment or precision placement — and this section describes each one and when to use them.

Here are the tools that Visio provides for positioning shapes:

- **Rulers** are horizontal and vertical bars around the drawing page that show distances based on the measurement units for the drawing. They are ideal for diagrams drawn to scale, but work equally well for positioning shapes using paper dimensions.

- The **grid** is a collection of horizontal and vertical lines that shapes snap to for fast and accurate alignment. The grid is perfect for lining up shapes with each other when you don't care about the shapes' exact position on the drawing page.

- The **dynamic grid** is different than the drawing grid, but even better for aligning shape with one another. When you drag one shape near another shape, the dynamic grid leaps into view with horizontal and vertical reference lines for aligning the shapes.

- **Guides** are nonprinting lines that provide reference points for positioning shapes. You can glue guides to shapes to align other shapes quickly or position guides at specific dimensions on the page.

- **Drawing aids** are like the dynamic grid on steroids and are best-suited to engineering or other types of technical drawings that require geometric constructions. For example, drawing aids can show you the tangent to a curve or the midpoint of a line.

- The **Layout Shapes** command (and its side-kick, Configure Layout) automatically positions shapes and connectors according to a layout style, such as the hierarchical trees of organization charts or the right-angle connections between steps in flowcharts.

Positioning with Rulers, Grids, and Guides

Rulers, grids, and guides tend to stay out of the limelight, but they are always available to help you position shapes, whether you want shapes merely aligned or placed in a precise location. You can use any or all of these tools to position your shapes, as demonstrated in Figure 4-6. If these features distract you from your diagramming task or you want to capture a clean screenshot, you can hide them from view, as described in the following sections.

FIGURE 4-6

Rulers, grids, and guides help position shapes.

Position a shape at a specific measurement on the drawing page.

Align shapes by positioning them on a grid intersection.

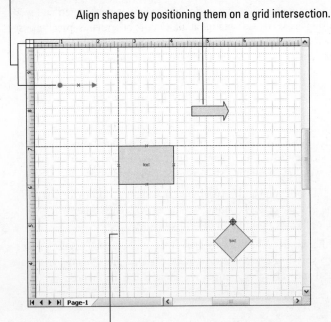

Align shapes by snapping them to a guide glued to another shape.

About Rulers, Grids, and Guides

Visio rulers, grids, and guides are similar to the real-world tools used by drafters and graphic artists. More powerful than their real-world counterparts, rulers, grids, and guides can act as magnets, so you can quickly snap shapes into place. Here's what each one does:

- **Visio rulers** — In Visio, vertical and horizontal rulers appear along the side of the drawing window and show full-size measurements based on the drawing scale, just like a wooden ruler or architectural scale ruler would when you place it on a paper drawing.

- **Visio grid** — This background grid is like drawing on a sheet of grid paper. You might not know the exact coordinates of a grid intersection, but they make it easy to align different components on a drawing.

- **Visio drawing guides** — These non-printing shapes are like light pencil lines you might sketch on a sheet of paper before you begin inking a drawing. You can glue them to shapes to simplify aligning additional shapes. When Visio prints the drawing, the guides don't appear.

Working with Rulers

The intervals on Visio rulers correspond to the measurement unit you specify (in the Page Setup dialog box on the Page Properties tab). As you move the pointer on a drawing page, dotted lines on the rulers indicate the current pointer position, which is helpful for any editing task, whether you are moving existing shapes, drawing lines, or specifying tabs in a text block. For example, adding a wall 40 feet from the building shell is as easy as dragging the wall to the appropriate marker on the ruler and dropping it into place. Choose View ⇨ Ruler to toggle the ruler's visibility on and off.

NOTE   The ruler subdivisions also determine the distance an object moves when you nudge a shape into position. When you press an arrow key with a shape selected, Visio moves the shape by one tick mark on the ruler.

The length of ruler subdivisions and the position of the ruler origin are not fixed. Suppose you want to place shapes relative to the first shape on a diagram. You can reposition the ruler origin to a corner of that first shape and position other shapes the same number of inches from the origin. Similarly, you can use one-foot ruler subdivisions for working on the exterior of a building and then change the ruler subdivisions to inches when working on window details. To specify the subdivisions that appear on the rulers, use one or both of the following methods:

- **Set Measurement Units** — To change the units that appear on the rulers, choose File ⇨ Page Setup and select the Page Properties tab. Choose the units you want in the Measurement Units drop-down list.

NOTE   When you choose units such as Inches, or Feet and Inches, Visio sets the units based on eighths of an inch. If you choose Inches (decimal), Visio divides an inch into tenths.

- **Set Ruler Subdivisions** — To change how many subdivisions Visio displays, choose Tools ⇨ Ruler & Grid. Select Fine, Normal, or Coarse in the Horizontal and Vertical Subdivisions boxes.

You can reposition the origin for rulers, known as the zero point, to align with an element on your drawing. Visio uses the zero point as the center of rotation when you rotate the drawing page. The zero point is usually located at the lower-left corner of the page. To move the zero point for the rulers, use one of the following methods:

- **Change the zero point on both rulers** — Hold the Ctrl key and drag from the blue cross at the intersection of the two rulers to a position on the drawing page. As you drag, Visio displays dotted lines that represent the x and y axes. When you release the mouse button, Visio moves the zero point to that location.

NOTE   Be sure to press the Ctrl key before you click the blue cross at the ruler intersection. If you press the mouse button before you press the Ctrl key, Visio drags a guide point onto the drawing page.

- **Change the zero point on one ruler** — Hold the Ctrl key and drag from the either the horizontal or vertical ruler.

- **Reset the zero point to the lower-left corner** — Double-click the intersection of the two rulers.

Working with a Grid

The Visio grid — horizontal and vertical lines that crisscross the drawing page — help you position shapes. Depending on the settings you choose, the grid can act merely as a visual reference or you can snap shapes to the grid intersections. For example, you can quickly position structural columns every 20 feet by defining a 20 foot grid for the drawing and dropping column shapes onto grid intersections. Choose View ➪ Grid to toggle the grid on and off.

TIP   By default, the drawing grid doesn't print, but you can print it with the drawing page by choosing File ➪ Page Setup, selecting the Print Setup tab, and checking the Gridlines check box.

Most Visio drawing types use a *variable grid,* which means that Visio determines the best grid spacing based on how far you are zoomed in or out. When you zoom in, the grid intervals represent smaller distances, and switch to larger distances as you zoom out. If you work with specific distances, such as the spacing for ceiling tiles, you can specify a fixed grid, in which the grid lines remain the same distance apart no matter how you zoom. Whether you are zoomed in or out, you can be sure that the shapes you snap into place are positioned at the distance you want.

You can adjust the coarseness and origin of the grid to facilitate drawing. Use one or more of the following techniques to set the grid up for your drawing needs:

- **Reposition the Gird Origin** — You can reposition the grid origin, for example, to a corner of a shape so you can easily draw other components relative to that shape. To reposition the grid origin, choose Tools ➪ Ruler & Grid, type the x and y coordinates for the new grid origin, and click OK.

NOTE   By default, the grid originates at the ruler zero point and moves when you move the ruler origin. However, when you move the grid origin, it remains at that location even when you change the ruler zero point.

- **Set Variable Grid Spacing** — To vary the grid spacing based on your zoom level, choose Tools ⇨ Ruler & Grid. Select Fine, Normal, or Coarse in the Grid Spacing Horizontal and Vertical lists.

- **Set Fixed Grid Spacing** — To specify an interval for the grid spacing, choose Tools ⇨ Ruler & Grid and select Fixed in the Grid Spacing Horizontal and Vertical lists. Type the distance for the grid interval in the Minimum Spacing boxes and click OK.

Working with Guides

Guides are like reference points or guidelines that help you position or align shapes. For example, if a building has walls at different angles, you can add guides at those angles to align furniture with the walls. In addition to snapping to guides, you can glue one or more shapes to a guide and move the shapes by moving their associated guide. Choose View ⇨ Guides to toggle guide visibility on and off.

TIP By default, guides are nonprintable objects, but you can print a guide by modifying the guide's ShapeSheet. To do this, select the guide you want to print and choose Window ⇨ Show ShapeSheet. Scroll to the Miscellaneous section and type False in the NonPrinting cell.

To create or modify guides, use one of the following methods:

- **Create a guide** — Drag from the horizontal or vertical ruler (not the intersection of the two) onto the drawing page. Visio displays a blue dotted line for the guide.

- **Create a guide point** — Drag the intersection of the rulers onto the drawing page. Visio displays a blue circle with crosshairs to indicate a guide point.

- **Use a shape as a guide** — Any Visio shape can act as a guide, including arcs and splines. To create a guide from a shape, right-click the shape, from the shortcut menu, choose Format ⇨ Style, and then, in the Line Style box, select Guide. The shape looks and acts like a guide but still has selection handles so you can edit it. However, guides created in this way remain visible when you hide guides and guide points.

- **Delete a guide or guide point** — Select the guide or guide point and press Delete.

NOTE After defining guides for your drawing, you might want to turn off the grid so it doesn't interfere with snapping to your guides and guide points. To turn off the grid, choose View ⇨ Grid.

- **Move a guide** — Drag a guide to a new position. You can also select a guide and type an x or y value in the Size & Position window.

- **Rotate a guide** — Choose View ⇨ Size & Position, select the guide you want to rotate, and type an angle in the Angle box in the Size & Position window.

TIP If you want to add several guidelines at a specific angle, rotate the page to that angle (see Chapter 3) and then add the guidelines. After the guidelines are in place, rotate the page back to its original angle.

Placing Shapes with Precision

Despite the ease with which you can drag and drop shapes onto drawings, there's no need to give up precision in placing shapes. In Visio, the snapping feature positions and aligns shapes by pulling them to elements on a drawing like diagrammatic pheromones. For example, you can snap a desk to the corner of an office cubicle. When you know exactly where something belongs, you can specify exact coordinates and angles. Visio includes even more tools for specialized placement and alignment, such as distributing several shapes equidistantly. This section describes the different methods for placing shapes precisely where you want on a drawing.

Snapping Shapes into Position

When snapping is activated, Visio pulls the pointer to possible placement positions on the drawing page, such as connections points on a shape or a position on the drawing ruler. For example, snapping makes it easy to connect an arrow shape to a block on a block diagram, but you can also snap furniture shapes to cubicle or wall shapes as well. You have the power to choose which elements Visio snaps to as well as the strength of attraction exerted by those elements.

Whether you are dragging an entire shape, a selection handle, a rotation handle, a vertex, or another shape element, Visio finds the closest snap point for your editing action. Here's how snapping progresses:

1. As you move the pointer, crosshairs indicate the current pointer location.

2. When the pointer nears a snap point, Visio displays blue crosshairs at the snap point.

3. When the pointer snaps to a connection point, Visio highlights the connection point with a red square, indicating that you can glue to that point.

To activate snapping, choose Tools ➪ Snap & Glue and select the Snap check box. You can snap to the following elements:

- **Ruler subdivisions** — Intervals on the horizontal and vertical rulers.

- **Grid** — The intersections of lines on the drawing grid.

- **Alignment box** — The dotted, green box that appears around a selected shape or group.

- **Shape extensions** — Dotted lines or points that show how to draw a line in relation to a geometric point, such as the tangent to an arc or the midpoint of a line. You can specify which extensions Visio displays by selecting the Advanced tab and selecting the extensions you want.

- **Shape geometry** — The edges of a shape.

- **Guides** — Guides and guide points you create, as described in the "Working with Guides" section earlier in this chapter.

- **Shape intersections** — Points where two shapes intersect, shape extensions and shapes intersect, or shape edges and the grid are perpendicular.

- **Shape handles** — Green selection handles that appear when you select a shape.

- **Shape vertices** — Green diamonds that indicate the start and end points of line segments.

- **Connection points** — Blue Xs that indicate points to which you can glue a shape.

 To see examples of shape handles, vertices, and connection points, refer to Chapter 1.

Configuring Snapping Elements

Unrestrained snapping could make it almost impossible to position shapes, as elements pull the pointer in every direction. Not only can you specify which elements you want Visio to snap to, you can set how tenaciously each of those elements pulls the pointer. Here are the steps for choosing snap elements and snap strength:

1. Choose Tools ➪ Snap & Glue and then select the Snap check box.

2. To specify snap to elements, under Snap To, check the check boxes for those elements, such as Ruler Subdivisions, Connection Points, and Shape Geometry.

3. To specify the snap strength, select the Advanced tab.

4. On the Advanced tab, select the check boxes for shape extensions you want to snap to, for example, Segment Midpoint or Isometric Angle Lines.

5. To strengthen or weaken the attraction of an element, drag the slider for that shape extension to the right or left, respectively. As you drag the slider, the number of pixels required to attract the pointer appears in the Pixels box for that element. For example, if you use one pixel for snap strength, an element doesn't attract the pointer until it is less than one-eighth of an inch away. If the snap strength is set to 40 pixels, the pointer snaps when it is about half an inch away.

TIP Depending on what you are trying to do, snapping can be a hindrance instead of a help. For example, when you are drawing freeform curves, snapping can turn rolling waves into choppy curves. If snapping is causing trouble, clear one or more check boxes in the Snap & Glue dialog box. You can also change Snap & Glue settings quickly by clicking commands on the Snap & Glue toolbar. To display this toolbar, choose View ➪ Toolbars ➪ Snap & Glue.

Positioning Shapes with the Dynamic Grid

The horizontal and vertical, dotted lines of the dynamic grid stay hidden until the pointer approaches a good place to drop a shape. For example, the dynamic grid appears when the pointer gets close to the top, bottom, left, right, center, or any connection point of another shape on the drawing. To enable the dynamic grid, choose Tools ➪ Snap & Glue, and check the Dynamic grid check box.

Using Drawing Aids

Drawing aids are dotted lines that show the correct position for drawing a circle, square, or line at a specific angle. When you use the Line tool, you can snap to these aids to draw lines at a specific angle, such as 45 degrees. Drawing aids are also handy when you use the Rectangle or Ellipse tools to draw a square or circle, because the dotted lines appear when the sides of the rectangle or the axes within the ellipse are almost the same length. To enable drawing aids, choose Tools ➪ Snap & Glue and select the Drawing aids check box. In the Snap To column, select the Shape extensions check box.

Positioning Shapes by Coordinates

For the most accurate positioning, nothing beats typing coordinates—albeit the correct coordinates. As you might expect, the Size & Position window displays fields for coordinates, size, and angle, depending on the type of shape you select. For example, Visio draws 1-D shapes based on the start point, end point, and an angle. 2-D shapes include width, height, angle of rotation, and the coordinates for the shape's pin, which is the shape's center of rotation.

The x and y coordinates that you see in the Size & Position window represent the page coordinates for the selected shape, which are based on the drawing scale for the page and relative to the origin of the rulers. For example, with the origin of the rulers at the bottom left corner of the page, the center of a letter-size landscape drawing at one-to-one scale has an x coordinate of 5.5 inches and a y coordinate of 4.25 inches. For 2-D shapes, the x and y coordinates represent the position of the shape's pin: a green circle with crosshairs that appears when you pause the pointer over the rotation handle of a shape.

 Although the pin is set to the center of a shape by default, you can drag the pin wherever you want. When you rotate the shape, it will rotate around the new pin location.

Open the Size & Position window by choosing View ➪ Size & Position window and select a shape to view its coordinates. To modify the position of a 2-D shape, as shown in Figure 4-7, type values in any of the following boxes:

- **X**—Move the shape by positioning its pin horizontally at this x coordinate.

TIP Values in the Size and Position window can be explicit values or formulas. For example, if you want to move a shape 4 feet away from a wall, you can add "+4ft" to the end of the current value.

- **Y**—Move the shape by positioning its pin vertically to this y coordinate.
- **Width**—This field doesn't move the shape, but it's a surefire way to specify a precise shape width.
- **Height**—This field specifies the shape height.

FIGURE 4-7

You can move or resize shapes using the Size & Position window.

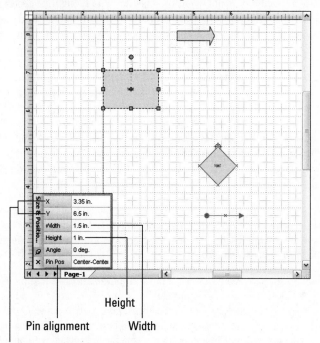

Height

Pin alignment   Width

X and Y coordinates of the pin

- **Angle** — To rotate the shape, type an angle in the Angle box. Angles start with 0 degrees pointing to the right, and increase as you move counterclockwise relative to the shape's alignment box.

- **Pin Pos** — To align the pin with one of the shape selection handles, select an alignment in the Pin Pos list, such as Center-Center or Bottom-Right. When you change the alignment of the pin position, the pin stays at the existing *x* and *y* coordinates, but the shape moves into alignment. For example, if you change Pin Pos from Center-Center to Bottom Right, the shape moves up and to the left until the bottom right corner of the shape is at the pin position.

TIP The Size & Position window locks to the bottom-left corner of the drawing window by default. You can drag and dock it at the bottom of the Shapes window to keep the entire drawing area visible.

Automatically Laying Out Shapes

Many diagrams use a specific type of layout, the most familiar being the vertical tree for organization charts or the right-angled top-to-bottom progression of a flowchart. You could resort to Visio positioning techniques to place shapes and connectors where you want them, but why spend the time when Visio can take care of shape layout for you?

Visio's automatic layout and routing tools arrange the shapes on your drawings according to the layout options you choose — as long as you use dynamic connectors to create shape-to-shape connections. This qualification is key, because dynamic, or shape-to-shape, connections know how to navigate around other shapes, jump over other connectors, and choose the connection points that create the best route between connected shapes. Harness this power and you can turn over layout, routing, and jumping over connectors to the Configure Layout and Re-layout Shapes commands. Although automatic layout and routing is typically well-behaved, understanding how layout and routing work helps you select the options that deliver the results you want.

NEW FEATURE In Visio 2003, the Layout Shapes command handled both the selection and application of the layout and routing options you selected. In Visio 2007, the Configure Layout command on the Shape menu handles the selection of layout and routing options. To apply those options, you choose Shape ⇨ Re-layout Shapes.

CAUTION Shape-to-shape connections are a must for automatic layout and routing to work properly. If connectors don't stay attached to shapes or connector end points are green, they aren't connected properly. To glue connectors to shapes, drag an end point on the connector until Visio highlights the shape with a red box. A connector end point changes to a solid red square when it uses a shape-to-shape connection. The end point is a smaller red square with an x if it uses a point-to-point connection.

If the connections are correct, but automatic layout still doesn't work properly, verify that the connector is drawn in the right direction. To do this, make sure the connector is not selected and then drag the connector away from its shapes. The starting point for the connector is a green square with an x in it. The end point is a green square with a +. Reattach the connector end points in the correct direction.

Specifying Layout Options

The Visio templates for diagrams that follow standard layouts come with default layout options. If you want to arrange your shapes differently, the Configure Layout command lets you control the shape layout on your drawing, as shown in Figure 4-8. For example, a work flow diagram might make more sense laid out horizontally or vertically. Besides the positioning of shapes, Configure Layout controls the style, direction, and appearance of connectors. It's smart enough to choose connector options that correspond to the placement options selected. For example, if you choose a Circular layout style, Visio changes the connector style to Center To Center, so that connectors radiate from a central shape to the center of outlying shapes. Although you're free to choose different options for connectors, drawings look cleaner when the placement and connector options match.

FIGURE 4-8

The Configure Layout command contains all the options for laying out and connecting shapes on a drawing.

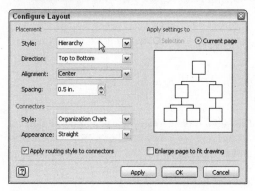

The Configure Layout dialog box provides a preview window so you can see what the settings you choose would look like before you apply them. To specify layout and routing options, choose Shape ➪ Configure Layout and then specify one or more of the following settings:

- **Placement Style** — Under the Placement heading, the Style setting determines the arrangement of shapes. Choose a setting from the Style drop-down list, such as Radial, Flowchart, or Hierarchy. Visio changes the options in the Connectors section to match the placement style you select.

- **Direction** — Select the direction you want shapes to flow on your drawing, for example, Top to Bottom for an organization chart, Left to Right for tournament results, or Bottom to Top or Right to Left.

- **Alignment** — Select how shapes align with each other. For example, choosing Left in a Top to Bottom diagram positions shapes in columns with the first shape on each level on the left side of the diagram.

- **Spacing** — Specify the space between shapes.

- **Connectors Style** — Under the Connectors heading, the Style setting determines how connectors travel through a layout. For example, Right Angle draws single straight lines at right angles between shapes. Center to Center draws straight lines that point toward the centers of the connected shapes.

- **Appearance** — Select Straight or Curved to determine what connectors look like.

- **Apply Routing Style to Connectors** — To apply new routing options to some or all connectors on a page, select the Apply Routing Style to Connectors check box. To rearrange shapes without changing the connectors, clear the check box.

- **Enlarge Page to Fit Drawing** — Sometimes, the settings you choose require more room than the current drawing page provides. To automatically increase the size of the page to accommodate your layout, clear this check box.

- **Apply Setting To** — By choosing the Selection option, you can apply a layout to only some of the shapes on a drawing, for instance, the subordinate who report to one manager in an organization chart. The Current Page option apples the layout to the entire page.

To apply the layout, click Apply. If you don't like the results, go back and choose other settings. Otherwise, click OK.

NOTE   If the new layout borders on disastrous, press Ctrl+Z to undo the layout.

TIP   If Visio creates an overly complicated route for a connector, you can modify its path manually by dragging the connector's green vertices to new positions. However, the Re-layout Shapes command or clicking Apply in the Configure Layout dialog box both overwrite those changes when applying layout rules.

Specifying Layout and Routing Spacing

Adjusting the spacing between shapes is helpful when you want to fit more shapes on a page or add white space around shapes to improve readability. In Visio 2003, spacing options were available in the Lay Out Shapes dialog box. In Visio 2007, the Configure Layout dialog box includes a setting for shape spacing, but the only place to change all aspects of spacing is within Page Setup. Choose File ➪ Page Setup. In the Page Setup dialog box, select the Layout and Routing Spacing tab, click the Spacing button, and then change one or more of the following settings:

- **Space Between Shapes** — Visio adds the distances in the Horizontal and Vertical boxes between each shape in a layout. The default values are .5 inches, but you can increase the dimension to add white space or decrease the dimension to squeeze more of a diagram on a page.

- **Average Shape Size** — These dimensions set the average height and width of the shapes in your drawing and also set the average shape size used by the dynamic grid. When a drawing has both very large and very small shapes, a smaller average shape size often produces better results.

- **Connector to Connector** — These dimensions control the minimum spacing between parallel connectors. For example, a horizontal spacing of one inch keeps at least one inch between the horizontal segments of connectors. For drawings with relatively few connectors, a large connector to connector spacing is easier to read. But, if you draw wiring diagrams, you often have no choice but to set small spacing between connectors.

- **Connector to Shape** — Specifies the minimum spacing between connectors and shapes. A horizontal spacing of 1 inch lays out shapes so that a shape and a vertical segment of a connector are no closer than 1 inch apart.

Specifying Line Jump Options

When connectors cross over each other like spaghetti, the relationships can be difficult to follow. Line jumps clarify the paths that connectors take by showing gaps or small arcs wherever two connectors cross. To specify line jumps, choose File ⇨ Page Setup and then select the Layout and Routing tab. Use the following settings to specify where to add line jumps and what you want them to look like:

- **Add Line Jumps To** — Choose the lines to which you want to add line jumps. For example, you can specify Horizontal Lines to add line jumps only to the horizontal lines on a drawing. Last Displayed Line adds line jumps to the line at the top of the stacking order.

- **Line Jump Style** — Choose a style for the appearance of a line jump, such as Arc or Gap. You can also choose from a number of multifaceted jumps.

- **Vertical Size** — This dimension sets the size of line jumps added to vertical lines.

- **Horizontal Size** — This dimension sets the size for line jumps added to horizontal lines.

Every once in a while, one connector might need a different type of line jump — for example, if a very complicated drawing presents an almost insoluble problem in one small area. To specify line jumps for a single connector, do the following:

1. Right-click the connector and, from the shortcut menu, choose Format ⇨ Behavior.

2. In the Behavior dialog box, select the Connector tab.

3. Under Line Jumps, in the Add drop-down list, choose a setting to specify whether the connector conforms to the line jump options for the page or uses a style other than the page default.

Configuring Placement Behavior

One aspect of shape behavior is how those shapes react when you use the layout and routing tools. Shapes that depend on connections, such as network diagrams or organization charts, are already configured to work with layout and routing. You can teach shapes to respond to layout and routing and change the layout and routing behavior of shapes. For example, you might want Visio to route connectors around a shape or you might want other shapes to move when you add the shape to a drawing page.

To configure placement behaviors for a shape, follow these steps:

1. Right-click the shape and, from the shortcut menu, choose Format ⇨ Behavior.

2. Select the Placement tab.

3. On the Placement tab, change one or more of the following behaviors:

    - **Placement Behavior** — The choices in this drop-down list tell Visio whether to include the shape in layouts and routing. Lay Out and Route Around enables the shape for automatic layout, which in turn makes the other placement options on the tab available as well. If you choose Do Not Lay Out and Route Around, Visio ignores the shape during layout.

▥ **Do Not Move During Placement** — To prevent Visio from relocating the shape when it lays out the drawing, check this check box. For example, if your computer center contains an old mainframe that no one can budge, you can simulate the same behavior in Visio by choosing Do Not Move During Placement.

▥ **Allow Other Shapes To Be Placed On Top** — Check this check box to allow Visio to place other shapes on top of the shape during automatic layout. If you want every shape to be completely visible, leave this check box unchecked.

▥ **Move Other Shapes Away On Drop** — Choose a setting to specify whether a shape plows other shapes out of the way when you drop it on a page. Choosing Plow Other Shapes means that a shape shoves nearby shapes out of the way to make room for itself, like one of your older relatives at the family reunion buffet line.

▥ **Do Not Allow Other Shapes To Move This Shape Away On Drop** — Select this check box to set up the antidote to another shape's plowing behavior. For example, with this check box selected, the shape remains where it is when you drop another shape near it no matter what.

▥ **Interaction with Connectors** — These check boxes specify whether connectors can route through the middle of the shape horizontally or vertically. Leave these check boxes unchecked if you want Visio to route connectors around the shape.

 Placement options apply only to 2-D shapes. If you select a 1-D shape and then choose Format ➪ Behavior, the Placement options are disabled.

Moving, Rotating, and Flipping Shapes

After a shape takes its place on a drawing, there's a good chance it won't stay where it is. Whether a step-by-step procedure gains a few additional steps, some additional causes crop up for a cause-and-effect diagram, or a computer room has to absorb new servers, you can move shapes around to accommodate changes. Moreover, rotating shapes often comes in handy for positioning furniture in oddly angled buildings or turning a valve to connect it to a vertical pipe. And in some cases, a mirror image of a shape is the only solution — such as when you use two different desk arrangements depending on whether workers are left- or right-handed. This section describes several methods for moving shapes as well as the steps for rotating and flipping shapes.

 The Action toolbar contains most of the commands described in this section with a few more for extra measure. If you are moving, rotating, or manipulating shapes, choose View ➪ Toolbars ➪ Action to display this handy toolbar.

Methods for Moving Shapes

The simplest way to move a shape is to position the pointer inside the shape and, when the four-headed arrow appears, drag the shape to its new location. This method is fine when you don't need precise placement and can provide precision when combined with Visio snapping. Another way to move a shape just a tad is to select the shape and then press one of the arrow keys to nudge the shape one interval on the ruler. The rest of this section discusses some of Visio's more robust tools for moving shapes.

CAUTION Moving a shape while it is selected can lead to resizing it instead. Before moving a shape, make sure it is not selected by clicking the page background or pressing the escape (Esc) key.

Aligning Shapes with One Another

Although snapping shapes into position is a convenient way to align shapes with one another, Visio provides other tools for arranging shapes neatly on a drawing. The Align Shapes command can line shapes up horizontally, vertically, or in two directions.

To align several shapes at once, follow these steps:

1. Select the shape that acts as the anchor for alignment — the one to which you want to align other shapes.

2. Press and hold the Shift key while you click each shape you want to align to the first shape. Visio outlines the first shape with a thick magenta line, as illustrated in Figure 4-9.

FIGURE 4-9

Align Shapes highlights the shape to which the others are aligned.

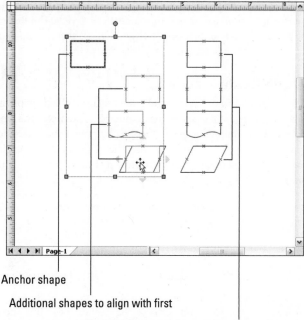

Anchor shape

Additional shapes to align with first

Result of aligning shapes vertically to the center of the first shape

3. Choose Shape ➪ Align Shapes.

4. In the Align Shapes dialog box, click the icon for the alignment you want: vertically to the top, center, or bottom of the primary shape; horizontally to the left, center, or right of the shape. (Click the red X to deselect an option.)

5. Click OK to align the shapes.

Distributing Shapes Evenly Over a Distance

In some instances, you want to spread shapes out evenly over a distance, for example, to position every step in a procedure the same distance from its neighbors. In this case, the Distribute Shapes command is what you're looking for. The Distribute Shapes command measures the distance between the outermost of the selected shapes and positions all the shapes equidistantly within that distance. To distribute shapes equally, follow these steps:

1. Select three or more shapes and then choose Shape ⇨ Distribute Shapes. The order in which you select the shapes doesn't matter.

2. In the Distribute Shapes dialog box, click a distribution option. For vertical distribution, the top and bottom shapes define the distances. For horizontal distribution, the shapes to the far left and far right define the distance.

3. Click OK. Visio repositions the shapes with an equal spacing between them.

 To distribute shapes at precise intervals, use the Offset command to create lines offset at specific distances from one shape. You can then snap shapes to these lines. To create off-set lines, select a line or curve and choose Shape ⇨ Operations ⇨ Offset. Type a value for the offset and click OK. Visio creates matching lines offset on either side of the original.

Moving Shapes with Guides

When shapes are glued to guides, moving the guide moves all the glued shapes at once without disrupting the arrangement of those shapes to one another. For example, you can create a guide through the center of a row of equipment. By gluing the equipment shapes to the guide, you can reposition the equipment on the plant floor by moving the guide. Because the guides are nonprinting objects, they won't appear when you print your drawing unless you modify the NonPrinting cell in their ShapeSheets.

NOTE You can also move several shapes at once by grouping them. When you group shapes, you can define separate settings and behaviors for the group in addition to the member shapes.

To glue shapes to a guide, follow these steps:

1. Choose Tools ⇨ Snap & Glue and make sure that both the Snap and Glue check boxes are selected.

2. In the Snap To and Glue To columns, check the Guides check box and then click OK.

3. To create a guide, drag from a ruler onto the drawing page.

4. Drag a shape to the guide and drop it when Visio highlights the connection point you want on the shape with a red box.

5. Repeat step 4 until all the shapes are glued to the guide.

You can also glue shapes to guides while aligning or distributing them by following these steps:

1. Select the shapes you want to glue to a guide and choose either Shape ➪ Align Shapes or Shape ➪ Distribute Shapes.

2. In addition to selecting the alignment or distribution options you want, check the Create Guide And Glue Shapes To It check box (or Create Guides And Glue Shapes To Them if you are distributing shapes), and click OK. Visio creates a guide or guides and glues shapes as follows:

   ▪ **Align Shapes** — Visio creates one guide and glues the shapes to it. You can move the shapes by dragging the guide to another location.

   ▪ **Distribute Shapes** — Visio creates a guide for each selected shape. You can redistribute the shapes by dragging one of the outermost guides. You can't drag an interior guide.

Moving Shapes Precisely with the Move Shapes Add-On

If you positioned shapes precisely in the first place, no doubt you expect to move shapes precisely as well. Available only in Visio Professional, the Move Shapes Add-On trumps the Size & Position window in several ways:

- **Move relative distances** — In the Size & Position Window, you can add a relative distance to the end of a value, such as **-5.5ft**. However, the Move Shape Add-On always moves shapes relative to their current position. If you want to move the shape 1 inch to the right, simply type **1 in** in the Horizontal box.

- **Use polar coordinates** — For geometry nuts, the Move Shapes Add-On accepts polar coordinates — a distance radially from the center in the direction specified by an angle. For example, you can move a shape 6 inches out at a 45-degree angle.

- **Move or copy shapes** — If you want to make a copy of the selected shapes instead of moving them, check the Duplicate check box.

To move or copy shapes with the Move Shapes Add-On, follow these steps:

1. Select the shape or shapes that you want to move or copy and then choose Tools ➪ Add-Ons ➪ Visio Extras ➪ Move Shapes.

2. Select the Horizontal/Vertical option to work with Cartesian coordinates or select the Distance/Angle option to work with polar coordinates.

3. Depending on the coordinates, specify the new position as follows:

   ▪ **Horizontal/Vertical** — Enter the distances you want the shape to move horizontally and vertically in the Horizontal and Vertical boxes, respectively. To move a shape down or to the left, use negative numbers.

   ▪ **Distance/Angle** — Type the radial distance (vector length) you want the shape to move into the Distance box. Type an angle to specify the direction you want to move the shape on the page, as shown in Figure 4-10. Angles start with 0 degrees pointing to the right, and increase as you move counterclockwise relative to the shape's alignment box. Ninety degrees points up; 180 degrees points to the left; and 270 degrees points down.

FIGURE 4-10

With the Move Shapes Add-On, you can specify a radial distance and angle.

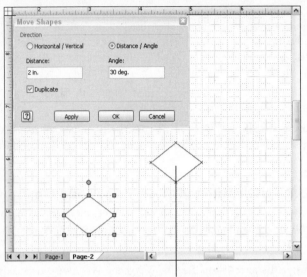

Center of copy positioned 2 inches away at 30 degrees

4. To copy the selected shapes instead of moving them, select the Duplicate check box.

5. To preview the move or copy action, click Apply. If the results are correct, click OK. Otherwise, click Cancel.

> **TIP** If you click OK without previewing the results, you can press Ctrl+Z to undo the move or copy.

Rotating Shapes

Visio shapes include rotation handles similar to those found in other Office products. The easiest way to rotate a shape is by dragging its rotation handle.

> **TIP** The Action toolbar includes icons for Rotate Left and Rotate Right and keyboard aficionados can press Ctrl+L or Ctrl+R to rotate to the left or right, respectively. Shape shortcut menus used to contain the Rotate Left and Rotate Right commands, but not in Visio 2007.

When you drag a shape's rotation handle, it's obvious that the status bar changes to show the current rotation angle. What isn't so obvious is how Visio snaps to specific rotation angles as you drag the pointer. The trick is that the snap angle gets smaller the further the pointer is from the shape's rotation pin. For example, if you drag in circles on top of the rotation handle, Visio snaps to angles in increments of one degree. On the other hand, dragging the rotation handle in a large arc snaps to angles within tenths of a degree.

TIP When you position the pointer over the rotation handle, Visio displays the Center of Rotation pin for the shape, a green circle with a cross inside it. Typically, the Center of Rotation pin is in the center of the shape and the shape rotates around that point. When you want to rotate several shapes at once, Visio adds a Center of Rotation pin at the center of all the shapes you select, so you can rotate several shapes the same way you rotate one. If you would rather have a shape or group of shapes revolve around a different point, simply drag the Center of Rotation pin to a new position.

Flipping Shapes

Sometimes you want the mirror image of an existing shape. For example, the layout for a workstation for a right-handed person might be the mirror image of a workstation for someone who's left-handed. In Visio, you can flip shapes and groups horizontally or vertically. Select the shape or group you want to flip and use one of the following methods:

- To mirror a shape about the vertical axis, choose Shape ⇨ Rotate or Flip ⇨ Flip Horizontal.

- To mirror a shape about the horizontal axis, choose Shape ⇨ Rotate or Flip ⇨ Flip Vertical.

- On the Action toolbar, click Flip Horizontal or Flip Vertical.

- Press Ctrl+H to flip horizontally or Ctrl+J to flip vertically.

Manipulating Shapes

In addition to positioning and relocating shapes, diagrams often require duplicating, resizing, and manipulating shapes in other ways. Regardless what sort of tweaking you want to perform to produce the perfect diagram, this section describes the techniques Visio offers.

CROSS-REF If you see padlocks where you would normally see selection handles when you select a shape, it is locked to prevent you from manipulating it. To learn how to lock or unlock shapes, see Chapter 7.

Undoing Actions and Deleting Shapes

When the editing you perform makes changes you neither expect nor like, you can delete or undo your actions with any of the following methods:

- **Undo one action** — To undo your last action, there's nothing faster than pressing Ctrl+Z.

- **Undo recent actions** — To undo several of your most recent actions, several fast Ctrl+Z keystrokes will do the trick. If mousing is more your style, on the Standard toolbar, click the arrow to the right of the Undo button and drag the pointer to select the actions you want to undo. When you release the mouse button, Visio undoes all the actions you selected.

- **Redo one action** — To redo one action that you undid, press Ctrl+Y.

- **Redo several actions** — To redo several actions, press Ctrl+Y multiple times. Or, on the Standard toolbar, click the arrow to the right of the Redo button. Drag the pointer to select the actions you want to redo and release the mouse button.

- **Delete shapes and other objects** — Select the shapes you want to delete and press Delete.

- **Cut shapes to the Clipboard** — To remove shapes from the drawing page and place them on the Clipboard, select the shapes and press Ctrl+X. To paste cut shapes, press Ctrl+V.

 You can also delete shapes on the Standard toolbar by clicking Delete. You can cut shapes by choosing Edit ⇨ Cut or, on the Standard toolbar, clicking Cut.

Duplicating Shapes

After you have molded a shape into exactly what you want, there's no reason to make those same changes to another shape. Visio makes it easy to duplicate shapes on a drawing. For shapes you use constantly or for those that you plan to use on other drawings in the future, saving the modified shape to your Favorites stencil or another custom stencil is even better. True, it takes a few steps to save the shape, but after that, dragging it onto any drawing is a breeze.

Copying Shapes

One shape or many, you can copy selected shapes once or multiple times with the following methods:

- **Copy one or more shapes on a drawing** — Select the shapes that you want to copy and press Ctrl+C to copy them to the Clipboard. Press Ctrl+V to paste them onto the drawing, another drawing, or a document in another program. To paste the selected shapes, press F4 or Ctrl+V.

- **Copy shapes on a layer** — To copy shapes that reside on one or more layers, choose Edit ⇨ Select By Type and select the Layer option. Select the check box for each layer you want to copy and click OK. Use Ctrl+C and Ctrl+V to copy and paste the selected shapes, respectively.

- **Copy a shape from a stencil** — To copy a shape from a stencil several times, you can use the Stamp tool. The Stamp tool doesn't work with shapes on a drawing page, only masters in stencils. The Stamp tool doesn't appear on a menu or toolbar by default. To add it to the Standard toolbar, click the downward pointing arrow on the very right of the toolbar (Toolbar Options) and choose Add or Remove Buttons ⇨ Standard ⇨ Stamp Tool. The button image looks like a real-world ink stamp. To stamp a shape, do the following:

  1. On a stencil, click the master you want to duplicate.

  2. Click the drawing page where you want to add an instance of the master to the drawing. Continue clicking to place additional instances.

  3. To specify the size of the shape, click and drag on the drawing page while the Stamp tool is active.

  4. To stop the Stamp tool, choose another tool, such as Selection.

Creating an Array of Shapes

The Array Shapes command is the king of duplication. It creates copies of one or more shapes across a number of rows and columns, separated by the distances you specify, as illustrated in Figure 4-11. The applications of Array Shapes are limited only by your imagination. For example, you can use Array Shapes to copy structural columns at specific distances from one another, reflected ceiling panels abutting each other, or office cubicles separated by the width of walls.

FIGURE 4-11

Array Shapes creates rows and columns of copies.

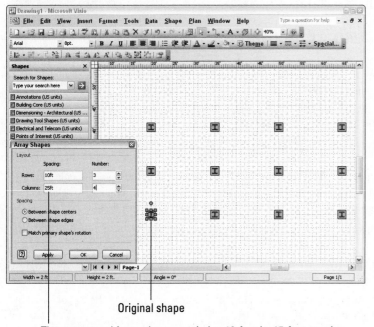

Original shape

Three rows and four columns copied at 10-foot by 15-foot spacing

Here are the steps for setting up Array Shapes to copy shapes:

1. Position a shape that you want to copy at the bottom left of the array and choose Tools ➪ Add-Ons ➪ Visio Extras ➪ Array Shapes.

2. To specify the distance between rows of the array (vertical spacing), in the Rows Spacing box, type the separation. In the Number box, type the total number of rows you want in the array (not the additional rows after the one that contains the original shape).

3. To specify the distance between columns of the array (horizontal spacing), in the Columns Spacing box, type the separation. In the Number box, type the total number of columns you want in the array.

4. Choose a spacing option to determine what the spacing distance applies to:

 ▓ **Between Shape Centers** — Separates the centers of the shapes by the spacing distance. This option is great for shapes such as I-beams, when you usually know the spacing from center to center.

 ▓ **Between Shape Edges** — Separates the edges of the shapes by the spacing distance. This option is better when you know the dimensions of the aisles between shapes, for example, when you separate grocery store shelving units by 4-foot aisles.

5. To rotate the copied shapes to match the primary shape, select the Match Primary Shape's Rotation check box.

6. Click Apply to preview the results. If they are correct, click OK. Otherwise, click Cancel.

Resizing and Reshaping Shapes

The shapes you drag from a stencil onto your drawing aren't always the right size for your purposes. For example, enlarging a shape helps to emphasize it or fit its text within the shape's boundaries. At other times, reducing dimensions of shapes comes in handy for fitting more shapes on a page. Visio provides several methods for resizing shape dimensions. You can also drag vertices and eccentricity handles to change the outline of a shape.

To resize a shape, select the shape and then use one of the following methods:

■ **Resize 1-D shapes** — Drag a selection handle at either end of the shape to lengthen or shorten the shape. For 1-D shapes with thickness, such as a 1-D Single Arrow, you can also drag the selection handles at the midpoints to modify the thickness.

■ **Resize 2-D shapes** — You can resize 2-D shapes horizontally, vertically, or in both directions at once:

 ▓ **Resize Horizontally** — Drag the selection handle at the midpoint of the left or right of the shape alignment box.

 ▓ **Resize Vertically** — Drag the selection handle at the midpoint of the top or bottom of the shape alignment box.

 ▓ **Resize Proportionally** — Drag a corner selection handle to change the horizontal and vertical dimensions in proportion to each other.

> **NOTE** When a shape belongs to a group, you have choices about how the shape resizes when the group resizes. A shape can scale up or down as the group size increases or decreases, or it can remain the same size but reposition itself relative to the new group boundaries. The third option is to resize the shape based on the group's resize settings, which ensures that every member of the group behaves the same way. To set this resizing behavior, subselect the shape (click it twice) and, from the shortcut menu, choose Format ⇨ Behavior. In the Behavior dialog box, on the Behavior tab, choose one of the options under Resize Behavior.

■ **Specify Shape Dimensions** — The Width and Height fields in the Size & Position window are another option for resizing shapes with utmost precision. Refer to the "Positioning Shapes by Coordinates" section earlier in this chapter for more information.

When you work on a scaled drawing, you can change the dimensions of a shape as you would on any other drawing. You can scale shapes to make them appear smaller on the drawing page without changing their real-world measurements, for example, to fit a floor plan on the same page after a new wing is added to the building. To set the scale for a drawing page, choose File ➪ Page Setup and select the Drawing Scale tab. Specify the scale you want and click OK.

■ **Reshape a shape** — You can change the form of a shape by dragging vertices to new positions. For example, you can turn a rectangle into a keystone by dragging the vertices at the top left and top right a few grid intersections to the left and right, respectively. You can even bend the sides of shape, for example to turn a rectangle into a barrel. The trick is displaying a shape's vertices and eccentricity handles. By default, when you select a shape, you see its selection handles. To view vertices and eccentricity handles, you must choose the Pencil tool, which resides on the Drawing toolbar.

Reordering Overlapping Shapes

Sometimes, one shape hides another shape that you want to see. For example, if you add a 3-D bar graph to a drawing and then drag the 3-D axis onto the page, the axis shape hides the bar graph completely. To correct this, you can rearrange the order in which Visio overlaps shapes. Visio provides four options for reordering shapes. Bring to Front and Send to Back appear on the Action toolbar. To access all four reordering commands, right-click the shape and, from the shortcut menu, choose Shape and then choose one of the following commands:

■ **Bring to Front** — Places the selected shape in front of all other shapes

■ **Bring Forward** — Brings the selected shape forward one layer in the stacking order

■ **Send to Back** — Places the selected shape underneath all other shapes

■ **Send Backward** — Sends the selected shape back one layer in the stacking order

Choosing Between Faster Display and Higher Quality

When you're manipulating shapes, the Visio display options can help or hinder. When you are making changes to beat a deadline, faster display might be all that matters. On the other hand, for meticulous editing, higher quality is more important. To control how Visio displays shapes, choose Tools ➪ Options and then specify the following options:

■ **Smooth Drawing** — To prevent your drawing from flickering as you stretch a bitmap or other non-Visio object, click the View tab and then select this check box.

■ **Higher Quality Shape Display** — Anti-aliased drawing displays smooth lines even at angles but is slower than aliased drawing. To draw shapes with anti-aliased lines, click the View tab and select this check box.

■ **Enable Live Dynamics** — To view shapes instead of only the alignment box as you transform shapes, click the General tab and select this check box.

 By clicking several times where a shape is located, you can cycle through selecting the shapes stacked on top of each other even when you can't see the shapes.

Grouping Shapes

Quite often, several shapes belong together and you would like to work with them as if they were one shape. At the same time, you want them to retain their individuality. For example, a table of data in Visio includes many individual boxes and text blocks. The table should move as a single unit, but you want to edit, format, and resize individual table cells. Creating a group from several shapes provides exactly this behavior. Another reason to group shapes is when you want to apply more than one color, line pattern, text block, or other characteristics to an entity. Individual shapes can have only one of each of these attributes. By grouping shapes, you can create a multi-hued element that behaves like a single shape. In fact, many shapes in Visio stencils are actually groups, such as title blocks. You can create your own groups, ungroup and modify built-in groups, and even nest groups within other groups.

 Groups come in handy when you want to apply the same edits to several shapes. For example, suppose you want to rotate all the shapes on a landscaping plan so that the patio tiles run horizontally on the drawing. By grouping all the shapes on the plan, you could then select the group, and, in the Size & Position window, type the angle you want for the group rotation. After the group is rotated, you can ungroup the shapes and continue working on individual shapes.

When an element on your drawing doesn't behave as you would expect, it could be a group rather than an individual shape. Shapes made from multiple geometric constructions can be quite elaborate, but they are still shapes. Groups behave differently. To determine whether an element is a shape or group, select it and then, from the shortcut menu, choose Format ➪ Special. The Type field displays Shape or Group.

 To access Format ➪ Special, you must run Visio in developer mode. Choose Tools ➪ Options and select the Advanced tab. Click the Run In Developer Mode check box.

Groups are separate objects in Visio, so each group has its own text block and ShapeSheet separate from the text blocks and ShapeSheets for each shape in the group. You can add text to the group text block, configure the group's options and behaviors, and define formulas for the group. For example, you could add text to the group text block to label the assemblage or create a formula that applies a fill color to shapes in a group based on the value of shape data.

CAUTION Think twice before ungrouping shapes; doing so eliminates the group ShapeSheet, which takes with it any associated data and formulas.

Making and Breaking Up Groups

Grouping several shapes makes them move around as one; whereas ungrouping shapes splits the group into its individual shapes and eliminates the ShapeSheet, shape data, and formulas associated with the group itself. Visio includes commands to add or remove shapes from an existing group, so you don't have to break apart a group to add or remove shapes. Use the following methods to create, modify, or eliminate a group of shapes:

- **Create a group** — Select the shapes you want to group and then choose Shape ⇨ Grouping ⇨ Group.

- **Break up a group** — Select the group and then choose Shape ⇨ Grouping ⇨ Ungroup.

- **Add a shape to a group** — Select the group and the shape you want to add and then choose Shape ⇨ Grouping ⇨ Add to Group.

NOTE A group alignment box is usually a rectangle that encloses the boundaries of all the group's component shapes. Sometimes the group's alignment box doesn't reflect changes you've made, particularly when you resize the shapes in a group. And when that occurs, snapping to the group might position shapes in unexpected places — to the alignment box and not the borders of the shapes. To reset the alignment box, select the group and then choose Shape ⇨ Operations ⇨ Update Alignment Box.

- **Remove a shape from a group** — Select the shape within the group and then choose Shape ⇨ Grouping ⇨ Remove From Group.

TIP You can select a shape within a group by double-clicking it.

CROSS-REF In addition to the behaviors that make shapes smart, there are behaviors for groups as well as special group behaviors for shapes that belong to groups. For example, you can control the order in which you can select groups and shapes or whether you can add shapes to groups by dropping them onto the group. To specify group behaviors, select the group and choose Format ⇨ Behavior. Chapter 33 describes in detail the options for controlling group and shape behaviors.

Working with Locked Groups

Some groups resist editing and warn you that shape protection prevents execution of the command, which occurs when a group is locked. Many groups in built-in stencils are locked so that you can't inadvertently blow away built-in behaviors or formulas. For example, the 3-D Bar Graph on the Charting Shapes stencil is a group of bar shapes. The group has control handles so you can adjust the height of the tallest bar and the width of all the bars. If you experiment with the 3-D Bar Graph, you'll find that you can't reposition, resize, or edit the shapes in the group. You can only change them by dragging the group's control handles or by changing values in the group's custom properties.

The group is locked so you can't ungroup it and eliminate the features provided by its control handles and ShapeSheet formulas. In addition, the group is configured so that you can't reposition shapes in the group. Because the power of many grouped shapes depends on their group features, you should think twice before removing locks and other protection settings. Visio makes these settings a bit harder to change by placing them on the group ShapeSheet.

TIP If you choose to unlock a group, consider making a copy of the group before you unlock or break it up.

To reset group protections, select the group and then choose Window ➪ Show ShapeSheet. Use one of the following options:

- **Unlock a group** — Scroll down until you see the LockGroup cell in the Protection section of the ShapeSheet. Click the LockGroup cell (which contains the number 1), type **0**, and then press Enter.

NOTE   In the Protection section of the ShapeSheet, typing 1, which represents True or On, in protection cells turns on protection. Typing 0, which represents False or Off, turns it off. The Protection section of a group ShapeSheet includes special group protection settings as well as the settings available in the Protection dialog box that appears when you choose Format ➪ Protection.

- **Configure a group so you can reposition its shapes** — In the group ShapeSheet, scroll down until you see the Group Properties section and the Don'tMoveChildren cell. Type **False** in this cell to enable repositioning of the shapes in the group.

CROSS-REF   To learn more about working with ShapeSheets, see Chapter 34.

Using Visio Drawing Tools

Most drawing content comes from dragging and dropping shapes and connectors. For unique graphics and complicated geometries, the Visio drawing tools let you build shapes from scratch. Although there are only six drawing tools, deft amalgamations of these tools with drawing aids and snapping and gluing techniques can quickly produce lines, curves, and closed shapes on any drawing.

CROSS-REF   To take your drafting to another level, apply one or more Shape Operations command to your drawing elements, described in Chapter 33.

To access the Visio drawing tools, display the Drawing toolbar by choosing View ➪ Toolbars ➪ Drawing. On the Standard toolbar, you can open the Drawing toolbar by clicking the Drawing Tools icon, which appears to the left of the new AutoConnect icon. The drawing Tools button is a mixture of a square, a circle, and a pencil.

TIP   Additional geometry handles for shapes drawn with the Visio drawing tools do not appear by default in Visio 2007. If you want to see all the handles on the shapes you draw, choose Tools ➪ Options and then select the General tab. Select the Show More Handles On Hover check box.

Drawing with Snap To Tools

Snapping helps you position lines and shapes by attracting a point to another shape, a ruler subdivision, a drawing guide, the drawing grid, or other elements on your drawing. Snapping is particularly helpful when you draw lines, arcs, and other constructions, because you don't want to specify the coordinates for every endpoint of every line you draw.

TIP   Snapping is usually helpful, but it can pull the pointer to places you don't want, particularly when you're trying to draw flowing freeform curves. To draw smoother freeform curves, choose Tools ➪ Snap & Glue and then clear the Snap check box in the Currently Active column. When you are finished, select the Snap check box to restore snapping.

CROSS-REF   The section "Placing Shapes with Precision" earlier in this chapter discusses snapping, drawing aids, and other positioning tools.

Drawing aids are dotted lines that indicate where to click to draw a circle, square, or a line at a particular angle. They are indispensable when creating lines and closed shapes. With the Ellipse and Rectangle tools, a dotted line shows you where to click to create a circle or square, as demonstrated in Figure 4-12. When you use the Line tool, guides appear when you approach an increment of 45 degrees. During edits on a line segment, drawing aids extend at 45-degree increments as well as the line's original angle. To display drawing aids, choose Tools ➪ Snap & Glue and check the Drawing Aids check box in the Currently Active column.

FIGURE 4-12

Drawing aids show you where to move the pointer to create circles, squares, and angled lines.

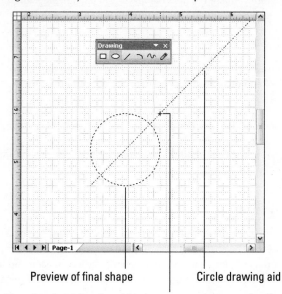

Preview of final shape   Circle drawing aid

Drag this point along drawing aid to increase or decrease the diameter of the circle

Drawing Lines

The Line tool can create individual line segments, a series of connected line segments, or closed shapes, as long as you know how to click and drag the pointer for each. When the Line tool is active, the pointer changes to crosshairs with a short angled line to its right. To draw lines, follow these steps:

1. On the Drawing toolbar, click the Line tool.

2. Drag the mouse button from the starting point of the line to the end point of the line, and then release the mouse button. Visio displays the new line segment and selects it.

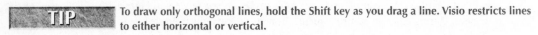 To draw only orthogonal lines, hold the Shift key as you drag a line. Visio restricts lines to either horizontal or vertical.

3. To connect another line segment to the one you just drew, without moving the pointer from the end of the last segment, drag to the next end point, and release the mouse button.

CAUTION If you click more than once before drawing a new line segment, Visio deselects the previous segment and the two segments won't be connected. To prevent this problem, make sure the previous segment is selected before you add the next one. If it's too late for prevention, join separate line segments by Shift-clicking each segment and then choosing Shape ➪ Operations ➪ Join.

4. To add another line segment, repeat step 3.

NOTE When you create a single line segment with the Line tool, the start and end points are green squares. In a series of connected lines, Visio indicates the vertices at the ends of each line segment with green diamonds.

5. If you want to close the shape by adding another line, click and drag to the starting point of the first line segment and release the mouse button. You can tell that the shape closed successfully when Visio applies a solid white fill to the shape, hiding the drawing grid behind the enclosed area.

Drawing Arcs and Curves

The Arc tool draws one arc at a time, and each arc represents no more than one-fourth of an ellipse or a circle. The curve of the arc depends on the angle between the start and end points you choose and can vary between a straight line and a circular curve. Visio arcs have vertices at each end and at their midpoints and also contain eccentricity points control the offset and curvature of the arc, as illustrated in Figure 4-13.

 The Arc tool can't create a complete circle or ellipse. You must use the Ellipse tool to draw a closed ellipse or circle.

FIGURE 4-13

Vertices and eccentricity handles modify the characteristics of an arc.

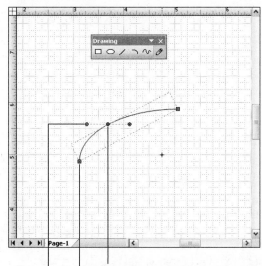

Drag this eccentricity handle to change the offset and curvature

Move vertex to lengthen, shorten, or change the direction

Drag this eccentricity handle to change the offset and lean of the curve

Use the following techniques to draw and modify arcs:

- **Draw a clockwise arc** — Click one point and then sweep the pointer clockwise to the next point.

- **Draw a counterclockwise arc** — Click one point and then sweep the pointer counterclockwise to the next point.

TIP   The direction of the arc remains fixed after the initial sweep. After the direction is set, dragging the pointer further defines the arc's curve and orientation.

- **Move an arc** — Select the Pointer tool, and click somewhere on the arc away from any vertices or control points. When the pointer changes to a four-headed arrow, drag the arc to a new location.

- **Extend or shorten an arc's length** — Drag one of the end points to a new location.

- **Change the depth of the curvature** — Turn on the Pencil tool. Select the arc and then drag the middle eccentricity handle.

NOTE   Depending on where you drag the end point of an arc, you can also change the shape of the arc in addition to changing its length.

- **Change the shape of an arc** — Turn on the Pencil tool. Drag the outside eccentricity handles to change the offset of the curve from the center of the arc and the sharpness of the curvature.

- **Modify direction or rotation** — Right-click an arc, from the shortcut menu, click Shape, and then click one of the Rotate or Flip commands.

The Freeform tool draws multiple arcs and splines — for instance, to mimic handwriting. This tool senses changes in the direction of the pointer and automatically adds vertices and control points as you draw. The result is a series of curves that you can tweak section by section. To draw a freeform curve, on the Drawing toolbar, click the Freeform tool and then drag the pointer slowly on the drawing page. Several factors contribute to the success of your freeform drawing efforts. To improve your results, try the following:

- **Take your time** — Drawing more slowly provides greater control over the curves you create, as Visio better recognizes your direction changes.

- **Modify the freeform precision option** — Precision controls how Visio switches between drawing straight lines and curved splines. To set precision, click Tools ➪ Options, click the Advanced tab, and drag the precision scroll bars to the left or right. Dragging to the left tightens the tolerance so that Visio draws splines unless you move the mouse in a very straight line. Dragging to the right loosens the tolerance, so Visio draws straight lines until you move the mouse in an obvious curve.

- **Modify the freeform smoothness option** — Smoothing controls how much Visio smoothes out your curves — in effect, how sensitive Visio is to changes in direction. Tighter settings add more control points, as shown in Figure 4-14, whereas looser settings add fewer. More control points provide greater control over the angles of arcs in a freeform curve.

FIGURE 4-14

The Smoothing setting influences the number of control points added to a curve.

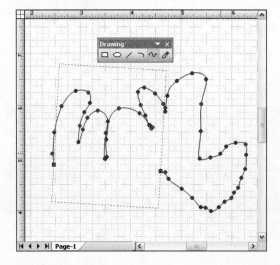

■ **Disable snapping** — If your freeform lines are erratic, choose Tool ⇨ Snap & Glue and uncheck the Snap check box in the Currently Active column.

TIP If you change your Snap & Glue settings frequently, save time by displaying the Snap & Glue toolbar to keep Snap & Glue commands handy. Simply choose View ⇨ Toolbars ⇨ Snap & Glue.

Drawing Closed Shapes

Although the Line tool can produce closed linear shapes, ellipses and rectangles have their own Visio drawing tools. With the Ellipse and Rectangle tools, you can draw ellipses, circles, rectangles, and squares by clicking just two points.

You can draw quadrilateral shapes (shapes with four sides) using the following methods:

■ **Rectangle** — Click and drag the pointer from one position on the drawing page to the opposite corner of the rectangle.

■ **Square** — Hold the Shift key while dragging from one corner to the opposite corner.

■ **Use drawing aids to draw a square** — Click the first corner, drag the pointer close to a 45-degree angle, and click on the drawing aid that appears to select the opposite corner.

Many diagrams use rectangles with rounded corners. To round the corners of rectangular shapes, choose one of the following methods:

■ Right-click a shape and, from the shortcut menu, choose Format ⇨ Line. Select the rounding you want and click OK.

TIP You can also specify the rounding in Format Line dialog box. In the Rounding box, type a value for the radius of the corner.

■ Choose Format ⇨ Corner Rounding. Select the rounding you want and click OK.

■ Choose View ⇨ Toolbars ⇨ Format Shape. Click the Corner Rounding button on the Format Shape toolbar.

■ To define a style with rounded corners, choose Format ⇨ Define Styles and click the Line button. Select the rounding you want and then click OK.

CROSS-REF To learn more about defining styles, see Chapter 35.

You can draw ellipses and circles using the following methods:

■ **Ellipse** — Click and drag the pointer from one corner to the opposite corner. The ellipse is circumscribed by the rectangle you defined with your two points.

■ **Circle** — Hold the Shift key while dragging from one corner to the opposite corner.

■ **Use drawing aids to draw a circle** — Click the first corner, drag close to a 45-degree angle, and click on the drawing aid that appears to select the opposite corner.

Using the Pencil Tool

The Pencil tool is quite versatile, working equally well for drawing new lines and arcs as it does for reshaping existing ones. In addition, you can use the Pencil tool to construct a polyline made up of a combination of straight lines and arcs, as demonstrated in Figure 4-15. The Pencil tool interprets your pointer movements to determine whether you want to draw a line or arc, and switches to either Line mode or Arc mode, respectively. In addition, the Pencil tool, unlike the Arc tool, can draw arcs that are almost complete circles.

FIGURE 4-15

The Pencil tool interprets the movements of the mouse to decide whether to draw a straight line or an arc.

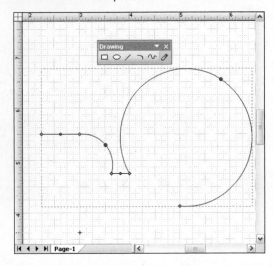

Here are the techniques for drawing connected lines and arcs with the Pencil tool:

- **Draw a straight line** — Drag the pointer straight in any direction. Visio indicates that it is in Line mode by changing the pointer to crosshairs with an angled line below and to the right.

- **Draw an arc** — Sweep the pointer in a curve. Visio indicates that it is in Arc mode by changing the pointer to crosshairs with an arc below and to the right. You can move the pointer to define the radius of the arc, the angle that the arc circumscribes, as well as the position of the arc on the drawing page.

- **Switch to Line mode when you are in Arc mode** — Move the pointer back to the last vertex. When the Arc next to the crosshairs disappears, drag the pointer straight to switch to Line mode.

- **Switch to Arc mode when you are in Line mode** — Move the pointer back to the last vertex. When the angled line next to the crosshairs disappears, sweep the pointer to switch to Arc mode.

Summary

Drawings are made up entirely of shapes, so Visio provides plenty of tools for finding, adding, and manipulating the shapes on your drawings. When you start building a drawing, finding the right shapes is your first task. When you add shapes to a drawing, you can position shapes in several ways and select shapes to modify them further. Visio provides tools for snapping shapes into position or positioning them with precision.

For editing, Visio offers techniques and specialized tools for moving and copying shapes, resizing them, and changing the order in which they appear. Similarly, you can simplify work on several related shapes by turning them into a group.

If no shape exists to satisfy your need, you can build your own with Visio drawing tools. In addition to constructing basic forms, the Pencil tool stands out by enabling you to reshape the appearance of lines, arcs, and closed shapes. Visio provides even more tools for customizing and creating your own shapes, which are covered in Chapters 33 and 34.

Chapter 5

Connecting Shapes

Some shapes stand on their own, such as title blocks or legends, but, in most diagrams, the relationships between shapes constitute much of what you're trying to communicate. Imagine an organization chart without reporting relationships between managers and employees, or a data flow diagram without flows between processes. In Visio, connectors show the relationships between shapes.

One of the great new features in Visio 2007 is AutoConnect — it takes care of adding shapes to drawings, connecting them to other shapes, and aligning the connected shapes all at once. For example, you can click one of the blue arrows on shapes on the drawing to add, connect, and align another shape to the shape you clicked.

AutoConnect also speeds up adding several copies of the same shape to a drawing. When you select a shape in the Shapes window, you don't have to drag and drop it to add it to your drawing. Simply click a blue arrow on the drawing and AutoConnect adds the shape to the page. AutoConnect can even simplify connecting shapes already on a drawing page. Suppose you added shapes to a drawing before you discovered AutoConnect. As long as the shapes are next to each other, you can connect two adjacent shapes by clicking one of the blue AutoConnect arrows on the side that faces the other shape.

In this chapter, you'll learn about all the different methods Visio offers for connecting shapes, from the AutoConnect feature to dragging and gluing shapes and connectors. When you connect shapes in Visio, the program can choose the best spot for the connection. However, you can take charge of connection locations by attaching connectors to specific points on shapes.

Understanding Connectors

Most diagrams show relationships, whether they are the computers in a network, the steps in a process, or the people in an organization. Visio represents the relationships between the 2-D shapes on a drawing with connectors, which are 1-D shapes made from one or more line segments.

Connector Basics

Connectors come in all shapes and sizes, from unadorned lines to 1-D shapes specialized to suit the different drawing types that Visio supports, as illustrated in Figure 5-1. Yet, no matter how fancy the formatting, connectors all boil down to lines that attach to 2-D shapes at each end. The stencils associated with each drawing type usually include connectors appropriate for your drawing. In addition, the Connectors stencil (choose File ➪ Shapes ➪ Visio Extras ➪ Connectors) contains many of the most popular connectors for all drawing types.

FIGURE 5-1

Visio provides all kinds of connectors from plain lines to connectors customized to suit specific types of diagrams.

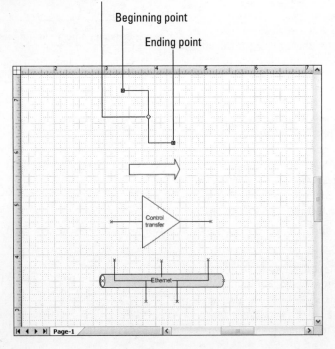

Control handle to reposition connector text

Beginning point

Ending point

Unlike the connecting lines you draw in a paint program or even a CAD application, Visio connectors are smart. When you glue the end points of connectors to shapes, the connectors know how to keep those shapes connected, regardless how you move the shapes on the drawing. For example, a dynamic connector that runs straight across between two shapes automatically adds line segments to itself if you move the shapes out of alignment.

Connectors have start and end points that are important for defining the predecessor and successor for two connected shapes. All connectors store direction information, whether they display visual cues, such as arrowheads, or not. For example, in an organization chart, the direction of a connector differentiates the manager and the employee. When sequence is important, you must connect the correct end of the connector to the predecessor and successor shapes. If order doesn't matter, connect whichever end of the connector you want.

Connectors also respond to the layout and routing settings you choose, described in Chapter 4. For example, connectors change their appearance to match the connector style you choose, switching from multi-segmented orthogonal lines to straight lines between the centers of shapes. Connectors also can route around shapes on a drawing or include line jumps to make crossing connectors easier to follow.

CROSS-REF For an overview of connection concepts, see Chapter 1.

An Introduction to Connecting Shapes

Visio provides visual cues that indicate where you can connect shapes and connectors. For example, 2D shapes often include *connection points*, which are likely connection spots on those shapes—midpoints of sides are common sites for connection points. Visio indicates a connection point on a shape with a blue x. On connectors, each end point looks a little different, so you can tell the beginning and end apart. The *start point* for the connector is a green square with an x inside the square. The *end point* is a green square with a + inside the square, as shown (without the color, of course) in Figure 5-1.

Visio recognizes other types of points as connection spots for gluing shapes and connectors. To learn more about specifying glue points, see the "Specifying Points for Glue" section later in this chapter.

The point you select on a shape determines whether you create a dynamic or static connection. When you drag an end point inside a shape, Visio highlights the entire shape with a red box, indicating that you're creating a dynamic, or shape-to-shape, connection. Shape-to-shape connections tell Visio to choose the best sides to use for connecting the shapes. For example, if the shapes are one above the other, Visio connects the shapes from the bottom of one to the top of the other. But, should you move the shapes to be side by side, Visio automatically changes the connection to be between the facing sides.

On the other hand, static, or point-to-point, connections remain on the sides to which you glue. Static connections are perfect for connections in which the location communicates information, such as a piping diagram. Dragging an end point on a connector to a connection point on a shape first highlights the connection point with a red box and then glues the two together with a static connection.

Depending on the template you use and the shapes you want to connect, you can choose from a variety of methods for connecting shapes. With some templates, such as the Organization Chart, you can simply drop a shape onto another shape to connect them. You can also drag connectors from stencils or use the Connector tool to create connectors by dragging from shape to shape or even to create a sequence of connected shapes. The rest of this chapter describes these methods and the best time to use them.

CROSS-REF   Layout and routing settings affect the appearance of dynamic connectors. See the sections "Setting Up Pages" in Chapter 3 and "Specifying Layout Options" in Chapter 4 to learn more about layout and routing settings.

Working with AutoConnect

Some shapes in Visio 2003 were smart enough to connect and align themselves when you added them to a drawing. For example, dropping an organization chart shape on top of one on the drawing page would add the shape to the page, connect the shape to its superior, and position the shapes according to the organization layout you selected. In Visio 2007, AutoConnect brings this type of convenience to more shapes.

NOTE   Not all shapes support AutoConnect. To see whether a shape works with AutoConnect, position the pointer over the shape on the drawing page. If blue arrows appear on each side of the shape, you can use AutoConnect to connect other shapes.

Although AutoConnect doesn't prevent you from using other connection methods, it could become your favorite Visio tool. You can connect shapes with AutoConnect in three different ways, which satisfy the majority of connections you make on diagrams:

- **Drag and drop shape onto an AutoConnect blue arrow** — Drag and drop the first shape on a drawing as you normally would. With no other shapes on the drawing, there are no blue arrows to click. Then, drag another shape from the stencil and position the pointer on top of the shape to which you want to connect it. As illustrated in the shapes at the top of Figure 5-2, blue arrows appear on each side of the shape already on the drawing. Position the pointer on top of the blue arrow on the side you want to connect and release the mouse button. Visio adds the new shape, connects it to the first shape, and lines the two shapes up, shown at the bottom of Figure 5-2.

FIGURE 5-2

AutoConnect rolls adding, connecting, and aligning into one operation.

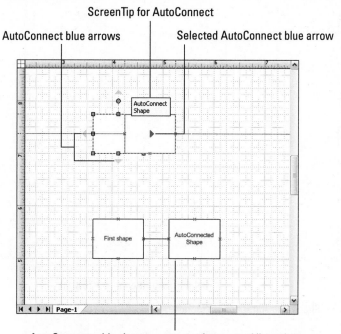

ScreenTip for AutoConnect

AutoConnect blue arrows · Selected AutoConnect blue arrow

AutoConnect adds shape, connects shapes, and lines them up.

- **Speed up adding the same shape several times** — In Visio 2003, you could add several copies of the same stencil shape to a drawing, but those copies weren't connected. Dragging a shape onto the drawing, copying and pasting, or using the Stamp tool all had the same limitation. With AutoConnect, when you select a shape in a stencil, you can AutoConnect as many copies as you want. Drag the shape onto a blue arrow on the drawing page to AutoConnect the first copy. After that, simply click additional blue AutoConnect arrows on the shapes on the drawing to add more of the selected stencil master. To add a different kind of shape, select it in the stencil, AutoConnect it to one shape on the drawing page, and start clicking blue arrows to add more copies of that shape.

- **Connect two adjacent shapes on a drawing** — AutoConnect can also connect shapes that are already on a drawing, as long as the shapes are near one another. As illustrated in Figure 5-3, clicking a blue AutoConnect arrow between two adjacent shapes connects them automatically.

TIP To use a specific type of connector, select it in the stencil in the Shapes window before you click a blue arrow on the drawing page.

FIGURE 5-3

Clicking a blue arrow between two unconnected shapes connects them.

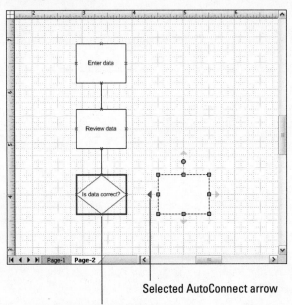

Selected AutoConnect arrow

Shape that AutoConnect found

Other Methods for Connecting Shapes

AutoConnect can handle almost all the connections you make on drawings, but there are still situations that require old-fashioned methods for connecting shapes. For example, connecting existing shapes on a drawing that aren't adjacent to each other is one connection task that AutoConnect doesn't cover. This section describes several methods for connecting shapes. Use them all and pick your favorite.

Using Stencil Connectors

You can connect shapes with connectors from stencils, whether you drag the connector onto a page and attach it to shapes or use the connector with AutoConnect. Because templates include stencils with the most appropriate connectors, opening a new stencil is rarely necessary. In addition, the connectors are often designed to communicate information specific to the type of drawing your building, such as the Ethernet connector with connection points for network connections.

The Dynamic connector and the Line-curve connector — the workhorses of the Visio connection world — appear on many stencils. For the widest selection of connectors appropriate for numerous types of drawings, open the Connectors stencil by choosing File ⇨ Shapes ⇨ Visio Extras ⇨ Connectors.

Regardless of the connector you choose, connecting shapes with a connector from a stencil follows the same steps:

1. Drag the connector you want from its stencil onto the page.

2. Drag one end point of the connector to a connection point (for static connection) or the inside of one shape (for dynamic connection).

3. Drag the other end of the connector to a connection point or the inside of another shape.

NOTE **If the direction of the connection is important, glue the start point of the connector to the predecessor or superior shape. Glue the end point of the connector to the successor or subordinate shape.**

4. To label the connector, make sure the connector is selected and then type the text for the label. To reposition the text, drag the yellow control handle for text on the connector to a new position.

NOTE **With a few settings in place, you can drop a shape onto a connector and Visio will split the connector in two and glue the dropped shape to the adjacent shapes with the newly spawned connectors. To enable this behavior, you must enable connector splitting and set up the shape to split connectors.**

First, choose File ⇨ Page Setup. In the Page Setup dialog box, on the Layout and Routing tab, check the Enable Connector Splitting check box. If a shape doesn't split connectors, right-click it and, from the shortcut menu, choose Format ⇨ Behavior. On the Behavior tab, select the Shape Can Split Connectors check box.

Using the Connector Tool

The Connector tool is at its best when you want to create connectors by drawing lines — with this method, you drag the pointer from the first shape to the second. This method makes it easier to create connectors with a specific direction, because most people naturally draw from the predecessor shape to its successor. To draw connectors with the Connector tool, from the Standard toolbar, click the Connector tool and then use one of the following methods:

- **Create a Static (Point to Point) Connection** — Position the pointer over a connection point on the first shape. When Visio highlights the connection point with a red box, drag to a connection point on the second shape.

- **Create a Dynamic (Shape to Shape) Connection** — Position the pointer inside the first shape. When Visio highlights the shape with a red box, drag to a position inside the second shape.

- **Create a Connection with Both Static and Dynamic Behavior** — This connection adjusts the side to which it is attached at the end that uses dynamic glue. The end that uses static glue remains glued to the same point regardless how you rearrange the connected shapes. To create the static connection, position the pointer over or drag to a connection point on one shape. Visio highlights the connection point with a red box to indicate a static connection. Create the dynamic connection by positioning or dragging the pointer inside a shape. Visio highlights the shape with a red box to indicate the dynamic connection.

- **Automatically connect a new shape to one of the drawing** — When the Connector tool is active, Visio automatically connects each new shape you add to the previous shape with a dynamic connection. Select an existing shape on the page. On the Standard toolbar, choose the Connector tool and drag a shape from a stencil onto the drawing page. Visio adds the new shape and a connector and glues the connector to both the existing shape and the new shape.

 When shapes are small or contain connection points in inconvenient locations, dynamic connections can be tough to create. To create a dynamic connection no matter where you position the pointer on a shape, hold the Ctrl key as you drag the end point of the connector or Connector tool over the shape.

The Connector tool creates dynamic connectors by default, but you can convince the Connector tool to create any kind of connector by following these steps:

1. On the Standard toolbar, click the Connector tool.

2. In the Shapes window, click a connector in a stencil.

3. Draw a connector from the first shape to the second. Visio creates the type of connector you selected in the stencil.

Dragging Points to Connect Shapes

Connectors sometimes have behaviors and features tailored to a specific type of drawing. These connectors are actually shapes with connectors built in and typically contain control points that you drag to attach the connectors to other shapes. For example, by dragging the control point on the trunk of the Multi-Tree Sloped shape on the Blocks stencil, you can connect branches on the tree to several shapes, as shown in Figure 5-4.

TIP To find out what a control point can do, position the pointer over the control point. Visio displays a tip on the screen, as demonstrated in Figure 5-4.

Connecting a Sequence of Shapes

When you're creating a drawing that shows a sequence, such as the steps in a procedure, you can add the shapes to your drawing, and instruct Visio to create all the dynamic connections at once. To connect a sequence of shapes, select the shapes in the order you want them to appear in the sequence and then choose Shape ➪ Connect Shapes. Visio draws dynamic connectors from each predecessor to each successor. Connect Shapes creates shape-to-shape connections so the connections adjust as you reposition the shapes.

FIGURE 5-4

Drag control points on specialized connectors to create and configure connections.

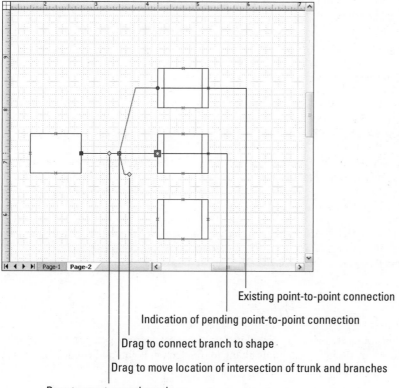

Existing point-to-point connection

Indication of pending point-to-point connection

Drag to connect branch to shape

Drag to move location of intersection of trunk and branches

Drag to create new branch

Organizing Connections with Glue

The paths that your connectors take depend on how you glue them to shapes. Visio glue comes in two types: static and dynamic. *Static glue* (also called *point-to-point glue*) connects specific points on two connected shapes. As you move those shapes around, the connector remains attached to the points you glued, no matter how convoluted the connector path becomes. Static glue is best used when the connection point is important. For example, in a flowchart, you might not want Visio to reattach a process to a different decision point when you move the process on the page. With *dynamic glue* (also called *shape-to-shape glue*), Visio changes the connection points it uses as you move shapes around, connecting the two shapes by the shortest route. Dynamic glue is more convenient but suitable only where the specific connection point doesn't matter.

Gluing Shapes

Connections are limited to one type of glue. You can combine glues in any way when you connect two shapes. However, for the automatic layout tools to work, you must connect shapes using shape-to-shape, or dynamic, glue at both ends of the connector. Connections between shapes behave according to how the connector ends are glued. For example, if one end of the connector uses static glue and the other uses dynamic glue, the connection point that is glued dynamically can change as you move the shapes. The end point that is glued statically remains glued to the same point no matter where you position the shapes.

> **TIP**   Use dynamic glue while you develop a diagram, so Visio automatically adjusts connec-tions as you move shapes around. When you finalize the layout, you can change the connections between shapes to static glue if you want to fix the connection points that Visio uses.

Unfortunately, glue can get you into some sticky situations. For example, when you're dragging connectors onto a crowded page, it's difficult to drop a connector onto the page without gluing it to a shape. To prevent Visio from gluing connectors to shapes, choose Tools ➪ Snap & Glue, clear the Glue check box under Currently Active, and click OK.

> **TIP**   The Snap & Glue toolbar includes buttons that control which points Visio uses for snap-ping or gluing. If you want to switch the options for snap or glue frequently, choose View ➪ Toolbars and then choose Snap & Glue. Click one of the buttons to toggle a snap or glue option on or off.

Specifying Points for Glue

Although connection points are natural choices for gluing connectors to shapes, you can glue to other points on shapes. By default, Visio limits gluing to Connection Points and Guides, but you can use other types of points for glue by following these steps:

1. Choose Tools ➪ Snap & Glue.
2. In the Snap & Glue dialog box, under Glue To, check one or more of the following check boxes:

   - **Shape Geometry** — Glues connectors anywhere on the visible edge of a shape
   - **Guides** — Glues connectors or shapes to guides
   - **Shape Handles** — Glues connectors to shape selection handles
   - **Shape Vertices** — Glues connectors to shape vertices
   - **Connection Points** — Glues connectors to shape connection points

Adding, Moving, and Deleting Connection Points

Many shapes come with connection points already built in, but you can add more connection points to shapes if that makes your connecting tasks easier. For that matter, connection points don't have to reside on or within the shape to which they belong, although connection points too far removed from their parent shapes are usually more confusing than convenient.

The Connection Point tool must be active before you can add, move, or delete connection points. To activate the Connection Point tool, on the Standard toolbar, click the arrow to the right of the Connector tool button and, on the drop-down menu, choose Connection Point Tool. Visio displays a green, dotted line around the selected shape and changes the button on the Standard toolbar to a blue X. To revert to the Connector tool, click the arrow to the right of the Connection Point tool and choose Connector Tool.

CROSS-REF Visio provides three types of connection points: Inward, Outward, and Inward & Outward. Most built-in masters already contain connection points. To learn how to specify connection point types when you create your own customized shapes, see Chapter 33.

To add, delete, or move connection points on a shape, activate the Connection Point tool, select a shape, and then use one of the following techniques:

- **Add a connection point** — Hold the Ctrl key and click the shape where you want to add the connection point. Visio displays the new connection point as a small purple x, which changes to blue when you complete another action.

 It's easier to add connection points when Snap & Glue is activated. To activate Snap & Glue, choose Tools ⇨ Snap & Glue, check the Snap and Glue check boxes, and click OK.

- **Move an existing connection point** — Drag the connection point to a new location.
- **Delete a connection point** — Select the connection point you want to delete and press the Delete key.

Summary

Connections between shapes convey a great deal of information on drawings. Visio provides several methods for adding connectors to shapes, as well as dozens of different types of connectors. Some shapes, such as trees, are shape with connectors built in. The new AutoConnect feature makes connecting shapes much faster, whether you are adding new shapes to a drawing or connecting shapes already on the drawing page.

Connections come in two flavors: dynamic and static. When you use static glue, Visio keeps connectors glued to the points you select. However, you must use dynamic glue if you want to automatically lay out and route your diagram, or you want Visio to create optimal routes between shapes.

Chapter 6

Working with Text

Pictures might be worth a thousand words, but usually, drawings need words, too. Text pops up in many forms in Visio: text that appears within shapes, shapes specifically for annotation, and stand-alone text blocks. All Visio shapes, including connectors, have associated text blocks perfect for labeling those shapes. Although these text blocks belong to their parent shapes, you can improve the readability of your drawings by relocating, rotating, or formatting shape text.

The built-in annotation shapes include boxes with callouts for adding notes to diagrams, as well as popular annotation devices, such as title blocks that display the attributes of drawings. You can also configure text into tabular form by using a built-in Table shape or by formatting the text in a text block. In addition, Visio add-ons can automate labeling and numbering shapes on your drawings.

If you're like most people, you probably devote a significant amount of time working with text, because drawings often contain as much text as they do graphics. Besides the text that you type, Visio can accommodate handwriting with digital ink. Whatever its form, text in Visio works like any other Visio shape. Visio makes it easy to add, edit, and format text and annotations. Visio, like other Microsoft Office applications, includes familiar commands for searching and replacing text as well as checking spelling.

Collaborating with teammates often results in questions and comments about the documents you work on together. Comments can communicate this type of information, and they're easy to remove when drawings are final.

Adding Text to Drawings

No matter which type of text medium you use, adding text to drawings is often as simple as clicking a shape or text block, typing text, and pressing Esc when you're done. Go ahead and type without worrying about whether the text will fit. Many Visio text shapes automatically resize to accommodate the text inside them. But you can always apply Visio's text tools to edit and format text after your words are in place.

If you're used to working with CAD programs, Visio text behaves differently, but is simpler to use. Text mimics the behavior of shapes on Visio drawings—it appears larger or smaller as you zoom in and out. In Visio, the drawing scale has no effect on the appearance of text. Text always appears at its actual size relative to the printed page. If you change the scale of a drawing, you might have to change the size of the text as well.

> **TIP** Options control whether Visio displays aliased or antialiased text, which is important when you want faster display times or higher quality. To set these options, choose Tools ⇨ Options and select the View tab. Select the Text Quality option you want.

Another way to improve performance on drawings with loads of text is called *greeking*. This feature displays a wavy line in place of text when the text is smaller than a specified point size on the screen. You can turn on this feature while you work on your drawing and then turn it off when the editing is finished and you want to display all text. On the View tab, in the Greek Text Under box, choose the point size.

Adding Text to Visio Shapes

To add text to most shapes, including connectors, simply select a shape and type the text you want to display. For example, in a flowchart box, you might type the name of the process. On a connector to a Decision shape, you might type **Yes** or **No**. As soon as you begin to type, Visio zooms in to 100 percent so you can see the text you're entering. Visio underlines spelling errors with a red, wavy line, just like Microsoft Word does, but unlike in Word, the misspelling indicator disappears when the shape or text block isn't selected.

> **TIP** Although automatic zooming makes text easier to see, it can slow you down when you have to continually change your view. To disable automatic zooming when you edit text, choose Tools ⇨ Options ⇨ General. In the Automatically Zoom Text When Editing Under drop-down list, choose 0 for the point size. Because text point size must be greater than zero, this choice ensures that Visio never zooms when you edit text.

To add text directly to a shape or connector, follow these steps:

1. Select a tool other than the Text tool, such as the Pointer tool, and click a shape or connector.

2. Begin typing.

3. To complete your text entry, press Esc or click outside the text block.

Pressing Enter inside the text block inserts a carriage return and moves the insertion point to the next line in the text box. For some shapes that use fields, such as the Organization Chart Position shape, an extra carriage return might appear at the end of the text. This occurs because the shape expects a pair of text strings separated by a carriage return. The first segment is assigned to the name and the second to the title.

When you add more text than a shape can hold, the text usually overflows the shape's boundaries. (Some shapes automatically resize to contain the text you type.) You can resize a shape to contain its text by dragging its selection handles. If you don't want to resize the shape, you can apply a smaller font or edit the text to be more concise.

Adding Text to Groups of Shapes

Some shapes look like they include more than one text block. For example, the X-Y Axis shape on the Charting Shapes stencil includes a text block to label each of the axes. The X-Y Axis shape performs this magic because it's actually a group of shapes. Each shape in the group has its own text block and the group has one as well.

Accessing text in grouped shapes sometimes requires a few additional clicks. The order in which Visio selects the text blocks within a group depends on the group's behavior settings. The default method is Group first, but you can also set text selection to Members first or Group only.

To add text to a shape in a group using the default behavior, click that shape once to select the group to which it belongs. Then click the shape a second time to select the shape. If you select the shape with the Text tool (on the Standard toolbar) Visio positions an insertion point in the text block of the shape. If you select the shape with any other tool, Visio selects the shape text block and replaces its contents when you begin typing.

Adding Text-Only Shapes

Drawings can contain text that does not have a visible shape. (Changing the line pattern from none to solid will make the rectangular shape visible.) The Text tool focuses on text, so it's the tool to use to create a shape that contains only text. When you activate the Text tool and drag to define a text block, Visio creates a separate text block shape and selects it so that you can begin typing text. All the standard editing and formatting features apply to the text in these blocks as they do to text in Visio shapes.

When the Text tool is selected, Visio positions an insertion point where you click in a shape text block.

To add a text-only shape to a drawing, follow these steps:

1. On the Standard toolbar, click the Text tool.
2. Drag between two points on the drawing to create a rectangular text block.

With the Text tool selected, you can click a point in the drawing and Visio creates a text block at that point using default dimensions.

3. Type the text.

4. To complete your text entry, press Esc or click outside the text block. By default, Visio creates a text-only shape with 8-point Arial text centered in the block.

NOTE To insert special characters into text, choose Insert ➪ Symbol, which opens the Symbol dialog box. For the most common special characters, select the Special Characters tab, select the symbol you want, and click Insert. For other special characters, select the Symbols tab, select the symbol you want, and click Insert. If you don't see the symbol you want, choose another font from the drop-down list.

Displaying Field Information in Text

Visio stores information about shapes and documents in fields. For example, Visio fields track who created a document, the name of a page, the angle of a shape, and the values in shape data. Fields can display the results of a formula that uses the values from other fields. This sort of information isn't visible on drawings by default, but you can add it by inserting fields into shape text. For example, a built-in Visio title block shape, shown in Figure 6-1, uses fields to automatically display information, such as drawing scale, file name, the author, and page number, for the drawing on which it is located. A Text block can contain as many fields as you want.

FIGURE 6-1

With fields in shape text, Visio automatically retrieves and displays drawing, page, shape, or custom property information.

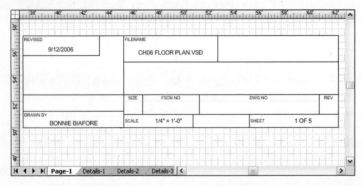

As with other Microsoft applications, Visio includes several categories of fields and numerous fields within each category. Here are categories of fields and the type of information the fields represent:

- **Shape Data**—When you choose this category, the fields that appear in the Field Name list correspond to the shape data fields for the selected shape. For example, if a Process shape is selected, the Field Name list contains Cost, Duration, and Resources.

- **Date/Time** — These fields include the current date and time or the date and time that the file was created, printed, or edited. For drawings under change control the field Last Edit Date/Time automatically shows when the drawing was edited last.

- **Document Info** — These fields include information from a file's Properties box, such as creator or keywords.

- **Page Info** — These fields relate to settings that apply to pages in a drawing file, such as the page number or number of pages in a drawing.

- **Geometry** — When you select the text box in a shape, these fields display the shape width, height, or angle of rotation.

- **Object Info** — These fields represent information from a shape's Special dialog box, such as the shape's internal ID or the master used to create it. To view these fields, right-click a shape and choose Format ⇨ Special from the shortcut menu. (You must turn on developer mode to access the Format ⇨ Special command. Choose Tools ⇨ Options and select the Advanced tab. Then, select the Run In Developer Mode check box.)

- **User-Defined Cells** — If you enter formulas in the User-Defined Cells section of the ShapeSheet, you can display the results of those formulas with these fields.

- **Custom Formula** — These fields are the results of a ShapeSheet formula you define in the Custom Formula box.

CROSS-REF For an introduction to ShapeSheet formulas, see Chapter 34. To learn more about fields and shape data, see Chapter 33.

Fields can go anywhere in a text block. To insert fields in a shape text block, follow these steps:

1. If you want to insert a field as the only text, select the shape and choose Insert ⇨ Field. To insert the field within existing text, double-click the shape to open its text block, click in the text to position the insertion point, and choose Insert ⇨ Field to place the field at that position in the text.

2. In the Field dialog box, under Category, select the field category you want, such as Geometry.

3. Under Field Name, Visio displays the fields for the selected category. Select the field you want, such as Width.

4. To apply a specific data format, click Data Format, select the format you want, and then click OK.

To change the field used in a text block or the format for the data, follow these steps:

1. Select the Text tool and select the shape.

2. Deselect the text and then select the field you want to modify by clicking within the field. (You can't position an insertion point within a field.) Visio highlights the value for the field.

3. Choose Insert ⇨ Field. The Field dialog box appears with the current field and settings selected, which is a handy way to determine what field and settings are used in a text block.

4. Choose a new field or click Data Format to modify the formatting.

5. Click OK.

Displaying Information in Headers and Footers

Headers and footers in Visio can be misleading. If you're new to Visio, you might think about using Visio headers and footers in the same way that you use headers and footers in Microsoft Word, Excel, and PowerPoint documents. It turns out that Visio has two features for displaying reference text, and it's important to understand when to use each one.

- **Background pages** — As described in Chapter 3, background pages contain supporting information for the diagrams on foreground pages. For example, a background page could contain a background design, the name of the diagram, the date it was created, and the copyright information. If the drawing page and the physical paper are the same size, a background page can perform all the functions of a header and footer and more. For example, you can place text and objects anywhere on a background page, whereas headers and footers are restricted to the top and bottom margin of a page. Another strength of background pages is that you can associate one background page with several foreground pages, which saves you time in adding reference information and ensures consistency.

- **Headers and footers** — So why does Visio have headers and footers if background pages are so great? In Visio, headers and footers are the solution to the problem that arises when a drawing prints on several pieces of paper. With a header or footer, you can include a page number field to identify the physical pages you print. Because a background page corresponds to one foreground page, it can't number physical printed pages. Headers and footers are limited. The only place you can add header and footer information is in the top and bottom margin of a drawing page. In addition, headers and footers each contain only three positions for adding information: Left, Center, and Right. Another drawback to headers and footers is that you must add that information to each drawing page in a drawing file.

Headers and footers appear only when you print or preview your drawings. Unlike in Microsoft Word, which displays headers and footers in the Print Layout view, you must choose View ➪ Header and Footer to add or change header and footer information. Choose File ➪ Print Preview to only view header and footer information.

To include text and fields in the header or footer of a drawing, follow these steps:

1. Choose View ➪ Header and Footer.
2. To place information in a header or footer, in one of the position boxes, either type text or choose a field from the drop-down list. You can add text or fields to the Left, Center, or Right box in either the Header or Footer column. The fields you can add include Page Number (the printed page number), Page Name, Total Printed Pages, Current Time, Current Date (Short), Current Date (Long), File Name, File Extension, and File Name and Extension.

3. To format the header and footer, click Choose Font, select a font, a style, a size, a color, and effects, and then click OK. Unfortunately, the formatting you select applies to *all* the text in the header and footer.

4. In the Header and Footer dialog box, click OK.

 Fields and text play well together within any of the six header and footer text blocks. For example, you can enter Page &p of &P in any of the fields to obtain Page 1 of 2.

Adding Comments to Drawings

The casual diagram can often perform its duties without being scrutinized by a cast of thousands. But many diagrams are the result of collaboration or require approval before being published. For drawings that must pass a meticulous review process, reviewers often provide feedback about a drawing with *redlining* — adding text to a separate layer reserved for comments, and typically in the color red, from which the process gets its name. For less stringent review processes, adding comments to a drawing is faster.

Visio supports both types of review processes. Markup features let each collaborator add comments and drawing modifications to their own markup layer. In addition, you can insert comments directly on a page. You can also modernize the redlining process by creating a layer to hold reviewer comments.

Comments also come in handy when you want to remind yourself about changes and additions to make before you call the diagram done. This section describes how to add comments to drawings. To learn about the Visio collaboration tools, see Chapter 11.

To add a comment to a drawing page, do the following:

1. Choose Insert ➪ Comment. Visio opens a popup comment box and flags the comments with your initials, a comment number, and the date, as illustrated in Figure 6-2.

2. Type your comment and press Esc or click anywhere in the drawing. A comment icon appears approximately in the center of the visible drawing area.

3. To position the comment near the shape to which it refers, drag the comment tag to the new position.

4. To add another comment, repeat steps 1, 2, and 3.

Comment tags are nice and compact, but the initials and comment number don't tell you much. You can view, edit, or delete comments on a page. To view a comment, click its comment tag. To edit or delete a comment, right-click the comment tag and choose Edit Comment or Delete Comment from the shortcut menu. You can also click the comment and type within the comment box or press the Delete key to delete it.

FIGURE 6-2

Comments are great for adding reminders, questions, or suggestions.

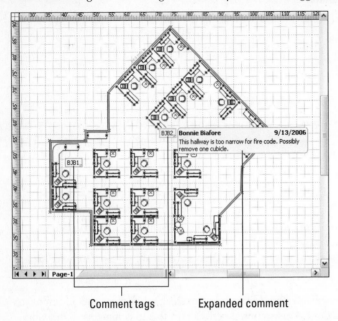

Comment tags   Expanded comment

Selecting and Editing Text

Changes are ongoing events in the life of any document and Visio drawings are no exception. Text often must move to a new position to make way for additional shapes. The words themselves usually change and, sometimes, you decide to change the formatting to make text more attractive. With text in Visio, you can reposition, format, and edit text blocks with most of the same tools that you use for shapes. Moreover, the cut, copy, and paste shortcuts from other Microsoft Office applications work in Visio as well. But before you can modify text in any way, you have to know how to select the text you want to work with.

Selecting Text

To replace or format the *entire* text block for a shape, you simply select the shape and then retype the text or click a formatting button on the Formatting toolbar. Editing or formatting part of a text block requires different but equally simple techniques.

To edit only portions of a text block, you either position the insertion point at the point in the text block where you want to work or you select the portion of text to change. The techniques you use and the results you obtain vary depending on whether you use the Pointer tool, Text tool, or Text Block tool. After you review the methods in Table 6-1, you might want to choose your favorite to keep things simple.

TABLE 6-1

Methods for Selecting Text

Selection Task	Tool to Use	What to Do
Select an entire text block for editing or formatting	Any	Select the shape in which the text block resides.
Select an entire text block for repositioning or resizing	Text Block tool	Select the shape in which the text block resides. Drag the text block to a new position or drag its selection handles to resize it.
Select all the text within a text block	Pointer tool	Select the shape and press F2 or double-click the shape. By selecting the text within a text block, you can work on the entire text block, select portions of text, or move the insertion point.
Select some of the text in a text block	Text tool	Within the shape, drag to select the text you want to work on.
Position the insertion point in a text block	Text tool	Click within the text where you want to position the insertion point.
Select a paragraph	Text tool	Select the shape and then triple-click the paragraph.
Select a single word	Text tool	Select the shape and then double-click the word.

Selecting Text in Groups

Grouping related shapes and text blocks makes them easier to work with. All the shapes and text blocks move and resize as a group. With grouped shapes, the group has a text block in addition to the text blocks for each shape in a group. For example, a group of shapes for an office cubicle might use text block to label each piece of furniture, whereas the group text block identifies the model of cubicle.

Even if shapes are grouped, you can still add, edit, or format the text in each shape. Plus you can work on the group text block, too. The main difference when working with text in grouped shapes is that it sometimes takes a few more clicks to select the text you want.

The selection techniques described in the previous section mostly work with text in groups. However, selecting a shape in a group with the Pointer tool could require more than one click. For example, if a shape belongs to a group that in turn belongs to a larger group, the first click selects the largest group. Clicking a second time in the same place selects the subordinate group. Clicking a third time in the same place selects the shape, at which point you can use the regular techniques to select the text.

Copying Text from Other Applications

The trick to copying text into Visio from another application is to first create a text-only shape on your drawing and then paste the text from the other application into the text-only shape. To do this, select the text in an application such as Microsoft Word and press Ctrl+C to copy it to the Windows Clipboard. In Visio, activate the Text tool and drag to create a text block in your drawing. Press Ctrl+V to paste the text into the text block.

Embedding text from another application makes it easier to modify the text should you need to change it in the future. When you double-click embedded text, Visio opens the application in which you created the text originally. To embed text, copy the text from the other application and paste it directly onto the drawing page *without* creating a text block.

Editing Text

You can edit text in Visio using techniques similar to those used in other Microsoft Office applications. For example, you can use the Cut, Copy, and Paste commands to modify or duplicate Visio text. As with other applications, text that you cut or copy moves temporarily to the Windows Clipboard so you can paste it several times. Unlike other Microsoft Office applications, Visio automatically zooms in to 100 percent when you edit text to make the text easier to see. If you want to eliminate that behavior, choose Tools ➪ Options ➪ General. In the Automatically Zoom Text When Editing Under drop-down list, choose 0 for the point size.

Deleting Text

Deleting text works differently depending on whether the text is within a Visio shape or a text-only shape. When you delete the text in a Visio shape, only the text goes away, not the shape's text block. You can always select the shape and type text to insert new text into the shape's text block. In a text-only shape, deleting all the text deletes the text-only shape as well. To delete text, use one of the following methods:

- **Delete text in a shape** — Select the Text tool and then select the shape. Select the text you want to delete and press Delete. Press Esc to close the text block.

- **Delete some of the text in a text-only shape** — Select the Text tool and then select the text-only shape. Drag to select the text you want to delete and press Delete. Press Esc to close the text block.

- **Delete a text-only shape** — Select the Pointer tool, select the text block, and press Delete.

Finding, Replacing, and Correcting Text

In Visio, the commands for finding, replacing, and checking spelling work similarly to those in other Microsoft Office applications. Visio offers more options for specifying where to search to accommodate the different elements that Visio drawings contain, as shown in Figure 6-3.

FIGURE 6-3

The Visio search for text feature includes several check boxes for specifying which Visio elements to search.

Finding Text

Visio can search for text within shapes, shape data, the name of shapes, and user-defined cells in ShapeSheets. Here are the check boxes that specify which text you want Visio to search:

- **Shape Text** — Searches for text in shape text blocks and text-only text blocks.
- **Shape Data** — Searches for text in the values in shape data fields.
- **Shape Name** — Searches for text in the Name field of shapes or masters. You can use this option to locate instances of a master in a drawing or to find a master in a stencil you are editing.
- **User-Defined Cell** — Searches the Value and Prompt cells in the User-Defined Cells section of ShapeSheets for all shapes in a drawing.

To find text on a drawing, follow these steps:

1. Open the Find dialog box by choosing Edit ➪ Find or pressing Ctrl+F.
2. In the Find What box, type the text you want to find. To specify a special character, click Special and, from the drop-down menu, choose the special character.
3. To specify how much of the drawing file to search, select one of the options (Selection, Current page, All pages).
4. Select the check box for each of the components you want Visio to search (Shape Text, Shape Data, Shape Name, User-defined Cell).
5. If you want Visio to match results in a specific way, under Options, check the check boxes.

The check boxes under Options control whether the results returned match the case of the text in the Find What box or match whole words, not just a portion of a word. The Match Character Width check box limits the results to characters of the same width, such as wide or narrow characters in the Katakana alphabet.

6. Click Find Next to find the next occurrence of the text. Visio highlights matching text on the drawing page.

7. To edit the text, close the Find dialog box and edit the text.

If Visio finds the text in a shape name or user-defined cell, it highlights the shape on the drawing and displays the name in the Found In section of the Find dialog box, as illustrated in Figure 6-2. If you want to edit the text in a shape name, close the Find dialog box, right-click the highlighted shape, and choose Format ➪ Special (you must run Visio in developer mode) to access the shape's name. To edit text in a user-defined cell, open the ShapeSheet for the highlighted shape and edit the text in the appropriate cell.

Finding Shapes by Searching for Text

The Search for Shapes box in the Shapes window will find shapes on stencils, which you can then add to a drawing. However, when shapes are on the drawing, finding them by name is a whole different story. Suppose you add one type of cubicle to an office layout and want to replace all of them with a different shape. The cubicles on the drawing all start to blur together, but searching for them by name finds them for you. Likewise, you can search for shapes based on the values in their shape data. It's easy to locate employees in an organization chart by searching the Shape Data fields for their names.

Replacing Text

To replace text on a drawing, follow these steps:

1. Choose Edit ➪ Replace.

2. In the Find What box, type the text you want to find. To specify a special character, click Special and, from the drop-down menu, choose the special character.

3. In the Replace With box, type the new text.

4. Select one of the options (Selection, Current page, All pages) to specify how much of the drawing to search.

5. If you want Visio to match results in a specific way, select the check boxes in the Options section.

6. Click Find Next to find the next occurrence.

7. Click Replace to replace the occurrence that Visio highlights or Replace All to replace all occurrences.

Checking Spelling

Visio can check spelling like its Office counterparts. You can perform all of the following tasks with Visio:

- Check spelling with the built-in dictionary.
- Create your own dictionary of words.
- Instruct Visio to correct entries automatically as you type.
- Add your own AutoCorrect entries.

CROSS-REF To learn how to create your own dictionary and AutoCorrect entries, in Visio Help, refer to the Check and Correct Spelling topic.

To check spelling, follow these steps:

1. Choose Tools ➪ Spelling ➪ Spelling or press F7. Visio opens the Spelling dialog box for each word not found in the dictionary.

2. If the word is spelled correctly, you can click Ignore or Ignore All. To prevent Visio from identifying the word as a misspelling in the future, click Add to include the unrecognized word in the spelling dictionary.

3. If the word in the Change To box is spelled correctly, click Change to replace the misspelled word with the Change To text. If it is not spelled correctly, edit the contents of the Change To box and then click Change or Change All.

NOTE If you want to turn off the spell checker, choose Tools ➪ Spelling ➪ Spelling Options. Uncheck the Check Spelling As You Type check box and click OK. You can turn spell checking back on by pressing F7.

Positioning Text

Text-only shapes, callouts, and other annotation shapes can go wherever you want them on a drawing. In addition to moving text shapes around, the text within a shape can change position. By default, a shape's text block is the same size as the shape and centered within it. Relocating or rotating shape text blocks helps remove overlapping text or makes your drawing more readable. For example, by moving the built-in axis label for a graph, you can make room for numeric labels along the axis.

In addition, text blocks are rectangular. If you want to wrap text within a circle or constrain the text to the outline of its shape, you must center the text and place carriage returns judiciously to vary the length of the lines appropriately. Visio also can't string text along a curve, although you can resort to creating fancy text in another program and inserting it as a graphic into Visio.

Repositioning Text in a Shape

The text block in a shape is a part of the shape but it doesn't have to reside in the same location on the drawing. You can move, resize, or rotate a shape's text block. To reposition text in a shape, follow these steps:

1. On the Standard toolbar, click the Text Block tool. If the Text Block tool isn't visible, click the arrow next to the Text tool and, from the drop-down menu, choose Text Block tool.

2. Select a shape. Because the Text Block tool is active, Visio displays the text block selection handles, as shown in Figure 6-4.

FIGURE 6-4

When the Text Block tool is active, clicking a shape selects the text block and displays the text block selection handles.

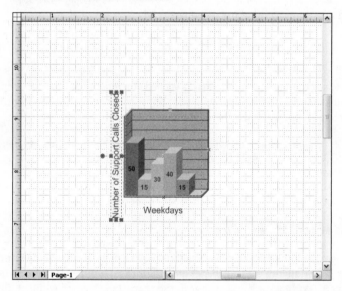

3. When the text block selection handles are visible, you can perform the following tasks:

    ■ Move the text block to another location by dragging it to the new location.

    ■ Rotate the text block by dragging the rotation handle until the text is rotated to the angle you want.

 **TIP**   Another way to rotate only a shape's text block is by clicking the **Rotate Text 90 Degrees tool on the Action toolbar.**

    ■ Resize the text block by dragging a selection handle.

TIP The properties of a shape text block are in the Text Transform section of a ShapeSheet. Some functions for text are only available through the ShapeSheet, such as preventing text from rotating. For an introduction to ShapeSheets, see Chapter 34.

Editing Locked Shapes

Some Visio shapes and connectors are locked so that text remains right side up no matter how you move a shape or line. When you select a locked shape with the Text Block tool, Visio indicates the protection by changing the handles to gray. To unlock a shape, right-click the shape and, from the shortcut menu, choose Format ⇨ Protection. Then, use one or more of the following settings to unlock the characteristics that you want to change:

- To unlock the shape rotation, clear the Rotation check box.
- To unlock a shape so you can resize it, clear the Width and Height check boxes.
- To unlock a shape so you can move the text block, clear the X Position and Y Position check boxes.

After you clear the protection check boxes and click OK, proceed with the editing you want to perform.

For grouped shapes, you can set the group's behavior to prevent selecting the text of the group shape or member shapes. To do this, choose Format ⇨ Behavior and change the group options.

Creating Special Annotations

Visio comes with predrawn shapes for displaying and emphasizing information in your drawings. Callout shapes emphasize important information on drawings. Built-in title blocks identify the drawing file and its contents. Legends identify the symbols used on a drawing. This section describes the techniques for putting these annotation shapes to work.

CROSS-REF You can produce bulleted and numbered lists by formatting the text in annotation shapes, which is described in Chapter 7.

Creating Tables

The Charting Shapes stencil contains several built-in shapes for different types of tabular information. Built-in table shapes make it easy to marshal text into rows and columns. Each text block goes into its own cell in the table. Showing information in tabular form in one shape's text block requires the use of tabs to position text in columns.

Using Built-in Table Shapes

It goes without saying that the tabular shapes on the Charting Shapes stencil offer myriad advantages over using tabs to create columns. The built-in tables are groups of shapes, so you can work with each cell individually or modify the entire table. For example, to resize all the cells in a table shape, select the group and drag a selection handle. These shapes also automatically assume the theme you apply to a drawing.

Here are the shapes for tables on the Charting Shapes stencil:

- **Deployment Chart** — This shape comes with labels for departments and phases, but there's no reason you can't change those labels to represent anything you want. Because this shape already has labels for every row and column, it's often easier to edit the shape's labels than to build a table with the Grid shape.

- **Feature Comparison Chart** — This shape includes labels for products and features but you can modify the labels to represent anything you want.

- **Grid** — The grid is a plain array of cells. When you drag it onto the drawing page, its Shape Data dialog box appears. Specify the number of rows and columns you want in the table and click OK. Visio configures the grid and adds it to the page.

- **Row Header and Column Header** — To label rows and columns of a Grid shape, drag the Column Header or Row Header shapes and snap them into place next to a row or column in the Grid shape. Each row and column needs its own header shape, which can be time-consuming for a large table. In this case, consider modifying a Deployment Chart.

To use the Grid shape to create a table, follow these steps:

1. To open the Charting Shapes stencil, choose File ⇨ Shapes ⇨ Business ⇨ Charts and Graphs ⇨ Charting Shapes.

2. Drag the Grid shape onto your drawing. The Shape Data dialog box opens.

3. Select the number of rows and columns you want in the drop-down lists and click OK.

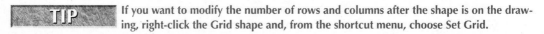 If you want to modify the number of rows and columns after the shape is on the drawing, right-click the Grid shape and, from the shortcut menu, choose Set Grid.

4. To label a row, drag the Row Header shape and snap it to the leftmost cell in a row. With the Row Header selected, type the label text.

5. To label a column, drag the Column Header shape and snap it to the topmost cell in a column. With the Column Header selected, type the label text.

6. To add text in the table, click a table cell and type the text.

Formatting Shape Text into Columns

It isn't elegant, but applying tab stops to one or more paragraphs in a text block does align the text into columns. Tabs align to the left, center, or right; a decimal tab aligns columns of numbers by their decimal points.

CAUTION If you resize a shape after applying tab stops, the shape margins can impinge on the text block and affect the alignment of your tabs. As you narrow the width of the shape, the lines of text begin to wrap around, ruining the column layout.

Tab stop positions are relative to the left edge of the text block. Because of this, it's easiest to begin formatting by adjusting the origin of the ruler to the left edge of the text block. To add tab stops to a shape, follow these steps:

1. To relocate the origin of the horizontal ruler, press the Ctrl key and drag the vertical ruler to the left edge of the shape. The change is subtle, but you should see the zero point on the horizontal ruler lined up with the left side of the text block

2. In the text block, select the text you want to format. To select it all, press F2.

3. Choose Format ⇨ Text and select the Paragraph tab.

4. To set the text block alignment, in the Horizontal Alignment drop-down list, select Left, and set the values in the Before Text, After Text, and First Line boxes to zero.

5. To add a tab stop, select the Tabs tab. In the Tab Stop Position box, type the distance from zero for the tab stop. Select the alignment you want and click Add.

NOTE You specify tab stop indentations in unscaled units even if you're working on a scaled drawing.

6. Repeat step 5 for each tab stop you want to add and click OK when you are finished.

7. To add text in columns, type a value in the text block and then press Tab to move to the next column.

Highlighting Information with Callouts

Built-in Visio callout shapes are great for calling attention to information on your drawings, as illustrated in Figure 6-5. All you do is drag these shapes onto a drawing, type your annotation, and drag the callout line or arrow to the area you want to highlight. If you glue a callout to the shape it documents, the callout moves when the shape moves.

Callouts from any stencil work regardless of which type of drawing you use. Simply open the stencil with the callout you want and drag the shape onto the drawing. Some stencils include one or two callout shapes, but you can find the widest variety of callouts on the following stencils:

- **Callouts** — Contains callouts, balloons, tags, notes, and other shapes for adding text to drawings. To open this stencil, choose File ⇨ Shapes ⇨ Visio Extras ⇨ Callouts.

- **Charting Shapes** — Contains callouts, balloons, and the annotation shape. To open this stencil, choose File ⇨ Shapes ⇨ Business ⇨ Charts and Graphs ⇨ Charting Shapes.

- **Annotations** — Contains callouts in addition to other reference shapes, such as north arrows. To open this stencil, choose File ⇨ Shapes ⇨ Visio Extras ⇨ Annotations.

FIGURE 6-5

Callout shapes often include instructions as default text and yellow control handles for positioning the callout arrow.

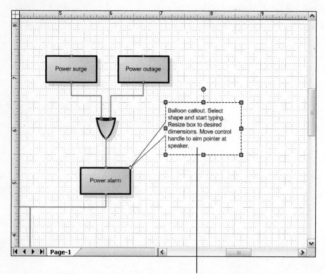

The default text explains how to use the callout shape.

Adding Callouts to a Drawing

Each callout has its own special attributes, but they work about the same way. Callouts often include instructions in the text block, as illustrated in Figure 6-5. To add a callout to a drawing, follow these steps:

1. Drag the callout from the stencil and drop it near the shape you want to annotate.

2. Select the callout shape and type text as you would for any kind of shape.

3. To point the callout at a shape, drag the selection handle or control handle at the end of the callout line or arrow.

4. To glue the callout to a shape, drag the selection handle or control handle to a connection point on the shape. A red box highlights the connection point when the callout and shape are connected.

Displaying Properties in Callouts

The text blocks within shapes can display information stored in the shape data, such as the person who occupies an office, their department, and their phone number. If the shape isn't large enough for the data you want to show, Custom Callout shapes know how to grab values in shape data from other shapes. In addition, you can specify which shape data fields to display and how to display them.

To display shape data in a custom callout shape, follow these steps:

1. Open the Callouts stencil.

2. Drag a Custom Callout shape (Custom Callout 1, Custom Callout 2, or Custom Callout 3) onto the drawing.

3. Drag the yellow control handle on the Custom Callout to the shape with shape data that you want to display. The Configure Callout dialog box appears, listing every shape data field for that shape.

4. Select the check box for each property you want to display in the callout. If you select more than one property in the list, in the Separator drop-down list, choose whether you want to separate data with commas, tabs, or returns.

> **TIP**   The shape data doesn't have to appear in the order it appears in the Configure Callout dialog box. To reorder the shape data in the callout, select a shape data field in the list and then click Move Up or Move Down until the field is where you want it.

5. To show only the shape data value and not the field name, clear the Show Property Name check box.

6. To move the callout when you move the shape, make sure the Move Callout with Shape check box is selected.

Documenting Drawings with Title Blocks

Title blocks are commonly applied to architectural and engineering drawings to identify and track their contents and revision history. Visio automatically opens the Borders and Titles stencil with many of the built-in templates, but title blocks from any stencil work with any type of drawing. If you want your title block just so, as most architects do, you can create your own title block with exactly the information and look you want. Find Visio's built-in title block shapes on either of the following stencils:

- **Borders and Titles** — Contains dozens of title block shapes, some of which use fields to automatically display information such as the date, scale, and page number. To open this stencil, choose File ➪ Shapes ➪ Visio Extras ➪ Borders and Titles.

- **Title Blocks** — Available only in Visio Professional, this stencil contains several title block and revision block shapes, along with shapes that display fields automatically. To open this stencil, choose File ➪ Shapes ➪ Visio Extras ➪ Title Blocks.

When none of the built-in title blocks suit your needs, a creative compilation of shapes from the Title Blocks stencil can become your customized title block. The Title Blocks stencil contains shapes that display fields such as the current date, author of the drawing, file name, page number, and scale. The Frame shape is ideal for the title block border and you can arrange the other shapes inside it. To create your own title block, follow these steps:

1. Drag the Frame shape onto your drawing and then drag other title block shapes to build the title block.

2. To group the shapes so you can move them as one, select all the title block shapes and choose Shape ➪ Grouping ➪ Group.

Working with Title Blocks

The title blocks built into Visio are often groups of shapes, so you can add the title block to your drawing in one step. Quite often, these groups are locked to prevent you from inadvertently changing part of the group. Depending on the protection applied, you might be able to resize, delete, format, or annotate the shapes that make up the title block.

If protections on a shape prevent you from modifying the title block, you can unlock the shape and then make the change you want. To do this, select the shape within the group and choose Format ➪ Protection. Clear the check boxes for the type of change you want to make and click OK. If some residual protection remains, choose Insert ➪ Behavior to see the group behavior or take a look at the ShapeSheet to turn off any remaining protection.

If you want to rearrange the shapes in a group, the Ungroup command is so tempting. However, ungrouping shapes eliminates group data and behavior that you might have trouble recreating. Instead, choose Edit ➪ Open *<shape name>* where *<shape name>* is the name of the master for the group or the word *Group* if the shape has not been saved. By opening the group, you can modify the shapes within it, even adding or deleting shapes without losing any of the features Visio built into the group.

 Creating several nested levels of grouping can reduce Visio's performance. An alternative is to modify a group by using Shape ➪ Grouping ➪ Add to Group or Shape ➪ Grouping ➪ Remove from Group.

 If you want to add your custom title block to a stencil, see Chapters 32 and 33 for instruction on creating custom stencils and new masters.

Identifying Drawing Symbols with Legends

Any time you use symbols on a drawing, it's a good idea to include a legend so everyone knows what the symbols represent. Legends are essential when you use symbols that aren't standardized, such as stars, exclamation points, or smiley faces. Your audience needs to know whether an exclamation point is good or bad, and what it means.

In Visio, a Legend shape displays the symbol shapes included on a page, the symbol descriptions, and the number of times each one appears on that page. The Legend shape updates its rows and fields automatically as you add symbol shapes to your drawing, so you can add the Legend shape before or after you add symbols.

The Legend Shapes stencil opens automatically for only the Brainstorming, Network Diagram, and Building Plan templates, but you can use it with any type of Visio drawing. If the symbols on the Legend Shapes stencil aren't what you need, you can turn any shape into a symbol, as described later in this section.

TIP Symbols have little room for text and no shape data for storing information. To add a reference to additional information associated with a symbol, select the symbol and type an alphanumeric ID, such as A1. Then, with the Text tool, add a text block or annotation shape to the drawing and add the ID and the more detailed information.

Creating New Symbol Shapes

As long as a Legend Shape is present on the drawing page, you can turn any Visio shape into a symbol. For example, on a site plan diagram, you could turn different types of signage shapes into symbols. To create symbols out of shapes, follow these steps:

1. Choose Shapes ➪ Business ➪ Brainstorming ➪ Legend Shapes.

2. Drag the Legend shape from the stencil onto the drawing page.

3. Drag the shape you want to use as a symbol onto the Legend shape.

4. When Visio asks whether you want to convert the shape into a symbol and add it to the legend, click Yes. Visio adds a row for the symbol to the Legend shape and adds a control handle to the shape on the drawing page, as shown in Figure 6-6.

FIGURE 6-6

Any shape can be a symbol for a legend.

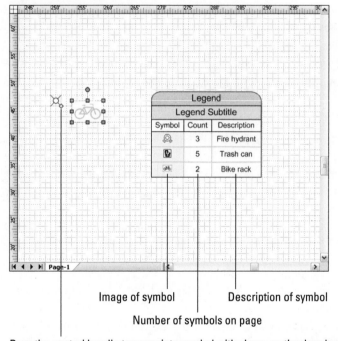

Image of symbol   Description of symbol

Number of symbols on page

Drag the control handle to associate symbol with shape on the drawing.

5. To associate the symbol with a shape on the drawing, such as a telephone sign with the light pole on which it is hung, drag the yellow control handle and glue the symbol to the shape.

Configuring Legends

A Legend shape catalogs the number and types of symbols on a drawing page, not the entire drawing file. To create a legend on a page, drag the Legend shape from the Legend Shapes stencil onto the drawing page. To add a title and subtitle, double-click the Legend Title or Subtitle text and type text.

Each Legend shape can have its own unique configuration and formatting. The Configure Legend command includes options for this customization, but you can edit and format text directly in Legend shapes. The easiest way to configure a legend is to right-click it and, from the shortcut menu, choose Configure Legend. In the Configure Legend dialog box, modify any or all of the following options:

- **Show or hide the subtitle** — Check or clear the Show Subtitle check box to show or hide the legend subtitle.

- **Show or hide the quantity column** — Check or clear the Show Count check box to show or hide the Quantity column.

- **Show or hide the column names** — Check or clear the Show Column Names check box to show or hide the column names.

- **Show or hide specific symbols** — To hide symbols in a Legend shape, clear the check boxes in the Visible column. By default, Visio displays all the symbols on the page.

- **Change the sort order for symbols in the Legend shape** — Select a row in the Legend shape for the symbol you want to move and click Move Up or Move Down to reposition it in the list.

NOTE Before you rearrange the sort order, clear the check boxes in the Visible column for each symbol you want to sort and then use the Move Up and Move Down buttons to re-sort the symbols.

Aside from the Configure Legend dialog box, you can modify Legend shapes in these additional ways:

- **Modify the width of the Legend shape** — Drag a selection handle on one of the sides of the Legend shape to the width you want.

- **Edit Legend text** — Subselect the title, subtitle, or column text (other than the Count column) you want to edit by clicking the Legend shape and then clicking the text. Type the new text.

- **Format Legend text** — Subselect the text you want to format and choose Format ➪ Text. Choose the formatting options you want and then click OK.

NOTE To delete a legend, select it and press Delete. Visio deletes only the Legend shape, not the symbol shapes associated with shapes, on the drawing page. You must delete symbol shapes separately on the drawing page.

Labeling and Numbering Shapes

Many types of drawings number or label their elements for identification. For example, blueprints number the columns on a structural plan so the construction crew knows how to assemble the steel. The Number Shapes Add-On can number shapes as you add them to your drawing or after all the shapes are in place. By default, Visio increments numbers moving from left to right and from top to bottom on the drawing, but you can choose the order you want, as shown in Figure 6-7, or click each shape to number it.

FIGURE 6-7

The Number Shapes Add-On adds numbers in sequence to the shapes you select.

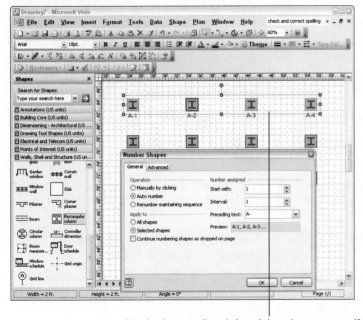

Numbering runs from left to right unless you specify otherwise

To number shapes on a drawing, follow these steps:

1. If you want to number specific shapes on the drawing, select the shapes in the order you want them to be numbered. Otherwise, select all the shapes to number.

2. Choose Tools ➪ Add-Ons ➪ Visio Extras ➪ Number Shapes.

3. In the Number Shapes dialog box, on the General tab, specify how you want to number the shapes:

  ▪ **Manually By Clicking** — Choosing this option adds the next number in a sequence to a shape.

  ▪ **Auto Number** — This option numbers shapes based on the order you specify on the Advanced tab or starting from the top left, working across, and then proceeding to the next row.

  ▪ **Renumber Maintaining Sequence** — When you choose this option, select the Advanced tab and specify whether you want to renumber shapes with unique numbers or allow duplicates in the sequence.

4. Under Number Assigned, choose the characteristics of the sequence in the following boxes:

  ▪ **Start With** — The first number in the sequence.

  ▪ **Interval** — The gap between each number in the sequence. For example, an interval of 3 would create a sequence that begins 1, 4, 7.

  ▪ **Preceding Text** — Text that precedes the number. For example, for steps in a process, you can choose Step in the list. For columns on a floor plan, you might choose A or a.

 Sometimes, you might want two numbering sequences, such as numbers for the columns and different letters to identify each row. To apply two-part numbering schemes, select one row, in the Preceding Text box specify the letter for that row, and number the shapes. Repeat this process for each row on the page.

5. To number shapes that you add later, check the Continue Numbering Shapes As Dropped On Page check box. Visio continues to add numbers in sequence as you add shapes to the page.

6. For total control over the numbering options, select the Advanced tab.

7. Select an option to specify where the number appears relative to the shape text.

8. If you are using the Auto Number option, specify the order you want. You can number from left to right and then top to bottom, top to bottom and then left to right, from back to front, or in the order that you selected the shapes.

NOTE To hide the shape numbers, on the Advanced tab, check the Hide Shape Numbers check box.

Summary

Drawing annotation comes in many forms and shapes — literally. Every shape on a drawing has its own text block for displaying text or field values. In addition, text-only shapes or built-in tabular shapes are better when you want to show large amounts of textual data. Callout shapes can include notes to highlight information on a drawing or to display the values of custom properties. Title blocks contain fields that automatically display the values of file or page properties.

Chapter 7

Enhancing Diagram Appearance with Formatting

These days, most audiences want both information and good looks—from TV news anchors to the Visio diagrams in presentations. Formatting can make the difference between dreary drawings that go unread and documents that make everyone take notice. Visio 2007 has taken some giant steps in making diagrams look good and communicate clearly with a minimum of effort on your part.

One new feature, Themes, can format an entire drawing with consistent and good-looking formatting in about the same number of clicks it took to open a Format dialog box in Visio 2003. Visio comes with predefined Theme Colors and Theme Effects, which are amalgamations of the formatting tools you know and love—line styles, fill styles, text formatting, and shadows. Visio themes are designed by professionals to look good, which is particularly helpful if you have to ask your spouse whether your clothes match.

For additional formatting, such as emphasizing shapes, you can use predefined styles, similar in function to the ones in Microsoft Word, to format lines, text, and fill. If a few shapes need even more detailed tweaking, such as those with large amounts of text, you can apply unique formatting to shapes, lines, or text blocks.

This chapter covers the Visio formatting features by order of convenience, starting with Themes. Then, you'll learn how to apply styles to Visio elements to fine-tune formatting. And, finally, you'll find out how to apply individual formatting for unique situations.

NEW FEATURE Data Graphics, another new feature in Visio 2007, turns data into pictures. For example, instead of displaying the number of weeks that a task in a project schedule is delayed, you can show a stoplight that turns green, yellow, or red to represent on time, a bit late, or in big trouble. In addition, Data Graphics can color-code shapes, display icons for status, and more — all based on the values in shape data. Chapter 10 describes the new Visio 2007 data features, including Data Graphics.

Fast Formatting with Themes

Themes, new to Visio 2007, give your productivity a big boost when it comes to making your diagrams look good. Now, you can apply formatting to an entire diagram with fewer clicks than it used to take just to open a formatting dialog box. Themes also make *you* look good, because the themes and effects that Visio offers were built by professionals with a good eye for color and design.

NOTE Visio 2007 Themes replace the Color Scheme feature offered in earlier versions of Visio. Color Schemes applied coordinated formatting to shapes — but only in templates that supported the feature. In effect, the Color Schemes used a set of styles that developers applied to shapes. If you chose a different color scheme for your drawing, the shapes that used those color styles changed color to conform to the new color scheme. With the Theme feature in Visio 2007, any shape can take on theme-based formatting, unless you specifically tell it not to.

Similar to the library of built-in shapes, Visio 2007 comes with a collection of ready-to-use Theme Colors and Theme Effects, from soothing pastel combos to lurid fluorescent colors that no one will miss. Theme Effects range from basic formatting — a thin border line to a combination of rounded edges on shapes with the illusion of a bevel and shading inside the shapes. If your organization already has its own unique look, you can define Theme Colors and Theme Effects to match. Table 7-1 shows the types of formatting that Theme Colors and Theme Effects apply.

TABLE 7-1

The Formatting That Theme Colors and Theme Effects Apply

Theme Colors	Theme Effects
Line and connector color	Line and connector weight, pattern, rounding, and transparency; for connectors, arrowhead style and size
Text color	Text font
Fill and fill pattern color	Fill pattern and transparency
Shadow color	Shadow style, offset, direction, magnification, and transparency
Accents colors (five accent colors that coordinate attractively with the foreground color)	Not applicable
Background color	Not applicable

Applying Themes

Themes pack a lot of punch, but they are incredibly easy to use. Turning a drab drawing into an attractive diagram with built-in themes takes four clicks. To apply a Theme, do the following:

1. To open the Theme task pane, on the Formatting toolbar, click Theme.

2. In the Theme - Colors task pane, which opens by default, choose a color theme by clicking a thumbnail, as shown in Figure 7-1. Or, right-click a thumbnail and choose the Apply to All Pages or Apply to Current Page option.

FIGURE 7-1

Coordinated colors and formatting are as easy as a couple of clicks in the Theme task pane.

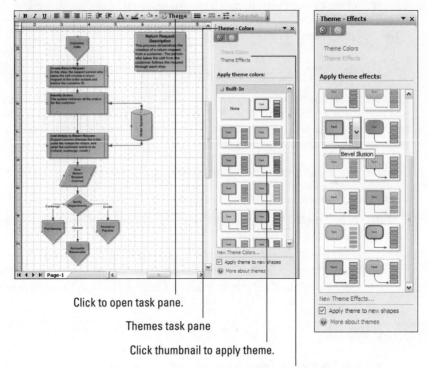

Click to open task pane.

Themes task pane

Click thumbnail to apply theme.

Click to choose colors or effects.

3. At the top of the task pane, click Theme Effects.

4. In the Theme - Effects task pane, choose the effect by clicking a thumbnail (or right-clicking and choosing the Apply to All Pages or Apply to Current Page option).

 Some of the shapes in Figure 7-1 use accent colors to highlight steps in the procedure. See Working with Colors later in this chapter to learn how to apply accent colors to shapes.

Protecting Shapes from Theme Colors and Effects

Some shapes come with colors and outlines that mean something, and, for these shapes, themes would jumble what the shapes are supposed to say. For example, warning signs are usually yellow triangles, so you wouldn't want those signs to take on a theme's attractive but misleading teal tint. To prevent themes from applying their formats to some shapes, you can choose from either of the following techniques:

- **Do Not Allow Themes** — Right-click the shape and, from the shortcut menu, point to Format and click Allow Themes to turn off the check mark that precedes it.

- **Apply Theme Protection** — Select the shape and, from the shortcut menu, choose Format ➪ Protection. In the Protection dialog box, select the From Theme Colors and the From Theme Effects check boxes.

Creating Custom Themes

Most people approach the paint chip section of the hardware store with a heady mixture of fear and loathing. But for the intrepid souls who devise daring color combinations, Visio provides tools for creating additional custom themes. As you can see in Figure 7-2, custom themes appear in their own section of the Theme task pane and are almost as easy as applying themes to drawings.

To set up your own custom theme, do the following:

1. On the Formatting toolbar, click Theme to display the Theme task pane.

2. In the Theme - Colors task pane, click the thumbnail of the theme that is closest to the collection of colors you want.

3. Below the thumbnails of built-in themes, click New Theme Colors. Visio opens the New Theme Colors dialog box, also shown in Figure 7-2, and creates a new custom theme using the colors of the currently selected theme. The New Theme Colors dialog box contains buttons for each type of Visio element it formats.

 The thumbnails for custom themes appear in the task pane above the built-in themes.

4. In the Name box, type a name for your custom theme, such as Gotta Be Green.

5. Under Theme Colors, click the arrow to the right of a color sample and choose the new color you want. Visio presents colors from the current theme color palettes, which work well together. You can also choose one of the standard Visio colors or click More Colors to choose your own.

FIGURE 7-2

Creating a custom theme is almost as easy as using a built-in theme.

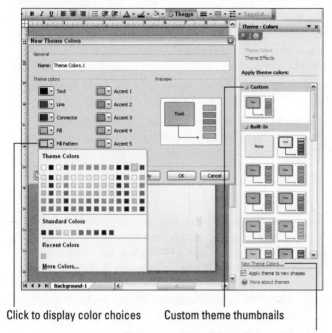

Click to display color choices   Custom theme thumbnails

Click to open New Theme Colors dialog box

6. Repeat step 5 for every theme color you want to change, basic elements colors or colors for accounts. As you modify colors, the preview shows the selected colors so you see whether they are breathtaking or heinous.

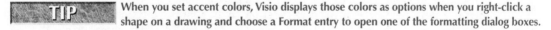

TIP   When you set accent colors, Visio displays those colors as options when you right-click a shape on a drawing and choose a Format entry to open one of the formatting dialog boxes.

7. Click OK to save the theme colors.

Defining custom theme effects works similarly to setting up custom color themes. In the task pane, click Theme Effects and then in the Theme - Effects task pane, click New Theme Effects. The New Theme Effects dialog box includes tabs for text, lines, fills, shadows, and connectors. Simply click each tab and choose settings as you would if you were applying formatting to an individual element (described in Applying Formats later in this chapter). Click OK to save the custom theme effect.

Working with Custom Themes

Besides creating custom themes, you can perform the following tasks:

- **Copy custom theme to another drawing** — To use a custom theme in another drawing, select a shape with the custom theme applied and then press Ctrl+C. In the drawing to which you want to copy the theme, press Ctrl+V and the theme appears in the Themes task pane. Delete the shape you copied.

- **Delete a theme** — In the task pane, right-click the theme and then, from the shortcut menu, choose Delete.

- **Duplicate a theme** — If you want to use one of your custom themes as the basis for a new theme, in the Theme task pane, tight-click the theme and then, from the shortcut menu, choose Duplicate. You can then edit the new theme.

Working with Colors

If the Visio 2007 themes whet your appetite for color, this section shows you how to do more with color in drawings. For example, after you apply Theme Colors to a diagram, you can add accent colors to specific shapes to emphasize them, such as the shapes for decisions, departments, and systems in Figure 7-1. The color picker in Visio 2007 now includes a Theme Color section, so you can pick colors that go with the other colors in the currently selected Theme. But the color picker still has options for choosing colors — and creating new ones — as it did in previous versions.

> **TIP**   Although Visio offers 16.7 million colors and 100 levels of transparency, drawings look more professional with a few well-coordinated colors. For example, Visio Color themes are sets of six colors. Even if you opt for custom colors, stick to half a dozen colors or fewer.

Using the Color Picker

By using Theme Colors and the Color Picker, adding harmonious colors to drawings is easy. The Color Picker appears whenever you click the arrow on the right side of a Color box in any of the formatting dialog boxes or next to the Text Color, Line Color, or Fill Color buttons on the Formatting toolbar. It displays the main colors for the current color theme, coordinated accent colors, and options for obtaining just about any other color you want.

To apply a color to a Visio element, click the arrow to the right of the corresponding Color box, as demonstrated in Figure 7-3. For example, if you right-click a shape, and, from the shortcut menu, choose Format ➪ Fill, the Fill dialog box includes several boxes with arrows that open the Color Picker.

FIGURE 7-3

The Color Picker provides color chips that you click to choose a color.

Click arrow to display Color Picker

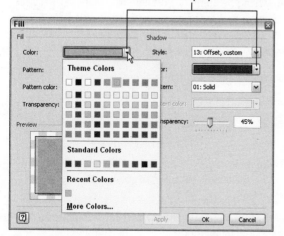

> **TIP**   With transparent fill colors, the colors mix when you overlap shapes. In the Format Fill dialog box, drag the Transparency slide to set the color from totally transparent to completely opaque.

Here is what each section of colors does:

- **Theme Colors** — Picking colors within the Theme Colors section of the Color Picker guarantees that the selected colors work well together — at least in the eyes of professional designers. Moreover, the colors in the Theme Colors section change automatically if you choose a different set of Theme Colors.

- **Standard Color** — The colors in this part of the Color Picker remain the same regardless of the Theme Colors in effect. These colors are ideal if you want to apply colors that carry a meaning, such as red to show debt or green to show progress.

- **Recent Colors** — This section shows custom colors that you have recently selected.

- **More Colors** — Click More Colors to open the Colors dialog box, so you can pick the exact color you want.

> **CAUTION**   If you apply custom colors to drawing elements, Visio stores the RGB or HSL values in the ShapeSheet so the colors won't change when you change theme colors.

Using the Color Palette

The color palette isn't quite as important in Visio 2007, because color themes are so much easier. But the color palette does contain the standard colors that Visio displays in the color picker. The color palette is a set of 24 indexed colors and when you apply one of them to a shape, Visio stores in the ShapeSheet the index value of the color, such as 4 for bright blue in the default color palette. One caveat for this behavior is that shapes can change color unexpectedly if someone redefines a color on the color palette or you copy shapes to a drawing that uses a different color palette.

If you want to edit a color in the color palette, follow these steps:

1. Choose Tools ⇨ Color Palette.

2. Click the color you want to edit and then click the Edit button to open the Edit Color dialog box.

> **CAUTION** Do not edit black (index 0) or white (index 1) in a color palette. Visio uses index 0 for the default line color and index 1 for the default fill color, so changes to index 0 or 1 can affect more than you might expect.

3. Choose a standard color or define a custom color and click OK when you are done. Define a color by using either of the following methods:

    - To use one of the colors on the Standard tab, click the hexagon with the color you want.

    - To define a custom color, select the Custom tab and choose RGB or HSL in the Color Model box. Type the values in the RGB or HSL boxes, depending on which color model you selected. You can also click a color in the Colors preview area. When the color is the hue you want, you can drag the arrow up or down to lighten or darken the tint.

4. Repeat step 3 to redefine other colors, and click OK when you have finished modifying the color palette.

Specifying Color Settings for Drawings and Stencils

Besides the colors for lines, text, and fills, you can also specify colors for the drawing page, drawing page background, stencil text, stencil background, print preview background, and full-screen background. To do this, follow these steps:

1. Choose Tools ⇨ Options, select the Advanced tab, and click the Color Settings button.

2. To change a color, click the arrow for the color you want to change and select a new color.

> **NOTE** The drawing window and stencil window each have two entries for background colors. With adequate screen resolution and your monitor set to display 32-bit color, Visio will grade the background from one of the colors into the other from the top to the bottom of the screen.

3. Repeat step 2 for each color you want to change. Click OK in the Color Settings dialog box and then click OK to close the Options dialog box.

NOTE   Most templates include the Backgrounds and Borders and Titles stencils, so you can easily add interesting backgrounds and borders to your drawing. If you start a drawing from scratch, you can open these stencils by choosing File ⇨ Shapes ⇨ Visio Extras and then clicking the stencil name that you want. When you drag a background shape onto your foreground page, Visio automatically creates a background page for you and assigns the background page to the current foreground page.

Formatting with Styles

Formatting with styles comes in handy when you want to apply a set of formatting options that differs from what the current Theme applies. For example, you can emphasize the business process steps targeted for streamlining by making the shape's border heavier, changing the fill color, and increasing the font size. Of course, you could apply each of these formats separately with individual formatting commands. But, by defining a style with those formats, you can apply them all at once and reuse that set of formats quickly on other shapes. Much like a style in Microsoft Word, a Visio style compiles several formatting options into one handy package. Because Visio works with more than text, a Visio style does much more than its Word cousin.

NOTE   Because the new Themes feature might satisfy most casual Visio users, styles are available in Visio 2007 only if you run Visio in developer mode. Choose Tools ⇨ Options and then select the Advanced tab. Check the Run In Developer Mode check box and click OK.

As you'd expect, Visio comes with built-in styles and the easiest way to format with styles or create your own styles is by starting with what Visio offers out of the box. To view a style to determine the formatting it applies, choose Format ⇨ Define Styles to open the Define Styles dialog box, shown in Figure 7-4. The check boxes in the Includes area indicate whether the style applies text, line, or fill formatting. To view the specific formatting options for the style, click the Text, Line, or Fill buttons in the Change area, but make sure you don't change any of the settings.

FIGURE 7-4

View the type of formatting a style applies in the Define Styles dialog box.

Templates often include specialized styles for the shapes they contain. For example, in the building plan template, Visio includes line styles with end points for building dimensions, fill styles that indicate new or to-be-demolished walls, and text styles for several purposes. In most cases, you don't have to think about applying styles; Visio assigns them automatically as you drag shapes onto the drawing. However, when you create a blank drawing or use drawing tools to add content, you can apply styles to format your shapes.

CROSS-REF   To learn how to create and modify styles, see Chapter 34.

Visio inserts five default styles for lines, fill, and text, even if no template is present. Some of the default style names sound similar, so it's helpful to know what each one does:

- **Guide** — For a drawing guide, the line style is a dashed, blue line. Text is Arial 9-point blue. There is no fill format.

- **No Style** — Despite its name, this style does apply some basic formatting: the line style is a solid black line, text is Arial 12-point black centered in the text block with no margins, and fill is solid white with no shadow.

- **None** — Removes lines and fill so a shape has no boundaries and is totally transparent. Uses default text options of Arial 12-point black but includes 4-point margins in the text block.

- **Normal** — By default, Normal uses the same settings as No Style. However, you can redefine normal if you want.

- **Text Only** — This style has the same settings as None except that it aligns the text to the top left of the text block, with no margins.

CAUTION   When you use styles to format shapes in a drawing, shape formatting can change when you copy the shapes to another drawing. This occurs when the destination drawing contains styles with the same names used in the source drawing but with different formatting options. To prevent your shapes from assuming the formatting in the destination drawing, rename the styles in the source drawing before copying the shapes.

Applying Styles

Visio includes style lists on both the Format Shape and Format Text toolbars. If you choose Format ➪ Style, Visio opens the Style dialog box with drop-down lists for text styles, line styles, and fill styles. One subtle but important change in Visio 2007 is that a style appears on a list only if it applies the corresponding type of formatting. For example, if a style applies only to line formatting, you won't see the style name in the text style and fill style lists.

To format using styles, right-click a shape or shapes and, from the shortcut menu, choose Format ➪ Style. In the Style dialog box, choose a style from one or more of the style lists depending on whether you want to apply text, line, fill, or multiple styles.

When you tweak a shape's formatting options after you apply a style, it's usually to resolve a readability issue for that shape. You can retain these individual formatting options even when you apply a different style to that shape. To do this, choose Format ➪ Style and select the Preserve Local Formatting check box.

TIP Applying styles has no effect if a shape is protected against formatting. Refer to the Protecting Shapes section in this chapter for instructions on removing formatting protection.

Restoring Default Styles

Sometimes, you want to remove the local formatting you applied to a shape — for example, when the pattern assigned locally to a shape clash with the drawing's theme. You can restore the default styles associated with the shape's master by selecting the shape and choosing Format ➪ Style. In any of the style lists, choose Use Master's Format, which is the first entry in the list, and click OK.

NOTE To restore the default style for a shape you created with a drawing tool, select the shape and apply the Normal style.

Applying Individual Formats

Applying styles is easier than specifying umpteen formats, but eventually you will end up applying specific formatting options to shapes, connectors, and text. And if you are going to define your own styles or theme effects, you have to know how the Visio formatting options work.

Visio provides methods for formatting just one item or several shapes at once. Whether you decide to use styles or apply formatting options individually, the options you use most frequently are available on the Formatting, Format Text, and Format Shape toolbars, shown in Figure 7-5.

TIP If you want to apply a collection of formats to new shapes you add to the drawing, start by making sure that no shape is selected, for example, by clicking an empty area of the drawing page. Then, select the formats you want from any of the formatting toolbars. When you add shapes to the drawing, they will use the formats you set.

FIGURE 7-5

The Formatting, Format Text, and Format Shape toolbars include common formatting options.

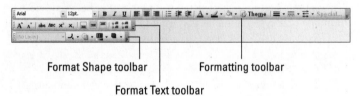

Format Shape toolbar   Formatting toolbar

Format Text toolbar

Applying Formats to Lines

Line formatting includes setting line weight, pattern, color, and end options and the Formatting toolbar includes buttons for all of them. Near the left end of the Formatting toolbar, click the option arrow next to the weight, pattern, or end buttons (which show examples of what they do) and then, from the drop-down menu, choose the format you want. The button to set line color is grouped with the buttons for specifying text and fill colors.

If the choices on the drop-down menu don't appeal to you, choose the More command (More Line Weights, More Line Patterns, or More Line Ends) at the bottom of the menu. In the Line dialog box that opens, you can apply any of all of the formatting options that lines accept. The Preview section shows a sample of the line with all the line formatting options you've applied.

The Line dialog box overflows with options, each described in the following list:

- **Pattern** — Patterns of dotted or otherwise broken lines. Visio includes 23 patterns in addition to a solid line, but you can also define your own.

- **Line Weight** — The thickness of the line, no matter what pattern you use.

CROSS-REF To learn how to define your own patterns, see Chapter 35.

TIP When you produce drawings for a wide audience, use patterns instead of line color so that your drawings don't lose their meaning, even when printed on grayscale printers.

- **Line Ends** — Line ends are symbols such as arrowheads that you can place at the end points of lines. The Formatting toolbar includes separate options for symbols at each end and both ends of a line. In the Line dialog box, you can specify the symbol and symbol size for each end of the line, such as a medium-sized arrowhead at the end of the line and a small circle at the beginning.

TIP If an arrow points the wrong way, you can correct this problem by switching the line end to the other end of the line or by reversing the line. To reverse a line, select the line and then choose Shape ➪ Operations ➪ Reverse Ends.

- **Color** — The color of the line.

- **Cap** — The Cap option, which is available only in the Line dialog box, specifies whether the ends of very thick lines are squared or rounded.

- **Transparency** — This option specifies how transparent the line is.

- **Round Corners** — Click a sample button to apply a radius to the corner between line segments. Or, in the Rounding box, type the radius you want to use.

Applying Fill Formats

Closed shapes are the only type of shape that includes fill, which represents the color and pattern applied to the interior of a shape. You can apply fill colors by clicking the Fill Color button on the Formatting toolbar. However, if you want to specify patterns, transparencies, or shadow

Highlighting Elements

Visio doesn't provide a highlighting feature like the one in Microsoft Word, but formatting can take its place in a pinch. For example, if you want to emphasize a shape on your drawing, you can change the fill color.

You can also highlight lines and connectors — for example, to show the redundant paths in a high-availability network. To do this, on the Drawing toolbar, click the Line tool. Choose Format ⇨ Line, select a wide line weight, a bright color, a high transparency percentage, and click OK. Draw lines using the same vertices as the paths you want to highlight. If you can't see the connectors, right-click the lines you added and, from the shortcut menu, choose Send to Back. If you want to be able to remove these highlight lines easily at a later time, you can add them to a separate layer and delete the entire layer to remove them. To learn more about layers, see Chapter 26.

formatting, select the shape you want to format and then choose Format ⇨ Fill. You can specify the following options for fill formatting:

- **Color** — The color for the interior of the closed shape.
- **Pattern** — Visio provides 40 patterns, including cross-hatching, stippling, and gradients. To define custom patterns, see Chapter 35.
- **Pattern Color** — The pattern color is the second color used if you select a pattern other than None or Solid, for example, for cross-hatching lines and patterns or for gradients.
- **Transparency** — By default, fill is opaque (0%). Drag the transparency slider to the right to increase the transparency of the fill color.
- **Shadow** — Shadows can make diagrams look more dramatic. By default, the Shadow style is set to None, but you can choose the offset and direction of shadows, as well as their color, pattern, and transparency.

Applying Formats to Text

The most frequently used text formatting options are conveniently located on the Formatting and Format Text toolbars, shown in Figure 7-5. The Formatting toolbar includes familiar buttons for specifying font, font size, font style, horizontal alignment, and text color. The Format Text toolbar goes further with options for increasing or decreasing font size, strikethrough and subscript, vertical alignment, and paragraph spacing. The Formatting toolbar appears by default when you create a drawing. To display the Format Text toolbar, choose View ⇨ Toolbars ⇨ Format Text.

Alternatively, the Text dialog box is the comprehensive source for text formatting options, such as language used for spell checking, transparency, character spacing, paragraph spacing, indents, margins around text blocks, tabs, bullet styles, and bullet characters. If you are unusually fastidious about formatting, you might prefer this dialog box over toolbars, because it contains every text formatting option available. In addition, the text formatting features are grouped on tabs and, in some cases, include visual clues about the results you obtain from choosing an option. To open the Text dialog box, choose Format ⇨ Text.

The text you can format depends on how you select it:

- **Select a shape** — This selection method sets up the shape for text formatting applied to all the shape's text.

- **Select text with the Text tool** — Select precisely the text you want to format by activating the Text tool, selecting the text you want to format, and then choosing the text formatting options.

TIP   You can copy text formatting from shape to shape by using the Format Painter. With the Text tool activated, select the shape that has the text formatting you want to copy. On the Standard toolbar, select the Format Painter, and then click the shape you want to format. Visio copies only the text formatting to the second shape.

Formatting Text Blocks and Paragraphs

The text blocks in many shapes contain no more than one paragraph, so it's easy to assume that formatting text blocks and paragraphs means the same thing. In reality, you can specify one set of options that apply to the text block itself and other options for each paragraph within the text block. The formatting options themselves are similar to those in Microsoft Word, including indentations, margins, and alignment.

For a text block, you can specify the vertical alignment of the text within the text block, the margins between the text and the text block boundaries, and color and transparency of the background in the text block, as illustrated in Figure 7-6. To apply text block formatting, select the shape, on the shortcut menu, choose Format ⇨ Text, and then choose the text block formats you want.

NOTE   Most Visio shapes use a transparent background color so you can see what lies behind the text. If elements behind your text make it unreadable, make the text background color opaque so that other shape components don't show through. To do this, right-click the shape, on the shortcut menu, choose Format ⇨ Text, and select the Text Block tab. Type 0% in the Transparency field. The opaque background blocks out components only when the shape contains text because the background color only fills the area around text.

For each paragraph within a text block, you can specify the horizontal alignment of the paragraph, the indentation for the first line of the paragraph, the indentation on each side of the paragraph, the spacing before and after the paragraph, and the spacing between the lines in the paragraph, also shown in Figure 7-6. To format a paragraph, do the following:

1. On the Standard toolbar, choose the Text tool.
2. Select the paragraph you want to format.
3. Right-click the selected paragraph and, from the shortcut menu, choose Format ⇨ Text.
4. Apply the paragraph formatting you want and click OK to close the Text dialog box.

FIGURE 7-6

Formatting options for text blocks and paragraphs are different.

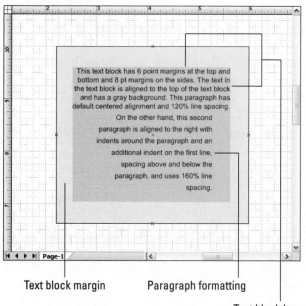

Text block margin　　　Paragraph formatting

Text block boundary

NOTE The Text Ruler is a handy tool for setting, modifying, or removing tab stops and indents. To display the Text Ruler, double-click a shape, right-click the text, and choose Text Ruler from the shortcut menu. To create a tab stop, select the text you want to format and then click the position on the ruler where you want to place the tab. To insert a different type of tab stop, click the Tab icon until the tab stop you want appears. You can also drag tab stops to other positions or remove them by dragging them off the ruler. To adjust indents, drag the top or bottom of the hourglass on the Text Ruler to another position.

Creating Bulleted and Numbered Lists

To create bulleted lists, you proceed as you do for formatting a text block or paragraph and then choose bullet options on the Bullets tab. Creating numbered lists is a bit more challenging, because you must specify every aspect of the numbering, spacing, and indents.

NOTE When you use the built-in bullets, you can change the font for a bulleted list without affecting the appearance of the bullet. However, when you change font size, bullet size adjusts to match.

To create a bulleted list, follow these steps:

1. Select the text block or the portion of the text block that you want to format as a bulleted list.

2. Right-click the selected text and, from the shortcut menu, choose Format ➪ Text. Select the Bullets tab.

3. To choose a different bullet symbol, click a bullet option or, in the Bullet Characters box, type the character that you want to use as a bullet.

4. To adjust the font size for the bullet and the space between each bullet point, in the Font Size list, increase or decrease the percentage.

5. To change the hanging indent between the bullet and the list text, in the Text Position box, type the distance you want.

6. To preview the formatting you have chosen, click Apply. If the list is formatted the way you want, click OK. Otherwise, click Cancel or try other settings.

Numbered lists work differently in Visio than they do in other applications, such as Word. In Visio, you have to add numbers, tabs, and indentation manually. If you plan to include a long numbered list in a Visio diagram, creating a link from your Visio drawing to a numbered list in a Word document is an easier approach.

CROSS-REF To learn how to create a link to a Word document, see Chapter 8.

To create a numbered list in Visio, follow these steps:

1. Make sure that text justification is left-justified. On the Formatting toolbar, click Align Left.

2. Double-click the shape in which you want to create the numbered list.

3. Type the 1, press Tab, and then type the text for the first entry.

4. Repeat step 3 for each entry in the numbered list (change the number accordingly).

5. To align the numbers and text, first display the Text Ruler. To do this, select the text, right-click the text, and, from the shortcut menu, choose Text Ruler.

6. In the Text Ruler, drag the bottom of what looks like an hourglass to the position where you want the text of each numbered step to begin.

7. In the Text Ruler, drag the top of the hourglass to the position where you want the step numbers to appear.

8. When you are done, press Esc or click outside the shape to close the text block.

Formatting Shapes

The formatting options you can apply to shapes depend on the shapes. For example, you can apply line formats to any shape containing lines whether they are open or closed. When you format closed shapes, such as rectangles and circles, you can also specify the fill formatting for those shapes. If you format text-only shapes created with the Text tool or other shapes containing text, you can format the text in those shapes.

Applying Formatting to Shapes

It's easy enough to format one shape by selecting that shape and then formatting with the techniques listed in Table 7-2. It's no harder to format several shapes or groups of shapes. Plus, you can copy formatting from one shape to another.

If you can't apply formatting options to a shape, the shape could be protected against formatting or it could belong to a group. See the "Protecting Shapes" section later in this chapter to learn how to undo formatting protection. To format a shape in a group, subselect the shape (click the shape within the group until Visio displays the shape's alignment box) and then apply the formatting you want.

NOTE To remove a shape's border, right-click the shape, and from the shortcut menu, choose Format ↪ Line. In the Pattern drop-down list, choose None.

TABLE 7-2

Applying Formatting to Shapes

Formatting Task	How to Accomplish It
Format a shape	Select a shape and then choose the formatting options you want.
Format several shapes at once	Select all the shapes and then choose the formatting options you want.
Format a group of shapes	Select the group and then choose the formatting options you want.
Copy formatting from one shape to another	Select a formatted shape, on the Standard toolbar click the Format Painter button, and click the shape to which you want to copy the formatting.
Copy formatting from one shape to several others	Select a formatted shape, double-click the Format Painter button on the Standard toolbar, and click each shape to which you want to copy the formatting. To stop copying formatting, click the Format Painter button or press Esc.

Applying Shadows to Shapes

Shadows behind shapes can add punch to your diagrams. Visio 3-D shapes already have shadows set up, but shadows are nothing more than formatting that you can apply to any shape.

NOTE An easy way to add shadows to shapes is with a default shadow for a drawing page. To do this, display the page, choose File ↪ Page Setup, and select the Shadows tab. Specify the shadow style, offset dimensions, magnification, and direction for the shadow, and then click OK.

To format and apply shadows to specific shapes, follow these steps:

1. Right-click a shape and, from the shortcut menu, choose Format ⇨ Fill. Or select several shapes and then choose Format ⇨ Fill.

2. In the Fill dialog box, in the Style list, choose the shadow style you want. Each shadow style includes offset dimensions, magnification, and direction for the shadow.

 To use the default shadow specified in Page Setup, choose Page Default in the Style list.

3. To change the shadow color from the one assigned by the current color scheme, in the Color list, choose a color.

4. To change the shadow pattern, in the Pattern list, choose the pattern you want. If you choose a pattern other than None or Solid, in the Pattern Color list, choose the secondary color for the pattern, such as the color of the hatch lines in a hatch pattern.

5. To specify the transparency of the shadow, drag the Transparency slider to the right. By default, the shadow is opaque.

Protecting Shapes

The Format menu is an unlikely place for the command for protecting shapes from inadvertent changes, but choosing Format ⇨ Protection is where you start to protect shapes against resizing, moving, rotation, text editing, and formatting. You can also prevent someone from selecting or deleting shapes. To protect shapes, select the shapes you want to protect and choose Format ⇨ Protection. The Protection dialog box enables you to protect a shape in the following ways:

- **Resizing** — Select the Width, Height, or Aspect Ratio check boxes to prevent users from changing the width or height of shapes or from modifying the proportions of a shape.

- **Moving** — Select the X Position and Y Position check boxes to prevent users from moving a shape to a new location.

- **Rotation** — Select this check box to prevent users from rotating a shape.

- **Moving Endpoints** — Select the Begin Point and End Point check boxes to lock the end points of 1D shapes in place.

- **Editing Text** — Select the Text check box to prevent users from editing shape text.

- **Formatting** — Select the Format check box to prevent users from modifying the formatting of a shape.

- **Selection** — Select this check box to prevent users from selecting a shape.

- **Deletion** — Select this check box to prevent users from deleting a shape.

- **Group and Theme Formatting** — Select the From Group Formatting, From Theme Colors, or From Theme Effects check boxes to prevent the shape from taking on formatting from its parent group or the theme applied to the drawing.

 In the Protection dialog box, quickly add or remove protection by clicking the All or None buttons.

 Some shapes are protected against changes with the GUARD function. To learn how the GUARD function works, see Chapter 34.

Summary

Visio provides formatting tools so you can make your drawings look exactly the way you want. You can choose options to format text, lines, fill, and shadows. The new Themes feature is the easiest way to get started with good-looking formatting, but you can use styles to highlight other shapes with consistent sets of formatting options. You can also modify the formatting for all shapes using a specific style by modifying the formatting options of that style. In addition, you can apply special formatting to specific shapes on your drawing and preserve those options as you apply themes or styles.

Part II

Integrating Visio Drawings

Chapter 8

Inserting, Linking, and Embedding Objects

Microsoft Visio diagrams typically don't exist in a vacuum, because the very point of Visio is to communicate information in a visual way to others out there. Sometimes, Visio drawings stand alone, but most of the time, they become a part of larger communications, in reports or presentations, for example.

To facilitate this kind of information sharing, you can create hyperlinks to Visio drawings from Word documents, Excel workbooks, PowerPoint presentations, and other programs. Click the hyperlink, just as you do in a web site, and Visio and the drawing are launched. Taking this concept of linking a step further, you can insert Visio drawings as links in other application files, ensuring that any edits to the Visio drawings automatically appear in the linked documents. A third method is to insert, or embed, separate copies of Visio drawings in other application files, where you can view the drawings and use Visio tools to edit them at will. And finally, you can always save Visio drawings in another file format, such as JPEG graphic files, to share the drawings while preventing any editing by others.

These methods work in the other direction as well. Hyperlinks can take you from Visio to documents in other applications. You can link or embed text from Word documents, tables from Excel workbooks, or the contents of other application files into Visio drawings and make changes to those documents without ever leaving Visio. Graphics such as clip art and digital photographs are common elements for embedding in Visio drawings.

In this chapter, you'll learn how to create hyperlinks to and from Visio drawings as well as how to link and embed information from one application into another. This chapter also explains the finer points of embedding graphics files in Visio and provides tips for adding Visio drawings to PowerPoint presentations to the best effect.

CROSS-REF Another way to exchange information between applications is to export data from one program and import it into another. To learn about importing and exporting with Visio, see Chapter 9.

Navigating Drawings and Documents with Hyperlinks

Hyperlinks are the ultimate in easy and convenient navigation. Regardless which application holds a hyperlink, you can tell the link is there with visual clues like underlined text, an icon, or a different mouse pointer. When you click the hyperlink, the application to which the hyperlink points launches in a separate window, and the file or web page opens. For example, in a Visio drawing, hyperlinks associated with process shapes could open the Word documents containing detailed business process descriptions. On multipage Visio drawings, hyperlinks can move you from one page to another. Of course, hyperlinks in other applications can jump to Visio drawings as well.

To enable hyperlink navigation, you have to define the destinations to which the hyperlinks jump. Although hyperlinks can belong to a drawing page, in Visio, they're more commonly associated with shapes, especially Hyperlink buttons or circles. This section describes how to add hyperlinks that navigate to other pages in Visio files, to other application files, and from other application files to Visio drawings.

NOTE Creating hyperlinks in other applications, such as Microsoft Excel, to Visio drawings works similarly to the reverse process. You choose the words, worksheet cell, PowerPoint bullet point, or other element, and use the technique for creating hyperlinks in that application to point to a Visio drawing file.

Hyperlinks at Work

The behavior of hyperlinks on a Visio drawing depends on how you view your drawing:

- **Full-screen mode** — If you choose View ⇨ Full Screen or use a Visio drawing as a web page, when the mouse pointer is over a shape containing a hyperlink, the pointer changes to the pointing hand icon. To follow the hyperlink, click the shape.

- **Normal mode** — If you are not in full-screen mode, when the mouse pointer is over a shape containing a hyperlink, the pointer changes to an arrow with the hyperlink globe icon. To follow the hyperlink, right-click the shape and then, from the shortcut menu, choose the link.

When you follow the hyperlink, the web page or file appears in its own window. If the hyperlink jumps to another Visio page, that page replaces the current page.

Inserting Hyperlinks

The first step to inserting a hyperlink in a Visio drawing is deciding where to place the hyperlink. You have three options:

- **Associate the hyperlink with a Visio Hyperlink shape** — Visio offers specially designed Hyperlink shapes, which give readers visual cues that hyperlinks exist, as demonstrated in Figure 8-1. To add a Visio Hyperlink shape, open the Borders and Titles stencil, drag one of the hyperlink shapes to the drawing page, and insert hyperlinks into those shapes. With Hyperlink shapes on a drawing, anyone viewing the drawing knows that clicking the shape will navigate somewhere else.

FIGURE 8-1

Hyperlink shapes provide visual cues that hyperlinks exist, but hovering the mouse over any shape with a hyperlink displays the Hyperlink icon.

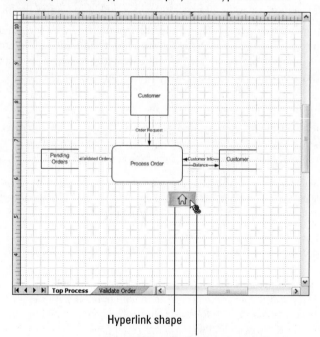

Hyperlink shape

Hyperlink icon indicating shape is "hot"

- **Associate the hyperlink with a diagram shape** — This method isn't as obvious as using Hyperlink shapes, because you must move the mouse over a shape before the pointer changes to one of the hyperlink indicators, identifying the shape as *hot*. When you click a hot shape, the linked application and file or Web page launches in a separate window. This approach works well when you want to navigate from the last step on one page to the continuation on another.

- **Associate the hyperlink with a drawing** — You can make the entire page hot, for example, to navigate from a project calendar in Visio to the Microsoft Project schedule that the calendar summarizes. For hot drawings, the mouse pointer changes to the hyperlink icon whenever you move the mouse over a blank part of the page.

TIP Visio Hyperlink shapes use icons to communicate their meaning. For example, you can indicate a web site home page with an icon of a house, as shown in Figure 8-1. However, you can change the icon on a hyperlink shape to any of the 12 icons Visio offers, such as Back, Forward, and Help. To change the icon for a specific Hyperlink shape, right-click the shape and, from the shortcut menu, choose Change Icon and then, in the Icon Type drop-down list, choose the icon you want.

To insert a hyperlink in a Visio drawing, follow these steps:

1. Click the shape or page in which you want to insert the hyperlink.

2. Choose Insert ➪ Hyperlinks.

3. In the Address box, enter the address for the hyperlink using one of the following methods:

   - **Linking to a web page** — Type the full address starting with the protocol — for example, **http://**. You can also click the Browse button and then click Internet Address to launch your web browser. Navigate to the web page to which you want your Visio hyperlink to point. In the Hyperlinks dialog box in Visio, the Address box should display the full web address of the page in your browser. If not, copy the address from the web browser and paste it into the dialog box.

   - **Linking to a file on your system** — Enter the full file path. You can also click the Browse button and then click Local File. Navigate to the location of the file to which you want your Visio hyperlink to point. Select the file and then click the Open button.

NOTE If you're entering a web address, remember to use front slashes (/) where necessary. If you're entering a file path, remember to use back slashes (\) to separate the folder names.

4. To display text when the pointer pauses over the hyperlink, in the Description text box, enter the text.

5. Click OK.

TIP By default, hyperlinks use a relative path — that is, the path relative to the location of the Visio drawing itself. If you prefer to use the absolute path to specify file location, in the Hyperlinks dialog box, clear the Use Relative Path For Hyperlink check box. Because Visio uses relative paths when creating hyperlinks, this check box is dimmed until you save the Visio file and thus define its path.

A single element can contain multiple hyperlinks — for example, to provide links to each detail drawing page for a high-level process. In the Hyperlinks dialog box, click New, enter a new hyperlink address as usual, and then click OK. When you right-click the hyperlinked shape or page, a shortcut menu of hyperlinks appears and you can choose the hyperlink you want. Multiple hyperlinks work whether you work in Normal mode or Full Screen mode.

NOTE   When you use Internet Explorer 5.0 or later and right-click a shape with multiple hyperlinks, the shortcut menu shows all the associated hyperlinks. For browsers or output formats such as SVG that don't support multiple hyperlinks, right-clicking a shape displays only the
default hyperlink or, if there is no default hyperlink, the first hyperlink in the list.

Using Hyperlinks to Navigate Between Drawing Pages

Suppose your Visio drawing starts with an overview process, which spills over onto other pages
that contain the details of the entire process. You can create hyperlinks from the overview page to
drill down to see the details on the other pages. Page-to-page hyperlinks work just as well to move
from one page to the next in a sequence of drawing pages, for example, to follow a flowchart from
its first step to the last.

To create hyperlinks that navigate between multiple pages in a Visio drawing, follow these steps:

1. Click the shape in which you want to insert the hyperlink.

   ■ On an overview drawing, click the shape that represents the overview of the detail
     drawings to come.

   ■ For the first page of a sequence of pages, add a navigation shape, such as a Hyperlink
     shape, to indicate navigation to another page.

2. Choose Insert ⇨ Hyperlinks.

3. Next to the Sub-address box, click the Browse button to open the Hyperlink dialog box.

4. In the Page drop-down list, select the Visio drawing page you want, as shown in Figure 8-2.

FIGURE 8-2

When you select the page for the hyperlink, its reference appears in the Sub-address
box.

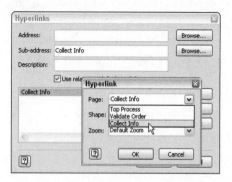

5. To focus on a specific shape on the destination page, in the Shape box, type the name of the shape. When you follow the hyperlink, Visio centers this shape in the drawing window. This comes in handy, for example, when a drawing page contains the details for two higher-level processes, the hyperlink can navigate to the beginning shape of one of the processes.

NOTE   To find the name of a shape, you must turn on Developer mode. To do this, choose Tools ➪ Options and select the Advanced tab. Select the Run In Developer Mode check box and click OK. On the drawing page, right-click the shape and, from the shortcut menu, choose Format ➪ Special. The shape name appears in the Name box — for example, Circle, Sheet.1, or Manager.18.

6. To change the default size of the destination page, in the Zoom drop-down list, select the percentage you want.

7. Click OK to return to the Hyperlinks dialog box.

8. To display text when the mouse pointer pauses over the hyperlink, enter the text in the Description text box.

9. Click OK.

Modifying Hyperlinks

If your Visio drawing has been around for a while, it's a good idea to periodically verify that your hyperlinks still work. This is especially important if hyperlinks point to web pages or files that you don't control. A company revamps its web pages or a colleague moves the files to which your hyperlinks point and, voilà, your hyperlinks no longer work. When you find broken hyperlinks, you can edit them to hook them back up. To modify a hyperlink, follow these steps:

1. Select the shape or page with which the hyperlink is associated and choose Insert ➪ Hyperlinks.

2. If there are multiple hyperlinks for the selected shape, in the Hyperlinks dialog box, in the list at the bottom, select the name of the hyperlink you want to change.

3. In the Address text box, edit the web address or path name. You can also click the Browse button and navigate to the web page or file to automatically update the address.

4. Make any other changes you want to the hyperlink and then click OK.

Understanding Linking and Embedding

The term *linking* often brings to mind hyperlinks between web pages, and rightly so, because hyperlinks in Microsoft Office applications navigate between files or to other locations within the same file. For example, you could add a hyperlink to a status report in Word that displays the Visio diagram of the current workflow for a business process. Hyperlinks also move within documents, so you can use hyperlinks to jump from the main page in a Visio drawing to the page that contains further detail.

Another type of linking that you're likely to employ is Object Linking and Embedding technology, or OLE. Applications that use OLE technology, including the Microsoft Office applications, can easily swap elements with other OLE applications — from entire files to individual items within those files. For example, you can include an entire Excel worksheet in a Visio drawing or only a few cells.

Swapping objects between applications comes in two flavors: linking or embedding. With linking, the object appears in the target, or container, application, but the object actually exists in another file on the computer. Linking makes updating objects easy, because the linked objects change whenever the original object changes in the container application. When you double-click linked objects, the source application's menus and tools open within the container application so you can make changes to the source file right there. Consider a PowerPoint slide that displays a linked Visio diagram. Suppose you notice a small error as you practice your PowerPoint presentation. You double-click the Visio object and Visio windows, menus, and tools appear, and your diagram from its separate Visio drawing file opens as well. After you make the changes to the diagram and save it, those changes are made to the separate Visio drawing file and become visible in any document that links to it.

When an object is *embedded,* rather than linked, the container application gains a separate copy of the object. The source file still exists, but the object in the container application is totally separate and independent. You can access the source application's menus and tools within the container application and make changes there, but the changes update only the copy of the object in the container application and do not affect the source file.

Whether you're linking or embedding, double-clicking the object in the container application opens the source application's menus and tools so you can make changes on the spot. Moreover, you can create new objects from scratch in a container application to link or embed.

Linking and embedding both have advantages and disadvantages, so the best technique depends on what you want to do. The bottom line is that linking is preferable when the source file tends to change frequently and you don't want to make changes in two places. Embedding works better if you want to distribute the container file to others without also providing linked source files. Table 8-1 lays out the pros and cons of linking and embedding so you can choose the best technique.

TABLE 8-1

Linking versus Embedding

Factor	With Linking	With Embedding
Ease of updating	Linking makes updating easier, because changes to a linked object update to both the source and container application, regardless of where you do your editing.	Changes to the source file do not affect the embedded object in the container file and vice versa. If you make changes in one place, you must repeat those changes in the other file.
File management	There's only one copy of the object to track. However, if you move the source file, the link breaks and you must recreate it.	You must manage two separate and independent versions of the object. On the other hand, embedding makes the object portable because the embedded object doesn't rely on the source file.
Linking individual elements	You can't create a link to an individual element within a file. You must link to the entire file.	You can embed individual elements within a file.
File size	Linking produces smaller files because the object isn't actually within the container file.	The container file size can grow quite large, because it actually stores the content of the object itself, not just the link to the source.
Slow file launch	Large or complex files containing several links can take a while to open, especially across a large network. This occurs because OLE checks the source files for changes and updates the linked objects, either manually or automatically.	Embedded objects can lead to slow file launch as well because the embedded files take time to load and draw completely.

CROSS-REF   Visio can also link shapes to data stored in databases or other data sources. To learn about linking shape data to a database, see Chapter 10.

Linking Visio Drawings with Other Files

Insert an OLE link when you want to show information from a file, but you don't want that file incorporated into the target document in its entirety. For example, suppose a Visio diagram illustrates the business problem discussed in a Microsoft Word document. You update the Visio drawing occasionally and you want to see those updates in the Word document. Creating an OLE link to the drawing from Word is the perfect solution.

This works just as well in the opposite direction. A Visio drawing might refer to a Microsoft Excel chart that is dynamically updated as data is entered. By linking that chart to your drawing, you can include the chart as an integral part of your Visio drawing and know that you're always looking at the latest version of the data.

The container application shows the link either as an application icon or a representation of the source file. Either way, when you double-click an OLE link, the source application launches in a separate window and opens the file.

Linking Visio Drawings to Other Documents

To link a Visio drawing to another Microsoft or other OLE application file, follow these steps:

1. Open both the container file and the source file that contains the content you want to link to using OLE. For example, open a Word document as a container and a Visio drawing as the source of the diagram you want to link into your report.

2. In the Visio drawing file, display the page that you want to link.

3. Make sure that no shapes are selected and then choose Edit ➡ Copy Drawing. (If anything is selected, the Copy Drawing command does not appear on the Edit menu.) Copy Drawing grabs the entire drawing page including background pages.

4. Switch to the container file without closing Visio and position the insertion point where you want to insert the link.

5. To create a link in Word 2007, Excel 2007, or PowerPoint 2007, follow these steps:

   On the Home tab, in the Clipboard group on the left side of the ribbon, click the arrow below Paste and then, from the drop-down menu, choose Paste Special.

 The command you use to create links varies depending on the application.

   In the Paste Special dialog box, in the As box, select Microsoft Visio Drawing Object.

   Select the Paste Link option.

   To show an icon rather than the drawing itself, in the container file, select the Display As Icon check box.

   Click OK. The drawing appears in the container application, as illustrated in Figure 8-3.

6. When you double-click the linked object, the source application launches in a separate window and opens the source file for editing.

FIGURE 8-3

The linked object retrieves information from the source Visio drawing.

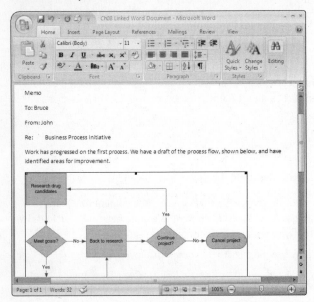

Linking Other Documents to Visio Drawings

To create a link from another document into a Visio drawing, follow these steps:

1. Open the Visio drawing container file and the source file, such as an Excel spreadsheet.

2. Choose Insert ➪ Object.

3. In the Insert Object dialog box, select the Create From File option.

4. Click Browse and in the Browse dialog box, select the file you want to link into the Visio drawing and then click Open.

5. In the Insert Object dialog box, select the Link To File check box.

6. Optionally, if you want to show an icon rather than the content, select the Display As Icon check box.

7. Click OK. The object appears as a shape on the Visio drawing, including selection handles that you can drag to resize.

8. When you double-click the linked object, the source application launches in a separate window and opens the source file for editing.

Moving and Resizing Linked Objects in Visio

Linked objects respond to editing methods like other elements in your container application. For example, in Visio, you can use any of the following methods with a linked object:

- **Select a linked object** — Click the object. A selection box similar to the selection box for a Visio shape appears around the object.

- **Resize the object proportionally** — Drag one of the selection handles, shown in Figure 8-3, to change the size of the linked object.

- **Stretch or condense the object along one side** — Drag one of the selection handles at the midpoint of a side.

- **Move the object** — Drag the middle of the object to the position you want. Take care to drag from the center of the object, and not along any of its edges.

Editing the Content of Linked Objects

With a linked object, you can edit its content from the most convenient place. Either way, you're working with the same file. If you have the object's source application open, open the source file and make the changes. When you save the file, the changes appear in the container application.

Alternatively, if you spot needed changes while looking at the file in the container application, double-click the object. The source application menus and commands appear so you can edit the linked object, as shown in Figure 8-4. Make whatever changes you want and then save and close the file. The source application menus disappear and you see the changes reflected in the object in the container application. If someone changes the source file while you're working with the linked object, you'll see the changes the next time you open the container file.

 To edit the content of a linked object, you must have read-write privileges.

If you receive an error when you try to edit a linked Visio drawing, the most frequent culprit is a broken link. Make sure the Visio drawing file is located where it's supposed to be — or recreate the link. If the link is correct, make sure that the file isn't open in another application. You can't edit the file in two places at once. And of course, to edit a Visio drawing even within another application, you must have Visio installed on your computer.

FIGURE 8-4

Double-click a linked object to access its source application's menus and commands for editing.

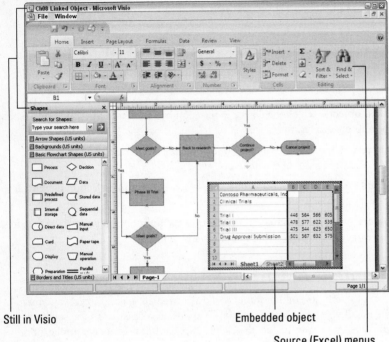

Still in Visio

Embedded object

Source (Excel) menus

Managing Linked Objects

When you have linked objects, you sign up for managing those links — whether it's making sure that the links don't break to specifying whether you want the links to update automatically or only when you say so. Managing links includes the following actions:

- Review details for all linked objects in the file.

- Review the path name for a linked source file and change the path if the file has moved. If you link to a file that is subsequently moved or deleted, the link breaks. The next time you open the container file, instead of a picture of the linked content, an error message appears, such as "Error! Not a valid link."

- Specify whether you want updates from the source to be automatic or manual (they're automatic by default).

- Break the link between the object and source, transforming the linked object into an embedded object or a static picture.

Change Links to Embedded Objects or Pictures

Suppose you linked to a file that was getting a major overhaul so that you would see the most up-to-date version. Now the file is static and you would prefer to store the copy completely within the container document so you can send the document to others without worrying about maintaining the link. You can break the link while keeping the object in the container, essentially converting a linked object to an embedded object. To do this, follow these steps:

1. In the container file, choose Edit ⇨ Links, and then select the link in the Links dialog box.

2. Click Break Link, and then, in the confirmation message box that appears, click Yes. The link information disappears from the Links dialog box.

3. Click Close. The object is still exactly where it was in your container document and looks just the same. Yet, it no longer is linked to the source and has become an independent copy existing within the container document — that is, it's now an embedded file or a static picture.

To manage links, follow these steps:

1. Choose Edit ⇨ Links to open the Links dialog box.

2. To change the path for a broken link, click the link whose path you need to edit. Click Change Source and browse to the new location of the source file. Select the file name and click Open to recreate the link.

3. To change an automatic link update to manual update, for example, to improve performance while you are performing heavy editing on the drawing, select the link and then select the Manual option, and click Close.

4. To update the manual links in your container file, open the Links dialog box again, select a manual link, and then click Update Now.

5. When you're finished making changes in the Links dialog box, click the Close or OK button.

 If you decide you no longer need a linked object at all, you can remove it from the container document by selecting the object and then pressing Delete.

Embedding Objects

Embedding an object from another application inserts an independent copy of the source file into the container application file. For example, if you embed a worksheet from an Excel workbook into a Visio drawing, a copy of the Excel worksheet becomes an integral part of the Visio drawing. Editing that copy while working in Visio has no effect on the original worksheet. Conversely, embedding a Visio drawing as an object in another application, such as Word, creates a copy of the Visio drawing in the Word document. As with linked objects, when you double-click the embedded object in the container application, the menus and tools for the source application appear, so you can make changes on the spot.

Embedding has some advantages over linking:

- You don't have to worry about the location of the source file, because the copy of the file exists inside the container file.

TIP **The disadvantage to stuffing the copy into the container file is that the container file gets bigger. If you embed several objects into a file, the file size could easily grow to several megabytes.**

- You can embed an entire source file or just a piece of it. For example, you can embed an entire Excel workbook within a Visio drawing, or just a single chart from that workbook. Likewise, you can embed all pages of a Visio drawing in a PowerPoint presentation, a single page, or even a single shape in another application.

- The embedded object displays completely and correctly in the container file even if the source application is not installed on the computer you use to view the file. This is incredibly important when you plan to distribute the container file to others, who might not have Visio on their computers.

Embedding Entire Files into Visio

Embedding another document into a Visio drawing is almost exactly like linking a document to a Visio drawing. The main difference is that you don't select the Link to File check box. To embed an entire document into a Visio file, follow these steps:

1. Open the Visio drawing into which you want to embed another file.

2. Choose Insert ➪ Object.

3. In the Insert Object dialog box, select the Create From File option.

4. Click Browse and in the Browse dialog box, select the file you want to link into the Visio drawing and then click Open.

5. In the Insert Object dialog box, make sure that the Link To File check box is not selected.

6. Optionally, if you want to show an icon rather than the content, select the Display As Icon check box.

7. Click OK. The object appears as a shape on the Visio drawing, including selection handles that you can drag to resize.

8. When you double-click the embedded object, the menus and tools for the source application appear so you can edit the source file.

Embedding Portions of Files

You might want only a single page or portion of a Visio drawing in a Word report. Similarly, you might need just a table from Word or a single PowerPoint slide in your Visio drawing, rather than the entire file. You can select and copy the portion of a file you want and embed only that much in the container file. Along with including only the necessary information, this can also help keep the size of the container file from ballooning larger than it needs to be.

To embed a part of a file in another application, follow these steps:

1. In the source application, open the file and go to the page that contains the portion you want to embed in the other application.

2. Using the application's tools, select the portion you want to embed. For example, if you're working in Visio, activate the Pointer tool if necessary, and then drag across the shapes you want to embed. The selection area must fully enclose all elements you want to embed. If you're working in Word, Excel, or another application, drag over the text, cells, or other elements.

> **TIP** If you're working with a multipage Visio drawing and you want to select the contents of a single page to embed in another application, display that page and make sure no shape is selected. Choose Edit ➪ Copy Drawing to copy the entire page into memory so that you can paste it into the other application.

3. Press Ctrl+C to copy the selection to the Windows clipboard.

4. Switch to the container application, open the file, and go to the page or worksheet in which you want to embed the object you just copied. If necessary, click the location where you want to embed the object.

5. Choose Edit ➪ Paste Special.

6. In the Paste Special dialog box, in the As box, select the type of object you copied, such as Microsoft Visio Drawing Object or Microsoft Excel Worksheet.

7. Click OK. The copied object appears in the container file.

If you need only a single Visio shape in the other application, you can copy and paste it. In Visio, show the shape you want in the Shapes window. Right-click the shape and then choose Copy or simply press Ctrl+C. In the container application, select the location you want and then use paste to add the shape to the file. For example, in Word, click the Home tab, and in the Clipboard group on the ribbon, click Paste. You can also arrange the Visio and container applications side by side and then drag the shape from Visio to the other application.

Positioning and Formatting Embedded Objects

As soon as you insert an object into the container application, you can perform the following actions:

■ Move it to the location you want

■ Resize it to the dimensions you need

■ Crop one or more edges of the object

■ Adjust the space surrounding the object

Creating an Embedded Object from Scratch

If you want to create a file whose only purpose is to augment information in your Visio drawing, you can create a brand-new file from within Visio. For example, suppose you want to create a bulleted list for a Visio drawing and would rather format the list using Word.

To create a new embedded object file, follow these steps:

1. In Visio, choose Insert ⇨ Object and then select the Create New option.

2. Under Object Type, select the application with which you want to create the new embedded file.

3. Click OK. A blank file is inserted in your container file, and you can start creating the new embedded file using its application's tools.

4. To return to Visio, simply click anywhere on the Visio drawing.

Moving and Resizing Objects

To move an embedded object, drag the middle of the object to the position you want. Take care to drag from the center of the object, not along any of its edges.

To resize an object proportionally, drag one of the selection handles in any of the four corners. To stretch or condense the object along one side, drag one of the selection handles at the midpoint of an edge.

Cropping Objects

If you need to trim extraneous space from the edges of an object, crop the object. To crop a Visio object in another application, follow these steps:

1. In the other application, click the Visio object and then choose Format ⇨ Object. The Format Object dialog box appears. Make sure the Picture tab is showing.

2. Under Crop From, enter the amount you want to crop from the Left, Right, Top, or Bottom.

3. When you're finished, click OK. Because you can't tell what the crop dimensions do until you see the results, you might have to repeat these steps a few times to crop the Visio drawing the way you want.

To crop an object in Visio, follow these steps:

1. If it's not already showing, display the Picture toolbar by choosing View ⇨ Toolbars ⇨ Picture.

2. Select the object you want to crop.

3. On the Picture toolbar, click the Crop button.

4. Drag a selection handle in the object in the direction you want to crop the object.

Adjusting the Space Surrounding Objects

To adjust the space surrounding a Visio object in another application, follow these steps:

1. In the other application, double-click the Visio object to open the in-place editing window.

2. Drag one of the selection handles to change the shape surrounding the object.

3. Click outside the in-place editing window. The editing window closes, and the container file reflects the new space surrounding the object.

To adjust the space surrounding an object in Visio, follow these steps:

1. If not already showing, display the Picture toolbar by choosing View ⇨ Toolbars ⇨ Picture.

2. Select the object.

3. On the Picture toolbar, click the Crop button.

4. Drag a selection handle *outward* from the object in the direction in which you want to add space around the object.

Editing the Content of Embedded Objects

With embedded objects, you can edit their content without leaving the comfort of the container application. In effect, you can work with the tools of two applications in one.

To edit an embedded object, simply double-click it. This opens in-place editing window and the menu and toolbars change from those of the container application to those of the embedded object's application, as shown in Figure 8-5. Use whatever commands you need on those menus exactly as if you were working with the source application. When you're finished editing, click outside the in-place editing window. The editing window closes and the embedded object reflects your edits.

FIGURE 8-5

Use the source application menus to edit the embedded object. Adjust the space around the object by dragging the selection handles in the in-place editing window.

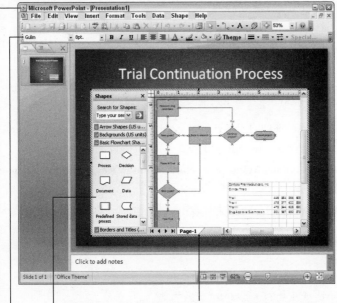

Drag selection handles to adjust space around the object.

Double-clicking an embedded object opens the in-place editing window.

The menus switch to those of Visio, the source application.

In this example, PowerPoint is the container.

If you prefer, you can open an object in a separate window. For example, in PowerPoint, right-click the object and, from the shortcut menu, choose Visio Object ➪ Open. The object and its source application open in a separate window where you can make the changes you want. When you're finished, choose File ➪ Update to update your changes to the container application. Close the Visio window to return to the container application. You can now see the object with your edits.

Perking Up Visio with Pictures

Inserting graphics such as clip art, photographs, or other types of images is a specialized form of embedding. Any kind of graphics file will do, including graphics from a digital camera or scanner. The Clip Art Task Pane helps you search a variety of sources for just the right piece of clip art. Because embedding graphic files is a frequent part of building diagrams, Visio includes tools to make the process easy and versatile.

Inserting Graphics Files into Drawings

Inserting graphics files on your hard drive or network drive into a Visio drawing couldn't be easier. Just follow these steps:

1. In the Visio drawing, display the page in which you want to insert the graphic.
2. Choose Insert ⇨ Picture ⇨ From File. The Insert Picture dialog box appears.
3. Browse to the location of the graphic, select the file, and then click Open. The graphic appears on the page.

> **NOTE** In the Insert Picture dialog box, the Files Of Type box is set to All Pictures, which means that you'll see any kind of graphics file, including JPEG, Tag Image File Format (TIFF), windows bitmaps, and more. If you want to look for a specific type of file, in the Files Of Type drop-down list, select the type.

4. Now, you can use standard Visio tools to resize, move, and crop the graphic as needed.

Searching for and Inserting Clip Art

You can't complain about the quantity of clip art available in Microsoft Office applications. Even if you end up looking elsewhere for the perfect picture, Visio and other Microsoft Office applications can search a number of places for images. To embed a piece of clip art in Visio, follow these steps:

1. In your Visio drawing, select the page in which you want to insert the graphic.
2. Choose Insert ⇨ Picture ⇨ Clip Art. The Clip Art Task Pane appears. Alternatively, if the Task Pane is already visible, click the Task Pane down arrow and then choose Clip Art.

> **NOTE** If you haven't installed the clip art feature, you'll see a prompt to do so. The CD is not required to install clip art.

3. In the Search For text box, type a key word or phrase that describes the type of image you want to add to your drawing. For example, if you want show clip art of engineers to make an engineering flowchart look friendlier, you could search for "engineer."

> **TIP** Searches execute faster if you use specific key words or phrases. However, you can use general key words to obtain a wider selection of clip art.

4. In the Search In box, specify where you want Visio to search. You can select the check box for a particular folder on your hard drive, such as Office Collections. You can also have Visio search the Internet in a particular web collection. Be aware that the wider the search and the larger the collections, the longer the search might take.
5. In the Results Should Be box, select the type of media you want to find: Clip Art, Photographs, Movies, or Sounds. To further specify file formats you want, click the plus sign under the media. This can help narrow your search, especially when searching web collections.
6. When you're finished defining your clip art search criteria, click Go. Results of your search appear in the Clip Art Task Pane as thumbnails.

7. When you find the clip art you want to use in your drawing, drag it into position in your drawing.

8. Use standard Visio techniques to resize the art if necessary.

NOTE In Visio 2007, the Insert ⇨ Picture menu no longer contains the From Scanner or Camera entry. To insert a picture coming from a digital device such as a scanner or digital camera, copy the file from the device onto your computer and then insert the graphic file as described in the section, "Inserting Graphics Files into Drawings."

Fine-Tuning Visio Drawings for PowerPoint

PowerPoint presentations and Visio drawings go together like bread and butter. Visio is great for illustrating processes, concepts, or other information, while PowerPoint provides a platform for explaining those items to your audience. Embedding Visio drawings in PowerPoint adds a world of clarity to presentations. Themes in Visio 2007 make it even easier to ensure that your Visio diagrams look good within your PowerPoint slides, because the same set of Themes Colors are available in Visio 2007 as well as the other Microsoft Office applications.

Formatting Visio Drawings in a Presentation

The Theme Colors you choose for your Visio drawing might look great on their own, but truly heinous when the drawing becomes part of a PowerPoint presentation. With Microsoft Office 2007, all you have to do to realign the colors is choose the same Theme Colors in your Visio drawing and the PowerPoint presentation, as described in Chapter 7. But, colors aren't the only formatting issue when you plan to project a presentation in front of an audience. The colors, text size, and other elements must be appropriate for projection on a screen and for viewing at a distance by an audience. For example, a smaller font size for body type and titles might be readable in a paper report or in a web page, but project the presentation to an audience and you need to be sure that the folks in the back of the room can read the text.

TIP Print your drawing, place it on the floor, and stand up over it. If you can read all the text comfortably from your height, your audience will probably be able to read all the text comfortably during the presentation.

When a Visio diagram is destined for a presentation, it's a good idea to use a font size at least two to four points larger than you typically use for reports or web pages. In addition, pay attention to the contrast of letters and lines against the background. What looks snazzy on a web site might look washed out or busy in a presentation. The most readable colors use high contrast between foreground and background elements, such as dark blue text and lines on a white background, or white text and lines on a black background.

CROSS-REF If you're using styles in your drawing, you might be able to adjust the drawing by simply changing the text, line, and fill styles. This way, you don't have to adjust each element individually. For more information about working with styles, see Chapters 7 and 35.

Building Drawings in a Presentation

If you use PowerPoint to present information to audiences, you already know that displaying all the bullet points at once is a sure way to lose your audience's attention until they've read the entire slide. PowerPoint includes features to feed pieces of slide in digestible portions, which work equally well for bullet points and parts of Visio diagrams. By displaying steps within a flowchart individually, you break the process down into manageable chunks and keep the audience focused on the current topic.

In PowerPoint, you can use animation to build a Visio diagram within a single slide or you can build a drawing incrementally across multiple slides. Either way, the elements of the drawing appear in sequence as you click the mouse.

Animating a Drawing on One PowerPoint Slide

Animation on a single slide means you can control how each element enters the slide. To build multiple elements in a drawing using animation on a single PowerPoint slide, follow these steps:

1. In Visio, select the part of the drawing that makes up the first element you want on the PowerPoint slide. Either drag across the shapes or Shift-click each shape.

 NOTE Connectors might bend in wrong directions after you separate and paste them in PowerPoint. In Visio, choose View ➪ Toolbars ➪ Layout & Routing. Shift-click the connectors in your drawing and then, on the Layout & Routing toolbar, click the Never Reroute tool.

2. Choose Edit ➪ Copy or press Ctrl+C.

3. Switch to PowerPoint and add or show the slide where you want to build the elements from the Visio drawing.

TIP The Blank or Title Only slide layouts are ideal for this purpose.

4. Choose Edit ➪ Paste or press Ctrl+V.

5. Repeat steps 1 through 4 for each succeeding element you want to build on the slide. After pasting the next element, drag the grouping into the position you want on the slide.

6. When all elements are showing in the PowerPoint slide, select the Animations tab.

7. Select the element you want to appear first in the slide show and then, in the ribbon, choose Custom Animation.

8. In the Custom Animation Task Pane, choose Add Effect ➪ Entrance, and then choose the entrance effect you want for the first element.

9. Under Modify, specify how you want the effect to behave. For example, you can specify what action triggers the effect, the direction, and the speed.

10. For each succeeding element in the order you want to introduce them in the animation, repeat steps 7 through 9. You can choose the same or different effects for each element.

11. When you're finished, at the bottom of the Custom Animation Task Pane, click Play to show the animation in the current window. You can also click Slide Show in the Custom Animation pane to show your animation in the full Slide Show screen. When you're finished, click a final time or press Esc to return to the normal window.

 To remove an animation effect, right-click it in the Custom Animation Task Pane and then choose Remove.

Building Sequences of PowerPoint Slides

If you don't care whether shapes come flying from all directions or bubble up onto the screen, it's easier to build the elements of a drawing over several sequential PowerPoint slides. To build a sequence of slides, follow these steps:

1. In PowerPoint, select the Insert tab and, in the ribbon, click Add Slide. Choose the slide layout and PowerPoint inserts a new slide. The Blank, Title Only, or Title and Content slide layouts are ideal for this purpose.

2. Repeat step 1 for each element you want to add until the Visio drawing is displayed in its entirety in PowerPoint.

3. In Visio, select the part of the drawing that makes up the first element you want on the PowerPoint slide. Either drag across the shapes or click each shape while pressing Ctrl or Shift, and then choose Edit ⇨ Copy.

4. Switch to PowerPoint, show the first slide of the sequence you're building, and then choose Edit ⇨ Paste.

5. Switch back to Visio again, and, with the first element still selected, Shift-click the second element and choose Edit ⇨ Copy.

6. To paste the elements for the second slide, switch to PowerPoint, show the second slide of the sequence, and choose Edit ⇨ Paste.

7. Repeat steps 5 and 6 for each additional element you're adding to the sequence.

 Be sure to paste the element in the same location on each slide. Zooming into the slide and using the ruler can help position the element precisely so that the elements transition smoothly from one slide to the next.

8. When you've added all Visio drawing elements to their PowerPoint slides, choose Slide Show ⇨ View Show and click through the slides to preview the construction of the sequence.

9. After the last slide, click or press Esc to return to the normal PowerPoint view. Make any necessary adjustments to the drawing sequence.

To add a transition effect to the sequence of building your Visio drawing, follow these steps:

1. In PowerPoint, in the Slides pane, Shift-click to select all the slides in the drawing sequence for which you want the same transition effect.

NOTE If the Slides pane is not showing, choose View ⇨ Normal (Restore Panes) and make sure the Slides tab is selected.

2. Select the Animations tab, and, in the ribbon, click Transition Scheme.

3. In the Transition Picker that appears, click the effect you want when the slides move from one to the next.

4. In the ribbon, specify how you want the transition effect to behave. For example, you can specify how fast the effect should occur and whether any sound effect should be associated with the transition.

5. Under Advance Slide, specify the action that moves from one slide to the next.

6. In the ribbon, click Preview to show the transition effect for the current slide.

7. When you're finished, you can view the slide show in its full glory by selecting the Slide Show tab and then, in the ribbon, choosing From Beginning.

Summary

You can link information between Visio and other applications in a variety of ways. Hyperlinks between Visio drawings and files in other applications or between pages in a drawing file make it easy to navigate to the information you need. You can create dynamic links between a Visio drawing and another document, so that edits instantaneously update both source and target. You can also embed an independent copy of information from another application.

Chapter 9

Importing, Exporting, and Publishing to the Web

V isio drawings are akin to team players—they generally work in conjunction with documents in other applications or include information from other applications to convey the full story. For example, Visio diagrams often find their way into reports written in Word, presentations built in PowerPoint, or web sites. Conversely, Visio drawings are equally likely to include images such as CAD drawings, or files such as notes in Word or numerical analysis in Excel.

Depending on your goal and the capabilities of the other applications, you can embed, link, import, or export information between Visio and other Office applications such as Excel, Word, and PowerPoint, as well as other applications such as databases or CAD programs. Visio drawings can reach even broader audiences through publishing Visio drawings to the Web or by exporting them to Adobe PDF format or XPS (XML Paper Specification).

Linking and embedding documents (see Chapter 8) provides a great deal of control over document appearance and the changes you can make, but sometimes, linking and embedding isn't enough. You can export Visio data, perhaps information from a prototype you built in Visio, to store it in Excel spreadsheets, AutoCAD drawings, or ODBC-compliant databases. Similarly, exporting is the solution when you want to incorporate Visio data in documents that don't use OLE or to include drawings in HTML files for publication to the Web. Conversely, importing into Visio usually comes in the form of inserting images into Visio drawings, but you can also import Outlook appointments into Visio calendars. In this chapter, you learn how to accomplish all these tasks.

CROSS-REF If you want Visio shapes to display data from other sources, read Chapter 10 to learn about Visio data features.

IN THIS CHAPTER

File formats that Visio imports and exports

Importing files and images into Visio

Importing Outlook appointments into a Visio calendar

Exporting drawings and shapes to other formats

Saving Visio drawings to PDF and XPS format

Choosing the right output format for Visio web pages

Saving drawings as web page files

Embedding Visio drawings into web pages

Making web page navigation easier

File Formats That Visio Imports and Exports

Visio supports numerous graphics formats for importing and exporting images and data. Visio inserts the files directly or as metafiles, depending on the type of graphic file, as shown in Table 9-1. For example, bitmaps inserted into Visio drawing come in as individual pictures, which you can resize, reposition, or crop. However, with inserted metafiles, you can not only resize, reposition, and crop, but also convert the metafiles to Visio shapes by selecting the metafile and choosing Shape ➪ Grouping ➪ Ungroup. Visio 2007 supports exporting to PDF and XPS when you install the Publish as PDF or XPS add-on. In addition to graphics formats, Visio can export data to XML files, Excel spreadsheets, and ODBC-compliant databases, such as Access.

TABLE 9-1

Graphic Formats That Visio Imports and Exports

Supported Formats	Type of File Visio Inserts
AutoCAD Drawing (.dwg)	Vector-based metafile
AutoCAD Interchange (.dxf)	Vector-based metafile
Compressed Enhanced Metafile (.emz)	Vector-based metafile
Enhanced Metafile (.emf)	Vector-based metafile
Graphics Interchange Format (.gif)	Bitmap
Joint Photographic Experts Group (JPEG) File Interchange Format (.jpg)	Bitmap
Portable Network Graphics (.png)	Bitmap
Scalable Vector Graphics Drawing (.svg)	Vector-based metafile
Scalable Vector Graphics Drawing – Compressed (.svgz)	Vector-based metafile
Tag Image File Format (.tif)	Bitmap
Web Page (.htm, .html)	
Windows Bitmap (.bmp, .dib)	Bitmap
Windows Metafile (.wmf)	Vector-based metafile

CROSS-REF   To learn more about importing and exporting CAD drawings in Visio, see Chapter 26.

NOTE   Beginning in Visio 2003, the following formats are no longer supported: Adobe Illustrator (.ai), ABC Flow Charter 2.0, 3.0, and 4.0 (.af2, .af3), CorelDRAW! 3.0 through 7.0 (.cdr), CorelFLOW (.cfl), Computer Graphics Metafile (.cgm), Corel Clipart Format (.CMX), Bentley Microstation Drawing (.dgn), MicroGrafx Designer 3.1 and 6.0 (.drw, .dsf), Encapsulated Postscript (.eps), Interchange Graphics Exchange Standard (.igs), ZSoft PC Paintbrush (.pcx), Macintosh PIST (.pct), PostScript (.ps).

Using Template Tools to Import and Export

Some of the more specialized Visio templates include tools for importing data to create new Visio drawings or for exporting Visio data by using database connections. For example, the following templates offer tools for importing and exporting data:

- Project Schedules, including Gantt Chart, Timeline, and Calendar (see Chapter 17)
- Building Plans, including Space Plan and other types of floor plans (se Chapter 26)
- Brainstorming Diagram (see Chapter 18)
- Organization Chart (see Chapter 14)
- Database Model Diagram (see Chapter 20)
- UML Model Diagram (see Chapter 22)

Although Visio supports the SVG graphics format, there are some limitations, as summarized in Table 9-2.

TABLE 9-2

SVG Format Features and Limitations

SVG Support	SVG Limitations
You can open and insert both uncompressed (SVG) and compressed (SVGZ) files, as well as save Visio drawings in both formats.	Visio does not support scripting, animation, sound, XSLT style sheets, CSS cascading rules, masking and compositing, and metadata.
When you open an SVG or SVGZ file, Visio translates SVG symbols into Visio masters and transforms SVG uses and paths into Visio shapes.	SVG parameterized linear gradients map to one of four predefined linear gradients in Visio.
Inserting SVG files as pictures inserts the SVG drawing into a Visio group.	Visio maps SVG formatting elements such as fill and line patterns, fonts, and markers to the closest corresponding Visio formats or creates custom elements.
Saving a drawing with multiple pages as a web page in SVG format creates a separate SVG file for each page and stores the files in the folder for the web page.	SVG filter effects are only supported for raster image elements and are limited to brightness, contrast, gamma, transparency, and blur.

Importing Data into Visio

For applications that support OLE, linking or embedding files in a Visio drawing (see Chapter 8) means you can both see and edit the file from within Visio. In many situations, editing isn't important and, in fact, might be something you don't want others to do. Inserting a picture of documents or data from other sources into Visio drawings is the perfect solution. What's more, a picture also displays documents from applications that don't support OLE. This section explains how to import graphic files into Visio and also describes one specialized import method — importing appointments from Outlook into a calendar you create in Visio.

Importing Graphic Files into Visio

Importing graphic files saved in other formats can be as easy as opening them in Visio. Choose File ⇨ Open, in the Files of Type box, choose the file format you want to import, and double-click the file you want to import. Visio creates a new drawing file and inserts the graphic file in its native format on the drawing page — complete with selection and rotation handles like built-in Visio shapes. Opening a graphic file is a better approach for importing vector-based graphic files (such as AutoCAD DWG or SVG formats), because the quality of the image is as high as possible.

To import a graphics file onto an existing drawing page, choose Insert ⇨ Picture.

 **If the file you want to import isn't in a format that Visio supports, use a graphics program to convert it to a supported format and then import the file.**

Cropping Imported Graphics

After you import a picture, you can crop the image to modify how much of the picture is visible on the drawing. For example, suppose you insert a full screen shot, but want to show only the dialog box in the middle. To crop a picture, do the following:

1. Display the Picture toolbar by choosing View ⇨ Toolbars ⇨ Picture.

2. On the Picture toolbar, click the Crop tool.

3. Drag selection handles on the picture to change the size of the border on the drawing page and hide some of the picture. For example, to show only a dialog box in the middle of a screen shot, you can drag the selection handle in the middle of each side of the picture to the corresponding borders of the dialog box, as illustrated in Figure 9-1.

4. If you want to display a different area of the picture, position the pointer over the picture and then drag the hand symbol until the area you want to display is in view.

5. When you finish cropping the picture, turn off the Crop tool by selecting another tool such as the Pointer tool.

Cropping a picture does not resize it. The cropped area is a border that reduces the amount of the picture that you see, but the entire picture is still there. If you want to delete the areas outside the cropped border to reduce the size of the image, choose Format ⇨ Picture, select the Compression tab, check the Delete Cropped Areas of Pictures check box, and then click OK.

FIGURE 9-1

Dragging selection handles when the Crop tool is selected hides a portion of the picture.

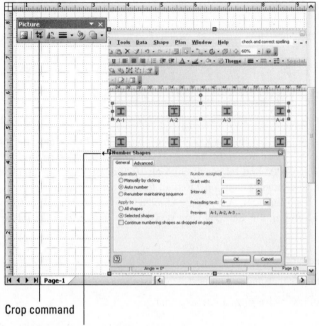

Crop command

Drag selection handle to hide part of picture

Formatting Imported Graphics

After you insert a picture into a Visio drawing, you can tweak its appearance by formatting its graphic characteristics, for example, brightening a dark picture or heightening the contract. To format a picture, do the following:

1. Select the picture on the drawing page and choose Format ➪ Picture.

2. In the Format Picture dialog box, modify one or more settings to adjust the image. Choose from the following controls:

   ▪ **Brightness** — Lightens or darkens all the colors in the entire picture.

   ▪ **Contrast** — Adjusts the contrast between the lightest and darkest areas in the picture. The higher the contract is, the more difference there is between colors.

   ▪ **Gamma** — Adjusts the brightness of gray tones in a picture, which tends to emphasize detail in shadows.

   ▪ **AutoBalance** — This control adjusts the brightness, contrast, and gamma based on the contents of the selected picture.

■ **Transparency** — 100 percent makes the picture completely transparent, whereas 0 percent makes the picture opaque so nothing in the background shows through.

■ **Blur** — Decreases the sharpness of edges and boundaries of areas.

■ **Sharpen** — Increases the sharpness of edges.

■ **Denoise** — Removes noise (stray speckles), which tend to appear in scanned images.

3. Click OK.

Importing Outlook Appointments to a Visio Calendar

Microsoft Outlook is a fabulous tool for keeping track of your appointments, but publishing and distributing your calendar is another story. You don't have much control over which parts of your calendar Outlook prints. With Visio 2003 and Microsoft Outlook 2000 or later installed, Outlook appointments import into a Visio calendar, which you can then format, print, share, or publish. When appointments reside in a Visio calendar, you can use familiar Visio calendar tools to format and customize the calendar. With the Import Outlook Data Wizard, you can perform the following actions:

■ Specify a date and time range

■ Include only appointments that match a subject

■ Create a one-week, multiweek, or one-month calendar

NOTE You can create Visio calendars based only on the Gregorian calendar. If your Microsoft Outlook calendar format is set to Arabic, Hebrew, Chinese, Japanese, Korean, or Thai, you can change your Outlook calendar format to Gregorian, use the wizard to import your appointments into Visio, and then change your Outlook calendar format back.

To create a Visio calendar from Outlook appointments, follow these steps:

1. Choose File ➪ New ➪ Schedule ➪ Calendar.

2. Choose Calendar ➪ Import Outlook Data Wizard.

3. In the Import Outlook Data Wizard dialog box, select the option to specify whether you want to create a new calendar or add appointments to the selected calendar and then click Next.

NOTE The Selected Calendar option is not available if you do not select a calendar before you start the wizard.

4. Specify the start and end dates and the start and end times that you want to scan for appointments.

5. To limit the appointments added to the calendar, click Filter. In the Filter Outlook Data dialog box, check the Subject Contains check box. Type the subject text for the appointments you want to import, click OK to close the dialog box, and then click Next.

6. If you chose to create a new calendar, specify the calendar type you want to create. You can specify a week, month, or multiple-week calendar, and select the day that begins the week. If you want to shade weekends, select the Yes option. If you want to display dates in a specific format, choose the language for that format in the Language list and click Next.

7. On the last wizard screen, review the calendar properties that you selected. If you want to change any properties, click Back or Cancel. To create the calendar, click Finish.

 If appointments overlap, you can resize the calendar or delete some appointments.

CROSS-REF Several people can combine their schedules onto one Visio calendar. To learn how to do this, see Chapter 17.

Exporting Shapes and Drawings

There are plenty of reasons you might want to export shapes or drawings as graphics files, for example, including a Visio diagram in a report, using a Visio diagram on a web page, or creating an instruction manual that shows how to use custom Visio shapes in a template you developed. Visio exports shapes and drawings to several different graphics formats, as well as to HTML. The format you choose depends on what you plan to do with the graphic file.

When you export shapes or drawings by choosing File ➪ Save As, an Output Options dialog box might appear (depending on the export format you select) in which you can specify the settings for the exported file. The settings vary depending on the file format to which you want to export your drawing. To find out more about the different export settings, in the Output Options dialog box, click the Help button.

Because files go through transformations at every export and subsequent import into another application, exporting to other formats might alter the appearance of your Visio drawings. If you find that a special fill pattern or other format causes problems, the best solution is simply to apply a different format and then try exporting your drawing again.

CAUTION Visio gradient fill patterns might not transfer accurately when you save them in a non-Visio graphics format. If gradient fills don't look the way you want, replace them with plain fill patterns and then export your Visio drawing.

Exporting Visio shapes and drawings isn't much different from saving Visio drawings with another file name. To export Visio shapes or drawings, follow these steps:

1. Display the drawing page you want to export in Visio. If you want to export specific shapes on a page, select them.

2. Choose File ➪ Save As and, in the Save As Type drop-down list, select the export format you want.

3. In the File Name box, type the name you want.

4. Click Save.

5. Depending on the type of format you chose to export in step 2, an Output Options or Filter Setup dialog box appears, as shown in Figure 9-2. If it does, specify the options you want, such as compression to reduce file size or the number of bits for color, and click OK. Visio exports the page and selected shapes to a file with the format you chose.

Different types of graphics files offer output options or filters for fine-tuning the exported image.

> **TIP**   If you export Visio shapes but they don't appear in the exported file, the shapes might be metafiles, such as Visio Network Equipment shapes or some objects linked or embedded in a Visio drawing. To export metafiles, apply the Ungroup command until Visio converts all the components of the metafile to shapes, and then export the drawing.

You can specify export options for the following formats:

- **TIFF** — Data compression, color format, background color, color reduction, transformation, resolution, and size

- **JPEG** — Baseline or progressive, color format, background color, quality, transformation, resolution, and size

- **GIF** — Data format, background and transparency color, color reduction, transformation, resolution, and size

- **PNG** — Same options as GIF

Saving Visio Drawings for Distribution

In Office 2007 programs including Visio 2007, you can save any file from your Microsoft application to either Adobe PDF or XPS (XML Paper Specification) format. These formats offer the best of both worlds when you want to distribute documents to a broad audience. The files can't be edited and they appear with exactly the formatting you originally applied.

To save a file to PDF or XPS format, you must first install the Save as PDF or XPS add-on. Navigate to www.microsoft.com/downloads and search for "PDF or XPS". Follow the instructions on the web page for downloading and installing the add-on.

After the add-on is installed, to save a Visio drawing as a PDF- or XPS-formatted file, do the following:

1. Choose File ➪ Publish as PDF or XPS.

2. In the Publish as PDF or XPS dialog box, navigate to the folder in which you want to store the file. In the File Name box, type the name of the file.

3. In the Save As Type drop-down list, choose either PDF or XPS Document.

4. By default, the Optimize For option selected is Standard, which creates a document suitable for both online and print publishing. To create a smaller file for online publishing, select the Minimum Size option.

5. To create the file, click Publish.

Publishing to the Web

When you want to share information with a large or widely distributed audience, the Web is frequently your first choice. You can publish Visio drawings to the Web by saving Visio drawings as their own web pages, as graphics files that you can embed in web pages, or as Visio XML files that you can open in a web browser. Visio drawings saved as web pages enable anyone with a more recent web browser to view your diagrams, including shape data. Visio takes care of almost every aspect of creating the web pages — you choose the Save As Web Page command and, in some cases, select a few publishing options. This section explains the different ways to publish Visio drawings to the Web, the benefits of each approach, and how to make the completed web page easier to navigate.

Choosing an Output Format for a Visio Web Page

The best format for a Visio drawing published to the Web depends on what you're trying to accomplish with web publication as well as the browsers your audience uses. The first step to choosing an output format is to decide what you want to publish:

- Saving Visio drawings as web pages is effective when you want to:
  - Maintain navigational links between Visio shapes when you publish them to a web page
  - Include reports that users can view easily
  - Export several pages of a multiple-page drawing at once

- Saving drawings in the JPG, GIF, PNG, or SVG format is preferable when you want to:
    - Insert a Visio drawing into an existing HTML web page
    - Publish only a portion of a drawing

The second aspect of choosing output formats is picking one that works well with the browsers people use to view your web pages. Although it's often difficult to determine which browsers your audience is likely to use, Table 9-3 shows which output format works best with different types of browsers.

TABLE 9-3

Browser and Output Format Compatibility

Output Format	Earliest Browser Supported	Internet Explorer 5.0 or Later Behavior	Other Browser and Version Behavior
VML (Vector Markup Language)	Any HTML 2.0–compliant browser with alternate browser support that supports frames	Drawing is displayed in VML and scalable with browser window. Left frame includes Go To Page, Pan and Zoom, Details, and Search Pages.	Drawing is displayed in JPEG, GIF, or PNG, but is not scalable. Left frame shows pages as links.
VML only	Internet Explorer 5.0	Drawing is displayed in VML and is scalable with the browser window. Left frame includes Go To Page, Pan and Zoom, Details, and Search Pages.	Not supported.
SVG with alternate browser support	SVG Viewer and any HTML 2.0–compliant browser that supports frames	Drawing is displayed in SVG with SVG Viewer or SVG-compatible browser; otherwise, as JPEG, GIF, or PNG. Left frame includes Go To Page, Details, and Search Pages.	Drawing is displayed in SVG with SVG Viewer or SVG-compatible browser. Left frame shows pages as links.

Output Format	Earliest Browser Supported	Internet Explorer 5.0 or Later Behavior	Other Browser and Version Behavior
SVG only	SVG Viewer, Internet Explorer 5, Mozilla Firefox 1.0	Drawing is displayed in SVG with SVG Viewer or SVG-compatible browser; otherwise, as JPEG, GIF, or PNG. Left frame includes Go To Page, Details, and Search Pages.	Drawing is displayed in SVG with SVG Viewer or SVG-compatible browser. Left frame shows pages as links
JPEG or GIF	Internet Explorer 3.0, Netscape Navigator 3.0, Mozilla Firefox 1.0, and any HTML 2.0–compliant browser that supports frames	Drawing is displayed in JPEG or GIF and is not scalable. Left frame includes Go To Page, Details, and Search Pages.	Drawing is displayed in JPEG or GIF and is not scalable. Left frame shows pages as links.
PNG	Internet Explorer 4.0 or later, Netscape Navigator 6.0 or later, Mozilla 1.0	Drawing is displayed in PNG and is not scalable. Left frame includes Go To Page, Details, and Search Pages.	Drawing is displayed in PNG and is not scalable. Left frame shows pages as links.

NOTE   VML (Vector Markup Language) represents graphics as vectors instead of pixels, so they look good no matter how far you zoom in. Vectors also speed up web navigation because they download faster than images made up of pixels.

Saving Drawings As Web Pages

Whether Visio drawing files contain one page or several, it's easy to transform them into web pages, despite the plethora of files it takes to represent each web page. When you use Microsoft Internet Explorer 5.0 or later and save Visio drawings as web pages, you can navigate around the drawing, zoom in and out, display shape data for shapes, search the drawing for shapes, or view reports associated with the drawing, as illustrated in Figure 9-3. When you save a Visio drawing as a web page, by default Visio generates HTML source code and creates HTML frame pages with frames for drawing pages and controls.

FIGURE 9-3

Visio includes controls and other information in the left frame of a web page.

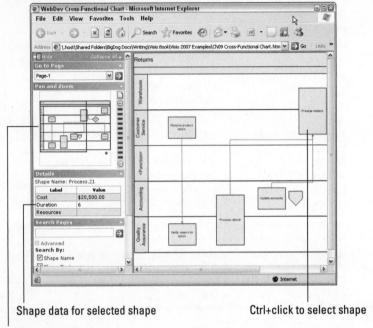

Shape data for selected shape   Ctrl+click to select shape

Drag to zoom into page

NOTE   In order for colors to appear properly when you display a Visio drawing as a web page, your system must have more than 256 colors. Even then, colors don't always match exactly when you display a Visio drawing on the Web.

To save a Visio drawing as a web page, follow these steps:

1. Open the Visio drawing that you want to publish to the Web and then choose File ⇨ Save As Web Page.

2. In the Save As dialog box, navigate to the folder in which you want to save the file. In the File Name text box, type the name for the web page file.

3. To specify the title you want to appear in the browser title bar, click Change Title, and then, in the Set Page Title dialog box, type the title you want. Click OK. The title you typed appears below the File Name and Save As Type boxes.

4. To specify the web page publishing options (described in detail in the next section), click Publish and choose the options you want.

5. To publish the Visio drawing to the Web, choose one of the following methods:

   - If you clicked Publish to specify web publication options, click OK.
   - If you didn't specify publishing options, in the Save As dialog box, click Save.

6. Visio creates the drawings as HTML files and opens the first page in Internet Explorer.

Specifying Web Page Publishing Options

When you save a Visio drawing as a web page, Web Page Publishing options control which pages are published, which controls appear in the left frame of the web page, and the output format and additional display options, such as the target resolution. To specify web page publishing options, in the Save As Web Page dialog box, click Publish.

The General tab contains options for configuring the following features in published web pages:

- **Specify pages to publish** — You can publish one page in a drawing file, all pages, or a range of consecutive pages. To publish all pages, select the All option. To publish one or more, select the Pages: From To option and type the page numbers for the first and last pages in the range.

- **Display search, navigation, and data controls** — In the Publishing Options list, select the controls that you want to appear in the left frame of the web page:

  - **Details** — Displays values for shape data for the selected shape. To view shape data in the left frame, Ctrl-click a shape on the web page or click the shape and then press Ctrl+Enter.

  - **Go to Page** — Displays a drop-down list of the pages, which you choose to navigate between pages.

  - **Search Pages** — Includes check boxes to specify what to search for on the pages.

  - **Pan and Zoom** — You can drag in the Pan and Zoom window to change the area that appears in the right frame of the web page or click to re-center the page. Pan and zoom controls are available only when you use the VML output format in Internet Explorer 5.0 or later.

  - **Report** — Select each report you want to publish in the Publishing Options list. Reports appear on separate pages that you can access by using the Go to Page control in the left frame of the web page. If no shapes in your drawing match the report query, Visio creates the web page without the report.

 If you don't want a left frame in your browser window and don't need Go to Page to navigate between web pages, uncheck Go to Page, Search Pages, Details, and Pan and Zoom.

- **Open web page in browser** — To open the web page in your browser immediately after you save it, under Additional Options, select Automatically Open Web Page in Browser.

- **Organize supporting files in a folder** — This option creates a subfolder for storing the supporting files for the web page. When you move your web page to another folder, the supporting files folder moves with it automatically.

CAUTION Visio looks for the supporting files for a web page based on the root file's name. If you rename a root HTML file, Visio displays a warning that renaming the file will break the links to the supporting file folder. Instead of renaming a root file, open the published diagram in Visio, save it with the new name (which also creates a new subfolder), and delete the old _files folder.

The Advanced tab includes options for the output formats, resolution, and other display options:

- **Specify output format** — In the Output Formats drop-down list, choose the format you want. VML produces the best results for displaying controls in the left frame of the web page. VML and SVG support scalable graphics, so the web page output resizes if the browser window resizes.

NOTE You must install an SVG viewer to view web pages in SVG format. If you open a web page in SVG format without an SVG viewer, the page appears in a format such as GIF.

- **Specify an alternate format for older browsers** — Because VML and SVG require more recent browser versions, it's a good idea to specify an alternate format so that the web page still opens in older browsers. Select the Provide Alternate Format for Older Browsers check box and select the format you want, such as JPG or GIF.

- **Specify the target resolution** — If you are creating a file in JPG, GIF, or PNG format, specify the smallest resolution that you expect people to use to view your web pages.

NOTE You don't have to specify the Target Monitor resolution for scalable formats such as VML and SVG.

- **Embed a saved Visio web page in another web page** — In the Host In Web Page drop-down list, choose the web page in which you want to embed the saved web page, as described in the next section.

- **Specify color scheme** — If you want to apply the color scheme for your Visio drawing to the resulting web page, in the Style Sheet drop-down list, select the color scheme. This color scheme applies colors to the left frame and report pages that match the color scheme for the Visio drawing.

Embedding Visio Drawings in Web Pages

If your organization already has a template for its web pages, chances are you want to embed Visio web pages within your company's template to maintain the look and feel across all the pages in the web site. To embed a Visio web page in another HTML page, you add an <IFRAME> tag to the host HTML page where you want the Visio drawing to appear, and specify the host page in the Save as Web Page dialog box. Visio provides a sample template, Basic.htm, in the Host in Web Page list if you need an example of how to use the <IFRAME> tag.

NOTE Of course, you can also publish a Visio drawing as a graphic image, if you don't need to show shape data or provide navigation tools.

Files That Visio Creates for Web Pages

Web pages are made up of a lot of files, which Visio creates for you and stores in a convenient location. Most of the time, you don't need to give these files a second thought. However, if you publish Visio drawings as web pages often, it's a good idea to understand what files Visio creates and where they're located, so you can modify or delete them if need be.

When you save a drawing as a web page, Visio creates a root HTML file for your Visio drawing in the folder that you specify. In addition, Visio creates a subfolder in the same location using the same name as the root HTML file but with _files appended to the end. Visio stores the files required for the web page in the subfolder. Because of this naming convention, the subfolder moves with its root HTML file; if you delete the HTML file, the subfolder is deleted as well. However, if you move HTML files, you might have to edit them to update pointers to graphics files.

For each Visio drawing, Visio creates the following files:

- A root HTML file with the name you specified, such as Flowchart05.htm

- Graphics files of the output format you specified for each published page in the Visio drawing, such as vml_1.htm, and the alternate format, such as gif_1.gif, if you chose to include an alternate format for older browsers.

- Other files that support the publishing options you selected when you published the web page, such as graphics (.gif) for controls, style sheet (.css), script (.js), and data files (.xml)

To use an existing HTML template to display a Visio web page, follow these steps:

1. Edit your HTML template to include the following HTML tag:

   ```
   <IFRAME src="##VIS_SAW_FILE##">
   ```

NOTE This tag embeds the Visio web page into your HTML template. It is case-sensitive and refers to the HTML output file that Visio creates when you create a Visio web page. The `<IFRAME>` tag for a Visio web page contains the information about the drawing. If you forget to include this tag, you'll see the HTML code instead of the drawing.

2. In Visio, choose File ⇨ Save As Web Page and, in the Save As Web Page dialog box, specify the folder and name for the web page.

3. Click Publish and select the Advanced tab.

4. Under Host in Web Page, browse to the HTML template file that you want to use as the host page and click OK.

NOTE Visio stores the Basic.htm template in the Visio path, usually C:\Program Files\Microsoft Office\Office12\1033\. Your customized HTML templates appear in the Host in Web Page drop-down list if you store copies of them in this same folder.

Other Ways to Reference Visio Drawings in HTML

Links to reference images of Visio drawings or Visio web pages within other web pages are another way to view Visio drawings on the Web, if you don't need special features such as shape data. To include a link to your Visio web page, you can add a tag to your HTML template, such as the following:

```
<a href="##VIS_SAW_FILE##">My Drawing</a>
```

To reference a graphics file in a web page, save your Visio drawing in the JPEG, GIF, or PNG format. Then add an `<IMG>` tag to your HTML code, such as `<IMG SRC="process_flow1.gif" >`. Visio drawings saved as graphics do not include shape data or navigation controls, such as hyperlinks.

To keep links intact, you can save a Visio drawing in PDF format and include that file in the HTML file.

Adding Hyperlink Navigation Shapes to Drawings

Navigating between shapes and pages in Visio is easy when you add hyperlinks to shapes on your drawings. When you create a web page from a Visio drawing, Visio saves these hyperlinks in shapes, so people viewing the drawing on the Web can navigate to the information they want as though they were viewing the drawing in Visio. However, no indication of a hyperlink appears until you position the pointer over a shape with a hyperlink. To make navigation options more obvious, you can add hyperlink navigation shapes to your drawing before you save it as a web page. People viewing the drawing can navigate simply by clicking the navigation shapes on the web page.

CROSS-REF To learn about creating hyperlinks in your Visio drawings, see Chapter 8.

Although you can add hyperlinks to any shape you want, the Borders and Titles stencil includes several built-in navigation shapes, Hyperlink Button, Hyperlink Circle 1, and Hyperlink Circle 2. These shapes automatically open the Hyperlinks dialog box when you drag them onto a page. After you add one of these hyperlink shapes to a drawing, you can change the icon that appears for the shape to indicate the function of the associated hyperlink. Table 9-4 includes the names of the icons you can use.

TABLE 9-4

Hyperlink Icons

Back	Forward
Up	Down
Home	Help
Directory	Info
Search	Mail
Photo	None

To add a navigation shape to a drawing, follow these steps:

1. Drag one of the hyperlink shapes from the Borders and Titles stencil onto a drawing page.

2. In the Hyperlinks dialog box, specify the address or sub-address for the hyperlink and click OK.

3. To change the icon in the shape, right-click the hyperlink shape and choose Change Icon from the shortcut menu. Select the icon you want in the Icon Type drop-down list and click OK.

Summary

Visio can import and export to a variety of formats, from AutoCAD drawings and ODBC-compliant databases to graphic files and web pages. Many of the built-in Visio templates include tools that simplify importing and exporting data to their specific types of drawings. However, to import data from another source into Visio without a specialized tool, you select the type of file in the Open dialog box and choose the file you want. To export data to another format, you use File ➪ Save As, selecting the type of format you want in the File of Type drop-down list. When you want to publish Visio drawings to the Web, the Save As Web Page command provides numerous options for specifying the contents and functionality of the resulting web page.

Summary

Chapter 10

Linking Shapes with Data

There are plenty of Visio drawings that are nothing more than a pretty face — like the attractively formatted diagram that you send to customers to guide them through the maze of your phone menu system. The shapes and connectors say all there is to say.

However, Visio drawings can put brains behind those good looks when you link shapes on Visio drawings with data. Combining shapes with data can be as simple as adding values manually to shape data, for example, when you build a small organization chart by typing employee names and other information in the organization chart shapes. Visio can also link shapes directly with data sources, so your drawings can act as visual and dynamic representations of data stored in data sources. For example, dynamic links between a Visio drawing of rentable office space and a database of facilities management data could generate a visual status report of occupancy and a text report of rental income.

Perhaps the most significant features in Visio 2007 Professional are related to data. Data Link is a new feature that greatly simplifies the process of connecting data sources and the data within them to shapes. When you have data linked to shapes, the new Data Graphics feature helps you present data graphically to highlight status, trends, or problems — without the customizations that earlier versions of Visio would have required.

This chapter describes the data features in Visio 2007. First, you'll learn the basics — how to add and view data in shape data. Then, you'll learn how to use the new Data Link feature to link shapes to data and refresh values when you want. The chapter continues with Data Graphics. You'll learn about the different types of Data Graphics that Visio 2007 Professional offers and how to set up shapes to use them. Finally, the chapter concludes by describing how to run reports and create customized reports.

An Overview of Data Features in Visio 2007

One of Visio's most powerful features is its ability to present information visually. Although Visio has done that in previous versions, Visio 2007 Professional greatly simplifies many data-related tasks. Data Link has streamlined connecting drawings to data sources; the Data Selector wizard provides a visual interface for connecting shapes and data; and Data Graphics display data as professional-looking graphics without hours of tedious customizing of shapes. This section summarizes the features in Visio 2007 for working with data.

> **NOTE** Both Visio Standard and Visio Professional include shape data and built-in reports. However, the more robust data-related features, including Data Link, Data Graphics, and the database wizards from previous Visio versions, are available only in Visio Professional.

Shape Data

One of Visio's basic concepts is that shapes can have properties that contain textual data, such as dimensions, cost, equipment type, or serial number. In Visio 2007 (both Standard and Professional), these properties are called shape data, although you probably know them as custom properties from earlier versions of the program.

Many of these properties never appear on a drawing page—they rest quietly in their shapes' ShapeSheets unless you summon them. For example, the Cost, Duration, and Resources properties found on the shapes in the TQM stencil might contribute to a report without ever appearing on a drawing.

Other shape data properties directly affect the appearance of their parent shapes. For example, engineering shapes often have configuration-oriented shape data, such as Width or Outside Diameter, which control the physical dimensions of the shapes you see on the drawing. Change the property and the size of the shape on the page changes, too. Shape data also can appear directly on the drawing page, such as the name and title properties, which show up as text within Organization Chart shapes.

One of the easiest ways to access shape data is within the Shape Data Window (choose View ⇨ Shape Data Window). The window includes two columns, one for shape data property names and the other for values. Viewing and entering data in the Shape Data Window is both easier and safer than opening the ShapeSheet to modify properties.

> **CROSS-REF** See the section "Storing and Viewing Shape," later in this chapter, for instructions on entering data in shape data fields and accessing those fields. To learn how to create new shape data properties and associate them with shapes, see Chapter 34.

Linking Shapes to External Data

Although you can add data directly to shapes and keep the data completely within your Visio drawing file, grabbing data from external sources is often a more efficient way to work. For example, with links between shapes and external data, you can work with a floor plan that automatically shows who's sitting where according to your facilities database. Or, a network diagram might show response times and system usage.

Previous versions of Visio had tools for linking shapes to external data, but the Visio 2007 Professional Data Link feature makes it easier to connect to data sources, link shapes to data, and refresh that data. For programming folks, Data Link includes an API, which provides a programmatic link to XML data and also means developers don't have to create shape data programmatically. Data Link supports more than one data source to the same drawing. For example, you might link to an employee database for the employees who occupy cubicles while also linking to an asset database to link to the equipment records for the computers and furniture in those cubicles.

CROSS-REF See the section "Linking Drawings to External Data," later in this chapter, for instructions on creating the links between drawings and data sources as well as between shapes and data records.

Presenting Data on Drawings

In Visio 2007 as well as earlier versions, add-ons help you display shape data within shapes or color-code shapes by shape data values (see Chapter 28 for the Label Shapes add-on and Color By Values add-on). Visio 2007 Professional has a new feature up its sleeve — Data Graphics display data graphically without all the customizing required in the past.

The built-in Data Graphics shapes can present data as text, but with nice formatting that makes the text stand out. In addition, Data Graphics can represent values as progress bars or icons. For example, instead of displaying the number of megabytes of free space on disk drives, you can have bars that grow shorter as you eat up hard disk space. Similarly, icon sets show different icons depending on the shape data values, so you can simulate a stop light with red, yellow, and green circles to represent dangerous, so-so, and good values.

CROSS-REF See the section "Visualizing Data with Data Graphics," later in this chapter, for instructions on attaching built-in Data Graphics shapes to drawing shapes and customizing your own Data Graphics.

Creating Reports

Visio also includes several built-in text reports, such as door schedules and equipment lists, which assemble data from built-in Visio shapes. You can also customize reports to build your own; although, for very specific reporting requirements, a dedicated reporting program is likely to be easier.

CROSS-REF See the section "Producing Reports" later in this chapter to learn how to run built-in reports or build your own.

Storing and Viewing Shape Data

Whether you manually type values for shape data or transfer data from an external data source through Data Link, shape data properties are the containers that store data (or the connections to external fields). The Shape Data Window is the easiest place to view the shape data for shapes and it doubles as a data entry window if you enter data manually.

 **If you want to keep the Shape Data Window from hiding part of the drawing area, dock the Shape Data Window within the Shapes Window by dragging and dropping it there.**

To view and store shape data, follow these steps:

1. To open the Shape Data Window, choose View ➪ Shape Data Window.

2. On the drawing page, select a shape to display its properties in the Shape Data Window. If the shape doesn't have any associated shape data properties, the Shape Data Window displays only the text "No Shape Data."

TIP **If you want to see shape data only occasionally, you don't have to keep the Shape Data Window open. The shortcut menus for many built-in shapes include the Properties entry, which opens the Properties dialog box. To access all the shape data properties for a shape through the Properties dialog box, right-click a shape and, from the shortcut menu, choose Properties. In the Properties dialog box, enter or edit shape data values and click OK when you're finished.**

3. To enter a property value in the Shape Data Window, click the value cell for the property you want to edit and type the value in the box. If the property includes a drop-down list, you can choose a value from the list.

TIP **If you aren't sure what a shape data property represents or what format you should enter, you can display the prompt for a property by pointing to the property label in the Shape Data Window.**

Linking Drawings to External Data

For all but the smallest data-related tasks, typing shape data values manually is tedious and time-consuming; and manually keeping data up-to-date is even worse. A much more effective approach is linking external data to shapes on drawings. The Visio Data Link feature helps you connect your drawings to data sources as well as link the records in your data sources to the shapes on your drawings.

NOTE **Data Link pushes data from external data sources to the shapes on your drawings, but it doesn't move data in the other direction. If you want a two-way communication channel, you can still use the Database Wizard, which is described in the section "Using the Visio Database Add-ons to Link Drawings and Databases" later in this chapter.**

In Visio 2003 and earlier, defining a link between a Visio drawing and a data source required steps that were often daunting to folks whose job titles didn't include database administrator. You started in Microsoft Windows by creating a data source — picking the correct driver for the data source and defining the data that the data source represented. Then, in Visio, you ran a wizard to connect the records to shapes. Data Link takes a lot of the mystery out of the process of linking a drawing to a data source, no matter which type of data source you use.

Linking a Drawing to a Data Source

The first part of Data Link is the Data Selector wizard, which sets up the connection between a data source and your drawing. To get started, choose Data ➪ Link Data to Shapes. In the Data Selector dialog box, from the following list of data source types, select the option for the type of data you have:

- **Microsoft Office Excel workbook** — An Excel workbook.

- **Microsoft Office Access database** — An Access database.

- **Microsoft Windows SharePoint Services list** — A list on a SharePoint site, such as a list of tasks.

- **Microsoft SQL Server database** — A SQL Server database.

- **Other OLEDB or ODBC data source** — Data stored in an ODBC-compliant database, such as Oracle, or a data source accessed through the Microsoft OLEDB API.

- **Previously created connection** — Choose this option if you want to link a drawing to a data source that already exists, for example, one that you created directly in Windows or, better yet, that your administrator set up for you. For example, Data Source Name connections end in the file extension .dsn.

When you click Next to continue the wizard, the steps you perform vary slightly depending on the type of data source. Here is an overview of the steps you're likely to perform:

1. On the first connection screen, select the data source you want to use and specify additional information about the data source. The following list describes the different steps to perform based on the type of data source:

   - **Excel workbook** — Choose the workbook file and click Next. In the next screen, from the drop-down list, select the worksheet that contains the data. To select a range of cells on the worksheet, click Select Custom Range. By default, the First Row Of Data Contains Column Headings check box is checked, which is almost always what you want. Of course, if your workbook doesn't include headings, clear the check box or the wizard will interpret your first row of data as headings.

   - **Access database** — Choose the Access database file. Then, in the What Table Do You Want To Import drop-down list, choose the table that contains the data and click Next. (If the database includes only one table, the wizard selects it by default.)

   - **Microsoft Windows SharePoint Services list** — In the Site box, type the URL for the SharePoint web page that contains the list you want to link and click Next.

■ **Microsoft SQL Server database** — For a SQL Server database, you must specify the name of the server and the credentials that give you permission to access the database. Windows authentication uses your Windows username and password, but you can also specify a user name and password specific to the database.

■ **Other OLEDB or ODBC data source** — For this option, you must first select the type of data source, such as ODBC DSN or Microsoft Data Access - OLE DB for Oracle. Then, you specify the file and authentication.

■ **Previously created connection** — In the drop-down list, choose the connection. Or, click Browse to navigate to the folder that contains your data connections.

2. On the Connect To Data screen, All Columns and All Rows are selected by default, as shown in Figure 10-1. If you want to link only specific columns and rows, which is wise for data sources with lots of data, click Select Columns. In the Select Columns dialog box, select or clear the column check boxes to include or exclude those columns. For example, if you use a data field as a unique identifier, you don't need to link an automatically incremented key field from a database. Click Select Rows to specify data records to link.

You can proceed with linking all columns and rows or click Select Columns or Select Rows to specify the data you want.

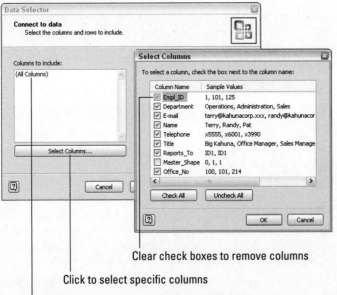

Clear check boxes to remove columns

Click to select specific columns

All columns and all rows are selected by default

3. Click Next to proceed to the Configure Refresh Unique Identifier screen. In this screen, select the check box or boxes for the fields that uniquely identify records, for example, an employee number field or a serial number. Select the Rows In My Data Are Uniquely Identified By The Value(s) In The Following Columns option. With a unique identifier, Visio can update the shapes in the diagram that are linked to records that have changed in the data source.

 **You can choose the other option if no fields uniquely identify records. In that situation, Visio updates the data based on the order of the rows.**

4. Click Next and then click Finish. The External Data window and the Data toolbar appear, which is your cue that the next major task in linking data is at hand. Read the section "Linking Data to Shapes on a Drawing" to learn how to link records to shapes.

Linking a Drawing to More Than One Data Source

Unlike the Database Wizard in earlier versions of Visio, the Data Selector Wizard can connect a drawing to more than one data source. To do this, follow these steps:

1. Right-click anywhere in the External Data Window and, from the shortcut menu, choose Data Source ⇨ Add. The Data Selector Wizard appears.

2. Follow the steps to connect a data source, as described in the section "Linking a Drawing to a Data Source," earlier in this chapter.

3. When you complete the steps, a new tab appears at the bottom of the External Data Window and the records for the new data source appear in the window.

Linking Data to Shapes on a Drawing

With a link between a data source and a drawing file, the data is available to the Visio drawing file, but it isn't connected to shapes. That's where the Data toolbar and the External Data Window come into play. The External Data Window shows the records in the data source that are available to link to shapes. The Data toolbar includes commands for working with the data.

To transfer data from a record to a shape's shape data, you can drop a row from the External Data Window onto the shape on the drawing page. That's easy enough for one shape, but dragging and dropping grows tiresome if the drawing contains hundreds of shapes or the data changes frequently. The Automatic Link Wizard can link shapes with records when they satisfy a condition (such as the employee ID value in both match). This section describes the different ways to link data and shapes.

Link Records Directly to Shapes

If you're working with an existing Visio diagram with only a few shapes, dragging and dropping records is probably the easiest approach. From the External Data Window, drag a record and drop it onto the corresponding shape on the drawing page. Before you drop the record, Visio displays a link icon to indicate that it's about to link the record and the shape. After you drop the record, the link icon appears in the first column of the External Data Window for that record, as illustrated in Figure 10-2, so you can tell which records you have already linked.

FIGURE 10-2

The link icons in the External Data Window indicate which records you've linked.

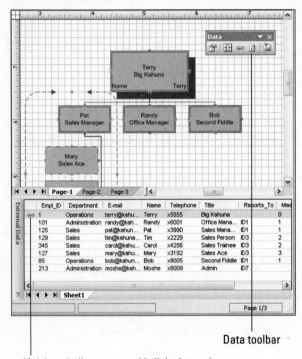

Data toolbar

Link icon indicates record is linked to a shape

Automatically Link Records to Shapes on a Drawing

When the volume of data or the frequency of changes is high, automatically linking shapes and records is a better way to go. The catch is that automatic linking works only when values in shape data match values in the fields in the data source. If you have a shape data property for the field that uniquely identifies records in the data source, you're all set.

To automatically link records to shapes, follow these steps:

1. On the Data menu, click Automatically Link, whose icon looks like a few links of chain. The Automatic Link wizard launches.

2. On the first screen, Visio selects an option based on your shape selection. If you did not select shapes on the drawing before starting the wizard, Visio selects the All Shapes On This Page option. If you selected the shapes you want to link automatically, Visio selects the Selected Shapes option, but you can select the other option as well. Click Next to proceed to matching up columns in the data source with shape data properties.

3. On the next screen, the label is "Automatically Link Row To Shape If," which is a big hint that this is where you define the criteria for matching shapes and records. In the Data Column drop-down list, choose a field in the data source. In the Shape Field drop-down list, choose the corresponding shape data property, as shown in Figure 10-3. If your data source uses more than one field to uniquely identify records, click And to define the next criterion. If you want Visio to replace any existing links (for example, a few you added manually), select the Replace Existing Links check box.

FIGURE 10-3

You must tell the Automatic Link wizard about the criteria that indicate matching records and shapes.

Matching data source field and shape data property

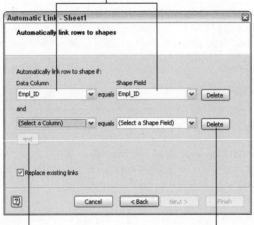

Click to add another criterion   Click to delete a criterion

4. Click Next to proceed to the summary screen. If the actions in the list meet your approval, click Finish to link the records and shapes. A message box with progress appears to show how many links the wizard has created and disappears when the automatic links are in place.

5. To double-check the results, look at the records in the External Data Window to verify that all the records that you want linked have the link icon in the first column.

NOTE When you link data to shapes, Visio stores the external data in shape data. Another way to verify that the links are working the way you want is to choose View ⇨ Shape Data Window. Select a shape and, in the Shape Data Window, verify that the data appears.

Adding Shapes with Built-in Data Links

If you are adding new shapes to a drawing, you can add the shapes and the data links at the same time. This works best when you are adding several instances of the same master. Here are the steps to follow:

1. In the Shapes window, in the stencil, select the shape you want to add to the drawing when you drag rows of data onto the drawing page.

2. In the External Data Window, drag one or more rows onto the drawing page. (To select several contiguous rows, Shift-click the first and last row. To select several separate rows, hold the Ctrl key and click each row you want to add.) Visio adds the shapes to the page with links to the records you dragged.

Refreshing the Data in Drawings

The last piece of the Data Link puzzle is the new Refresh Data wizard. You can specify how frequently you want to refresh data and Visio takes over the job of checking the data source for changes and updating shape data with those changes. If data conflicts arise, the Refresh Conflicts Task Pane appears, so you can tell Visio what to do to resolve them.

Setting Up Data Refresh

Whether you want to refresh data immediately or set up a schedule for automatic refreshes, you begin with the Data menu:

■ **Refresh the data on your drawing immediately** — Choose Data ➪ Refresh Data. In the Refresh Data dialog box, select the data source you want to refresh and then click Refresh. You're done. The linked shapes on the drawing page and the records in the External Data Window both reflect any changes.

■ **Set up a schedule for refreshing data** — For example, you can tell Visio to update the data every few minutes, when data changes are fast and furious. By automatically refreshing every few hours or even once a day, Visio performs more responsively, but you won't have to wait too long to see changes. Besides, you can always request an immediate refresh before you print your drawing for a big presentation. To schedule refreshes, do the following:

1. Choose Data ➪ Refresh Data.

2. In the Refresh Data dialog box, select the data source you want to refresh and then click Configure.

3. In the Configure Refresh dialog box, under Automatic Refresh, select the Refresh Every __ Minutes check box. In the drop-down list, choose or type the number of minutes between refreshes.

> **NOTE** If you want to change the data source to configure or specify different fields to uniquely identify your data, you can do so in the Configure Refresh dialog box.

4. If you want to overwrite any changes in shape data values, select the Overwrite User Changes To Shape Data check box. For example, if you inadvertently modified some shape data properties, selecting this check box pulls the data from the data source and fills in the shape data properties. With the check box cleared, the modified values remain in the shape data despite the refresh.

5. Click OK to close the dialog box, sit back, and let automatic refresh run.

Resolving Conflicts

When Visio can't uniquely identify shapes and records, or rows of data have been deleted, the Refresh Conflicts Task Pane appears, because Visio can't resolve these conflicts on its own. As illustrated in Figure 10-4, the Refresh Conflicts Task Pane presents options for the conflicts that it finds.

- **Rows of data were deleted** — If records in the data source were deleted, the shapes that correspond to those records appear in the Refresh Conflicts list. The choice boils down to deleting the shapes that have no corresponding records or keeping them — a better idea if you think the records were deleted in error. To process one shape, select it in the list and then click either Delete Shape or Keep Shape. To process all the shapes at once, click Delete All Listed Shapes or Keep All Listed Shapes.

- **Data that can't be uniquely identified** — If Visio couldn't match some of the shapes and records, you won't see the link icon in the External Data Window and the shapes won't display the data from the data source. To link shapes and records, you can drag a row in the data source onto a shape. The alternative is to make sure that the shapes and records have all the information required to match them up.

> **NOTE** The Refresh Data command on the Data menu is grayed out in two situations. If you did not use Data Link to attach data to a diagram, you can't refresh the data with the Refresh Data command. Instead, you must use the same feature that you used to initially import the data, such as the Database Wizard or one of the import commands on a template menu. To refresh data on a PivotDiagram, choose PivotDiagram ➪ Refresh Data.

FIGURE 10-4

The Refresh Conflicts Task Pane lists conflicts it finds and includes buttons for specifying the action you want to take.

Visualizing Data with Data Graphics

If you've watched a few crime movies, you know that one of the best ways to obscure information is to provide too much of it — demonstrated so craftily by cinematic legal teams delivering truck-loads of records to their opposition. Yet, one of the reasons you use Visio is to communicate information effectively. So, what do you do when you have lots of information to convey and want to make sure your message gets through? In Visio 2007 Professional, Data Graphics help solve this predicament.

Data Graphics are a combination of text, graphics, and formatting that can turn tedious text into telling images. For example, the length of progress bars quickly conveys differences in cost, duration, and other values. Likewise, up and down arrows effectively communicate increasing or decreasing trends. And, if you want to identify which aspects of a project are on track or falling behind, icons such as green check marks, yellow caution signs, and red flags usually get people's attention.

Data Graphics can show information in four basic ways, but your choices within each are almost unlimited:

- **Text callout** — This callout simply displays data values as text, although Data Graphics can format the text to make it stand out. For example, if you want to show the resources responsible for each process in a TQM diagram, a Data Graphic with a text callout can place the value from the Resource shape data property anywhere around the shape with your choice of formatting. For example, you can display the text with a simple underline or place it in a bubble, oval, or type of filled area (shown at the top of Figure 10-5).

- **Data Bar callout** — A data bar communicates data values visually by altering the callout to reflect the data value. For example, you can show relative values with data bars that lengthen or shorten based on the data values, illustrated in the second row of Figure 10-5. A Progress Bar shows a box and indicates progress by filling in part of the box. Similarly, you can choose a Speedometer to see the needle move as values change or a Thermometer, which pushes the mercury higher as values increase.

- **Icon Set callout** — With an icon set, the callout displays different icons based on conditional statements. For example, you could add stoplights to steps in a process to indicate whether quality results are acceptable, as shown in the third row of Figure 10-5. If the quality is within bounds, the green stoplight icon appears. If the number of defects is less than 10% over the limit, the yellow stoplight appears. Any other values make the red stoplight appear. Stoplights play a big part in the built-in icon sets, but you can also choose from flags, arrows, smiley faces, and more.

- **Color By Value** — The fourth method of displaying data isn't a callout. Color By Value, as its name implies, applies color formatting directly to shapes based on how values match the conditional formatting criteria. For example, instead of using a stoplight icon set, you could turn entire Process shapes green, yellow, or red depending on their defect rates, as illustrated in the bottom row of Figure 10-5.

Data Graphics Basics

A Data Graphic is a collection of specialized shapes designed for displaying data, including the three different types of callouts (Text, Data Bars, and Icon Sets) and Color By Value criteria. A shape on a drawing can have only one Data Graphic at a time. However, if you want to show several different types of callouts and Color By Value on the same shape, you can add them all to a single Data Graphic and apply it to the shape.

NOTE Data Graphics are designed to display data stored in shape data, so they won't display anything unless the shapes you apply them to contain at least one shape data property.

When you apply a Data Graphic to a shape on a drawing page, the original shape and the shapes within the Data Graphic become a group. You can see this clearly by trying to select a callout in a Data Graphic. The first time you click the callout, Visio selects the original shape. The second time you click, the handles on the callout appear, indicating that the callout shape is subselected. Because the original shape and the Data Graphic belong to the same group, when you move, copy, or delete the original shape, the Data Graphic follows along.

FIGURE 10-5

Data Graphics can turn textual data into visual cues or make values stand out.

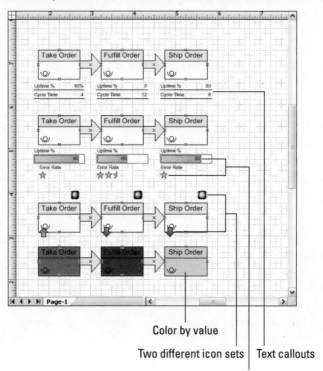

Color by value

Two different icon sets | Text callouts

Two types of data bar callouts

You can control the appearance of a Data Graphic in several ways. By combining these different settings, you can quickly display data in numerous ways. Here are the modifications you can make to Data Graphics:

- **Add additional callouts or Color By Value criteria** — A Data Graphic can contain several items, including the different types of callouts and Color By Value criteria. You can add more items to a Data Graphic, for example, to show the assigned resource as a text field, the duration as a progress bar, the change in duration as an icon, and the cost compared to budget as a color. You can also delete items or change the order in which they appear.

- **Specify the type of callout** — When you add an item, you choose Color By Value or the type of callout for that item. For each type of callout, you can also choose the appearance of the callout, for example, bar, progress bar, or speedometer, for a Data Bar callout.

- **Position the callout** — For each callout, you can tell Visio where to place it within or around the original shape. For example, you can position callouts above or below, to the left or right, inside or out — on any side of the shape. By default, Visio places all the items in one location on the shape, but you can position each one in a different spot.

- **Choose the shape data property that the callout presents** — Each callout can represent a different shape data property. For each callout or Color By Value rule, choose the field to display.

- **Specify the callout details** — Each type of callout has its own set of details, such as the Minimum Value and Maximum Value for a Data Bar.

- **Display a border** — You can tell Visio to place a border around the items at the default position within the Data Graphic.

- **Hide shape text** — Because Data Graphics often duplicate the data that already appears in the text blocks of shapes, you can tell Visio to hide shape text when a Data Graphic is applied.

Despite all the different looks and fancy formatting you can achieve, the great thing about Data Graphics is that Visio takes care of the hard-core formatting for you. Your job is to choose the types of callout and the conditions under which the callouts appear.

Adding a Data Graphic to Shapes

If you use Data Link to associate shapes to data records, Visio applies a basic Data Graphic to the shapes you link to external data. Whether you want to edit the Data Graphics that Data Link adds or apply Data Graphics of your own, the Data Graphics Task Pane is home base.

> **TIP** If the Data Graphics Task Pane isn't visible, choose View ➪ Task Pane. Then, in the Task Pane, click the title bar down arrow and choose Data Graphics. You can also display the Data Graphics Task Pane by choosing Data ➪ Display Data on Shapes.

To apply a Data Graphic to a shape, do the following:

1. On the drawing page, select the shape or shapes to which you want to apply a Data Graphic.

2. In the task pane, click the thumbnail of the Data Graphic you want (or the one that's closest to what you want). Initially, Visio displays a few basic Data Graphics with different types of callouts. As you edit these or create new Data Graphics, their thumbnails also appear in the task pane, as shown in Figure 10-6.

3. To edit the Data Graphic, for example, to choose the shape data fields to display, right-click the Data Graphic and, from the shortcut menu, choose Edit Data Graphic.

4. In the Edit Data Graphic dialog box, make any changes you want, such as the shape data field or the settings for the callouts, as described in the next section, Configuring Data Graphics. Click OK to close the Edit Data Graphic dialog box.

FIGURE 10-6

The Data Graphics Task Pane displays thumbnails of the Data Graphics that you've used recently. You can edit these or create new ones.

5. In the Data Graphics Task Pane, right-click the Data Graphic again and, this time, from the shortcut menu, choose Apply to Selected Shapes. The Data Graphics appear on all the selected shapes on the drawing page.

TIP If you always use the same Data Graphics for specific shapes, you can create a master that includes the Data Graphic. Then, when you drag the master onto the drawing page, the Data Graphic is already in place. To create a master with a built-in Data Graphic, apply the Data Graphic to a shape on the drawing page and then drag that shape onto an editable stencil. See Chapter 32 for instructions on adding shapes to stencils.

Configuring Data Graphics

Even if you use one of the Data Graphics that Visio provides out of the box, there's always some tweaking you have to do. At the very least, you have to tell Visio which shape data properties you want to highlight with the Data Graphic. You can choose which callout best illustrates the data or specify settings that control the appearance of the callout. Whether you want to make a simple edit or configure a Data Graphic of your very own, the configuration steps are almost identical.

Here are three ways to get started configuring Data Graphics:

- **Edit an existing Data Graphic** — If you want to make changes to an existing Data Graphic, in the Data Graphic Task Pane, right-click the thumbnail for the Data Graphic and, from the shortcut menu, choose Edit Data Graphic. The Edit Data Graphic dialog box appears, showing the items currently set within the Data Graphic.

- **Base a new Data Graphic on an existing one** — In the Data Graphics Task Pane, right-click the existing Data Graphic and, from the shortcut menu, choose Duplicate. A new Data Graphic thumbnail appears at the bottom of the task pane. Right-click the new Data Graphic thumbnail and, from the shortcut menu, choose Edit Data Graphic to open the Edit Data Graphic dialog box. The items and settings from the duplicated Data Graphic appear.

- **Create a brand new Data Graphic** — If you're convinced that it's easier to start from scratch, in the task pane, click New Data Graphic. The New Data Graphic dialog box appears, which is identical to the Edit Data Graphic dialog box except for the title and the empty item list. You must add callouts and Color by Value criteria, as described in the following sections.

Working with Items in a Data Graphic

In the Data Graphic universe, callouts and Color By Value criteria are know as items. Whether the New Data Graphic dialog box or the Edit Data Graphic dialog box (shown in Figure 10-7) is open, you can add new items to a Data Graphic, edit or delete existing items, or change the order in which they appear:

FIGURE 10-7

The New Data Graphic dialog box and the Edit Data Graphic dialog box are almost identical and let you add, edit, delete, or rearrange Data Graphic callouts.

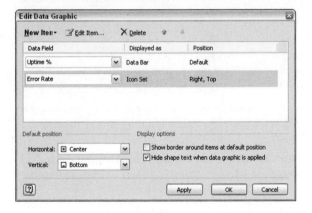

- **Create a new item** — In the dialog box icon bar, click New Item, and, in the drop-down menu, choose the type of callout you want (Text, Data Bar, Icon Set, or Color By Value). A dialog box for creating that type of item appears, for example, the New Data Bar dialog box, if you chose Data Bar.

- **Edit an existing item** — In the dialog box, click the row for the item you want to edit and then, in the dialog box icon bar, click Edit Item. A dialog box for editing that type of item appears, for example, the Edit Icon Set dialog box, if you selected an item that uses an icon set.

- **Delete an existing item** — In the dialog box, click the row for the item you want to delete and then, in the dialog box icon bar, click Delete. Visio doesn't bother with a request for confirmation and simply deletes the item. If you change your mind, press Ctrl+Z or click Cancel.

- **Rearrange items** — If you position each item in a different place around the shape, rearranging doesn't do anything for you. However, if you place all the items in the same location, you can determine the order in which they appear. Select an item and then, in the dialog box icon bar, click the up arrow or down arrow to place the item in the order you want.

Specifying the Default Position and Hiding Shape Text

A few settings apply to the overall Data Graphic rather than the items within the Data Graphic. Here are the settings that appear in the New Data Graphic or Edit Data Graphic dialog boxes and what they do:

- **Choose the default position** — At the bottom left of the dialog box is the Default Position area, which includes drop-down lists for specifying where the Data Graphics items appear by default. The Default Position makes it easy to collect every item at the same place relative to the original shape. When you create a new item, Visio automatically sets the position to the default, unless you choose to change it. You can specify the horizontal and vertical positions in the drop-down lists of the same names. The entries encompass seven positions, for example, Far Left, Left Edge, Left, Center, Right, Right Edge, and Far Right, for horizontal. These positions begin outside the shape on the left and move across until the position is outside the shape on the right, as illustrated by the icons that precede each entry.

> **TIP** Take care in choosing the location of your items. If you place them inside the shape, they might overlap text within the shape. If connectors tend to attach at the top and bottom of shapes, placing the items on the sides or at the corners keeps them in view.

- **Add a border** — Because many Data Graphics group callouts in one place, you can draw a border around all the items at the default position. Select the Show Border Around Items At Default Position check box.

- **Hide shape text** — If the text in the shape text block duplicates text that you display in a Data Graphic, you can hide the shape text while the Data Graphic is attached to the shape. Select the Hide Shape Text When Data Graphic Is Applied check box.

Specifying the Settings Common to All Types of Callouts

Whether you choose a text callout, a data bar, or an icon set, you choose the shape data field to show, the style of callout, and where you want it relative to the parent shape. Color By Value fills in the parent shape with a color, so it only requires a data field.

To configure a callout, in the Edit Data Graphic or Create Data Graphic dialog box, either select an existing callout and click Edit Item, or choose New Item and then choose the type of callout you want (or Color By Value). The settings are the same whether you are editing an existing callout or defining one from scratch.

In the dialog boxes for creating or editing items, choose settings in the following fields:

- **Data Field** — The Data Field is the setting you're likely to change most often. The drop-down list includes all the shape data fields associated with the shapes on the drawing page, but you can also choose More Fields to specify other shape data fields, page and document fields, user-defined properties, or custom formulas.

- **Callout** — In the drop-down list, choose the style of callout. For text callouts, the previews in the Callout drop-down list show a shaded area where the label and value appear. Some of the styles show only the data value, such as the headings, circles, and triangles. To show the shape data field name and its value, look for a callout that has two shaded areas, such as Bubble callout. The previews for data bars and icon sets help you choose the style you want.

- **Position** — To group callouts in one place, make sure the Use Default Position check box is selected, and the callout appears at the default position you chose for the entire Data Graphic. To position a callout somewhere else, clear the Use Default Position check box and then, in the Horizontal and Vertical drop-down lists, choose the position you want.

Emphasizing Textual Data with Text Callouts

A text callout takes a value from a shape data property and displays it in the vicinity of the parent shape. If seeing the data values is essential, a text callout can make them stand out from the other text on the drawing. If you've used the Label Shapes add-on, a text callout is the same basic idea with the addition of a dialog box for specifying the position of the text and the callout style, which in turn applies several types of formatting automatically.

To configure a text callout, in the Edit Data Graphic or Create Data Graphic dialog box, either select an existing text callout and click Edit Item, or choose New Item ➪ Text Callout. The settings for a text callout, shown in Figure 10-8, are the same whether you are editing an existing text callout or defining one from scratch.

The settings in the Details table vary depending on the type of text callout you choose. For text callouts, Value Format controls how the value looks, for example, showing a number as a percentage or the format for the date and time. Other settings control the border, fill, offset from the selected position, the label position, and the label value.

Visualizing the Magnitude of Numbers with Data Bar Callouts

Data bars are great for visually communicating magnitude, for example, the percentage complete on a project task, the number of support calls for a category, or the customer satisfaction rating for a product. Data bars come in several dramatically different styles. Simple bars show a value visually, whereas progress bars show a value relative to the maximum value, which is perfect for showing percentages. Star ratings, thermometers, and speedometers are other styles of data bars that transform numbers into interesting visual cues.

FIGURE 10-8

Configuring a text callout is a matter of choosing settings and values in a dialog box.

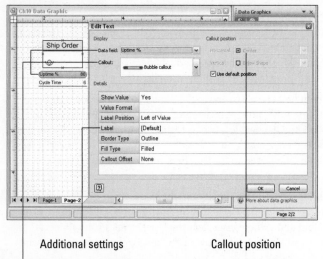

Additional settings

Callout position

Data field and callout style

The settings in the Details section shown in Figure 10-9 are the same for most Data Bar callouts. The Multi-bar Graph and Stacked Bar styles have extra settings for assigning multiple shape data fields to the callout. Here are the common settings and what they do:

- **Minimum Value** — The default for minimum is zero. However, if data values are within a range, assign the lower end to the Minimum Value field.

- **Maximum Value** — The default for maximum is 100, which sets data bars up perfectly for percentages. Otherwise, assign the highest value to the Maximum Value field. When you specify the maximum value for a data bar and a shape data value comes in larger than the maximum, the bar grows no farther than the maximum value.

CAUTION If you work with percentages, be sure to omit the percentage sign from your shape data values. Otherwise, the data bar displays the maximum value.

- **Value Position** — If you want the data bar to hide the value, choose Not Shown. Otherwise, choose from Left, Right, Top, Bottom, or Interior to specify where the data value appears relative to the data bar.

- **Value Format** — Choose a format from the drop-down list to control how the value looks.

- **Label Position** — If you want the data bar to hide the label, choose Not Shown. Otherwise, choose from Left, Right, Top, Bottom, or Interior to specify where the label appears relative to the data bar.

- **Label** — [Default] appears in this field initially and indicates that the name of the shape data field appears as the label. If you want to use a different label, type the text in the cell.

- **Callout Offset** — This setting is helpful if none of the callout positions are quite what you want or the callout overlaps a portion of the shape outside its alignment box. You can offset the callout to the left of right, or leave it at the designated position.

FIGURE 10-9

Configuring a data bar callout is a matter of choosing settings and values in a dialog box.

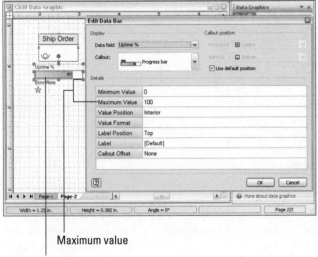

Maximum value

Shape data value

Showing Conditions with Icon Set Callouts

Unlike data bars, icon set callouts highlight numeric or text values that satisfy conditions. Whether you use colored flags, traffic signals, smiley faces, or other icons, an icon set callout displays an icon based on the rules you define. For example, if your company has set goals for accuracy on orders, you can show a green traffic light when the numbers exceed the goal, a yellow traffic light if the numbers are just short, and a red traffic light if the numbers are below the acceptable minimum.

As illustrated in Figure 10-10, a Rules For Showing Each Icon section appears for icon set callouts with a rule for each icon in the set. If you don't want to use all the icons in the set, in the icon's test drop-down list, choose [Not Used], as shown in the last icon in Figure 10-10.

If you use conditional formatting in Microsoft Excel, the way icon sets work should be familiar. Visio tests the shape data value against the first rule you define. If the value passes the test, the first icon appears. If the value doesn't pass the test, Visio checks it against the second test, and so on until it passes. The icon for the test that the value passes is the one that you see.

FIGURE 10-10

For icon set callouts, you must define a rule for each icon you want to display.

Position

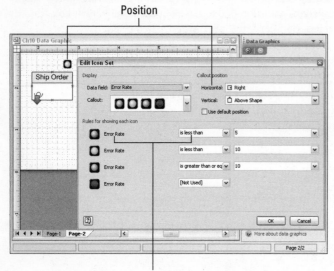

Rule that value must pass for icon to appear

You have to build rules carefully or icons won't display the way you expect. For example, if you have two "if less than" tests, you must test for the smallest value first. If you test for values less than 10 before you test for values less than 5, the second test never gets a chance to execute. Another potential flaw in tests is not covering every value that occurs. For example, suppose you want to test for values less than 5, less than 10, and then all other values. Be sure to use "is greater than or equal to" in the third test. If you use "is greater than," none of the tests pass for the value 10.

CROSS-REF If you're feeling energetic, you can develop your own callouts and icon sets for Data Graphics. Read the Building Custom Data Graphics for Visio 2007 article at `http://msdn2.microsoft.com/en-us/library/aa468596.aspx` to learn about customizing callouts and creating new icon sets.

Color-Coding Shapes with Color By Value Criteria

Color-coding shapes changes the fill color of the shapes based on their shape data values. Color-coding comes in two different styles:

- **Each color represents a unique value** — This approach is ideal when you want to show all the shapes with the same value in the same color, such as the highlighting all the offices occupied by the accounting department with green.

Copying Data Graphics Between Files

Data Graphics reside in the drawing in which you create them. However, you can share them with others or use them in other drawings that you create. The easiest way to reuse Data Graphics is to create a Visio template that includes the stencils you use for the type of diagram and the Data Graphics that you typically apply. When you create a new drawing with that template, the Data Graphics appear in the Data Graphics Task Pane. See Chapter 31 for instructions on creating templates.

To copy a Data Graphic from one drawing to another, apply the Data Graphic to a shape in the current drawing. When you copy the shape and paste it into another drawing, Visio copies the Data Graphic definition and the callouts it uses to the second drawing. The Data Graphic becomes a part of the second drawing, even if you delete the pasted shape.

- **Each color represents a range of values** — For values that cover a range, you can choose colors for each range. For example, you might use colors that go from pale green to emerald depending on the revenue for different product lines. However, you can also choose different colors for each range, for example, blue, green, yellow, orange, and red to represent different levels of drought.

In the New Color by Value dialog box or Edit Color by Value dialog box, in the Coloring Method drop-down list, choose one of the following:

- **Each Color Represents A Unique Value** — The Color Assignments area shows all the unique values in the drawing and assigns a fill color and text color to each value. If you want to use different colors or text colors, click the arrow for that entry and choose a color.

- **Each Color Represents A Range Of Values** — The Color Assignments area divides the values into equal ranges and assigns a gradually lighter fill color to each range as well as a text color to each range. If you want unequal ranges or any other customized range, type the minimum and maximum values in the Value 1 and Value 2 boxes. To choose a different color for the scheme, choose a different color for the first range only. Visio adjusts the remaining colors to lighter shades of the first color. If you want completely different colors for each range, select a color for each range.

Using the Visio Database Add-ons to Link Drawings and Databases

Although the new Data Link feature simplifies several aspects of linking data and drawings, the database features available in earlier versions of Visio are still around — and they still have value. The Data Link feature is a one-way street from data records to Visio shapes. If you want to set up two-way communication between your drawings and data sources, the Database Wizard and its relatives are the way to go.

By using the Database Wizard to link your Visio drawings to databases, you can update Visio drawings to show the current values in your databases or push changes made in Visio drawings to the corresponding records in your database. You can use either the Database Wizard, Link to ODBC Database, and other add-ons within the Visio Extras category to perform the following database-oriented tasks:

- Link shapes to records in databases
- Create Visio drawings that represent the data in database tables
- Create masters linked to database records
- Export Visio Shape Data to database records
- Synchronize shapes and database records

Understanding Two-Way Links Between Shapes and Databases

When you use Data Link to connect shapes and database records, the values from the database come over and sit in shape data properties. When you refresh the data, the Data Link takes care of the links for you and brings across updated values from the data source to the shape data. The two-way links you can build with Visio add-ons aren't as transparent, so it's helpful to understand how Visio keeps track of what goes where. In essence, Visio links shape data properties directly to fields in database records.

Similar to the setup for Data Link, two-way connections use a data source to connect Visio drawings with databases. Whether you use the Database Wizard, Link to ODBC Database, Export to Database, or Database Export Wizard add-ons, you can choose an existing DSN or create a new one.

Each shape in a drawing corresponds to a single record (or row) in a database table, and each linked field in the record corresponds to one property in the shape, which you can see via the ShapeSheet. With a link between a shape property and a database field, you can change the value in the property to update the field, or vice versa.

Although the most common link is between shape data and database fields, you can link other properties as well. For example, you can propagate equipment dimensions in a database table to the properties that define a shape's size. If a shape doesn't have properties suitable for linking to your database fields, the Database Wizard or Link to ODBC Database add-ons can create new shape data properties for that purpose.

When a shape is linked to a database, you can see the connection in the `User.ODBCConnection` cell in the User-Defined Cells section of the ShapeSheet. The connection is a concatenation of several lines of text separated by vertical bars. By separating the lines, you can see the components of the connection string:

```
ODBCDataSource=Visio Database Samples
ODBCQualifier=C:\Program Files\Microsoft
Office\Office12\1033\DBSAMPLE.mdb
ODBCTable=Airplane - Filled Seats
1
SeatNumber=Prop.SeatNumber
5
Baggage=Prop.Baggage=32
Class=Prop.Class=0
FrequentFlier=Prop.FrequentFlier=0
PassengerName=Prop.PassengerName=0
TicketPrice=Prop.TicketPrice=111
```

Here's how the components of the connection string translate in more human-friendly terms:

- **ODBCDataSource=<*name*>** — A simple text name for the data source to which the shape links. The example uses a sample data source installed with Visio.

- **ODBCQualifier=<*name*>** — This parameter specifies the direct path to the database to which the shape links.

- **ODBCTable=<*name*>** — This is the table to which the shape links.

- **<*number*>** — The fourth line in the example represents the number of key fields you specified to identify records. The example uses only one key field.

- **<*Field Name*>=<*name*>** — The fifth line in the example identifies the key field in the database and the ShapeSheet cell used to store the key.

- **<*number*>** — The sixth line in the example represents the number of database fields linked to shape data. In the example, five fields are linked.

- **<*Field Name*>=<*name*>=<*number*>** — For each field linked to shape data, the connection string identifies the linked field name and the ShapeSheet cell name to which it is linked. The number at the end of the line represents the constant for the Evaluate As setting. For example, 0 represents Value, whereas 111 represents Currency.

Limitations of Database Links

Links between Visio shapes and tables in databases have the following limitations:

- **String Length** — Visio cells and fields can hold up to 64K for strings.

- **Binary Field Length** — Visio cells and fields can hold up to 32K for ODBC binary fields.

- **Precise Numbers** — Visio stores numeric values as double floating-point numbers, truncating numbers with greater precision to 17 significant digits.

- **Database Key Field** — A field of the SQL_TIMESTAMP type can't be a primary key.

- **ID Replication** — You can't update replication IDs in Microsoft Access.

- **Timestamps** — You can't update Timestamp fields when you use an Informix database.

- **Row Deletion** — You can't delete rows in Excel when you use the Excel ODBC driver. The Database Wizard indicates deleted rows by setting text fields to #ROW DELETED# and numeric fields to 0. For wizard operations, such as Update, Select, and Refresh, the values 0 and #ROW DELETED# are invalid keys.

Creating Connections to Data Sources

Each of the Visio database add-ons includes the ability to create a data source. If you want to create a data source as a stand-alone step, from the Windows Start menu, choose Control Panel ➪ Administrative Tools ➪ Data Sources (ODBC) and follow the steps in the ODBC Data Source Administrator dialog box.

The steps for defining data sources in the various Visio database add-ons are similar. The following steps use the Link to ODBC Database add-on as an example:

1. Choose Tools ➪ Add-ons ➪ Visio Extras ➪ Link to ODBC Database.

2. In the Link to Database dialog box, click Create. This button is grayed out if no drawing is open.

3. In the Create New Data Source dialog box, choose one of the following options for type of data source and then click Next:

    - **File Data Source** — This option creates the most flexible data source. You can copy it to other computers or network locations and any user who can reach the data source can use it.

    - **User Data Source** — This option is the most limited, because it applies only to the current computer and is visible only to the current user.

    - **System Data Source** — This option creates a data source available only on the current computer but accessible to any user who logs on to the system.

4. In the Driver list, select the 32-bit ODBC driver that matches your data source, such as Excel, Access, or SQL Server. Click Next.

> **NOTE** If your data source is a specialized database, click Advanced. You can type keywords and values that your database driver uses in the box.

5. In the next screen, specify the location and name for the data source. Because the functionality of ODBC drivers varies, it's a good idea to include the type of data source in the name, along with a brief description of the data, such as `Employees_Excel2007`. After you specify the location and name for the data source, click Next and then click Finish.

> **NOTE** By default, Visio creates data sources in `C:\Program Files\Common Files\ODBC\Data Sources`. Unless you have a reason for storing data sources elsewhere, the default location has the added benefit of displaying all its data sources in the list of available data sources. If you don't see your data source in the list of available data sources, click Browse and navigate to the location of your data source.

6. Depending on the type of data source you're creating, there is some final cleanup to perform:

   - **Access data source** — In the ODBC Microsoft Access Setup dialog box, click Select to specify an existing database for the DSN. Select the database file you want to use and click OK. For example, to use the sample database included with Microsoft Office, in the Select Database Dialog box, navigate to `C:\Program Files\Microsoft Office\Office12\1033\DBSAMPLE.mdb`. In the Database Name list, select `Northwind.mdb`, and click OK. The ODBC Data Source Administrator creates the data source and adds it to the list of available data sources.

   - **Excel data source** — You must specify the version of Excel and then select the Excel workbook. After you select the file to use, click Options and clear the Read Only check box. In the Database Wizard on the Choose a Database Object to Connect to screen, you can create tables in Excel files by clicking Define Table. Visio stores tables created in Excel 5.0 or later as separate worksheets in the Excel file and creates a named range for the records.

When you finish these steps, in the Link to Database dialog box, the Name box shows the data source name and the Qualifier box shows the path and file name.

> **CAUTION** To use an Excel workbook as a data source, you must create a named range in a worksheet that includes all the rows and columns of data. Visio uses the values in the first row of the named range as table column names.

Linking Drawing Shapes to Database Records

The Database Wizard and the Link to ODBC Database add-ons both create links between your drawing and database records. For example, if you have a furniture plan in Visio, you can link the furniture shapes to records in a facilities management database. The Database Wizard feeds you steps one at a time, whereas the Link to ODBC Database add-on provides the tools you need to create a link in one dialog box.

NOTE You can link Visio shapes with a database without a database application installed on your computer. However, if you want to open the database directly to edit its records, you must have the corresponding database application installed.

To link shapes using the Link to ODBC Database add-on, follow these steps:

1. Select the shapes you want to link to a database and then choose Tools ⟹ Add-Ons ⟹ Visio Extras ⟹ Link to ODBC Database.

2. In the Name drop-down list, choose the data source. If you select Excel Files as the data source, select the Excel file (.xls) that you want to use in the Select Workbook dialog box, clear the Read Only check box, and click OK.

NOTE If your data source is absent in the Name drop-down list, don't worry. Click Browse, navigate to your data source folder, and select the data source. For example, to use the sample database included with Office, in the File Open dialog box, navigate to `C:\Program Files\Common Files\ODBC\Data Sources`, select the data source file you want, and click Open. You can click Create to create a new data source, as described in "Creating Connections to Data Sources" in this chapter.

3. If you can access multiple databases through your data source, in the Qualifier drop-down list, select the database you want to use. To filter the list of tables to those for a specific owner, in the Owner drop-down list, select an owner. For data sources with single databases, the Qualifier box shows the database path and name, and the Owner box is set to All Users.

4. In the Table/View list, select the database table you want to use. For Excel data sources, the Table/View list shows the worksheets within the Excel workbook. The Field Links section displays the fields in the data source as well as the cells in the ShapeSheet for the values.

NOTE You can create a new table or worksheet in the data source by clicking New.

5. If you want to change the default mapping between database fields and ShapeSheet cells, select a field and click Modify. Select the new ShapeSheet cell. If you want to change the data type for the field, select a type in the Evaluate As drop-down list. If you want to use a field as the primary key, select Yes in the Key drop-down list.

NOTE Click Add or Delete to create new links or delete existing links between fields and cells.

6. Click OK to create the links.

7. To associate a shape on a drawing to a specific record in the table in the data source, right-click the shape and choose Select Database Record. In the Key Value list, select the record you want and click OK to insert the record values into the linked ShapeSheet cells.

Linking Masters to Databases

By linking masters to database records, you can add shapes already set up with links to corresponding database records to drawings. This is useful when a master always uses the same database record, for example, to pull model information, such as manufacturer, model, cost, and color, from a catalog database. When you drag a master onto the page, the shape data is automatically populated.

What about masters that link to a database when each instance on a drawing page links to a different record? You can also create masters that link to any database record. When you drag the master onto the drawing page, the link to the database is in place, but you have to choose the record to which you want to link.

Generating Masters from Each Database Record

If you want to create a stencil of masters that link to a database, the Database Wizard is the add-on to use. To generate masters from a database table, follow these steps:

1. Choose Tools ⇨ Add-Ons ⇨ Visio Extras ⇨ Database Wizard. Click Next on the first screen.

2. Choose the Generate New Masters from a Database option and click Next.

3. In the Stencil box, select the stencil that contains the Visio master that you want to use as the basis for new masters. The Stencil drop-down list displays the stencils that are open. If you want to choose another stencil, click Browse, and select the stencil.

4. In the Masters list, choose the master and click Next. For example, if you want to create a stencil of chairs from your catalog database, you might choose a furniture stencil and then a chair master.

5. Continue through the wizard to select the data source, the table, and the primary key.

6. In the Choose the Database Link and Naming Options screen, select the Keep Database Links in New Masters check box. This setting creates a bi-directional link between the database and the masters. Select an option to either generate master names based on the values in the primary key field or on the original Visio master name. Click Next.

 **If the master you use doesn't show shape data in text fields, it's easier to identify the master you want when you name the master using the primary key.**

7. In the Events and Actions screen, review the settings that Visio selects by default and change any, if needed. The default settings work for most situations. Click Next.

8. Specify the cell in which you want to store the primary key field value and click Next.

9. In the next screen, you see a list of cells in the master ShapeSheet and fields in the database. To create links, click a ShapeSheet cell and then a database field, and click Add. If the names are the same in both lists, click Automatic. When the links are set up, click Next.

10. To choose the stencil for the new masters, select either the Create A New Stencil option or the Append To An Existing Stencil option. For an existing stencil, select the name or click Browse. Click Next.

11. Click Finish.

Linking Masters to Specific Records

When you want every instance of a master to use the same information, you can link masters to specific records in an existing database table. Perhaps the easiest way to create a master linked to a record is to create a link between a shape on a drawing page and a database record (by right-clicking the shape, and from the shortcut menu, choosing Select Database Record). Then drag the linked shape onto an editable stencil.

You can also link a master directly. Edit the master, as described in Chapter 34. Use the Link to ODBC Database add-on to link the master to a data source. Then, in the master drawing window, right-click the master and choose Select Database Record to specify the record for that master.

Creating Masters That Link to Any Database Record

To create a master that can link to any record in a database table, follow these steps:

1. Choose Tools ➪ Add-Ons ➪ Visio Extras ➪ Database Wizard. Click Next on the first wizard screen.

2. Select Link Shapes to Database Records and click Next.

3. Select the Master(s) on a Document Stencil option to link masters specific to a drawing, or Master(s) on a Visio Stencil to link masters that you can use on any drawing. Click Next.

4. Select a document stencil or click Browse to select a Visio stencil or one of your custom stencils. By default, Visio opens the File Open dialog box at the My Shapes folder.

5. Choose the master you want to link and click Next.

6. Continue through the Database Wizard screens to select a data source, the table to which you want to link, the primary key, events and shape shortcut menu commands, and the ShapeSheet cell that holds the primary key value.

7. On the Link ShapeSheet Cells to Database Fields screen, click a ShapeSheet cell in the Cells list, click the corresponding database field in the Database Fields list, and then click Add. Visio adds the shape-field link in the Links list. When the links you want exist, click Next and then click Finish.

Creating Drawings from Database Records

The Database Wizard has one additional trick up its sleeve. It can generate a Visio drawing that contains a shape for each record in an existing database table. For example, if you have a database or spreadsheet that delineates the computer equipment you're going to install at a client site, you can use the Database Wizard to create a layout drawing with shapes linked to each record. Then you can move the shapes into the layout you want.

To create a drawing based on a database table, follow these steps:

1. Follow the steps in the preceding section to create a master that links to the database table you want to use to generate the drawing.

2. Choose Tools ➪ Add-Ons ➪ Visio Extras ➪ Database Wizard. Click Next on the first wizard screen.

3. Select the Create a Linked Drawing or Modify an Existing One option and click Next.

4. Select the Create a Drawing Which Represents a Database Table option and click Next.

5. Select the Create New Drawing option and click Next.

6. Continue through the wizard to select a drawing template, options to use to monitor the drawing, and information about the data source to use.

7. When the Select a Visio Master Shape screen appears, click Browse and navigate to the stencil you saved in step 1. Select the linked master and click Finish to create a new drawing with an instance of the master for every record in the table.

Keeping Drawings and Databases in Sync

When you use the Database Wizard or the Link to ODBC Database add-on, you can take advantage of several methods for synchronizing the data between your Visio drawings and linked databases. For example, the shortcut menu for each linked shape can contain four commands for maintaining database links. In addition, Visio provides several add-ons that help you update many shapes at once.

To update a single shape or database record, right-click a shape and then choose one of the following commands from the shape's shortcut menu:

- **Select Database Record** — Specify the database record to which you want to link the shape.
- **Refresh Shape Properties** — Replaces the shape's properties with data from the linked database record.
- **Update Database Record** — Replaces the data in the database record with values from the shape's shape data.
- **Delete Shape and Record** — Removes the shape on the drawing page and the linked record in the data source.

If you want to refresh or update all the shapes on a drawing page, use one of the following add-ons:

- **Database Refresh** — Replaces each shape's properties with data from its linked database record
- **Database Update** — Replaces the data in each database record with values from the linked shape's properties

Adding Drawing Page Commands to Synchronize Shapes

The Database Wizard can add commands to synchronize links to shape or drawing shortcut menus. In addition, you can specify what happens when you drop a shape or copy and paste one on a page. If you use the Link to Database command to link your shapes, you can use the Database Wizard later to add synchronizing commands to the drawing page.

To add shortcut commands to a drawing page, follow these steps:

1. Choose Tools ➪ Add-Ons ➪ Visio Extras ➪ Database Wizard. Click Next on the first wizard screen.

2. Click the Create a Linked Drawing or Modify an Existing One option and click Next.

3. Click the Add Database Actions and Events to a Drawing Page option and click Next.

4. Select the drawing file and page to which you want to add actions and events and then click Next. Select one or more of the following actions or events:

   ■ **Refresh Shapes on Page** — Adds a command to refresh all shapes on the page with the data from their linked database records.

   ■ **Update Shapes on Page** — Adds a command to update the database records with values from the shapes on the page.

   ■ **Refresh Linked Shapes on Document Open** — Each time you open the drawing file, Visio refreshes ShapeSheet cell values to match linked database records for all shapes in the drawing file.

   ■ **Periodically Refresh Based on NOW Function** — Refreshes ShapeSheet cell values to match linked database records for all shapes in the drawing file at the interval specified by the NOW function. When you choose this option, Visio adds commands to the drawing page shortcut menu to start and stop the continuous refresh.

5. Click Next and then click Finish.

Adding Actions and Events to Shapes

By default, Visio sets Refresh Shape as the default Shape Drop Event. When you select this option and drop or paste a shape on a drawing page, Visio refreshes the values based on the database record linked to the original shape. If you want to select a different database record when you copy a shape, select the Select Record option. When you drag a master onto the page or paste a copy, Visio prompts you to select a database record.

To add or modify the actions and events after you've linked your shapes, follow these steps:

1. After you've linked the shapes to a database, select the shapes to which you want to add actions and events.

2. Choose Tools ➪ Add-Ons ➪ Visio Extras ➪ Link to Database and click Advanced.

3. Select the actions and events you want and then click OK.

Synchronizing Shapes and Database Records Automatically

If you would prefer to have Visio automatically refresh your drawings, you can have Visio monitor the database linked to a drawing at a regular interval and refresh the shapes in the drawing with the values from the linked database records. To refresh shapes automatically, follow these steps:

1. Choose Tools ➪ Add-Ons ➪ Visio Extras ➪ Database Settings.

2. Select the Automatically Refresh Drawing Page check box.

3. In the Refresh Drawing Interval (Secs.) box, type the number of seconds between every refresh and click OK.

Exporting Shape Data to Databases

The chances are slim that you assemble significant data in Visio that's not in a database somewhere. However, you might use Visio to input information initially, in which case, you can export shape data to databases using either the Database Export Wizard or Export to Database add-ons. The Database Export Wizard presents the process in smaller steps and includes an option to export all shapes on a layer. However, if you don't want to export by layers, you can follow a more stream-lined process by choosing Tools ➪ Export to Database.

To export shape data to a database with the Database Export Wizard, follow these steps:

1. Choose Tools ➪ Add-Ons ➪ Visio Extras ➪ Database Export Wizard and click Next.

2. Select the drawing file that you want to export and then select the page you want to export and click Next.

3. Click one of the following options for the shapes you want to export and then click Next:

   - **All Shapes on the Page** — This option exports all shapes on the selected page.

   - **Selected Shapes on the Page** — This option exports any shapes you selected before you started the wizard. To select additional shapes, click Select Shapes.

   - **All Shapes on One or More Layers** — Select the layers you want to export in the Layers list to export all shapes on those layers.

4. For each Visio item you want to export, select the item in the Visio Cells and Fields list and click Add. Visio adds the cell or field you chose to the Cells and Fields to Export list. Click Next.

 Cell names for shape data begin with Prop. followed by the shape data name. For example, the cell name for Duration is Prop.Duration.

5. Select the data source to which you want to export and click Next.

6. In the Table Name list, select the table to which you want to export the Visio data. If you want to create a new table in the data source, type the name of the table in the Table Name box.

7. To specify a key to uniquely identify each record, type a name in the Key Field box. By default, the key field is the ShapeID, which is the shape name with a sequential ID generated when you add the shape to a drawing page. To use the GUID as the unique identifier, select GUID in the Key Type drop-down list. Click Next.

 If the ODBC driver does not support primary keys, the Make Key Field the Primary Key for Table check box is grayed out.

8. To modify the default field mapping that Visio defines, in the Specify the Export Mapping Details screen, select an item in the Visio Data list, modify one or more of the following options, and then click Next:

- ■ **Evaluate Data As** — Specify the data type or units for the Visio item.
- ■ **Field Name** — Modify the field name to use in the export data source.
- ■ **Field Type** — Modify the field type, which is a broad category such as Number.

9. If you want to re-export data by right-clicking the drawing page, make sure the Add Export Right Mouse Action to the Drawing Page check box is checked.

10. Click Next and then click Finish to export the data.

TIP   When you export shape data with the Database Export Wizard, the wizard stores export-related information with the drawing page. If you want to export the data again after you have modified the drawing, right-click the drawing page and choose Database Table Export.

Producing Reports

If you link your Visio drawings to databases, the reporting tools you use with your database management system are more versatile than Visio's reporting feature, and might even seem easier to use. However, if you keep data only in Visio or aren't well-versed in your database reporting features, Visio does include predefined reports that you can run to view and analyze the data stored with the shapes in your drawings. Although you can create new reports from scratch, you can also modify existing reports to match your requirements and then save the modified report with a new name.

Running Reports with Shape Data

When you run a report, you can specify how to format the results: as an Excel spreadsheet, as an HTML file to display as a web page, as an XML file, or within a Visio shape. To run an existing report, built-in or custom, follow these steps:

1. Open the drawing on which you want to report and choose Data ➪ Reports.

2. In the Reports dialog box, select the report. If you don't see the report in the list, clear the Show Only Drawing-specific Reports check box.

3. Click Run.

Defining Custom Reports

When you choose Data ➪ Reports, Visio displays the reports specific to the type of drawing that is active. To see all report definitions, clear the Show Only Drawing-specific Reports check box. To create a new custom report, you have two methods to choose from:

- ■ In the Reports dialog box, click New to create a report from the bottom up.
- ■ If you want to base your report on an existing report, in the Report list, select the existing report and click Modify.

Either way, the Report Definition Wizard steps you through the process of defining a report. When you have completed the report definition, Visio saves your report where you specify and adds it to the list of available reports. As you step through the Report Definition Wizard, you specify the following features of your report:

- The objects on which you want to report

- The properties you want to display in the columns of the report and any criteria you want to use to limit the results shown in the report

- The format for the report, including how the contents are grouped and sorted and how numbers are formatted

- The Save options for the report, including the report definition name, a description, and where Visio saves the report

The following sections describe how to use the features in each step of the wizard.

Selecting Shapes on Which to Report

The first step to defining a report is to choose the shapes to scan. You can report on every shape in your drawing file, the shapes on the current page, or only selected shapes.

In the Advanced dialog box, you can define multiple criteria that shapes must meet for inclusion in the report. For example, to produce a report of the furniture for the Accounting department, you can specify that the Department property must equal Accounting for shapes only on the Furniture layer. Advanced settings limit the shapes used in the following ways:

- **Shapes on a layer** — Select the <Layer Name> property. You can include or exclude a layer from a report by selecting = or <> in the Condition box.

- **Shapes by name or ID** — You can use the <Master Name>, <Shape Name>, or <Shape ID> properties to specify the named shapes you want.

- **Shapes with specific shape data** — Select the shape, select Exists in the Condition list, and then select TRUE in the Value list.

- **Shapes with specific values in shape data** — Select the shape and then specify the condition and value for the property. For example, you can select Duration, >, and 5 to report on all processes whose duration is longer than five weeks.

- **AutoDiscovery shapes** — Select the <Autodiscovery Shape> property to report on shapes that result from using AutoDiscovery.

> **NOTE** The conditions and values you can choose depend on the property that you select. However, Visio offers a limited number of conditional operators, such as =, <>, and >=.

To add an additional criterion, define the criterion and then click Add. To delete a criterion, in the Defined Criteria list, select it and then click Delete. You can delete all the defined criteria by clicking Clear.

After you define criteria, click OK to close the Advanced dialog box and then click Next to continue selecting the properties for the report columns.

Choosing Report Columns

In the screen for choosing properties, Visio displays default shape properties, such as <Master Name> and <Height>, in addition to shape data and user-defined properties from the ShapeSheet User-Defined Cells section. Default shape properties appear within angled brackets. To see user-defined properties from the ShapeSheet, check the Show All Properties check box.

Select the check box for each property you want to use as a column in the report. Click Next to continue grouping, sorting, and formatting your report.

Grouping, Sorting, and Formatting Report Contents

As with most reporting tools, Visio reporting can group, sort, and format report results in several ways. For example, you can produce a report that groups the office furniture by department, calculating the number of desks for each department, and sorting the report by the average furniture cost per department.

- **Group results** — Click Subtotals. In the Group By drop-down list, which contains the properties you specified as columns for your report, choose the property you want to use to group your results. You can also specify which rows appear in your report by clicking Options. For example, you can show all values, only unique values, only subtotals, or grand totals. You can also prevent duplicate rows from appearing in your report.

- **Calculations** — For each property in the Properties list, select the check boxes for the calculations you want. Count is available for any type of shape, but the other calculations only work with numerical properties. If a calculation isn't valid for a property, the check box is dimmed. When you perform a calculation, it calculates the value for each group in the report. If you create an ungrouped report, the calculation represents all entries. You can choose from the following calculations:

  - **Count** — Calculates the number of shapes with the same value in a shape. For example, if there are 16 shapes with the value Accounting in the Department property, Count returns 16.

  - **Total** — Sums the value of all the entries in a group or all entries for an ungrouped report.

  - **Avg** — Calculates the average of property values in a group or the entire report.

  - **Max** — Returns the largest number from the list of property values.

  - **Min** — Returns the smallest number from the list of property values.

  - **Median** — Calculates the median for a list of property values. In a group of numbers, the median is the number with an equal number of values greater and less than it.

- **Rearrange Columns and Sort Rows** — You can arrange the order of the columns in your report as well as specify the sort order for rows by clicking Sort. To rearrange the columns in your report, select a column in the Column Order list and then click Move Up or Move Down to change its position in the column order. You can sort the rows in a report with up to three properties. Choose the shape data for the first sort in the Sort By list and select the Ascending or Descending option. Specify one or two additional sort properties in the two Then By lists, also specifying ascending or descending order. For example, you can sort a report of office furniture first by department, then by employee name, and finally by Shape Name if it indicates the type of furniture.

- **Format Values** — For shape data with numeric values, you can specify the number of decimals in numbers and whether to show units. Click Format and then choose the number of places to the right of the decimal point in the Precision box. To show units, select the Show Units check box.

 After you specify any grouping, subtotals, sorting, and formatting you want, click Next to continue to saving the report.

Saving Custom Reports

When you save a custom report, you name the report and choose a location in which to save the report definition file. To save a report, follow these steps:

1. In the Reports dialog box, type the name for the report. Use a brief but descriptive name for each report to identify your reports more easily. Type a detailed description in the Description box, in case the descriptive name you choose doesn't seem to be as descriptive when you see it later on. This description appears in the Description box in the Reports dialog box.

2. Choose an option to either save the report in the current drawing file or to a VRD file in a folder on your computer. Saving a report to a VRD file is more flexible, because you can run the report for any Visio drawing file. In addition, VRD files are XML files that contain the report specification, so you can modify them using any text editor. If you save the report in the current drawing file, you can access the report whenever the drawing file is open.

3. Click Finish. Visio adds the report to the Report list in the Reports dialog box.

 The final step in the save process only saves the report. You must click Run in the Reports dialog box to run a report and view the results.

Summary

You can link the shapes on Visio drawings with data in ODBC-compliant databases. The new Data Link feature simplifies creating one-way links from data sources to Visio drawings. If you want to create two-way links between shapes and database records, the Visio add-ons are still available.

You can set up connections that link shapes on drawings to database records. If you want to add shapes to drawings with links already in place, you can create masters with links to databases — to any record or to specific records. You can also generate drawings based on the records in a database table. After you link shapes to database records, you can update values by using shortcut menu commands or by specifying a refresh interval and letting Visio update your shapes or records automatically.

Part III

Using Visio for Office Productivity

Chapter 11

Collaborating with Others

I t takes a team to brainstorm, generate, and develop new ideas and processes. Together, you can all show your brilliance. Sometimes, however, you're toiling away as a team of one, and yet you probably still need to share your Visio drawings with others, whether they are reviewers, customers, or other outside resources.

Either way, with Visio, you can use e-mail to distribute drawings to others, from simple attachments to a sequential routing system. Collaboration comes in many forms, whether it's with non-Visio users, collaborating with team members in a shared workspace, or working with colleagues across continents and languages.

Another application of collaboration is exchanging and reviewing one another's Visio drawings, tracking markup to capture important ideas. Recognizing the importance of collaboration in its various forms, Visio includes many features that facilitate collaboration with others.

In this chapter, you learn how to distribute Visio drawings through e-mail, share drawings using Visio Viewer, and collaborate using document workspaces. You'll learn how to work with other languages in Visio. Finally, the chapter shows how to track changes in Visio and how to review and accept those changes into your drawings.

IN THIS CHAPTER

Distributing drawings

Sending drawings using e-mail

Routing drawings

Posting drawings to Exchange folders

Sharing drawings with the Visio Viewer

Working with document workspaces

Working with multiple languages

Tracking and reviewing changes

Distributing Drawings

Sending Visio drawings via e-mail is an easy way to get your drawings into the hands of colleagues or your target audience. Using e-mail features, you can send drawings as attachments and even route the e-mails sequentially to a series of reviewers. With a distribution list, you can deliver Visio drawings with a single e-mail. If you store the drawings in a public Microsoft Exchange folder, you distribute the drawing without clogging the e-mail system with multiple copies of attachments.

 Visio can distribute drawings using any e-mail program that supports the MAPI (Messaging Application Programming Interface) protocol.

Sending Drawings Using E-mail

You can send a Visio drawing from your e-mail program or directly from Visio, depending on your preferred method and the length of your attention span. If you're in and out of e-mail all day long, in a new message form in your e-mail program, choose Insert ⇨ File (or the equivalent command), and then select the drawing file you want to attach. For folks who tend to forget to attach attachments, which seems to be the rule rather than the exception, a better approach is to create an e-mail from within Visio with the Visio drawing attached automatically. To do this, follow these steps:

1. In Visio, open the drawing you want to send.

2. Choose File ⇨ Send To ⇨ Mail Recipient (As Attachment). Your e-mail program launches and a new e-mail message form appears with the current drawing as an attachment.

3. Enter the recipients for the drawing in the To and Cc boxes, revise the Subject if necessary, and then type your message in the message area.

4. When you're finished, choose Send or the equivalent command.

Routing Drawings

When several people must review or approve a drawing, the easiest solution is to route an e-mail to multiple recipients with the Visio drawing attached. You can route the drawing sequentially to one recipient after the other or to all recipients at once. After the recipients review or approve the drawing, the drawing routes back to the person who sent it in the first place. To route a drawing to multiple recipients, follow these steps:

1. In the Visio drawing you want to route, choose File ⇨ Send To ⇨ Routing Recipient. The Routing Slip dialog box appears.

2. Under the To box, click Address. The address book for your e-mail program opens.

3. Select the e-mail addresses for the recipients of the routed drawing, and then click OK to add the selected names to the To box.

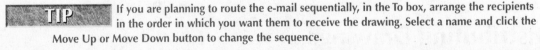 If you are planning to route the e-mail sequentially, in the To box, arrange the recipients in the order in which you want them to receive the drawing. Select a name and click the Move Up or Move Down button to change the sequence.

4. Revise the Subject if necessary and then, in the Message Text box, type your message.

5. Under Route to Recipients, select One After Another to route the drawing sequentially, which is the default. To route the drawing to all recipients simultaneously, select All At Once.

 When you route a drawing sequentially, each subsequent recipient can see the comments of the previous reviewers and add to them. If you want each person to provide comments without being influenced by others, select the All At Once option. With this option, you receive an e-mail from each recipient after he or she reviews the drawing.

6. To have the routing e-mail sent back to you after it has been routed, select the Return When Done check box.

7. To monitor who has the drawing at any given time, select the Track Status check box. This is particularly helpful if you're using the One After Another routing method.

8. When you're finished, click OK. Visio sends an e-mail with the current drawing attached to your routing recipients.

Distributing Drawings to Exchange Folders

Instead of sending a Visio drawing to a large number of recipients, posting the drawing to a public Exchange folder is a more effective way to distribute the drawing and keeps disk space consumption to a minimum. Then you can alert the recipients of the drawing's presence there so they can review it. To add a drawing to a Microsoft Exchange folder, follow these steps:

1. Open the drawing you want to add to an Exchange folder.

2. Choose File ➪ Send To ➪ Exchange Folder. The Send to Exchange Folder dialog box appears.

3. Select the folder, expanding your folders as needed. You can also create a new folder for your drawing by clicking New Folder, typing a name, and then pressing Enter.

4. Click OK. Visio saves the drawing in the Exchange folder, so people can open the file directly from that folder without copying it to their own computers.

CROSS-REF Another method for distributing drawings to others is to publish them on an intranet or Internet web site. For more information, see Chapter 9.

Sharing Drawings

Visio offers several methods for sharing drawings — even with people who don't have Visio installed on their computers. The Microsoft Office Visio Viewer opens and displays Visio documents, so people without Visio installed on their computers can still review Visio drawings. If your organization has set up a SharePoint document library, you and your team can employ Document Workspaces to share drawings and collaborate on their content.

Sharing Drawings with Colleagues Without Visio

If you're working with non-Visio users, one way to share drawings is to publish them to a web page (see Chapter 8). However, if you don't have a web site to publish to or want to restrict the audience that sees your drawings, you can send the drawings to the recipients you choose and they, in turn, can use the Microsoft Office Visio Viewer 2007 to open, view, and print Visio files. The Visio Viewer is an ActiveX control that displays Visio drawings in a Microsoft Internet Explorer (version 5.0 or later) window. With the Visio Viewer, someone can perform the following actions:

- View one drawing page at a time
- Navigate to another page in the drawing
- Zoom in and out
- View another area of the drawing
- Follow hyperlinks attached to shapes
- View shape data
- Print Visio drawings

As its name implies, the Visio Viewer is for viewing drawings, not editing them, so it doesn't show Visio elements, such as panes, stencils, rulers, guides, or guide points. The Visio Viewer doesn't support rotated pages, more than one hyperlink on a shape, drawing page hyperlinks, or drawing page properties. In addition, custom styles (including fills, line styles, and line ends) might not appear the same as they do in Visio.

The Microsoft Visio Viewer is available for download at the Microsoft Download Center (www.microsoft.com/downloads). To locate the download, in the Search box on the Microsoft Download Center site, type "Visio Viewer" and follow the instructions for downloading and installing the Viewer. In addition, Internet Explorer 7 has the Viewer built in.

You can use the Visio Viewer to open Visio drawings (.vsd files) saved in versions 2000 or later, or Visio XML format files (.vdx files) saved in version 2002 or later. Only Visio Viewer 2007 can open drawings created in Visio Professional 2007. Choose one of the following methods to open a Visio file using the Visio Viewer:

- If you do not have Visio installed on your computer, double-click a Visio file in Windows Explorer.
- If you do have Visio installed on your computer, right-click a Visio file in Windows Explorer, choose Open With ➪ Internet Explorer. If Internet Explorer is not listed, choose Program, select Internet Explorer, and then click OK.
- In your web browser, choose File ➪ Open and navigate to a Visio file.

NOTE If both Visio and the Visio Viewer are installed — for example, so you can see how your drawings appear in the Viewer or to test Visio Viewer capabilities — opening a file in your web browser launches Visio, rather than the Visio Viewer.

- Drag a Visio file from Windows Explorer into your web browser window.

To work with a Visio drawing in the Visio Viewer, use one or more of the following techniques:

- **Pan over a drawing** — Drag the drawing where you want it in the browser window.

- **Zoom in and out** — Use any of the following methods to zoom in and out:

  - To zoom in, right-click the drawing, and then click Zoom In. Or, on the Visio Viewer toolbar, click Zoom In.

  - To zoom out, right-click the drawing, and then click Zoom Out. Or, on the Visio Viewer toolbar, click Zoom Out.

  - To zoom using a zoom percentage, right-click the drawing, click Zoom, and then click a zoom percentage. Or, on the Visio Viewer toolbar, in the Zoom box, choose a zoom percentage.

  - To view the whole page in the Visio Viewer window, right-click the drawing, click Zoom, and then click Whole Page. Or, on the Visio Viewer toolbar, click Zoom Page.

- **Follow a hyperlink** — Position the pointer over a hyperlinked shape and click the shape. To return to the original place in your Visio drawing, click your browser's Back button.

- **Go to a different page** — Click the page tab for the page you want to navigate to. Or, press Ctrl+Page Down to move to the next page and Ctrl+Page Up to move to the previous page.

- **View shape data** — Double-click a shape. Shape data appears in the Properties and Settings dialog box on the Shape Properties tab. The tab is blank if the shape doesn't contain any shape data.

- **Print a drawing** — Display the drawing centered in the window as you want it to print and with the zoom factor you want. Choose the orientation you want and then, on the Visio Viewer toolbar, click Print.

- **Change drawing properties and settings temporarily** — If you want to view the drawing with different colors or with some layers hidden, on the Visio Viewer toolbar, click the Properties and Settings button. You can change the color of the drawing page, background, or layers. You can also hide layers, for example, to remove markup by selecting the Layer Settings tab and selecting or clearing layer check boxes.

Working with Document Workspaces

If your organization uses a SharePoint site, you can collaborate on Visio drawings and other Office documents through the Document Workspace feature. A Document Workspace provides a shared area with tools to share and work on files, and communicate with other team members about those files. Teammates can access this shared workspace through a web browser or the Shared Workspace task pane in Visio and can perform the following tasks:

- Share and work on drawings and related documents

- Exchange information
- Maintain lists and related links about the drawing
- Assign tasks regarding the drawing
- Update one another about drawing and task status

When you have the appropriate permission to create a Document Workspace, you can set one up to share drawings and invite the members you want to participate. Members work on their versions of the drawing, and update them periodically to the web server. The other members receive updates so that all Document Workspace members can see the changes that others have saved to the drawing so far.

As long as you have the appropriate permissions, you can create a Document Workspace for your drawing as a subsite of a SharePoint site. You become the administrator of any Document Workspace you create.

Creating a Document Workspace in Visio

If you are working on a drawing and want to create the workspace without launching another program, you can create a workspace from within Visio. To create a Document Workspace for a Visio drawing using the Shared Workspace Task Pane in Visio, follow these steps:

1. Choose Tools ⇨ Document Management. The Document Management Task Pane appears.

NEW FEATURE In Visio 2007, the name of the Task Pane for workspaces has changed from Shared Workspace to Document Management.

2. In the Document Workspace Name text box, type a descriptive name for the drawing workspace.

3. In the Location for a New Workspace box, enter the web address for the SharePoint site, as illustrated in Figure 11-1.

NOTE If you don't know the web address for the SharePoint site, check with your system administrator. If you are the system administrator, refer to your notes for the URL or check your Favorites list.

4. Click Create.

5. In the Document Management Task Pane, click Members. At the bottom of the task pane, click Add New Members.

6. Type the names of the members you want to add to your Document Workspace, separating them with semicolons. You might use e-mail addresses or Windows SharePoint Services user names, depending on how your system administrator has set up Windows SharePoint Services users.

FIGURE 11-1

Create a new Document Workspace using the Document Management Task Pane.

Creating a Document Workspace by Sending an E-mail

This approach is particularly efficient, because it creates the workspace, adds the members, and notifies them all at the same time. To create a Document Workspace for a Visio drawing using Microsoft Outlook 2007, follow these steps:

1. Choose File ➪ Send To ➪ Mail Recipient to create an e-mail with the current Visio drawing as an attachment.

2. Enter the e-mail addresses of all the individuals with whom you are collaborating on the drawing. These people become members of the Document Workspace.

3. Edit the subject and type a message as needed.

4. In the Attachment Options Task Pane, click Shared Attachments.

 If the Attachment Options Task Pane is not visible, click Attachment Options.

5. In the Create Document Workspace At box, enter the web address of your SharePoint site. As long as you have permission to create Document Workspaces for this web site, the Document Workspace will be created as a subsite of the SharePoint site, using the e-mail recipients as members.

6. Click Send.

To create a Document Workspace for a Visio drawing from within your Windows SharePoint Services web site, follow these steps:

1. In your web browser, go to the SharePoint site.

2. On the Site actions menu, click Create.

3. Under Web Pages, click Sites and Workspaces.

4. Type a title, description, and web address.

5. In the Permissions section, select a permission setting. By default the workspace uses the same permissions as the parent site. Choose Use Unique Permissions to assign permissions to the members of the workspace.

6. In the Template Selection section, select the Collaboration tab, and then click Document Workspace

7. Click Create.

Whenever you create a Document Workspace for a drawing, that drawing is automatically added to the SharePoint document library. Any time a member of the Document Workspace opens a drawing stored in the document library, the Shared Workspace Task Pane opens as well.

 **For drawings stored in a SharePoint document library, you can create a Document Workspace from the document library. In your web browser, go to the SharePoint site and then open the document library. Point to the name of the drawing, click the Edit arrow, and then click Create Document Workspace.**

Working with Drawings in a Document Workspace

Here are the steps to opening a drawing and beginning your collaboration work with it:

1. To open a file, use one of the following methods:

   ■ If you received an e-mail with a Visio drawing attachment, double-click the attachment to open it. A message indicates that the drawing is stored in a Document Workspace.

   ■ Choose File ➪ Open. In the Open dialog box, under Look In, click My Network Places, select the SharePoint site (or type the URL in the File Name box), and click Open. Select the library that contains the drawing and click Open. Finally, select the file and click Open.

NOTE **When the drawing opens in Visio, and the Document Management Task Pane appears, you know that the drawing is part of a Document Workspace.**

2. In the Document Updates Task Pane, click Get Updates to immediately update the content of your version of the drawing. As other team members update and save their version of the drawing, the Document Management Task Pane indicates that updates are available.

3. Make any changes you want to the drawing. If another member of the Document Workspace has specified a particular aspect of the drawing for you to work on, you might see a task assigned to you in the Document Management Task Pane.

> **TIP** The Document Workspace administrator can establish that drawing changes should be made with Track Markup turned on. If that's the case, then the changes made by each member of the workspace appear in a different overlay layer. If the workspace administrator has not turned on Track Markup but you want your changes to show as markup, choose Tools ⇨ Track Markup.

4. Save the drawing periodically, as usual.

5. If you must check in a file to the Document Workspace, choose File ⇨ Check In.

> **TIP** You can also edit the shared drawing in the SharePoint site for the Document Workspace if you are using Internet Explorer 6.0 or later. In the document library containing the drawing, point to the name of the document and then click the Edit arrow that appears.

Deleting Document Workspaces

When you and your team are finished collaborating on a drawing, you can delete the Document Workspace. When deleting a Document Workspace, keep the following principles in mind:

- Deleting a document workspace can be done only by the administrator of the Document Workspace — that is, the person who created it.
- It deletes all the data in the Document Workspace.
- It removes the associated document library, including all the documents stored there.
- It does not delete your own copy of documents stored on your computer.

To delete a shared workspace from the SharePoint site, follow these steps:

1. Use your web browser to go to the SharePoint site and the Document Workspace.

2. In the Site Actions menu, choose Site Settings.

3. Under Site Administration, click Delete This Site, and then click Delete.

Managing Shared Workspace Tasks

Members of a Document Workspace can create and assign tasks associated with the shared drawing to other members. You can assign to-do items with due dates to members of the shared workspace. To assign a task to another member, follow these steps:

1. With the shared drawing open, in the Document Management Task Pane, click Tasks and then click Add New Task.

2. Complete the fields in the Task dialog box. This includes the task title, current task status, priority, the Document Workspace member to whom you want to assign the task, any description, and the due date and time.

3. Click OK. The task is added to the Document Management Task Pane. All members of the workspace see the task assignment and associated information.

If another member has assigned a task to you, after completing it you can check it off in the Tasks list. When other team members open the Tasks list in the Document Management Task Pane, they can see that you have completed the task. To check off a completed task that has been assigned to you, follow these steps:

1. With the shared drawing open, in the Document Management Task Pane, click Tasks. The list of all tasks assigned to all Document Workspace members appears.

2. Select the task assigned to you.

3. In the Task dialog box, change the status to indicate that it's complete. Enter any information in the Description box.

4. Click OK. The check box is selected, indicating that your task is complete.

Working with Multiple Languages

If you share your Visio drawings with colleagues or customers in other countries, such as Hungary, Greece, or Japan, you might need to include elements of other languages and other language formats in a single drawing. You can also share and collaborate on drawings across multiple languages.

With multilanguage support, you can do all of the following:

- Flexibly format date, time, and number styles according to a specific region and language.

- Type characters for Asian languages using an Input Method Editor (IME).

- Link fonts automatically to find needed characters in other languages. If a selected font does not include all the required characters, Visio automatically links to a second font to find the missing characters. This is particularly useful in multilingual drawings that include East Asian and right-to-left text.

- Create multilingual web pages and intranet content in Visio.

To work with additional languages in Visio, you might need to adjust settings in the Windows Control Panel, in Microsoft Office, and in Visio itself. Some languages require additional resources installed.

NOTE Windows installs many files needed for multilanguage support. However, if the language you're using requires additional Windows resources, learn how to install that language by opening Windows Help and searching for "Installing languages." Pick the task for the language you want to install, such as East Asian language files, and follow the instructions. For some characters in some languages, you might also need to install a particular keyboard layout.

Visio multilanguage support includes the following:

- **Unicode** — A character encoding standard that enables almost all the written languages in the world to be represented by using a single character set. It uses more than a single byte to represent each character. Unicode makes it possible for multiple languages to appear in a single Visio drawing.

- **End User Defined Character (EUDC)** — A character set with which you can form Asian names and other Asian words using characters that are not available in standard screen and printer fonts.

- **GB18030** — A Chinese character-encoding standard that contains Chinese characters.

Configuring Office for Multiple Languages

To work with different languages in Visio, enable the appropriate languages to make additional language-specific options available. To do this, follow these steps:

1. Click Start and choose All Programs ⇨ Microsoft Office ⇨ Microsoft Office Tools ⇨ Microsoft Office 2007 Language Settings.

2. Select the Editing Languages tab.

3. Under Available Editing Languages, select the language you want to use for editing Office documents, including Visio drawings, and then click Add. The language appears in the Enabled Editing Languages box. If you haven't installed all the components required for that language, the words "limited support" appear next to the language.

4. When you're finished adding languages, click OK.

 If your organization has purchased the 2007 Microsoft Office Language Pack, you can also change the language of the user interface and Help.

Installing East Asian Language Files

To enter ideographic characters for Asian languages, you must use an Input Method Editor (IME). This feature is available only if support for Japanese, Simplified Chinese, Traditional Chinese, or Korean is enabled through Microsoft Office Language Settings as described above. To install East Asian language files and the Input Method Editor that you use to enter characters for those languages, do the following:

1. In the Windows Control Panel, click Date, Time, Language, and Regional Options.

2. Under Pick a Task, click Add Other Languages.

3. On the Language tab, under Supplemental Language Supports, select the Install Files For East Asian Languages check box.

4. Click OK. When you see the prompt to insert the Windows CD-ROM, do so or browse to a disk or share network folder in which the language files are located.

5. Follow the prompts to install the files and restart your computer.

After you have installed the IME for the language you're working with, you can access it from the Language bar that appears by default in the upper-right corner of the Visio screen.

Tracking and Reviewing Changes

One of the most powerful means of collaborating on documents is the capability to track changes using some type of markup. With the Visio Reviewing Task Pane and Reviewing toolbar, it's easy to track and review changes using separate colored overlays for each reviewer. When track markup mode is turned on, reviewers can add text comments, shapes, or use Ink to create freehand markups on a drawing. After the reviewers have finished their work, the person assigned to incorporate all changes and suggestions can review the comments and other markup, and make whatever changes are needed.

Turning Markup On or Off

When track markup mode is on, reviewers can add comments and other markup. Each reviewer's markup is kept separate from the original drawing and markups from all other reviewers. To turn on track markup mode, follow these steps:

1. Open the drawing that team members are going to review.
2. Choose Tools ⇨ Track Markup. A colored band appears around the drawing workspace, and the Reviewing Task Pane and Reviewing toolbar appear, as illustrated in Figure 11-2. In addition, a message box appears telling you that the document is ready for review.

FIGURE 11-2

Visual cues, including the Reviewing task pane, indicate that track markup mode is on.

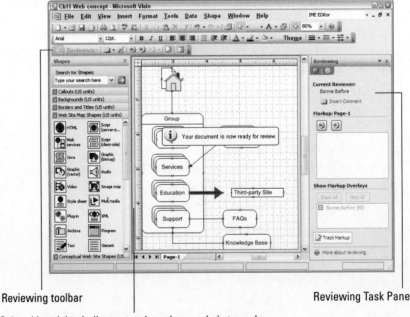

Reviewing toolbar

Colored band that indicates track markup mode is turned on

Reviewing Task Pane

 To edit markup, track markup mode must be turned on. However, to edit the original drawing, you must turn track markup mode off.

To turn track markup mode off, at the bottom of the Reviewing task pane, click Track Markup. You can also choose Tools ⇨ Track Markup, which acts as an on/off toggle. When markups exist and track markup mode is turned off, tabs appear on the right side of the drawing window, showing the original drawing, and the overlays of each reviewer's comments, each one in a different color.

 To print a drawing without your markup and comments showing, be sure to turn off track markup mode.

Marking Up Drawings

When you're marking up a Visio drawing with track markup mode turned on, Visio assigns you an overlay in a particular color. You can add markup, including comments, shapes, and Ink to your overlay without affecting the original drawing or other reviewers' markup.

CAUTION Markup appears only in the reviewer's assigned color. Although a reviewer can apply colors to shapes and Ink, those colors aren't visible until the shape is copied or moved onto the original drawing.

Inserting Comments

To add a text comment as drawing markup, follow these steps:

1. Make sure that track markup mode is on for the drawing.
2. If your comment is associated with a particular page in a multipage drawing, display the page.
3. In the Reviewing Task Pane, click Insert Comment. A comment bubble appears in the drawing.
4. Type your comment. When finished, click the drawing page away from the comment. The bubble disappears, but the comment marker with your initials remains, associated with the current page. A list of your comments builds in the Reviewing task pane, as illustrated in Figure 11-3.

Inserting Shapes

To add a shape to a drawing, simply drag it from the stencil into place. The shape appears in the color of your markup, and "Shape added" appears with your initials in the Reviewing task pane.

Although you can review other reviewers' markup, you can change or remove only your own. To remove a markup, select it in the drawing or in the Reviewing task pane and then press Delete.

FIGURE 11-3

The comments, shapes, and Ink that reviewers add to a drawing as markup appear in the Reviewing task pane preceded by the reviewer's initials.

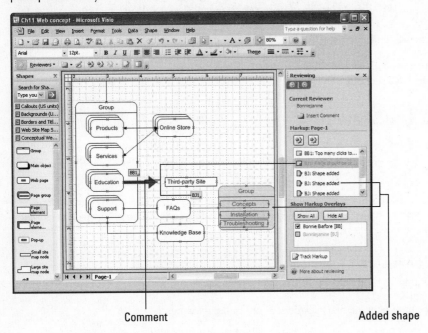

Comment   Added shape

Adding Freehand Markup Using Ink

Ink is the name of the freehand method of annotating a drawing in track markup mode. With Ink, you can draw shapes and add handwritten notes. These shapes behave like any other shape. You can even add them to custom stencils if they're shapes you want to reuse, such as your initials to indicate your approval of a change. While Ink facilitates tablet computer input with a stylus, you can use your mouse on a desktop or notebook computer to draw freehand markup. To use Ink to mark up a drawing, follow these steps:

1. Make sure that track markup mode is on for the drawing.

2. If you're adding Ink on a particular page in a multipage drawing, display that page.

3. On the Reviewing toolbar, choose the Ink tool. The Ink toolbar appears, showing different Ink colors, an Eraser tool, a Color tool, and a Line-width tool, as shown in Figure 11-4.

4. Choose the tools on the Ink toolbar to set up the markup you're adding.

5. Use your computer's pointing device (such as the mouse or stylus) to draw the markup shape or use handwriting to mark up the drawing. After you finish a shape, it is converted to a shape that can be manipulated as a unit like any other Visio shape. The message "Ink added" appears with your initials in the Reviewing task pane.

6. When you're finished using Ink, choose the Pointer tool on the Standard toolbar.

TIP You can set the speed of your Ink entry conversion to a shape. Choose Tools ➪ Options and then select the Advanced tab. Under Ink Tool, drag the slider in the direction you want to indicate how fast or slow you want an Ink entry to be transformed into a Visio shape.

FIGURE 11-4

You can add handwritten comments or shapes to markup with the Ink tool.

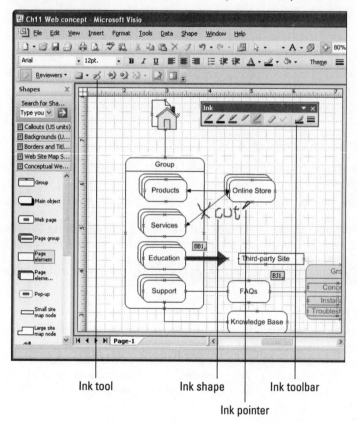

Ink tool   Ink shape   Ink toolbar

Ink pointer

Reviewing Markup

By default, every reviewer's markup is displayed, each in a different color. To see a reviewer's overlay, select the tab containing the reviewer's initials on the right edge of the drawing. To see the original drawing containing all reviewers' markups, select the Original tab.

To hide all markup, in the Reviewing Task Pane, click Hide All. To show all markup again, click Show All. To hide just the markup of selected reviewers, in the Reviewing Task Pane under Show Markup Overlays, clear the check boxes for those reviewers. To specify which reviewers' markup should show, select the check boxes for those reviewers.

Updating Drawings with Markup Changes

To incorporate markups into the original drawing, you first must turn off track markup mode. Then you can review markups and copy elements into the original drawing. To do this, follow these steps:

1. In the Reviewing task pane, click Track Markup to turn track markup mode off.

2. Select the tab for the reviewer whose markup you want to incorporate into the original drawing. Visio selects all the shapes on the overlap by default.

3. Select the shape(s) you want to copy, and then, on the Standard toolbar, click the Copy tool.

4. Select the Original tab at the lower-right edge of the drawing.

5. Click the Paste tool on the Standard toolbar. The copied shapes appear at the center of the drawing.

6. Drag the shape(s) to move them into place, using the markup overlay as a guide.

 To move from one markup to the next, in the Reviewing task pane or the Reviewing toolbar, choose the Next Markup tool. To delete a markup, select it, and then choose the Delete Markup tool in the Reviewing task pane or the Reviewing toolbar.

Summary

Visio helps you work as closely as you need to with colleagues, whether they're right next door or on the other side of the globe. Using e-mail, a Windows SharePoint Services site, or sophisticated layers of markup, you can discuss, experiment, and hammer out the most innovative ideas and processes. You can then effectively capture those ideas and processes in your Visio drawings.

Chapter 12

Building Block Diagrams

Visio block diagrams are the workhorses of the diagram world—easy to create and yet able to communicate an astounding variety of ideas. Block diagrams help illustrate flow within processes as well as the structure and relationships for all sorts of elements, such as ideas, concepts, designs, business units, manufacturing components, causes of problems, and more.

In Visio, you can create simple diagrams using basic shapes or produce eye-catching diagrams for presentations with 3-D shapes using perspective. In addition, the Visio Block Diagram stencil includes shapes to develop more specialized arrangements, such as hierarchical trees or onion diagrams. In this chapter, you'll learn how to create different types of Visio block diagrams and configure your Block Diagram shapes.

NEW FEATURE In Visio 2007, the three Block Diagram templates (Basic Diagram, Block Diagram, and Block Diagram with Perspective) have moved to the General category instead of residing in their own Block Diagram category as in Visio 2003.

Exploring the Block Diagram Templates

The Visio Block Diagram templates are some of the most popular templates because workers from any industry can communicate their ideas to their colleagues using Visio techniques they already know, such as dragging, dropping, connecting shapes, editing, and formatting. Block Diagram shapes are so simple that the Block Diagram templates don't contain any specialized menus, toolbars, or add-ons. You can build diagrams that satisfy many different requirements by dragging and dropping shapes from the Visio Block Diagram stencils.

You can modify and tweak Block Diagram shapes by dragging selection handles or control handles. Annotation is as easy as selecting a shape and typing. Some of the Block Diagram shapes and connectors include powerful yet simple-to-use features that speed up common diagramming tasks. For example, you can connect branches on one of the Visio Tree shapes to boxes on a drawing to construct a hierarchical tree.

Choosing the Right Template

When you review the three Block Diagram templates in the General category, their descriptions sound quite similar — they document structure, hierarchy, and flow using a combination of 2-D or 3-D shapes. The variety of diagrams that you can create with these templates boil down to three fundamental diagram types: blocks, trees, and onions, as illustrated in Figure 12-1.

- **Block diagrams** communicate steps in a process, such as how to set the alarm on your watch, or show the relationships between elements, for example, the interaction between departments handling a product return. Block diagrams often use geometric shapes such as rectangles and circles connected with arrows, but, in Visio, you can replace simple geometry with shapes that indicate the function of a department or pieces of equipment.

- **Tree diagrams** present hierarchical information, such as the ancestors and descendants in a family tree or the advancement of teams in tournament play-offs.

- **Onion diagrams** illustrate relationships that build from a core. For example, an onion diagram is the best way to show the layers that make up the earth from its core to the crust.

The three fundamental types of block diagrams don't correspond directly to the three templates that Visio provides. Use Table 12-1 to choose the template that contains the shapes you need for the diagram type you want to create. Each Block Diagram template automatically sets the page to a letter-size sheet with portrait orientation and uses inches drawn at one-to-one scale.

TABLE 12-1

Templates for Block Diagrams

Diagram Type	Template	Features
Basic Blocks	Basic Diagram	Opens the Basic Shapes, Borders and Titles, and Backgrounds stencils.
Blocks with Style	Block Diagram	Opens the Blocks, Raised Blocks, Borders and Titles, and Backgrounds stencils.
Tree	Block Diagram	Opens the Blocks, Raised Blocks, Borders and Titles, and Backgrounds stencils. Tree shapes show hierarchy.
Onion	Block Diagram	Opens the Blocks, Raised Blocks, Borders and Titles, and Backgrounds stencils. Concentric and Partial Layer shapes build onion diagrams.
High-Impact Blocks	Block Diagram with Perspective	Opens the Blocks with Perspective, Borders and Titles, and Backgrounds stencils. 3-D and Vanishing Point shapes show perspective.

FIGURE 12-1

Block diagrams can show flow, hierarchical structure, or concentric layers.

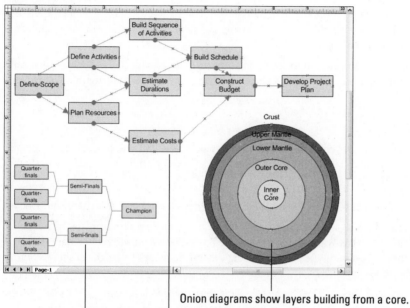

Onion diagrams show layers building from a core.

Block diagrams can show flow or relationships.

Tree diagrams present hierarchies such as family trees or playoff results.

CROSS-REF The General category also includes the Basic Flowchart template. See Chapter 15 to learn how to use that template.

Exploring Block Diagram Shapes

Every Block Diagram template sets up the same basic environment, so you can use Block Diagram templates almost interchangeably. Each stencil within the Block Diagram family offers some specialized shapes to address specific diagramming needs. By understanding the shapes available on each stencil, you can open the stencils as you need them, regardless of the Block Diagram template you start with.

Basic Shapes

The Basic Shapes stencil is a workhorse for simple block diagrams. It offers basic geometric shapes, arrow-like shapes that can act as connectors, as well as the standard Dynamic connector and Line-curve connector.

- **Geometric Shapes** — Drag and drop geometric shapes such as Rectangles, Circles, Stars, Rounded rectangles, Shadowed or 3-D boxes, and shapes for polygons from Triangles to Octagons.

- **Arrows** — Drag and drop Arrow shapes onto a drawing to connect the geometric shapes. You can choose from arrows with different arrowheads and tails, but you can modify only the length and width of the arrows.

- **Flexi-Arrows** — In addition to modifying the length and width of these shapes, you can drag control points on Flexi-arrow shapes to change the angles and lengths of arrowheads and tails.

Blocks

The Blocks stencil is the most versatile, with shapes for block, tree, and onion diagrams. It contains connectors with dozens of different end styles. Block shapes come with behaviors that help show relationships and flow.

- **Geometric Shapes** — Box, Diamond, and Circle shapes can represents steps or components.

- **Auto-sizing Boxes** — The Auto-height Box obligingly increases or decreases its height to accommodate text as you type. The Auto-size Box increases its width to contain the longest line of text you type before pressing Enter. It increases its height to show all the lines of text you type.

- **Open/Close Shapes** — The Open/close Bar and the Open/close Arrow shapes can display borders to represent a boundary, or hide borders so shapes appear to flow together.

- **Arrow Box** — This shape combines a box and an arrow to show both a process and the flow to the next step.

- **Arrows** — Drag the control points on the Curved Arrow shape to change the direction of the arrowhead and the curvature of the bend.

- **Onion Shapes** — Concentric Layer and Partial Layer shapes drop on top of each other to show relationships around a central core.

- **Tree Shapes** — Tree shapes show hierarchies with two to six branches.

- **Connectors** — In addition to the Dynamic connector and the Line-curve connector, the Blocks stencil includes several connectors with specialized styles, such as dots and arrows, at the end or midpoint.

Blocks Raised

The Blocks Raised stencil contains geometric shapes and arrows that appear three-dimensional, but the height and orientation of the third-dimension do not change as you move a *vanishing point*. These shapes can add visual interest to diagrams without the extra step of managing perspective with vanishing points. If perspective is what you want, use the shapes on the Blocks with Perspective stencil, described in the next section.

Blocks with Perspective

The Blocks with Perspective stencil contains geometric shapes and arrows that change their perspective in relation to a vanishing point. These shapes add visual impact to diagrams when you adjust the depth and angle of perspective by moving the Vanishing Point shape on the drawing.

- **Geometric Shapes** — Drag Block, Circle, Arrow, and Elbow shapes that adjust to the position of a Vanishing Point shape.

- **Holes** — Drag a Hole shape onto another shape to create the appearance of a hole.

- **Wireframe Blocks** — These shapes are three-dimensional boxes in which the edges are visible and the sides are transparent.

- **Vanishing Point** — Drag a second Vanishing Point shape onto a drawing to add more impact to diagrams.

Showing Structure and Flow

Block diagrams work equally well to represent static structural relationships or the dynamic flow between processes or steps. Boxes and other geometric shapes represent components, and arrows or connectors indicate order or hierarchy. For processes and procedures, geometric shapes signify each process or step, while arrows show the dependencies and sequence between them. No matter which type of relationship you're trying to communicate, you can use the same basic steps to create your diagram.

Creating Block Diagrams

You can use basic Visio techniques to create your Visio Block diagrams — begin by dragging shapes onto a drawing and typing the text you want to appear in each shape. Then, connect shapes as you go or attach arrows and connectors after the shapes are in place. If necessary, you can fine-tune the appearance of the diagram by rearranging the shapes and formatting them. To create a block diagram, follow these steps:

1. Choose File ➪ New ➪ General ➪ Block Diagram.

TIP To access other Block Diagram shapes, open the Basic Shapes stencil by choosing File ➪ Shapes ➪ General ➪ Block Diagram ➪ Basic Shapes.

2. Drag shapes from the Basic Shapes, Blocks, or Blocks Raised stencils onto the drawing (or use the new AutoConnect feature).

NEW FEATURE Visio 2007 introduces the AutoConnect feature, which simplifies connecting shapes whether you want to connect shapes as you drag them onto the drawing page, select them in a stencil, or connect shapes already on the drawing. See Chapter 5 to learn more about the new Visio 2007 connection methods.

3. Select a shape and type the text that you want to appear in the shape.

4. If you don't use AutoConnect, connect shapes by dragging an Arrow shape from one of the stencils and gluing it to a shape on the drawing. A red square highlights a shape connection point when the Arrow shape connects to another shape. After one end of the Arrow is connected to a shape, drag the end point at the other end and glue it to another shape.

Modifying Block Diagrams

Block diagrams don't include any specialized formatting or layout tools. You can use basic techniques to rearrange the shapes on a drawing or apply predefined backgrounds and color schemes to enhance its appearance. To adjust the overall appearance of a diagram, use any of the following techniques:

- **Rearrange Shapes** — Rearrange shapes in a diagram by dragging them to new locations. You can also specify the arrangement of shapes by choosing Shape ➪ Configure Layout. Then, you can reapply that layout by choosing Shape ➪ Re-layout Shapes.

- **Adjust Shape Location** — To make minor adjustments to the position of a shape, select it and then press one of the arrow keys to nudge it in that direction.

CROSS-REF   To learn about other ways to modify the location of a shape, see Chapter 4.

NEW FEATURE   In Visio 2007, you apply color schemes with the new Themes feature, described in Chapter 7. Themes replace the Visio 2003 Apply Color Schemes feature.

Modifying Block Diagram Shapes

Some Block Diagram shapes exhibit special behaviors, but you still use basic Visio techniques to modify all the shapes available in the Block Diagram stencils. Use any of the following techniques to modify shapes on a block diagram:

- **Add Text** — To add or modify text for an existing shape, select the shape and begin typing. You can also double-click a shape to edit its existing text.

NOTE   When you type text in a shape, Visio zooms in to make the text more legible.

- **Modify Relationships** — Select a connector or Arrow shape. Drag one of its end points and glue it to another shape or another connection point on the same shape.

- **Resize Shapes** — Select a shape and drag one of its square, green selection handles to resize it. Drag a corner to modify height and width proportionately. Drag a mid-point selection handle to change just one dimension.

- **Reshape Shapes** — Activate the Line, Arc, Pencil, or Freeform tool on the Drawing toolbar and select a shape. Drag a vertex (a green diamond) to reshape it.

- **Bend Shape Segments** — Activate the Line, Arc, Pencil, or Freeform tool on the Drawing toolbar and select a shape. Drag an eccentricity handle (a green circle) to bend one segment of the shape.

- **Reorder Overlapping Shapes** — To bring a shape to the very front of the stack or move to the very bottom of the stack, right-click it and choose Shape ➪ Bring to Front or Shape ➪ Send to Back, respectively. To move a shape one position in the stack, use Bring Forward or Send Back.

- **Format Shapes** — To apply formatting to a shape, right-click it, choose Format from the shortcut menu, and then choose one of the Format commands.

- **Specify Shadow Colors** — To set the shadow colors for Raised Blocks and Blocks with Perspective, right-click a shape and choose one of the shadow color options from the shortcut menu. You can choose from two options:

  - **Automatic Shadow** — Sets the shadow color based on the shape's fill color, which is controlled by the Theme that you applied (or the default shape color, if you haven't applied a Theme). This is the default setting.

  - **Manual Shadow** — Displays the shadow color you specified for the shape. To specify shadow color, select a shape, choose Format ➪ Fill or Format ➪ Shadow from the text menu, and select the color you want from the Color drop-down list.

Using Special Editing Techniques for Boxes

You can use special behaviors, control handles, and shortcut menu options that come with some of the Block Diagram boxes to modify their appearance. If you use these shapes, take advantage of the following editing shortcuts:

- **3-D box** — Drag the control point on a 3-D box to modify the depth and direction of the 3-D faces of the box.

- **Auto-height box** — Type text in an Auto-height box, and the height of the box changes automatically to accommodate the text you enter. To adjust the width of the box, drag one of the side selection handles.

- **Auto-size box** — Type text in an Auto-size box, and the height and width of the box changes to fit your text. Press Enter to start a new line. The box width is set by the longest line of text, whereas the height is set by the number of lines of text.

Modifying the Appearance of Block Diagram Arrows

You can use special behaviors, control handles, and shortcut menu options that come with some of the Block Diagram arrows to modify their appearance. If you use these shapes, take advantage of the following editing shortcuts:

- **Arrow box** — Drag the control point on the arrowhead to adjust its width. Drag the control point at the intersection of the arrow and the box to change the height of the box and the length of the arrow.

■ **Flexi-arrows** (in the Basic Shapes stencil) — Drag the control points on the arrowhead to change the width and shape of the arrowhead and the width of the arrow tail, as shown in Figure 12-2.

FIGURE 12-2

Dragging yellow control handles (shaped like diamonds) reshapes the arrowhead and arrow tail of the Flexi-arrow shapes.

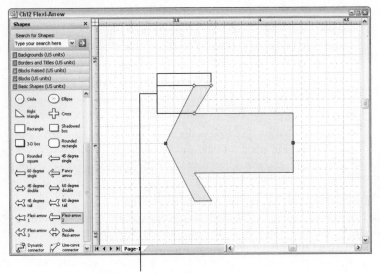

Drag to change arrow shape

■ **Curved Arrow** — Drag the control point on the arrowhead to reposition the arrowhead. Drag the control point at the curve to change the curvature.

Emphasizing Flow Between Shapes

Flow is easier to see on a diagram when there are no boundaries between shapes. Several Block Diagram shapes hide or show boundaries to emphasize flow. In the Blocks stencil, these shapes include the 1-D Single Arrow, 2-D Single Arrow, and Open/closed Bar shapes. In the Blocks Raised stencil, you can open and close Right Arrows, Up Arrows, Left Arrows, and Down Arrows, Horizontal Bars, Vertical Bars, and Elbow shapes. To open and close these shapes, drag a shape onto the drawing and then use one of the following methods:

■ To open the end of an arrow, right-click a closed shape and choose Open Tail from the shortcut menu. Drag the open end of the arrow to the flat side of a box or other block shape.

■ To close the end of an arrow, right-click the shape and choose Close Tail from the shortcut menu.

- To open or close Bar shapes from the Blocks or Blocks Raised stencil, right-click the bar and choose one of the following commands:
  - Open Left End Only
  - Open Right End Only
  - Open Both Ends
  - Close Both Ends

NOTE For vertical bars, the first two commands on the shortcut menu change to Open Top End Only and Open Bottom End Only.

Creating Hierarchical Trees

You can use tree diagrams to show hierarchies such as play-off standings or genealogy. As with other block diagrams, you can drag, arrange, and format shapes using basic Visio tools. Tree connectors include control points to help build your hierarchy. To build a hierarchical tree, follow these steps:

1. Choose File ➪ New ➪ General ➪ Block Diagram.
2. Drag boxes from the Blocks stencil.
3. Drag one of the four Tree shapes from the Blocks stencil onto the drawing.

NOTE You can choose from Trees with square or sloped branches. The Double-tree Sloped and Double-tree Square shapes provide only two branches. With the Multi-tree Sloped and Multi-tree Square shapes, you can draw from two to six branches.

4. To create a vertical tree, rotate the shapes by dragging one of the green selection handles on a horizontal Tree shape.

TIP Press Ctrl+L to rotate a Tree by 90 degrees. With horizontal Trees, you can also press Ctrl+H to flip the Tree from right to left, for instance to illustrate the ancestors in one show dog's pedigree rather than the offspring of two prize-winning hounds.

5. To connect a branch to a shape, drag the control handle at the end of a branch to a connection point on the shape. A red square highlights the connection point when the branch and the shape are connected.
6. To add text to the trunk of a tree, select the Tree and type the text you want.

NOTE You can add text to trunks only, not the branches of a tree. Use separate text blocks to add text at the branches.

Modifying Tree Shapes

Basic Visio techniques are all you need to modify and format the blocks and text in tree diagrams. You can use control points and built-in behaviors to modify tree trunks and branches. Use the following methods to modify trees:

■ **Add a Branch** — Drag the control handle on the trunk of a Multi-Tree shape to a position. With the control handle, you can drag out four extra branches. If you need more than six branches, drop a second or third Multi-Tree shape on top of the first. The distance perpendicular to the trunk controls the width of the branch, whereas the distance parallel to the trunk determines the length of the branch, as illustrated in Figure 12-3.

FIGURE 12-3

Drag Multi-Tree control handles to add branches to the tree and glue branches to other shapes.

Drag to control length and position of branch or to glue to box.

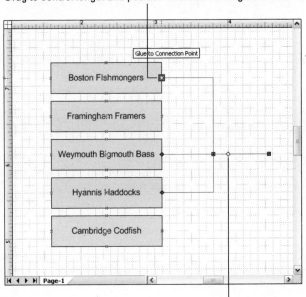

Drag to add a branch to tree.

■ **Remove a Branch** — Drag the control handle at the end of the branch on top of any other control handle on the tree.

■ **Adjust Branch Position** — Drag the control handle at the end of a branch to a new position.

■ **Modify Distance Between Branches** — Move the boxes attached to the branches to new positions. You can move these boxes by dragging them individually or as a group, or by applying the Align Shapes or Distribute Shapes commands from the Shape menu.

■ **Move a Tree Trunk** — Select the Tree and press an arrow key to move the trunk in that direction.

CAUTION When you move the shape connected to a tree trunk, the tree trunk rotates. One end of the trunk moves with the shape while the other end of the trunk stays fixed. To keep the structure of the trunk and branches of a tree orthogonal, select the tree and all the shapes connected to it and drag them all to a new position. If you want to move only the shape connected to the trunk, select the Multi-Tree shape when selecting the shape connected to the trunk.

Adding Impact with 3-D Block Diagrams

3-D block diagrams are visually appealing, so they're perfect for presentations. Although they look like they require hours of effort, they're just as easy to construct as regular block diagrams. When you create a block diagram using the Block Diagrams with Perspective template, the drawing includes a vanishing point that defines the perspective for the three-dimensional shapes. You can adjust the depth and orientation of the shape shadows by moving the Vanishing Point shape on the drawing.

CROSS-REF To learn how to change the color of the shadows for 3-D shapes, see the "Modifying Block Diagram Shapes" section earlier in this chapter.

NOTE Only shapes from the Blocks with Perspective stencil adjust to the position of the Vanishing Point shape. Shapes from the Blocks Raised stencil might look three-dimensional, but their depth and orientation remain fixed.

To create a 3-D block diagram, choose File ➪ New ➪ General ➪ Block Diagram with Perspective to open a drawing that contains a Vanishing Point shape. Drag 3-D shapes from the Blocks with Perspective stencil onto the drawing. You can drag, arrange, align, and format 3-D block shapes with basic Visio tools.

The depth and orientation of a 3-D shape changes as you move the Vanishing Point on a drawing. To add impact or emphasize specific parts of a diagram, you can change the perspective on the diagram, change the depth of a 3-D shape, or disconnect a 3-D shape from the Vanishing Point.

Modifying Perspective

Use the following methods to modify the perspective of a 3-D block diagram:

- **Change the Diagram Perspective** — Make sure no shapes are selected and then drag the Vanishing Point to another location.

- **Change a Shape's Perspective** — To change the perspective for one shape, select the shape and then drag the red control handle on the Vanishing Point to another location. The Vanishing Point's control handle that you move turns yellow, indicating that it and the selected shape are no longer connected to the Vanishing Point. However, when you select the Vanishing Point again, a red control handle still appears for the rest of the shapes connected to the Vanishing Point.

WARNING   If a shape on the drawing doesn't adjust its perspective as you move the Vanishing Point, it might not be a shape with perspective, such as a Box on the Blocks Raised stencil. However, the problem could be that a shape with perspective has become disconnected from the Vanishing Point. A yellow control point appears somewhere on the drawing page if you select a shape with perspective that is no longer connected to the Vanishing Point.

- **Connect a Shape to the Vanishing Point** — Select the shape. Drag the yellow control handle that appears on the drawing page and glue it to the connection point on the Vanishing Point shape.

- **Change a Shape's Depth** — Right-click a shape and choose Set Depth from the shortcut menu. Select a smaller percentage for a shallower shape, a larger percentage for a deeper shape.

TIP   You can hide the Vanishing Point shape — for example, to print the diagram or use it in a presentation. To hide the Vanishing Point for printing, choose View ⇨ Layer Properties and clear the check mark in the Print column of the Vanishing Point row. To hide the Vanishing Point on the drawing, clear the check mark in the Visible column of the Vanishing Point row.

Using Multiple Vanishing Points

You can create even more dramatic diagrams by adding additional Vanishing Point shapes to a diagram and associating shapes to those Vanishing Points. Visio doesn't support true two-point perspective. Although each shape connects to only one Vanishing Point, adding a second Vanishing Point can spice up your presentation graphics by linking different collections of shapes to different vanishing points. To work with an additional Vanishing Point, follow these steps:

1. Drag a Vanishing Point from the Block with Perspective stencil onto the drawing.

2. To associate a shape to the new Vanishing Point, first select the shape.

3. If the shape is associated with another Vanishing Point, drag the red control handle that appears in the first Vanishing Point and glue it to the connection point on the new Vanishing Point. A red square highlights the Vanishing Point, indicating that the shapes are connected.

4. If the shape is not associated with a Vanishing Point shape, drag the yellow control handle that appears on the drawing page when you select the shape, and glue it to the connection point on the new Vanishing Point.

NOTE   When you add new shapes, they connect automatically to the first Vanishing Point, which Visio adds by default to each block diagram that you create with the Block Diagram with Perspective template. You must change the connection for each shape you want connected to the other Vanishing Point.

Working with Onion Diagrams

Onion diagrams use concentric rings to illustrate concepts or elements that build up from a core, such as the layers that make up our planet. Although the objects represented on an onion diagram grow from the center, in Visio, you construct an onion diagram from the outside in.

Creating Onion Diagrams

The Blocks stencil contains Concentric Layer and Partial Layer shapes that you can use out of the box for up to four layers of an onion. If you require more than four layers, you can resize the largest layer and add additional rings. To create an onion diagram, follow these steps:

1. Choose File ⇨ New ⇨ General ⇨ Block Diagram.

2. To establish the outer layer of the onion, drag the Concentric Layer 1 shape onto the drawing page.

3. To add the next layer of the onion, drag the Concentric Layer 2 shape onto the drawing and drop it onto the center of the first concentric shape.

4. To add the third layer of the onion, drag the Concentric Layer 3 shape onto the drawing and drop it onto the center of the other concentric shapes.

5. To add the core of the onion, drag the Concentric Center shape onto the drawing and drop it onto the center of the other concentric shapes.

6. To add text to a ring, select the shape and type the text you want.

Modifying Onion Diagram Shapes

Whether you need additional layers or want to change a layer's size or thickness, it's easiest to modify the standard concentric rings after you add them to your drawing.

Adjusting the Dimensions of Onion Rings

You can resize Concentric Layer shapes or change their radius and thickness. However, you must realign the shapes after you make these adjustments. Use one of the following methods to adjust Concentric Layer shapes:

- **Resize a Concentric Layer shape** — Drag one of the selection handles to change the radius of the circle. The opposite selection handle remains fixed on the drawing, as demonstrated in Figure 12-4.

- **Change the thickness of a ring** — Drag the yellow control handle (the diamonds) on the inside edge of the shape.

> **TIP** To realign concentric rings after you modify their size or thickness, select all the Concentric Layer shapes and choose Shape ⇨ Align Shapes. Select the centered vertical alignment option and then click OK. If your concentric rings are still not nested properly, continue by choosing Shape ⇨ Align Shapes, selecting the centered horizontal alignment option, and clicking OK.

FIGURE 12-4

You can change the radius, thickness, and text position of a Concentric Layer shape.

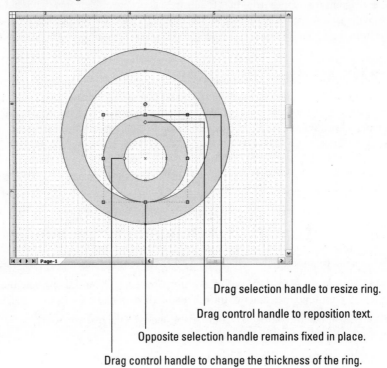

Drag selection handle to resize ring.

Drag control handle to reposition text.

Opposite selection handle remains fixed in place.

Drag control handle to change the thickness of the ring.

■ **Fit a smaller ring inside a larger ring** — Drag the selection handle on the left outside edge of the smaller ring and snap it to the connection point on the left inside edge of the larger ring. Then, drag the selection handle on the right outside edge of the smaller ring and snap it to the connection point on the right inside edge of the larger ring.

TIP   When you move the control handle for text in a Concentric Layer shape, Visio rotates the text to keep it within the ring. For example, if you move the control handle to 9 o'clock in the ring, the text rotates to be vertical reading from bottom to top. When you drag the control handle into the bottom half of a Concentric Layer shape, Visio automatically flips the text so that it still appears right side up in the ring.

Working with Text in Onion Diagrams

Text can be difficult to work with in onion diagrams because the shapes are curved and the text is straight. You have a few options when a long text string doesn't fit within a concentric ring.

- For text that *almost* fits, apply a smaller font or try a narrower font such as Arial Narrow.

- For long text, position the text outside the shape by dragging the control handle in the middle of the Concentric Layer to a position outside the shape.

- You can also annotate an onion diagram using Callout shapes. To open a stencil of Callout shapes, choose File ⇨ Shapes ⇨ Visio Extras ⇨ Callouts. AutoConnect is the fastest way to connect a Callout shape to a ring of the onion. Select the callout shape you want in the stencil. On the drawing page, position the pointer over the shape you want to annotate and click the blue arrow on the side on which you want to glue the Callout shape. Double-click the text area to add the text.

Dividing a Concentric Layer into Sections

You can divide a Concentric Layer shape into sections to show several components. The Partial Layer shapes on the Blocks stencil fit the Concentric Layer shapes. When you drop a Partial Layer shape onto a Concentric Layer shape, they connect and act as one, so you can drag a Concentric Layer shape's selection handles to resize it and its associated Partial Layer shapes. To divide a Concentric Layer into sections, follow these steps:

1. Drag a Partial Layer shape that matches the size of the Concentric Layer onto the drawing.

2. To rotate a Partial Layer, select it after dropping it onto the page and then press Ctrl+L as many times as necessary to rotate the shape into the correct quadrant.

3. If necessary, use the editing techniques described in the previous section to adjust the radius or thickness of the partial layer.

4. Drag the Partial Layer shape over the Concentric Layer shape you want to divide. When the Partial Layer snaps to the Concentric Layer, the red squares highlight the outside connection point and the center of the Concentric Layer shape to indicate that the shapes are glued, as shown in Figure 12-5.

5. To modify the length of the arc for a Partial Layer, drag the yellow control handle on the Partial Layer's outside edge.

6. To rotate a Partial Layer within the Concentric Layer, drag the red selection handle.

FIGURE 12-5

You can subdivide Concentric Layers by gluing Partial Layer shapes to them.

Partial Layer shape on top of Concentric Layer shape

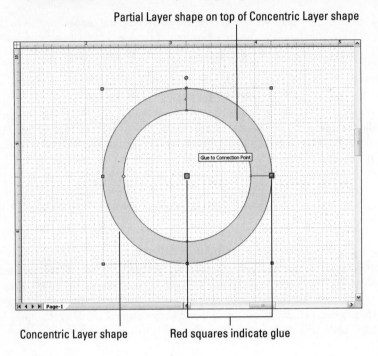

Concentric Layer shape   Red squares indicate glue

Summary

The templates and shapes for Block Diagrams are easy to use and include many helpful features. Using basic Visio techniques, you can document structure and flow or communicate hierarchy or concepts revolving around a central idea. Visio Block shapes can be as simple as one-dimensional arrows and two-dimensional geometric shapes, but you can also produce impressive diagrams for presentations by using shadows or shapes with perspective.

Chapter 13

Constructing Charts and Graphs

Charts and graphs illustrate quantitative results and trends, such as financial performance, marketing analysis, or statistical distributions. Microsoft Visio and Microsoft Excel both offer features for constructing charts and graphs, so your first decision is the program that offers the best tools for the charting job at hand. In most cases, Excel is better than Visio for developing and formatting charts.

If the data you want to present is already stored in a Microsoft Excel spreadsheet, there's no need to look any further than the Insert Chart command in Excel. You can format every component of an Excel chart to achieve exactly the look you want.

However, Visio 2007 has simplified connecting Visio shapes with data from one or more data sources (as described in Chapter 10), making it the perfect solution if you're developing a presentation that summarizes data from a variety of sources.

The Visio Charts and Graphs template includes shapes for common chart and graph styles, such as bar graphs, line graphs, pie charts, distribution curves, feature comparison tables, and other kinds of tabular information. In addition, you can use the Visio marketing shapes to analyze and communicate sales and marketing information, such as sales prospects, SWOT (strengths, weaknesses, opportunities, and threats) analysis, market share, and marketing mix.

NEW FEATURE In Visio 2007, the Business category is the new home for the Charts and Graphs template and the Marketing Charts and Diagrams template.

Exploring the Chart and Graph Templates

Chart and graph shapes are simple enough that the Chart and Graph templates don't contain any specialized menus, toolbars, or add-ons. Creating charts and graphs is as easy as connecting shapes copied from the Visio stencils and adding text and numbers to components within the shapes. You can change the configuration and arrangement of chart and graph components by dragging the shape handles or make the charts and graphs more informative by formatting them.

Choosing the Right Template

The Visio Business template category offers two templates for charts and graphs. The Charts and Graphs template has the more compelling name, but you're better off using the Marketing Charts and Diagrams template, because it opens all three of the Visio chart and graph stencils: Charting Shapes, Marketing Shapes, and Marketing Diagrams. Both templates automatically set the page to a letter-size sheet with portrait orientation, use inches (or millimeters for metric drawings) drawn at a one-to-one scale, and open the Backgrounds, Borders and Titles, and Charting Shapes stencils.

NOTE Because Visio eliminated the Forms template and stencil beginning with Visio 2003, turn to Microsoft InfoPath when you want to design and construct forms. Infopath forms are based on XML so you can collect data for XML-compatible databases and back-end business systems and incorporate handy form-filling features that Visio never provided.

Exploring Visio Chart and Graph Shapes

Chart and Graph shapes contain selection handles and control handles for resizing and reconfiguring their components. In tabular shapes, such as the Grid shape and the Feature Comparison chart, you can select and edit individual cells. When you drop some charting shapes on the drawing page, they give you a head start by displaying a dialog box with shape data for configuration. For example, the Bar Graph 1 shape displays a dialog box with a drop-down list to set the number of bars in the chart. After you specify properties or number of elements you want, click OK to close the dialog box and apply the configuration to the shape. To reconfigure one of these shapes after you add it to a drawing, right-click it and choose a configuration command from the shortcut menu.

NOTE You can open diagrams that were created in previous versions of Visio, although some shape behaviors from previous versions might no longer work. For example, the Table shape, which was replaced in Visio 2003 by the Grid shape on the Charting Shapes stencil, retains its previous formatting, but its formatting commands are no longer available to modify its formatting further. If you want to format the table, you must recreate your table by dragging a Grid shape onto the drawing page and reentering the table values.

Charting Shapes

The Charting Shapes stencil, shown in Figure 13-1, provides shapes to create standard charts and graphs. After you drop shapes onto the drawing, you can add text directly to them or in separate annotation shapes. Some of these shapes include control handles you can drag to modify the appearance of the chart and graph components, as described in the shape descriptions that follow.

FIGURE 13-1

The Charting Shapes stencil provides shapes for commonly used charts and graphs.

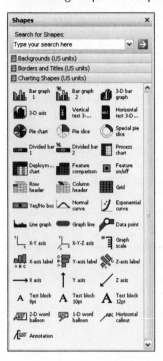

- **Bar graphs** — Visio bar graphs are standard bar graphs — great for comparing numbers or percentages, such as membership numbers from different regions. The Bar Graph 1 shape contains bars for displaying numerical values, whereas Bar Graph 2 is set up to show percentages. To communicate the information on bar graphs more easily, add the X-Y axis, Graph Scale, and axis label shapes to identify the numbers and measurements that you are presenting.

- **3-D bar graphs** — Three-dimensional bar graphs work well when you want to compare numbers across two different categories or measures, for instance, showing the trends in membership numbers for each region over several years. The 3-D Bar Graph and 3-D Axis shapes work together to document data that requires three-dimensions, such as revenue growth by quarter for the past five years. To build 3-D graphs from individual components, combine the 3-D bar shapes, the X-Y-Z axis shape, and the shapes that label the *x*, *y*, and *z* axes.

- **Pie charts** — Pie charts are perfect when you want to emphasize numbers or percentages that make up a total, because each slice represents a portion of the whole pie. The Pie Chart shape creates a whole pie with up to 10 slices. To create a pie with more slices, drop as many Pie Slice shapes as you need onto the drawing page and connect them to make up the whole pie. The Special Pie Slice shape shows a concentric ring that you can place on top of a pie slice — similar to the Partial Layer shape for onion diagrams (in the Block Diagram template described in Chapter 12). For example, Special Pie Slices might represent the percentage of sales for each business unit in the European operations of a company.

- **Divided bars** — Divided bars are like rectangular versions of pie charts. Each section of the divided bar shows a portion of the whole bar. You can put divided bars to work showing zones such as the suggested prices for buying, holding, and selling an investment. The Divided Bar 1 shape chops a rectangle into smaller boxes that show numerical values, whereas the Divided Bar 2 shape is set up to show percentages. You can add text to the divided bars or use the X-axis and Y-axis label shapes to annotate the divided bars.

- **Tabular charts** — The Process Chart shape includes up to 10 steps, with symbols to document the activities within each step. The Deployment Chart is a table for tracking the rollout of systems in an organization for up to six departments implemented over five phases. You can customize the shape data for larger organizations or more complex deployments. In the Shape Data dialog box, click Define and, in the Define Shape Data dialog box, modify the properties, as described in Working with Shape Data in Chapter 10.

- **Feature comparison charts** — Out of the box, the Feature Comparison Chart shape can compare up to 10 features across up to 10 products. You can indicate whether a product supports a feature completely, partially, or not at all by adding and setting the values for Feature On/Off shapes in grid cells within the Feature Comparison Chart shape.

- **Grids** — For generic tables, combine the Grid, Row Header, and Column Header shapes. The Yes/No Box can display a filled circle (to indicate Yes) hollow circle (to indicate No) or text (to further clarify the results).

- **Distribution and exponential graphs** — The Normal Curve shape displays a distribution curve. Control points change the shape and skew of the distribution. You can change the height and width of the Exponential Curve shape, but not its curvature.

- **Line graph shapes** — Line graphs are better than bar graphs when you have many data points, such as the temperature measured each second over a day-long test. The Line Graph shape displays a series of data points with the area under the graph filled. You can highlight lines on a graph with the Graph Line and Data Point shapes.

- **Annotation shapes** — You can annotate your chart or graph with labels, text blocks with different font sizes, balloons, callouts, or annotation shapes.

Marketing Shapes

Most of the shapes in the Marketing Shapes stencil look like clip art and are handy for developing sales and marketing presentations. A few of the shapes are extendable, such as People and Variable Building. When you drag the selection handle on the side of the People shape, Visio adds up to four people to the shape — but, sadly, not to your real-world team. Dragging the selection handle on the top of the Variable building shape adds up to 10 floors to the Visio skyscraper.

Marketing Charts and Diagrams

The Marketing Diagrams stencil, shown in Figure 13-2, provides charts and graphs typically used for marketing, such as market share, circle-spoke, or marketing mix. However, Visio won't complain if you borrow these shapes to illustrate other kinds of data.

FIGURE 13-2

The Marketing Diagrams stencil provides shapes for a variety of marketing-oriented charts.

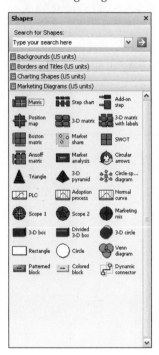

Constructing Basic Charts and Graphs

If you find that a predefined chart doesn't illustrate your data the way you want, you can build a graph from individual shapes. This section describes the steps that work for creating and formatting most Visio chart and graph shapes.

Constructing Bar Graphs

Visio provides predefined 2-D shapes that can display up to 12 bars. The 3-D bar graph shape includes up to five 3-D bars.

Creating 2-D Bar Graphs

Constructing a 2-D bar graph requires dropping a bar graph shape onto a drawing and adding text or annotation shapes to identify values. To create a 2-D bar graph, follow these steps:

1. Drag a Bar Graph shape onto the drawing, choose the number of bars from the drop-down list, and click OK.

2. To set the height for the tallest bar in the graph, drag the yellow control handle at the top left of the Bar Graph shape until the tallest bar is the height you want for the largest y-axis value, or 100% for graphs showing percentages, as demonstrated in Figure 13-3.

3. To set the width of all the bars, drag the yellow control handle at the bottom right of the first bar and drag it until the bars are the width you want.

4. To specify the value for a bar, click the graph, click the bar to select it, and type the value for that bar.

FIGURE 13-3

Drag selection and control handles to adjust the size of a bar graph.

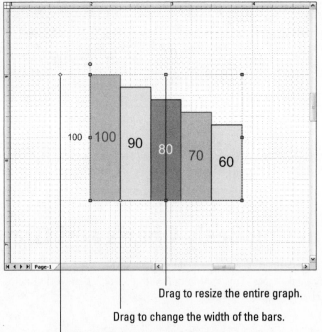

Drag to resize the entire graph.

Drag to change the width of the bars.

Drag to set the height of the tallest bar.

CAUTION Although the Bar Graph 2 shape is set up to show percentages, when you type a value for a bar, you must type % after the number. If you don't end a number with %, Visio converts the number to a percentage by multiplying by 100 and sets the bar to the resulting height. For example, typing 1 will create a bar the same height as 100%.

After you add a bar graph to a Visio drawing page, you can modify or format the bar graph in the following ways:

- **Change the number of bars in a graph** — Right-click the Bar Graph shape, choose Set Number of Bars from the shortcut menu, select the number of bars you want, and click OK.

NOTE To access the shortcut menu for the Bar Graph shape, make sure that none of the individual bars are selected and then right-click anywhere on the bar graph. Right-clicking a selected bar displays the shortcut menu for that bar. To deselect a bar and select the Bar Graph shape, right-click the dotted, green boundary line of the shape.

- **Change the color of a bar** — First, select the bar, then right-click it, and choose Format ⇨ Fill from the shortcut menu. Select the fill options you want and click OK.
- **Add the x and y axes to a 2-D bar graph** — Drag the x-y axis shape until the origin snaps to the bottom-left corner of the first bar in the graph.
- **Label the units for the axes** — Click the x-axis or y-axis text blocks and type the labels you want. For example, a revenue growth graph usually shows revenue in dollars on the y-axis and fiscal quarters along the x-axis.

Creating 3-D Bar Graphs

To create a 3-D bar graph, follow these steps:

1. Drag the 3-D Axis shape onto your drawing. The control handles on the 3-D Axis shape reposition the labels, change the number of grid lines, change the thickness of the wall, or change the depth of the third dimension, as demonstrated in Figure 13-4.

2. Drag and drop the 3-D Bar Graph shape onto the origin of the 3-D Axis shape. In the Shape Data dialog box that appears, select the number of bars from the drop-down list. You can also specify the values and colors for each of the bars. Click OK when you are finished.

TIP If you drag the 3-D Bar Graph shape onto your drawing before the 3-D Axis shape, the axis shape hides the bar graph. To display the bar graph in front of the axis, change the stacking order of these shapes by right-clicking the 3-D Axis shape and, from the shortcut menu, choosing Shape ⇨ Send to Back.

FIGURE 13-4

Drag control handles to configure all aspects of a 3-D Axis shape.

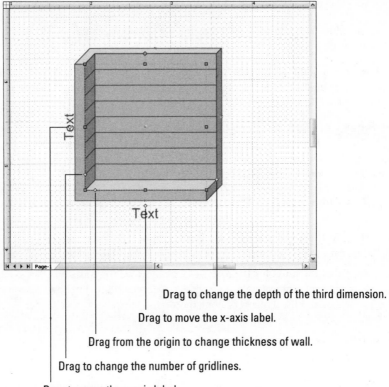

Drag to change the depth of the third dimension.

Drag to move the x-axis label.

Drag from the origin to change thickness of wall.

Drag to change the number of gridlines.

Drag to move the y-axis label.

3. To change the height of the graph, drag the green selection handle at the top or bottom of the shape.

4. To set the width of all the bars, drag the yellow control handle at the bottom right of the first bar and drag it until the bars are the width you want.

After you add a 3-D bar graph to a Visio drawing page, you can modify or format the bar graph in the following ways:

■ **Specify the value or color of a bar** — Right-click the 3-D Bar Graph shape, choose Bar Properties from the shortcut menu, and edit the values in the Shape Data dialog box. Click OK when you are finished.

■ **Change the number of bars in a graph** — Right-click the 3-D Bar Graph shape and, from the shortcut menu, choose Bar Count and Range. Select the number of bars you want in the Bar Count drop-down list and click OK.

■ **Change the height of bars in relation to the overall shape** — Right-click the 3-D Bar Graph shape and, from the shortcut menu, choose Bar Count and Range. In the Range box, type the value you want at the top of the y axis. For example, if you change the range from 4 to 8, the bars in the graph shorten by half.

Constructing Line Graphs

Unlike Microsoft Excel, Visio provides only one type of line graph. If you want to graph two lines or choose from different markers for data points, it's easier to add data to a spreadsheet in Excel and use its Insert Chart command. To create a simple line graph in Visio, follow these steps:

1. Drag the Line Graph shape onto a drawing. In the Shape Data dialog box that appears automatically, select the number of data points you want, and click OK.

TIP   To change the number of data points after adding the shape to a drawing, right-click the line graph and, from the shortcut menu, choose Set Number of Data Points. In the Data Points drop-down list, choose a number and click OK.

2. To change the length of the x or y axis, drag the control handle at the end of the axis to the length you want.

3. To change the value of a data point, drag the control handle for that data point to the appropriate value on the y axis.

4. To emphasize data points on a line graph, drag a Data Point shape onto the drawing and snap it to the control handle for a data point. To emphasize the lines between data points, drag a Graph Line shape onto the drawing. Glue each end to a pair of consecutive data points.

Labeling Axes

A steadily decreasing line graph might tell the story of a successful diet, but most people watching their weight want to know how many pounds they've lost and what they weigh now. Visio axis shapes include text blocks that you can edit to show the units for an axis, be they pounds, the number of support calls handled, or the time increments for measurements. But to also show numeric values for those units along each axis, you need axis label shapes as well.

To label the units on the axes, follow these steps:

1. To label the x and y axes with the units they represent, first select the Line Graph shape you added.

2. Click the y-axis text block to select it and then type the units represented by the values on the y axis, such as Pounds.

3. Click the x-axis text block to select it and then type the units for the x axis, such as Weeks.

To indicate the values at grid lines or simply at regular intervals along each axis, follow these steps:

1. Drag the first y-axis Label shape onto the drawing so that its bottom end lines up with the origin of the Line Graph shape.

 **Zooming in makes it easier to align labels. Press and hold Ctrl+Shift and click the graph to zoom in.**

2. With the y-axis Label shape still selected, type the value for that label.

3. To copy the first y-axis label, select it and then press Ctrl+D. Drag the second label so its horizontal line is even with the highest value you want to label. Type the value for the label.

4. Repeat step 3 to create labels for intermediate values along the y axis.

5. Repeat steps 1 through 4 with X-axis Label shapes to add labels along the x axis.

 **The Distribute Shapes command is an easy way to add all the shapes you need along an axis without worrying about positioning them accurately. Select the label shapes for an axis and choose Shape ⇨ Distribute Shapes. For y-axis labels, select the first option for Vertical Distribution, and click OK. For x-axis labels, select the first option for Horizontal Distribution and then click OK.**

Working with Pie Charts

Although Microsoft Excel provides more options and flexibility for pie charts, Visio pie charts are fine for simple pie charts for presentations. Create a pie chart by following these steps:

1. Drag the Pie Chart shape onto the drawing, select the number of slices you want up to 10, and click OK.

2. To specify different percentages for each slice, right-click the pie chart and, from the shortcut menu, choose Set Slice Sizes. In the Shape Data dialog box, type the percentage for each slice and then click OK.

NOTE **If the values you enter for the slices don't total 100 percent, an empty slice shows the left over percentage.**

To create a pie chart with more than 10 slices or to emphasize one or more of the slices, you can build a pie out of individual Pie Slice shapes. To do this, follow these steps:

1. Drag the Pie Slice shape onto the drawing.

2. To change the radius of the slice, drag the green selection handle at the outside edge of the slice.

3. Drag another Pie Slice shape and drop it close to, but not on top of the first slice.

4. Glue the green selection handle at the bottom right of the second slice to the vertex at the top left of the first slice, as shown in Figure 13-5.

5. Drag the green selection handle at the center of the second slice to the vertex at the bottom left of the first slice. When you are done, the radius of the second slice will match that of the first slice.

6. To modify the percentage of a slice, select the slice and drag the yellow control handle until the percentage you want appears in the text within the slice.

TIP To increase the size of a slice by 1 percent, right-click the slice and, from the shortcut menu, choose Add 1%. Alternatively, drag the control handle far from the center of the pie slice. The farther from the center you drag, the finer the control you have over the angle. (The same is true for the rotation handle.)

7. To change the color of a slice, right-click it and, from the shortcut menu, choose Format ⇨ Fill. Select the fill options you want and click OK.

8. Repeat steps 3 through 7 to add additional slices, always adding slices counterclockwise around the pie.

FIGURE 13-5

Drag selection handles to build a pie chart from pie slice shapes.

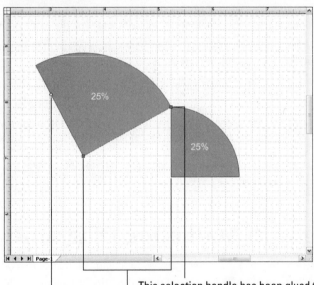

This selection handle has been glued to the first slice.

Drag to rotate the slice into postion.

Drag to change the percentage for the slice.

If you create a pie chart with individual slices, you can emphasize a slice by dragging the slice away from the center of the pie. If the slice tends to snap to another shape, choose Tools ⇨ Snap & Glue. Uncheck some of the check boxes, such as Shape Geometry, in the Snap To column, click OK, and then try to move the slice again.

Creating Feature Comparison Charts

Feature comparison charts illustrate which features products offer so that you can choose the product that best fits your requirements. The Feature On/Off shape includes three status options:

- **Blank** — Indicates that the feature doesn't exist for that product
- **Filled circle** — Indicates that the product provides the feature
- **Hollow circle** — Can indicate that the product provides the feature with some limitations

To create a feature comparison table, follow these steps:

1. Drag the Feature Comparison shape onto the drawing. Select the number of features and number of products to compare and click OK.

2. To enter a feature description, click a row header cell and type the name of the feature. To add a product name, click a column header cell and type the name of the product.

3. To add a status, drag the Feature On/Off shape onto a cell in the comparison chart, select the status option you want, and click OK.

4. Repeat step 3 for each cell in the chart.

CROSS-REF Visio doesn't provide a command for copying and pasting multiple Feature On/Off shapes over a grid of cells. To learn how to copy the On/Off shape over an array of columns and rows, see the section "Creating an Array of Shapes" in Chapter 4.

5. To change the number of features or products, right-click the chart and choose Set Fields from the shortcut menu.

Working with Marketing Diagrams

Many of the marketing shapes in Visio include shortcut menu commands to modify the shape configuration. For example, you can specify the number of arrows in a Circular Arrows shape when you first add the shape or specify the number later using the Set Number of Arrows command on the shape's shortcut menu. Some shapes also include control handles that for making other adjustments to configuration. This section describes some of the special features on marketing diagrams.

Building Circle-spoke Drawings

Circle-spoke diagrams illustrate the relationships between a single element and several satellites, such as the business units and headquarters for an organization. The Visio Circle-spoke Diagram shape includes up to eight circles arranged on spokes around a center circle. To create a circle-spoke diagram, follow these steps:

1. Drag the Circle-spoke Diagram shape onto the drawing.

2. In the Shape Data dialog box, select the number of outer circles you want and click OK. To change the number of circles later — for example, to add a new marketing region — right-click the shape and, from the shortcut menu, choose Set Number of Circles.

3. Resize the diagram by dragging a selection handle.

CAUTION Be sure to size the diagram before you reposition any circles. When you resize a Circle-spoke Diagram shape, Visio repositions the outer circles to be equally spaced from one another and at the same distance from the center.

4. To relocate or rearrange the outer circles, drag the yellow control handle in the middle of a circle to a new location, as shown in Figure 13-6.

FIGURE 13-6

Drag control handles to reposition and rearrange the circles in a circle-spoke diagram.

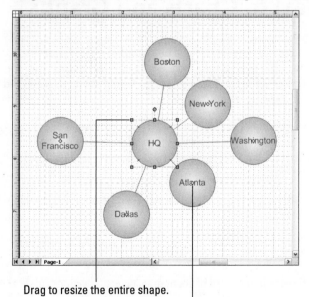

Drag to resize the entire shape.

Drag to reposition a circle.

Constructing Triangles and Pyramids

Triangular charts and 3-D pyramids show hierarchical relationships, such as the food pyramid that tells us how many servings to eat of different types of food. To create a triangular chart, follow these steps:

1. Drag a Triangle shape onto the drawing. In the Shape Data dialog box, select the number of levels you want, and click OK.

2. To separate the triangles in the chart with a small gap, right-click the shape and, from the shortcut menu, choose Set Offset. Type the number of inches (or decimal fractions of an inch) between the layers and click OK.

3. To switch between a flat triangle and a three-dimensional triangle, right-click the triangle shape on the drawing and, from the shortcut menu, choose either 2-Dimensional or 3-Dimensional.

4. To change the number of levels after you have added the shape to the drawing, right-click the shape and, from the shortcut menu, choose Set Number of Levels.

When you add a 3-D pyramid to a drawing, you can specify up to six levels in the pyramid and choose one color for the entire pyramid. To change these settings later, right-click the pyramid and, from the shortcut menu, choose Set Number of Levels or Set Pyramid Color.

Adding Text to Charts and Graphs

Depending on the chart and graph shapes you use, you can select a shape, a cell, or a text block within a shape and add text simply by typing. For example, you can select the text block on the left side of a Bar Graph shape to label the y axis. You can also select each bar in the Bar Graph shape to specify the height of the bar. If you select the entire Bar Graph shape and begin typing, Visio adds a text block below the x axis.

In addition to the text within shapes, the Visio Chart and Graph templates open stencils with title and callout shapes for annotating your drawings. To add a title to your drawings, you can use the Text Block shapes from the Charting Shapes stencil or choose one of the title block shapes from the Borders and Titles stencil.

Charts and graphs often include other types of annotation. To add a word balloon to a drawing, follow these steps:

1. From the Charting Shapes stencil, drag a 1-D or 2-D Word Balloon shape onto the drawing.

2. With the word balloon shape selected, type the text you want in the word balloon.

3. To change the size of the word balloon, drag a green selection handle to a new position.

4. To aim the Relocate Pointer, which protrudes from the side of the balloon, drag the yellow control handle to a new position. The Relocate Pointer automatically moves to the side of the balloon closest to the end of the pointer.

To add other annotation to a drawing, follow these steps:

1. Drag a Horizontal Callout or Annotation shape onto the drawing.

2. With the shape selected, type the text you want. The height of the shape adjusts to display the text you type.

3. To change the width of the annotation, drag the right, green selection handle to a new position.

4. To modify the orientation, length, or position of the callout line, drag one of its green selection handles to a new position.

 To learn more about annotating drawings, refer to Chapter 6.

Using Stackable and Extendable Shapes

Stackable and extendable shapes show more or fewer people, buildings, or smokestacks, as you lengthen or shorten the shapes. Changing the length of stackable shapes either tacks on or removes elements vertically or horizontally — for example, showing additional heads and shoulders to represent population. Extendable shapes stretch without distorting parts of the shape.

To work with stackable shapes in a chart or graph, follow these steps:

1. From the Marketing Shapes stencil, drag a stackable shape, such as the People shape or the Variable Building shape onto the drawing.

2. To extend a horizontal shape, drag either the left or right selection handle. Additional repeating elements will appear. For a vertical shape, drag the top or bottom selection handle.

3. To create a longer series, press Ctrl+D to copy the stackable shape and align the copy end to end with the first shape.

To work with extendable shapes in a chart or graph, follow these steps:

1. Drag an extendable shape, such as the Pencil shape, from the Marketing Shapes stencil onto the drawing.

2. To change the length of the extendable shape, drag the left or right green selection handle.

Summary

Visio chart and graph templates provide shapes to produce a wide variety of charts, graphs, and marketing diagrams. However, if you want to create commonly used charts and graphs from numeric data stored in a spreadsheet, the Insert Chart command in Microsoft Excel is easier and more flexible. The Visio Marketing Shapes stencil includes clip art shapes for sales presentations. If you want to communicate the results of marketing efforts, you can use marketing-oriented shapes, such as feature comparison charts or marketing mix shapes in the Charting Shapes and Marketing Diagrams stencils.

Chapter 14

Working with Organization Charts

Real-world organizations come in many shapes and sizes, with both formal and informal reporting structures. In strictly hierarchical enterprises, authority and communication mostly travels up and down official lines of command. But, companies that perform projects often use a matrix structure whereby workers report to project managers for the projects to which they are assigned in addition to functional managers for other assignments and administrative issues.

Organization charts document the formal structure and relationships within an enterprise, from top-level business units, down through functional groups, teams, and individuals. Visio 2007 simplifies the process of creating organization charts whether you build your charts one shape at a time or by importing organizational data from another data source.

Although Visio organization charts look like the hierarchies shown by tree shapes in Block Diagram templates (see Chapter 15) they are much more powerful. The Organization Chart template includes tools and features specifically designed to create and maintain organization charts. You can build organization charts by importing data from other sources or using standard Visio techniques to add and connect employee shapes. The template includes layout tools that automatically arrange shapes into typical organization chart layouts.

Visio organization charts are inherently hierarchical diagrams, so you can use the Visio 2007 organization chart features just as easily for other hierarchical diagrams, such as the genealogy of a family tree. In this chapter, you first learn how to create organization charts from scratch or by importing organization data from other sources. Then, you learn about different ways to arrange and format organization chart shapes until your diagram looks just the way you want.

NOTE   Regardless of your organization's official style, informal relationships are often the key to progress. For example, programming teams might share tips and tricks that improve everyone's output. These unofficial relationships often go unrecognized and even Visio 2007 can't help you document them.

Exploring the Organization Chart Template

The Organization Chart template is one of the more obliging templates in the Visio stable. It offers several specialized features to help you perform common organization chart construction tasks:

- If you have employee data stored in a database or other repository, the Organization Chart Wizard can import data to build a Visio organization chart.

- The Synchronize command keeps data in shapes synchronized when you expand the levels of a department or group on another page.

- The Visio organization chart layout options automatically arrange organization chart shapes, so you don't have to worry about placing shapes.

- The Organization Chart menu and toolbar provide fast access to the organization chart commands.

- As you add Organization Chart shapes to a drawing, they automatically position themselves in a hierarchy according to the layout you specify.

Exploring the Organization Chart Environment

As you can see in Figure 14-1, when you open an existing Visio organization chart or create a new organization chart drawing, the Organization Chart stencil and toolbar appear and the Visio menu bar gains an Organization Chart entry. This section introduces these specialized features and explains how to put them to best use.

The following are the Organization Chart tools you're likely to use most often:

- **Smart Organization Chart shapes** — If you're putting together an organization chart for a small group, the shapes in the Organization Chart Shapes stencil include behaviors that speed up the process. They know enough to connect themselves to the superior shapes on which you drop them and to position themselves based on the layout options you specified for those superior shapes. Visio adds connectors between the shapes, which creates reporting relationships between employees and their managers.

- **Organization Chart menu** — This menu contains all the commands specially designed for working with Organization Charts. Whether you want to run the Organization Chart wizard, specify layout options, or make other changes to a diagram, this menu has it all.

- **Organization Chart wizard** — Tedium is what comes to mind when thinking of manually adding, connecting, and annotating shapes for a large organization. If you store employee data in another program, the Organization Chart wizard can make short work of reading that data and building an organization chart diagram for you. On the Organization Chart menu, choose Import Organization Data to launch the wizard.

- **Organization Chart toolbar** — This toolbar includes six commands for fast access to the most common layout tasks. It includes three layout options, the Re-layout command to reapply the current layout option and two commands for moving shapes within the hierarchy.

- **Arrange Subordinates** — If the layout options on the Organization Chart toolbar aren't what you want, on the Organization Chart menu, choose Arrange Subordinates to choose from every layout option Visio provides.

Along with the tools here, the rest of the Organization Chart commands are described in detail in the remaining sections in this chapter.

 Many of the Organization Chart commands are also available on context-sensitive shortcut menus, displayed by right-clicking a shape or on the drawing page itself.

FIGURE 14-1

The Organization Chart template includes a menu, toolbar, stencil, wizard, and other tools.

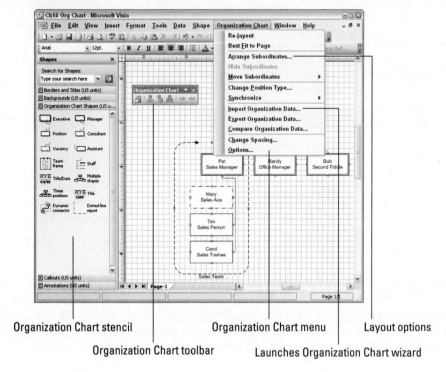

Organization Chart stencil   Organization Chart menu   Layout options

Organization Chart toolbar   Launches Organization Chart wizard

Exploring the Visio Organization Chart Shapes

The appearance of the shapes in the Organization Chart stencil denotes levels in a hierarchy or special conditions such as assistants, outside consultants, and vacancies. However, any employee shape in the stencil can act as either superior or subordinate, just as managers are superiors to the employees on their teams but subordinates to the higher-level managers to whom they report. The Organization Chart stencil also includes a few shapes that speed up the addition of frequently used groupings, such as several resources reporting to a superior. Here are the employee shapes that represent single positions at different levels in an organization:

- **Executive** — A larger box than other shapes with a heavier border and a shadow.

- **Manager** — A standard-sized box with a heavier border.

- **Position** — A standard-sized box with a single line border.

- **Staff position** — A shape that shows the name and title without a border.

- **Consultant** — A standard-sized box with a dashed-line border.

- **Assistant** — A box that looks like the Position shape but positions itself offset from the subordinates to the manager.

- **Vacancy** — A standard-sized box with a dotted-line border.

CROSS-REF To learn how to customize the appearance of Organization Chart shapes, refer to the "Formatting Organization Chart Appearance" section later in this chapter.

The following shapes simplify working with groups of employees in an organization chart:

- **Multiple Shapes** — When you drop this shape onto a superior shape, Visio connects the number of positions and connectors that you specify to the superior shape.

- **Three Positions** — When you drop this shape onto a superior shape, Visio connects three positions as subordinates to the superior shape.

- **Team Frame** — Indicates graphically that the employees within the frame are members of a team.

TIP An employee can have only one primary reporting relationship in Visio, but may report to a secondary manager in real life. Unfortunately, this reality of reporting isn't supported by the wizard. To show this relationship, drag the dynamic Dotted Line Report connector onto the secondary manager's shape and drag the other connector end to the employee shape.

Creating Organization Charts Manually

Although importing data into Visio is usually the most efficient approach, building an organization chart manually can make sense if you don't have an existing source of employee data or you're documenting a small group. Starting at the top of the hierarchy — with a company president or group manager, for example — you drag and drop shapes representing employees from the Organization Chart stencil onto the drawing page. By dropping a subordinate shape onto a superior, you not only connect the shapes but define the reporting relationship, as demonstrated in Figure 14-2.

FIGURE 14-2

Organization chart shapes are smart enough to create connections and reporting relationships when you drop them on other shapes.

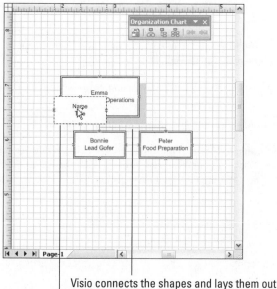

Visio connects the shapes and lays them out

Drag shape onto the shape it reports to

To create an organization chart manually, follow these steps:

1. Choose File ⇨ New ⇨ Business ⇨ Organization Chart.

2. Drag an Executive shape onto the drawing page.

NOTE   When you begin your first organization chart, Visio displays a message box that demonstrates how to connect shapes. To hide this reminder in the future, select the Don't Show This Message Again check box and click OK.

3. Because Visio selects the Executive shape as soon as you add it, fill in the name and title simply by typing the employee's name, pressing Enter, and then typing the employee's title. To complete the text entry, click the drawing page, press Esc, or drag another shape from the stencil.

Visio uses shape data to store the information you type within organization chart shapes. If you enter more than one line of text (by pressing Return), the first line is assigned to the Name shape data field and the second line is assigned to the Title shape data field. Alternatively, the Shape Data window displays all the shape data fields, so you can enter the information for the selected shape directly into the appropriate fields. Placing the right values in the right fields is essential, particularly if you plan to use the reporting feature.

4. To add a manager reporting to the executive, drag and drop a Manager shape from the stencil onto the Executive shape. Visio adds a connector between the shapes and assigns the manager as a subordinate to the executive. Visio also arranges the shapes based on the current layout (horizontally by default).

You can assign a name and title to shapes as you add them or add all the shapes to the chart and then select each one and type its values.

5. Drag and drop additional shapes from the stencil onto their superior shapes on the drawing page.

6. Select a shape and type the employee's name and title as you did in step 3.

If any shapes overlap each other, tell Visio to rearrange them by clicking the Re-layout button on the Organization Chart toolbar. Visio repositions all the shapes according to the layout options you've chosen. The section "Laying Out the Organization" describes how to select and apply layout options in detail.

7. To modify the layout of a group of subordinates, do the following:

   a. Select the superior shape to which the group reports.

   b. On the Organization Chart toolbar, click a layout option (Horizontal, Vertical, and Side by Side).

   c. From the drop-down menu, select a layout configuration (such as Center and Align Left).

8. To add additional information such as department, telephone number or e-mail address; right-click a shape and choose Properties from the shortcut menu. In the Shape Data dialog box, type values for the data you want to add and click OK to finish.

If your organization chart is cluttered or simply doesn't look the way you want, see the "Formatting Organization Chart Appearance" section later in this chapter.

Creating Organization Charts Using the Organization Chart Wizard

If your company is like most, you already have personnel data in some sort of database — an ODBC-compliant database such as Microsoft SQL Server, a Microsoft Exchange Server, or even a spreadsheet or text file. If personnel data is stored in a human resource system, that data can still be used to construct an organization chart by first exporting it into a spreadsheet or comma-delimited file.

The Organization Chart Wizard guides you through the steps of creating an organization chart whether you have organization data ready to import or you want to build a file of personnel data as you go. If personnel data is still in your head, the Organization Chart Wizard simply creates a sample file or spreadsheet for you to fill out and then import. The sections that follow describe in detail the steps you walk through in the Organization Chart wizard. The major tasks are as follows:

1. Specify the data source to use to build the organization chart.

2. Define the mapping between the fields in the data source and the fields in Organization Chart shapes.

3. Select the fields to display in the organization chart.

4. (Optional) Configure additional fields of data to import.

5. Specify the layout of the shapes in the organization chart.

Launching the Organization Chart Wizard

You can use one of the following methods to access the Organization Chart Wizard to create an organization chart:

- Choose File ⇨ New ⇨ Business ⇨ Organization Chart Wizard.

- Choose Tools ⇨ Add-Ons ⇨ Business ⇨ Organization Chart Wizard.

- If the Organization Chart template is active, choose Organization Chart ⇨ Import Organization Data.

What Visio Needs in a Data Source

Visio has only a few requirements for the data source you serve to the Organization Chart wizard. It doesn't care how you name the columns (data field names), because you tell the wizard which column corresponds to each shape data field for Organization Chart shapes (Name, Reports To, Department, Title, and Telephone). In fact, if employee names are unique, you need to provide only two fields: one for employee names and a second field for the names of the people to whom the employees report. To handle names that aren't unique, simply add a third column with a unique identifier for the employees, for instance, their employee IDs.

If your data source includes other information, the Organization Chart wizard is happy to import that into an organization chart as well. Data for department, title, and telephone slide into the existing shape data fields for Organization Chart shapes. For additional fields, such as building or office number, Visio simply creates additional shape data fields. However, this extra information is initially visible only through the Shape Data Window or in the Shape Data dialog box. See Chapter 6 to learn how to use the Label Shapes add-on to display other fields within the shapes on the drawing page or Chapter 10 to enlist Data Graphics for this task.

Building a File for Organization Data Using the Organization Chart Wizard

If you don't have an existing data source of employee information, the Organization Chart wizard can walk you through the creation of one and then use that source to construct an organization chart drawing. This section describes how to construct a data file of organization data using the Organization Chart wizard.

WARNING When you develop your data source within the Organization Chart wizard, you can enter data only for the default columns: Name, **Reports To**, **Title**, **Department**, and **Telephone**. To import additional data into a Visio organization chart, you must create your organization chart data file *before* you start the wizard. Then, when you start the wizard, select the Information That's Already Stored in a File or Database option, as described in the next section, "Building an Organization Chart from Existing Data."

1. On the first screen of the Organization Chart Wizard, select the Information That I Enter Using The Wizard option and click Next.

2. Select either the Microsoft Excel option to enter data in an Excel workbook or the Delimited Text option to create a comma-delimited text file.

3. In the New File Name text box, click Browse to specify the folder and name for the new data file.

4. In the dialog box that opens, navigate to the folder you want to use. In the File Name box, type the name for the data file and then click Save.

5. In the Organization Chart Wizard, click Next to open the file.

6. After reading the instructions for creating the data file, click OK. For the Excel option, a template file opens in Excel, as shown in Figure 14-3. For the Delimited Text option, Windows Notepad opens a sample delimited file. The file contains three sample entries that demonstrate how to define top-level and subordinate entries in the organization chart. Replace the sample text with your data and enter additional rows of data for each position in your organization chart.

7. After you complete your data entry, save the file and exit the other program. If you built an Excel data source, in Excel, click the File icon and then, at the bottom of the menu, click the Exit Excel button to save the file and return to the Organization Chart Wizard. For a delimited text file, in Notepad, choose File ➪ Exit.

8. To specify the layout of the diagram and create it using the data file you just created, jump to the Finalizing Your Organization Chart section later in this chapter and follow the steps outlined there.

FIGURE 14-3

Visio creates data files with instructions for entering data and sample entries, which you can overwrite with your employee information.

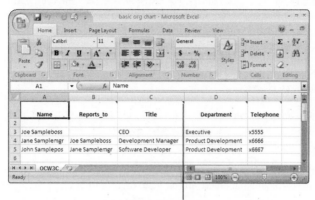

Click to read additional instructions

Building an Organization Chart from Existing Data

When you already have an existing data source of employee data, perform the steps in this section to select that data source as the foundation for building an organization chart:

1. To create an organization chart from existing data, on the first screen of the Organization Chart Wizard, select the Information That's Already Stored in a File or Database option and click Next.

2. Select the type of storage that contains your organization information and click Next. The options are a Microsoft Exchange Server directory, Org Plus text (*.txt) or Excel file, and an ODBC-compliant data source.

NOTE If your data resides in a Microsoft Exchange Server directory, the option to choose is self-explanatory. Likewise, choosing the option for an ODBC-compliant data source is easy once you determine that your database is ODBC compliant. However, if your data file doesn't fit either of those categories, you can transform it into a delimited text file or an Excel workbook and then choose the remaining option.

3. For the type of data storage you chose in step 2, specify the information required to identify the data file you want to use. Your options are as follows:

   - **Microsoft Exchange Directory** — You do not have to provide any information for this option.

   - **A text, Org Plus (*.txt), or Excel file** — Click Browse to specify the folder and file name.

■ **ODBC-compliant data source** — If you have already defined a data source in Microsoft Windows for your organization data, click the name of the data source in the list. If you select a generic type of ODBC-compliant data source, such as Microsoft Access Database or Excel Files, you must also specify the folder and file name of the database. After specifying the data source, you must also select the table in the database that contains your organization data.

TIP One advantage to using an ODBC-compliant data source is that you can link database records to shapes so that values displayed within shapes automatically update when the values in the data source change.

CROSS-REF If the ODBC-compliant data source list doesn't contain the type of data source you are using, click Create Data Source. See Chapter 10 for detailed instructions on setting up a data source using Visio tools.

4. After defining the data source, click Next to begin associating the columns or fields in the data file to the shape data fields in the Visio organization chart.

5. To specify the field that contains employee names in the data source, in the Name drop-down list, select the appropriate field name, as shown in Figure 14-4. The entries correspond to the column headings in an Excel workbook or delimited file, or to the field names in an ODBC-compliant database. Optionally, if you store employee first and last names in separate fields, you can associate a field or column heading with the First Name drop-down list.

FIGURE 14-4

The Organization Chart wizard maps fields in your data source to Organization Chart shape data fields.

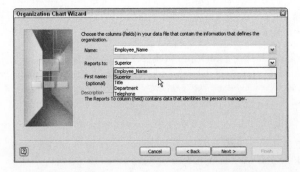

6. To specify a field that contains reporting relationships in a data source, select an entry in the Reports To drop-down list and click Next.

7. From the field names in the Data File Columns list, select the fields that you want to appear in each shape on the Visio organization chart. Use the following methods to build the list of fields:

   ▪ Click Add to move the selected fields to the list of fields that appear in organization chart shapes.

   ▪ To reorder the fields in Visio shapes, in the Displayed Fields list, select a field and click Up or Down to rearrange the order of fields within shapes.

   ▪ To remove a field from the Displayed Fields list, select it and click Remove.

8. Click Next to complete the mapping of the standard organization chart fields.

NOTE In the wizard, Organization Chart shapes contain four default properties: Name, Title, Department, and Telephone. If you remove these properties from the field lists in the wizard, the shapes still contain these shape data fields but the wizard does not transfer the values from your data file to the Visio shapes.

9. The next screen also includes a Data File Columns list, but this screen associates fields from your data source to shape data fields in organization chart shapes. In the Data File Columns list, select the fields whose data you want transferred to shape data in the organization chart shapes. You can use the Add or Remove buttons to build the list of fields. Click Next.

TIP Displaying a field in an Organization Chart shape automatically includes that field as a shape data field. By associating a field with shape data without displaying it in Organization Chart shapes, you can reduce clutter while still providing access to the data. To view shape data, right-click a shape and, from the shortcut menu, choose Properties.

10. *Execute this step only if you are using an ODBC-compliant data file.* Select either the Copy Database Records To Shapes option or Link Database Records To Shapes option. If you link database records to shapes, the values in the organization chart will change when the database values change. Click Next.

11. Proceed to the Finalizing the Organization Chart section to create the organization chart and lay it out.

Finalizing the Organization Chart

The entire organization chart for a large company would be one large black blob, if you added all the shapes to one 8 1/2" × 11" piece of paper. On the final screen of the Organization Chart wizard (the one that contains the Finish button) Visio provides all the options for specifying the allocation of organization chart shapes across multiple pages. For example, you can produce summary pages of executive levels for large organizations or show employees for different departments

Linking Database Records to Shapes

When you link database records to shapes by choosing the Link Database Records to Shapes option within the Organization Chart wizard, you can control how these links operate. In the wizard, on the same screen that includes the options for copying or linking records to shapes, click Settings. In the dialog box that appears, you can specify which database-related commands to include on a shortcut menu when you right-click a linked shape. Here's what each command does when you execute it in an organization chart drawing:

- **Select Database Record** — Opens a dialog box in which you select a database record to link to the shape.

- **Refresh Shape Properties** — Retrieves the corresponding record from the database and updates the shape data with values in the database record.

- **Update Database Record** — Uses the values in the shape's shape data to update the linked database record.

- **Delete Shape and Record** — Removes the shape in the organization chart as well as the linked record in the data source.

The dialog box also contains options for controlling when updates occur. Select the Refresh Shape Data On Page check box and the Update Database Records check box to configure Visio to update shape data and database records each time you display a page.

The check boxes under the Events heading tell Visio to refresh shape data either when you open a Visio document or periodically by using the NOW function. When you turn on the Periodically Refresh Based on NOW Function option, Visio adds the Start Continuous Refresh command to the shortcut menu that appears when you right-click a page. Choosing that command launches a link between the drawing and the database, which transfers data until you right-click the page once more and, from the shortcut menu, choose Stop Continuous Refresh.

on separate pages. The steps to spread shapes over multiple pages are the same whether you build your chart from existing data or create a data file within the wizard:

1. To create a hyperlink between shapes that represent the same employee on more than one page of your organization chart, select the Hyperlink Employee Shapes Across Pages check box. With this option enabled, you can jump to a copy of an employee shape by right-clicking the shape and choosing the hyperlink from the shortcut menu.

2. To synchronize all shapes that represent the same employee so that changes made to one copy propagate to all other copies, check the Synchronize Employee Shapes Across Pages check box. For example, if an employee changes her last name, you change her name in only one shape and Visio makes the changes in all other copies.

3. If you don't care how Visio allocates shapes to pages, select the I Want the Wizard to Automatically Break My Organization Chart Across Pages option. After choosing this option, click Finish to create the organization chart and skip the rest of the steps in this list.

4. To maintain control of the distribution of organization shapes across multiple pages, select the I Want to Specify How Much of My Organization to Display on Each Page option, and then click Next. The Organization Chart wizard displays a screen, shown in Figure 14-5, for creating and managing drawing pages, as well as specifying the portions of the organization to show on each drawing page.

FIGURE 14-5

When you retain control over how to place shapes on an organization chart drawing, you can create and modify organization chart pages and specify how many levels appear on each one.

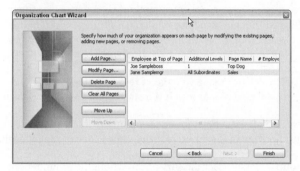

5. To specify the levels that you want to display on the first page, click Modify Page and perform the following tasks:

- To use the first page as an executive summary, in the Number Of Additional Levels box, select the number of levels that you want to appear on the summary page.

- In the Page Name box, type a name for the page, such as Executive Circle, and then click OK.

6. To add additional pages to the organization chart — for example, to create a page for a lower level department — click Add Page. When you create additional pages, you can specify their contents and then click OK to return to the wizard screen:

- **Choose the name that appears at the first level on this new page** — In the Name At Top Of Page list, select a name. For example, select the name of a manager shown at the lowest level of the executive summary page.

- **Show all subordinates who report to the person at the top-level of the page** — In the Number Of Additional Levels list, select All Subordinates.

- **Show only a few levels below the top-level on a page** — In the Number of Additional Levels list, select the number of levels to include.

- **Name the page** — Type a name for the page in the Page Name box.

7. When you have finished defining the pages for your chart, click Finish. Visio creates the organization chart on a new drawing page in the current Visio drawing file.

Laying Out the Organization

Organizational structure can be quite fluid, with people being promoted or reassigned and even business units being reorganized, acquired, or divested. Visio eases the challenge of keeping organization charts up-to-date with specialized tools for rearranging shapes.

Controlling the Layout of Subordinate Shapes

Organization charts use numerous layouts to show subordinates reporting to a superior and Visio can handle them all. Although Visio arranges subordinate shapes horizontally by default, you can specify different layout options for each person with subordinates or *direct reports*. For example, a vertical layout is more miserly in its space consumption when a manager supervises a plethora of subordinates. If a diagram contains synchronized copies of shapes on multiple pages (see the section "Spreading an Organization Over Multiple Pages"), you can apply a different layout to each synchronized copy.

 Although each superior shape with direct reports can use a different layout, your organization chart will be easier to read if you stick to only one or two layouts.

You can control the layout of direct reports for one or more managers or ask Visio to choose the best layout. To lay out shapes on an organization chart, choose one of the following methods:

- **Lay out a group** — First, select the superior to whom the group reports. On the Organization Chart toolbar, click one of the layout options, and then, from the drop-down menu, click the configuration you want.

- **Lay out multiple groups** — Select the superior for every group you want to lay out by clicking the first superior and then Shift-clicking the others. On the Organization Chart toolbar, click one of the layout options and click the configuration you want.

NOTE To see examples of each layout option, as shown in Figure 14-6, choose Organization Chart ➪ Arrange Subordinates. Click the layout you want and then click OK.

- **Optimize your layouts** — If you add employees, revise reporting relationships, or make other changes, you can reapply the layouts you chose. On the Organization Chart menu or toolbar, choose Re-layout. Visio moves the shapes on the page to conform to the layout options you selected for each superior.

- **Optimize overall layout** — To hand over layout decisions to Visio, choose Organization Chart ➪ Best Fit to Page. When you use this command, verify that ancillary shapes, such as Team Frames are still positioned correctly or reposition them, if necessary.

FIGURE 14-6

By choosing the Arrange Subordinates command, you can preview layouts before applying them.

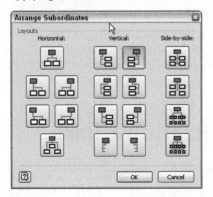

CAUTION   If your Organization Chart shapes explode over your drawing page when you apply a layout option, the connections between shapes could be broken. This can occur if you or another author positioned shapes manually and inadvertently broke the connections between them.

The Undo command is the easiest way to correct this problem. However, if someone was helpful enough to save the chart with the shapes in disarray, you can correct this chaotic behavior by reconnecting shapes, ideally with shape-to-shape glue. To connect shapes with shape-to-shape glue, select a connector and drag each end point to the center of a shape, or drag the disconnected subordinate shape over the supervisor shape.

Moving Organization Chart Shapes

Although the Visio organization chart layout options (Arrange Subordinates, Re-layout, and Best Fit to Page) are likely to handle most shape arrangement for you, from time to time, you want to move individual shapes. For example, when a subordinate joins the organizational hierarchy in a chart, Visio inserts the shape to the right of existing subordinates in a horizontal layout or below existing subordinates in a vertical layout. If you want the subordinates in a specific order, reordering the shapes is the only solution.

Change the order of subordinates by selecting a subordinate shape to move and choosing Organization Chart ⇨ Move Subordinates. On the submenu, choose Left/Up to move the subordinate left in a horizontal layout or up in a vertical layout. Choose Right/Down to move the subordinate to the right or down, for horizontal or vertical layouts, respectively.

CAUTION Visio's standard commands for moving shapes can cause problems in organization chart drawings. If you move Organization Chart shapes by dragging them to new positions, any subsequent application of Visio Organization Chart layout tools overrides your manual placements. You're better off applying layout options, adjusting shape spacing, and spreading the organization chart over several pages, as described later in this chapter.

In addition, moving shapes manually can break connections between shapes, which renders the layout tools powerless. This can happen if a connector isn't glued to both shapes. Another way that connections break is if you select a connector, but only one of its connected shapes. For Visio layout tools to work once more, you must reglue the shapes or be sure to select all the shape(s) you want to move.

Formatting Organization Chart Appearance

Visio 2007 offers Themes, which simplify the formatting of colors and shapes for many Visio templates. Visio's organization chart options offer formatting that applies specifically to organization chart shapes and help you wrangle shapes into just the look you want. Organization Chart options control the appearance of Organization Chart shapes and text and specify the data that appears within the shapes. To specify these settings, choose Organization Chart ➪ Options. To restore the Visio default settings, in the Options dialog box, click Restore Defaults. This section explains the layout and formatting tools that Visio offers for organization charts.

NEW FEATURE Visio 2003 included an option to apply an organization chart theme. In Visio 2007, you can obtain professional presentation results by clicking the Themes button on the Formatting toolbar. To apply a theme to an organization chart, in the Formatting toolbar, click Themes. In the Theme - Colors task pane, click the theme you want to apply. See Chapter 7 for the full description of working with Themes.

If the Visio drawing was created in a previous version of Visio, when you click Themes, a message box appears informing you that the shapes must be updated to work with the Themes feature. Click OK to apply the theme.

Setting Options to Customize Shape Appearance

Select the Options tab to customize shapes in the following ways:

- Set the height and width for all employee shapes.
- Show or hide employee pictures that have been inserted into shapes.
- Show or hide the divider line between the first and second line of text in shapes.
- Show drawing tips (when available).
- Use the organization chart options that you choose for all new organization charts and for all Organization Chart shapes used on other types of drawings.

Displaying Pictures of Employees

If your organization constantly adds new people or you simply like to personalize your organization chart, displaying employee photos in your Visio organization chart is one way to go. To attach a picture to a shape, right-click the shape and, from the shortcut menu, choose Insert Picture. Browse to the folder and file name for the picture and click Open.

To display pictures in the organization chart, choose Organization Chart ⇨ Options. Select the Options tab and then select the Show Pictures check box. If a picture doesn't appear in a shape, the picture might be too large to fit in the shape. Crop the picture to a smaller size.

Specifying the Information to Display in Shapes

Organization Chart shapes can display values for shape data in five locations within their boundaries. The center of a shape can show one or more fields and each corner of the shape can show the value for one field.

By selecting the Fields tab, you can perform the following actions:

- Select the fields that appear in the center of a shape by selecting the field check boxes for Block 1.

- Rearrange the order for the fields in the center of a shape by selecting a field and then clicking Move Up or Move Down.

- Select the fields that appear at each corner of a shape by choosing fields from the drop-down lists for Block 2 through Block 5.

Setting Options to Customize the Appearance of Text

By selecting the Text tab, you can customize the appearance of organization chart text in the following ways:

- Choose the field whose text style you want to set.
- Specify the font to use for the selected field.
- Specify the font size to use for the selected field.
- Specify whether the field's text should be bolded or italicized.

Improving Chart Readability

Sometimes, organization charts require more tweaking than organization chart layouts and options can provide. If your chart is hard to read no matter how you configure the diagram, adjusting the spacing between shapes or hiding some subordinates might do the trick.

Modify Shape Spacing

To modify the spacing between shapes, choose Organization Chart ⇨ Change Spacing. The spacing options control the spacing you use and the shapes to which the spacing applies, as follows:

- **Make spacing larger or smaller** — Select the Tighter or Looser option to decrease or increase the current spacing, respectively. For example, if a chart oozes beyond the border of the page, choose Tighter to shrink the area you need.

- **Define specific spacing** — Select the Custom option and click Values to define the distances for spacing. For each layout type, you can specify the spacing between subordinates, the spacing between superiors and subordinates or assistants, and the justification spacing for layouts aligned to the left or right.

- **Specify the shapes or pages for spacing** — Below the Apply Spacing To heading, select an option to apply the spacing changes to the selected shape or shapes, all shapes on the current drawing page, or all shapes on all pages.

> **TIP** Increasing the spacing around shapes makes charts more readable. But it can also expand an organization chart until it no longer fits on the drawing page. To distribute a large chart across multiple pages, see the "Spreading an Organization Over Multiple Pages" section later in this chapter.

Hiding Subordinate Shapes

For particularly tough readability problems, temporarily hiding subordinate shapes can help. To hide the subordinates for a shape, select the superior shape and choose Organization Chart ⇨ Hide Subordinates. When subordinates are hidden, the entry on the menu changes to Show Subordinates. After you have printed or saved a picture of the chart, you can choose Show Subordinates to restore the hidden shapes to the view once more.

Creating a Legend for an Organization Chart

Visio differentiates different organization positions with shapes of different sizes and border styles. The large box for the Executive shape is a good indicator of hierarchy, but it's hard to remember what a dotted or dashed line represents. To help your audience interpret an organization chart, add a Legend shape from the Legend Shapes stencil to the drawing and add the organization chart shapes you use to the legend. Chapter 6 includes detailed instructions for working with legends in Visio.

To display Organization Chart shapes in a legend shape, drag a copy of the shapes that you want in the legend from the Organization Chart Shapes stencil and drop them onto the legend shape. Visio displays the shape in the Symbol column, shows the shape name in the Description column, and calculates the number of each type of shape in the Count column. After you have converted the shapes to symbols, delete the copies that you dropped on the legend.

CAUTION The Re-layout and Best Fit to Page functions optimize the layout of your chart based on the shapes that *aren't* hidden. When you redisplay subordinates, re-run the Re-layout or Best Fit to Page command again.

Editing and Deleting Organization Chart Shapes

How you edit the text that appears in Organization Chart shapes or delete those shapes depends on how you built your organization chart. For organization charts built from scratch, standard Visio techniques work perfectly well, but for organization charts built from data in other programs, you must use Organization Chart tools instead.

NOTE Because organization chart shapes are so specialized, Organization Chart commands are usually better for editing than standard Visio techniques. For example, if shapes aren't large enough to display the employee information you want, consider specifying larger shape dimensions in the Organization Chart Options dialog box. To learn more about organization chart options, see the "Formatting Organization Chart Appearance" section earlier in this chapter.

Editing Shape Text and Data

The text that appears in Organization Chart boxes are actually text values stored in each shape's shape data fields. By default, the values for the Name and Title shape data properties appear in the center of a shape. (To learn how to select the fields that appear in shapes, see "Specifying the Fields to Display in Shapes" earlier in this chapter.)

You can modify organization chart shape data values in several ways:

- **Change values in a shape** — To change the value of shape data displayed in a shape, select the shape and type new entries for the fields you can see in the shape.

- **Change shape data values directly** — To change the values for any shape data field, right-click a shape and, from the shortcut menu, choose Properties. In the Shape Data dialog box, click a shape data box and enter the new value.

- **Update a shape's properties from a linked database record** — Copy values from the linked database to the shape data fields by right-clicking a shape and, from the shortcut menu, choosing Refresh Shape Properties.

- **Update a database record from the shape data values** — Transfer modified values in shape data to the linked database by right-clicking a shape and, from the shortcut menu, choosing Update Database Record.

Changing the Type of Organization Chart Shape

Sometimes, the type of shape you dragged onto a drawing is no longer appropriate — for example, when a person in the organization is promoted or a position becomes vacant. When this occurs, you can change the type of shape for a position by selecting a shape and choosing Organization Chart ⇨ Change Position Type. Select the new position type and click OK. Because each Organization Chart shape has the same shape data and behaviors, you don't have to make any other adjustments.

Deleting Organization Chart Shapes

To delete a shape that isn't linked to a database, simply right-click the shape and choose Cut, or select the shape and press Ctrl+X.

Cutting a shape that's linked to a database record does not delete the linked record in the database. To delete a shape *and* its linked database record, right-click the shape and, from the shortcut menu, choose Delete Shape and Record.

Spreading an Organization Over Multiple Pages

For large organizations, it's often clearer to present portions of the organization on different pages. For example, you can show top-level executives on a summary page and include additional pages to display the organization that each executive leads. When you do this, the same employee appears on more than one page, once as a direct report and again as the head of a group. By creating synchronized copies of these shapes, you can ensure that any text, shape data, or subordinates that you add to a shape on one page apply to the copies on other pages.

 Adding, deleting, or moving a shape, or modifying a shape's associated layout, does not affect its synchronized copies.

To create a synchronized copy of a group or department, follow these steps:

1. Select the shape that represents the head of the group that you want to move and choose Organization Chart ⇨ Synchronize ⇨ Create Synchronized Copy.

2. To create a copy on a new page, select the New Page option. To create a copy on an existing page, select the Existing Page option and choose the page from the drop-down list.

 If the Visio drawing has only one drawing page, the Existing Page option is grayed out.

3. If you don't want to see the subordinates on the original page after you create a synchronized copy, select the Hide Subordinates On Original Page check box.

4. Click OK to create a synchronized copy. If you create a copy on a new page, double-click the tab for the new page to rename the page.

Creating Hyperlinks Between Synchronized Copies

Hyperlinks to synchronized shapes make it easier to navigate between them. When you add a hyperlink to a synchronized shape, following the hyperlink pans and zooms to the synchronized shape. Follow these steps to add hyperlinks to shapes:

1. Select a shape and choose Insert ⇨ Hyperlinks.

2. Click the Browse button next to the Sub-Address box and choose the page that contains a synchronized copy of the shape.

3. To create a hyperlink, type the name of the shape to which you want to link. To obtain a shape's name, on the Tool menu, choose Options. In the Options dialog box, click the Advanced tab, and in the Advanced Options section, select the Run In Developer Mode check box. Right-click the shape, click Format, and then click Special. The Name box shows the shape's name.

If you add subordinates to a synchronized copy, those subordinates are initially visible only on that page. To update another synchronized copy to show the new subordinates, select the appropriate superior on the synchronized copy and choose Organization Chart ⇨ Synchronize ⇨ Expand Subordinates. Expanding subordinates applies to one level of subordinates for a shape, so you must expand each level that you added to a synchronized copy.

Comparing Versions of Organization Charts

When you build organization charts from imported data, you end up with more than one version of an organization chart. The Compare Organization Data command produces a report of changes, which is helpful for seeing what has changed, such as telephone extensions, office numbers, or employees' job titles. Follow these steps to compare two versions:

1. Open the more recent organization chart and choose Organization Chart ⇨ Compare Organization Data. If you would rather update the older version to maintain the layout and settings you have applied, open the older version. Regardless, in the My Drawing To Compare box, Visio displays the name of the drawing file you opened.

2. In the Drawing To Compare It With box, specify the file for the other version by clicking Browse, navigating to the folder that contains the files, and clicking Open.

3. To compare values only from specific fields, click the Advanced button, delete the fields you don't want to compare, and click OK.

4. If you opened the newer version first, select the My Drawing Is Newer option. Otherwise, select the My Drawing Is Older option.

5. Under Report Type, choose Sort By Change to produce a report that shows all changes made. Choose Sort By Position to see a report of positions that were added or removed.

6. Click OK. Visio compares the information in the two drawing files and, in your browser, displays the Comparison Report, which shows the differences between the two. To save the file, choose File ⇨ Save As, and specify the folder and file name.

Sharing Organization Chart Data

If your Visio organization chart is your source for information about an organization, you can share that data with others through reports or by exchanging data files. For example, you can produce a report showing employees along with other information stored in shape data. To provide organization chart data for someone to use in another application, export the data to other file formats, including Excel, HTML, Visio shapes, and XML.

Reporting Organization Data

To produce a report for an organization chart, follow these steps:

1. Choose Data ➪ Reports and, in the report list, select Organization Chart Report.

2. If you want to modify the report definition, click Modify and navigate through the wizard to make the desired changes. Specify the shapes to include in the report or define criteria that limits the report. Click Finish when you've completed the configuration of the report.

3. Click Run, choose the report format and location in which to save the file, and then click OK.

Exporting Organization Chart Data for Use in Other Programs

You can also export organization chart data to Excel spreadsheets, text files, or comma-delimited files. To do this, choose Organization Chart ➪ Export Organization Data. Specify the folder and file name for the export file. Choose a file format in the Save As Type box and click Save. Visio exports the values stored in shape data. It also exports the shape ID and the master shape designation so that you can use the exported file to build a new Visio organization chart.

Summary

With Organization Chart template features, maintaining documentation for an organization is quick and easy. Organization Chart shapes are smart enough to create reporting relationships when you drop one shape on top of another. However, if you already have organization chart data in another data source, you can build and update Visio organization charts automatically from that data.

To satisfy different documentation requirements, you can modify the appearance of organization charts in several ways:

- Specify the layout for subordinates in a group or instruct Visio to optimize layout for you.
- Specify the information fields that appear in organization chart shapes.
- Format the shapes and text styles in a chart.
- Distribute portions of an organization across multiple pages.

You can also share the information in your Visio organization charts by producing organization reports or data files based on Visio organization chart data. In addition, you can publish the organization chart by saving it as a series of web pages.

Chapter 15

Building Visio Flowcharts

All kinds of human endeavors proceed in a sequence or flow, such as the steps in a customer support procedure, the accumulation of information and change of responsibility during a change management process, or the movement of material through a manufacturing environment. Whether you are documenting business procedures and processes, analyzing the flow of data for a business re-engineering initiative, or showing the interaction of systems or departments, Visio flowchart templates help illustrate those relationships and connections.

Visio includes templates for building basic flowcharts with simple geometric shapes and arrows, which are all you need for many step-by-step procedures, checklists, sequences, and so on. But the program also offers specialized flowcharts for business process analysis, quality management, risk management, and other organizational initiatives. Even organization charts and project schedules incorporate the connections characteristic of flowcharts. Although Visio provides templates and shapes for specialized methodologies such as SDL and IDEF0, it doesn't provide tools to ensure that you construct your flowcharts in accordance with the rules of those methodologies.

Because flowcharts satisfy a broad range of uses and industries, flowchart templates appear within several of Visio's template categories. In addition to the self-explanatory Flowchart category, you can find templates for flowcharts within the Business, Engineering, General, and Software and Database categories.

This chapter begins with an overview of different types of flowcharting diagrams that Visio provides, both basic and specialized. The chapter then describes how to build, modify, and format flowcharts in Visio, regardless of the templates you choose. In fact, many of the methods for working with flowcharts are nothing more than standard techniques that work with any Visio drawing. Finally, because flowcharts are often large or complex, you will learn how to make them easier to read by continuing diagrams onto additional pages.

Choosing the Right Template

The Flowchart template category lists six templates, but you can find templates that show relationships and flow in almost every template category. Because Visio 2007 has recategorized many of the flowchart templates, Table 15-1 identifies Visio flowchart templates, what they do, their new Visio 2007 locations, their old homes in Visio 2003, and the chapter in this book that covers them.

NEW FEATURE Visio 2007 introduces a new ITIL template to support the creation of ITIL (IT Infrastructure Library) diagrams, which document management procedures that help organizations make the most of their IT operations. Chapter 16 covers the creation of ITIL diagrams in Visio 2007.

CROSS-REF Refer to Chapter 42 to identify which templates are available in Visio Standard and Visio Professional.

If you're familiar with basic Visio techniques you already know most of what you need to build flowchart diagrams. Flowchart shapes work with Visio's automatic layout tools, so you can add and connect shapes and then let Visio take care of arranging them. If you want to spiff up the appearance of your flowcharts, the standard Visio annotation and formatting tools do the job.

To create a basic flowchart, choose File ➪ New ➪ Flowchart and then choose one of the templates from the submenu, such as Basic Flowchart, Data Flow Diagram, or Work Flow Diagram. If the type of flowchart you want appears in a different template category, for example, the ITIL Diagram in the Business category, choose File ➪ New. On the submenu, point to the category you want, and on the template submenu choose the template. Flowchart templates automatically set the page to a letter-size sheet and use inches (or millimeters for metric templates) drawn at a one-to-one scale. Visio sets the orientation depending on the template or diagram orientation that you choose.

Adding and Connecting Flowchart Shapes

The best method for adding and connecting shapes depends on the type of flowchart and the shapes you use. Because flowcharts indicate order or flow, you must take extra care to connect shapes in the correct order or the flowcharting version of the cart might end up before the horse.

TABLE 15-1

Visio Flowchart Templates

Flowchart	Purpose	Visio 2007 Category	Visio 2003 Category	Book Chapter
Basic Flowchart	Document processes and procedures, show work or information flow, track cost and efficiency, or describe process improvements and process management in projects.	General, Flowchart, Business	Flowchart, Business Process	15
Cross-Functional Flowchart	Show the relationship between a business process and the departments or organizational units that are responsible for steps in that process.	Flowchart, Business	Flowchart, Business Process	16
Data Flow Diagram	Document structured analysis, information flow, data-oriented processes as well as data flow within a system or organization.	Flowchart, Business	Flowchart, Business Process	16
IDEF0 Diagram	Model decisions, actions, and activities based on the Structured Analysis and Design Technique (SADT). See nearby note for a definition of IDEF.	Flowchart	Flowchart	16
SDL Diagram	Document event-driven systems such as communication and telecommunication systems and networks using the Specification and Description Language (SDL).	Flowchart	Flowchart	16
Audit Diagram	Create auditing diagrams for accounting, inventory, financial and money management, tracking fiscal information, and decision-making.	Business	Business Process	16
Cause and Effect Diagram	Create cause-and-effect diagrams, also known as fishbone or Ishikawa diagrams, to categorize the factors that contribute to the problem being examined and identify the most significant factors for improving or resolving the situation.	Business	Business Process	16

continued

TABLE 15-1 *(continued)*

Flowchart	Purpose	Visio 2007 Category	Visio 2003 Category	Book Chapter
EPC Diagram	Use EPC (Event-driven Process Chain) from the SAP R/3 methodology to engineer business processes as chains of functions and events.	Business	Business Process	16
Fault Tree Analysis Diagram	Document events that can lead to failures to help prevent them. Fault tree analysis is commonly used in Six Sigma processes.	Business	Business Process	16
ITIL Diagram	Document management procedures that help organizations achieve business objectives and deliver quality, while making cost-effective use of IT operations.	Business	N/A	16
TQM Diagram	Document business process reengineering, continuous improvement, and quality solutions.	Business	Business Process	16
Work Flow Diagram	Show the flow of information or work for business process reengineering, business process automation, and management.	Flowchart, Business	Business Process	16
Brainstorming	Document ideas and concepts identified during brainstorming sessions	Business	Brainstorming	18
Organization Charts	Illustrate the hierarchy of organizations and other branching elements, such as genealogy.	Business	Organization Chart	14
Project Timelines	Show the chronology of work throughout the life of a project.	Schedule	Project Schedule	17

NOTE IDEF is an acronym of an acronym. IDEF originates from the acronym I-CAM Definition Methods. I-CAM is the acronym for Integrated Computer-Aided Manufacturing. The U.S. Air Force initiated the I-CAM project to develop methods for improving manufacturing productivity through a systematic application of rules enabled by computer technology.

Adding Flowchart Shapes in a Sequence

For sequential processes, you can create a sequence by dragging and connecting shapes in order or by defining the sequence after you add all the shapes to the diagram. Create a sequence of shapes using one of the following methods:

- **AutoConnect** — In Visio 2007, the new AutoConnect feature turns adding and connecting shapes into a two-click operation. To connect a new shape to one on the drawing, in the Shapes pane, select the shape you want to add. When you position the pointer over shapes on the drawing, Visio displays blue arrows on each side of an existing shape that is still available for a connection. Click the blue arrow on the side you want and Visio automatically aligns the new shape on that side of the existing shape and connects the two.

CROSS-REF To learn more about all the methods for connecting shapes, refer to Chapter 5.

- **Dragging** — When you already know the sequence of steps, the Connector tool is the easiest way to define that sequence. On the Standard toolbar, click the Connector tool. As you drag shapes onto the page, Visio connects the new shape to the shape you added previously with a shape-to-shape connection.

TIP Visio flowchart templates turn on the dynamic grid by default to make it easy to position shapes relative to shapes already on your diagram. When the dynamic grid is active, Visio displays dotted lines as you drag a shape around to indicate positions that align the shape with shapes already on the page. To align a shape, drag it and snap it to one of the horizontal or vertical dotted lines on the dynamic grid.

If the dynamic grid interferes with the placements you want, you can turn it off by choosing Tools ➪ Snap & Glue and, under the Currently Active heading, unchecking Dynamic Grid.

- **Selecting a sequence** — If you intend to work out the sequence by analyzing the shapes on your diagram, you can add shapes to the drawing page first and then use the Connect Shapes command to create the sequence. Follow these steps to create a sequence with shapes already on the page:

  1. Drag all the shapes for the sequence onto the page.

  2. After you determine the correct sequence, select each shape in order and then choose Shape ➪ Connect Shapes. Visio connects the shapes in order using shape-to-shape connections. Visio does not rearrange the shapes when it connects them.

  3. If the layout is difficult to read, rearrange the layout by choosing either Shape ➪ Re-layout Shapes to reapply the current layout options or Shape ➪ Configure Layout to specify different layout options (for instance, to switch a Right to Left layout to Top to Bottom).

CROSS-REF To learn more about layout options and laying out shapes, see Chapter 4. If you want to number steps in a process, for example, to reference steps in the diagram in your procedure instructions, use the Number Shapes command, described in Chapter 6.

Connecting Shapes Out of Sequence

Processes usually don't flow directly from start to finish. For example, one step in a process might trigger a notification to another department along with launching the next step in the process. Moreover, some steps can spawn several alternative paths. For example, Decision shapes connect to one other shape for each possible outcome of the decision and, sometimes, those shapes are earlier in the sequence.

When you want to connect shapes that aren't in sequence, you have to tell Visio which shapes to connect. The tools you use don't change, but the way you apply them does. To create connections between two nonsequential steps, use the following methods:

■ **Connecting out of sequence shapes** — As easy as AutoConnect is, it can't help you connect shapes that are out of sequence. Instead, turning on the Connector tool and then dragging from the first shape to the second is the only way to connect shapes that are out of sequence. To create a shape-to-shape connection, drag from inside the first shape to inside the second shape. Visio highlights the entire shape with a red box for shape-to-shape connections. To connect to specific connection points, drag between the connection points, which Visio highlights with small red boxes.

TIP If shapes are small, Visio usually highlights connection points instead of the entire shape. In this case, you can create shape-to-shape connections by pressing the Ctrl key while using the Connector tool.

■ **Creating additional connections on shapes** — To create additional connections on a shape, for example, to define alternate outcomes for a decision, you can use AutoConnect or the Connector tool. AutoConnect is better for creating additional connections to new shapes you add to the drawing, as demonstrated in Figure 15-1. In the Shapes pane, select the shape you want to add and then, on the drawing page, click a blue arrow that appears next to an existing shape to connect the new shape to that side of the existing shape. The Connector tool is preferable for creating connections between existing shapes. Use it as described in the previous bullet point.

FIGURE 15-1

AutoConnect can add initial connections as well as additional connections to existing shapes.

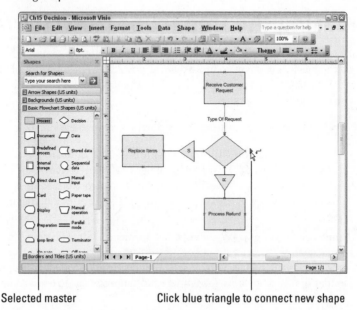

Selected master   Click blue triangle to connect new shape

Changing the Layout and Flow

From time to time, the sequence and connections in flowcharts change — a business process improvement initiative streamlines a process or quality initiatives add new steps to a procedure. When you modify a flowchart to add, remove, or reconnect shapes, chances are the layout needs some adjustment as well. In addition, despite your extra attention, shapes get connected in the wrong direction. Here are the standard Visio techniques that help you make these kinds of changes:

- **Rearrange Shapes** — You can manually rearrange individual shapes and connectors or use the automatic Visio layout tools:

  - **Move Individual Elements** — To manually rearrange shapes, drag shapes you want to move to new locations. To manually modify connector paths, drag green vertices on connectors to new locations.

  - **Automatically Layout Shapes** — To use the layout Visio tools to automatically lay out your diagram, choose Shape ➪ Configure Layout. Choose the options you want and click OK to apply them. To reapply layout options after you've manually positioned shapes, choose Shape ➪ Re-layout Shapes. Using automatic layout overwrites any manual arrangement you have performed.

■ **Reverse Direction of Flow** — To change the direction of flow between two steps, select the connector between the steps and choose Shape ➪ Rotate or Flip. Choose Flip Horizontal for a horizontal connector or Flip Vertical for a vertical connector.

Increasing the Impact of Flowcharts

Flowcharts communicate so many different types of information to so many different audiences that it's worthwhile learning how to make them as powerful as possible. By taking advantage of several of the standard Visio tools, including a few new to Visio 2007, you can make your flowcharts more persuasive and eloquent.

Here are several methods for increasing your audience's comprehension:

■ **Visualize information with Data Graphics** — In addition to the relationships between shapes, flowcharts usually have a lot of information to convey — be it the cost and duration of each step, the success rate, or where work is performed. Instead of overwhelming a diagram with tons of text, the new Data Graphics feature can highlight values with colors, icons, and other visual cues. Chapter 10 explains Data Graphics in detail.

■ **Annotate drawings** — When visual cues won't do, some descriptive text might be in order, such as labels on the connectors linked to Decision shapes to indicate the outcome that triggers a particular step. To add a border or title to identify your diagram, drag a border or title shape from the Borders and Titles stencil. To learn about other ways to annotate diagrams, see Chapter 6.

■ **Enhancing diagram appearance with Themes** — Whether the purpose of a diagram is to sell people on an idea or educate them, good-looking and easy-to-read diagrams are a must. With the new Theme Colors and Theme Effects, (described in Chapter 7), you can apply consistent and professional-looking formatting with one or two clicks.

■ **Add a background** — To add a background to a flowchart page, drag a background shape from the Backgrounds stencil onto the page. Visio automatically creates a new page for the background and associates it with the current foreground page.

> **NOTE** If the Flowchart template you're using doesn't open the Backgrounds stencil, choose File ➪ Shapes ➪ Visio Extras ➪ Backgrounds.

■ **Give diagrams enough room** — Flowcharts are often large or complex, so the default letter-sized paper that Visio chooses is usually too small to show all the information you want to communicate. If you have a printer or plotter that can produce larger printed diagrams, the easiest way to display flowcharts is on a larger piece of paper. In Page Setup on the Page Size or Print Setup tabs, you can specify a larger page or paper size, so the diagram doesn't have to continue across multiple pages. See Chapter 3 for instructions on specifying page and paper sizes.

■ **Make navigation easy** — Of course, some diagrams are so immense that there's no paper large enough to hold them on one sheet. If you must continue a diagram across several pages, you can simplify navigating from page to page by adding navigation shapes to your diagram, as described in the next section.

Creating Multi-Page Flowcharts

All but the simplest flowcharts quickly expand to more than one page. In many cases, the first page of a flowchart is a summary of the overall flow, while each additional page shows detailed steps. Although hyperlinks added directly to shapes jump to another location, another page, or even another file, they aren't visible unless you position the pointer over a hyperlinked shape. To make flowcharts easy to navigate, On-page and Off-page reference shapes are visible indicators that the process or flow continues elsewhere in the diagram.

Navigating to Other Pages

The Off-page Reference shape is so handy, it's located on the Basic Flowchart Shapes stencil, which means it's available in both Visio Standard and Visio Professional. It's designed to navigate to other pages in a diagram, so, as soon as you drop an Off-page Reference shape on a page, the Off-page Reference dialog box opens wanting to know the page to which the shape should connect. Whether you connect to a new page or an existing page, Visio adds another Off-page Reference shape to that page and hyperlinks the two shapes, as illustrated in Figure 15-2. To follow an Off-page reference, simply double-click the Off-page Reference shape.

FIGURE 15-2

Off-page Reference shapes jump to other pages in a diagram when you double-click them or follow their hyperlinks.

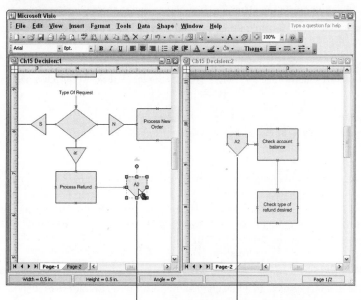

You can synchronize the labels on linked references

Double-click or follow hyperlink to jump to other page

To add an Off-page Reference to a diagram, do the following:

1. In the Shapes pane, select the Off-page Reference shape.

2. To connect the Off-page Reference shape to a step in the flowchart, position the pointer over the shape for the step and click a blue AutoConnect arrow.

3. In the Off-page Reference dialog box, the New Page option is selected by default and Visio fills in a name for the new page. To create a new page in the diagram, leave this option selected. If you want, in the Name text box, type a more meaningful name. If the diagram has more than one page and you want to link to another page, select the Existing Page option and choose the page.

4. To add another Off-page Reference shape to the linked page, make sure that the Drop Off-page Reference Shape On Page check box is selected. This tells Visio to add a second Off-page Reference shape to the destination page and hyperlink it to the first Off-page Reference shape.

TIP You can convey information about a reference by changing the shape of an off-page reference. To change the reference's appearance, right-click the off-page reference and, from the shortcut menu, choose Outgoing, Incoming, Circle, or Arrow.

5. If you want Visio to update the labels in the shapes to match one another, check the Keep Shape Text Synchronized check box. With this setting, changing the label in one reference shape automatically updates the linked reference shape.

6. If you plan to publish the diagram to the Web and want to navigate it there, select the Insert Hyperlinks On Shape(s) check box.

7. Click OK. Visio navigates to the destination page, whether it is a new page or an existing page, and selects the second Off-page Reference shape.

8. With the Off-page Reference shape still selected, type text to label the reference. If you selected the check box to synchronize text, Visio updates the label in the other reference shape.

Adding References to Other Steps on the Same Page

Sometimes, one step in a process jumps to another step located elsewhere on the same page. If the steps are at opposite ends of the page or separated by numerous shapes, a connection line between the two clutters the flowchart. The On-page Reference shape is the solution to this visual dilemma. An On-page Reference shape creates a visual link between two steps on the same page without a connection line. To create on-page references, follow these steps:

1. In the Shapes pane, select the On-page Reference shape.

2. To link the On-page Reference shape to a flowchart step, position the pointer over the shape for the originating step in the flowchart and click a blue AutoConnect arrow.

3. With the On-page shape still selected, type a number or letter to label the reference.

4. Position the pointer over the step in the flowchart where the procedure continues and click the blue AutoConnect arrow to add and link another On-page reference shape.

5. Type the same number or letter in the On-page reference shape.

Summary

Visio flowcharts are built primarily by adding and connecting shapes. With the new AutoConnect feature, connecting shapes in sequence is even easier than before. To fine-tune and format flowcharts, the standard techniques that apply to any kind of Visio diagram are all you need. For multipage diagrams, you can continue processes on other pages by adding Off-page reference shapes. Hyperlinks between the reference shapes enable you to quickly jump from the main process to the continuation.

Chapter 16

Documenting Processes, Workflows, and Data Flows

igher quality, increased customer satisfaction, lower costs, faster time to market, higher profits, competitive edge. Today, most organizations have to juggle so many conflicting demands that they have no choice but to look for ways to improve.

Business process improvement methodologies, such as Total Quality Management and Six Sigma, help streamline operations and reduce defects. These business methodologies model processes in different ways to highlight potential problems or opportunities for improvement. Other specialized diagrams focus on specific problems. For example, Cause and Effect diagrams help identify problem areas and their root causes. Sometimes, simply getting business processes onto a piece of paper — or a Visio diagram — makes potential improvements so obvious you say, "Well, duh!"

Whether you're documenting and analyzing current processes or engineering process improvements, this chapter shows you how you can use specialized templates to construct the diagrams you need. Although you have to know how to represent your business process information in a specific type of diagram before you begin your Visio session, you can still use standard Visio techniques to produce your documentation.

Visio 2007 introduced a few new business-related templates, including Value Stream Mapping, ITIL, and PivotDiagrams. In addition, the Work Flow template has been enhanced in Visio 2007. This chapter describes how to build almost all the business process diagrams Visio supports. Chapter 19 is devoted to PivotDiagrams.

CROSS-REF To learn about basic techniques for building flowcharts, see Chapter 15.

Showing Departmental Interaction with Cross-Functional Flowcharts

Cross-functional flowcharts are particularly well suited for showing the departments or functional areas that contribute to performing a process. Whether you're trying to streamline a process or simply need to know whose court the ball is in, the cross-functional flowchart is ideal. The flowchart includes steps, just as a basic flowchart does, but each department has a horizontal or vertical band on the diagram. You denote the participation of departments by stretching a shape for a step across the bands for all participating departments.

Setting Up Cross-Functional Flowcharts

When you create a new cross-functional flowchart, you can choose between a horizontal or vertical orientation. The Cross-Functional dialog box provides for only five functional bands, but you can create as many as you want when the diagram exists. To create a cross-functional flowchart, follow these steps:

1. Choose File ⇨ New ⇨ Business ⇨ Cross-Functional Flowchart. The Flowchart dialog box appears with options for setting up the diagram.

2. Select the Horizontal or Vertical option to specify the orientation of the bands that represent departments. If you choose the horizontal orientation, Visio changes the page orientation to landscape.

3. In the Number of Bands box, type a number from 1 to 5 for the number of departments.

CAUTION   Choose the orientation carefully, since it can't be changed after it has been set. If you chose the wrong orientation, create a new diagram with the correct orientation and copy the shapes from the old one.

4. To include a title bar for the diagram, select the Include Title Bar check box.

5. Click OK to create the diagram. Visio adds the number of bands you specified to the diagram and opens the Arrow Shapes stencil, the Basic Flowchart Shapes stencil, and the Cross-Functional Flowchart Shapes stencils for the orientation you chose.

6. To fill in the title bar, click within the title bar band. Although Visio selects the group that includes all the cross-functional shapes on the drawing so far, when you begin to type, Visio updates the title bar text.

7. To label the bands, select a band and type the label text.

Documenting Processes in Cross-Functional Flowcharts

Adding process steps to a cross-functional flowchart is no different than building a basic flowchart. You start by adding and connecting shapes that represent the process steps. Then, to indicate that a step crosses departments or functions, resize the shape for that step so that it spans the participating departments. Use the following methods to add processes and steps to a cross-functional flowchart:

- **Add Process Steps** — From the Basic Flowchart Shapes stencil, drag shapes such as Process and Decision onto the page as you would for a basic flowchart. Use your favorite method for connecting shapes to define the sequence and interactions within the process.

- **Label Steps** — Select a step shape and type the step description.

- **Associate Departments with Steps** — Drag a selection handle on a shape for a step until it spans the departments that participate in the process, as shown in Figure 16-1.

FIGURE 16-1

Steps span multiple departments to show which departments participate.

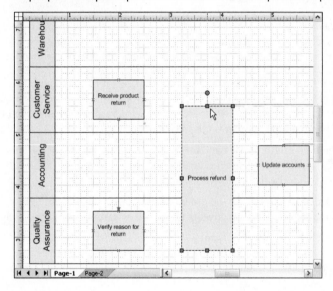

> **TIP** You can't split a step or process into multiple pieces to associate it with departments whose functional bands are separated in the diagram. However, you can show functional bands in a different order on each page. If several steps in a process relate to the same departments, show those steps on another page and rearrange the functional bands to co-locate those departments.

Working with Functional Bands

You can add, remove, edit, or rearrange the functional bands on a cross-functional flowchart to suit your needs. You wouldn't want the scientists performing the steps usually assigned to marketing, so, when you move or reorder the bands on a diagram, Visio moves the steps along with the bands. Unfortunately, this rearranging usually results in messy connections. To reorganize the paths between steps, drag the green vertices on connectors. The following methods for modifying the functional bands in a cross-functional flowchart use a Horizontal cross-functional diagram as an example:

- **Add a Functional Band** — Drag a Functional band shape from the stencil to an approximate vertical position on the page. Visio snaps the functional band into place, aligning the band horizontally with the other bands. To add a functional band above another band, position the pointer within or slightly above the existing band. Visio inserts the new functional band above that band. With the band selected, type the department or function name.

> **NOTE** If you insert a functional band between two bands that share a step, Visio resizes the step to span all three bands.

- **Remove a Functional Band** — Click once to select the label of a functional band and then press Delete.

> **CAUTION** When you delete a functional band, Visio also deletes any shapes within that band, without requesting confirmation. If you inadvertently delete shapes you want to keep, press Ctrl+Z to undo the deletion. Move the shapes to other bands and then delete the functional band.

- **Resize a Functional Band** — To resize a band, select the band and then drag one of the selection handles until the band is the width you want. For example, to change the width of a vertical band, drag a selection handle on either side. To change the height of a horizontal band, drag the top or bottom selection handle.

- **Change the Length of All Bands** — Select the border or title of the cross-functional flowchart to display the selection handles for the chart group. Drag a selection handle until the bands are the length you want.

- **Move a Band** — To change the order of bands, select a band and drag it to a new location. Visio moves or resizes any shapes wholly or partially contained within the band.

- **Move a Shape to Another Band** — Drag the shape to another band. If you want to associate a shape with multiple bands, drag a selection handle on the shape until it spans all the bands to which it relates.

Identifying Process Phases

The Separator shape is perfect for specifying phases within a process. When you add a Separator shape to a flowchart, Visio associates the steps that follow the separator with that phase. On vertical cross-functional flowcharts, separators are horizontal; for horizontal flowcharts, they are vertical.

To add a separator to a flowchart, drag the Separator shape from the stencil onto the page. Select the separator and type a description for the phase. When you move a separator, Visio moves all the steps in that phase. To move a separator without moving the steps, delete the separator and add a new one in the new location.

Making Good Decisions with Decision Trees

Sometimes, you face a seemingly overwhelming number of choices, each with their unique set of pros and cons. For example, in risk management, some risks are present only if other risks occur. For example, you don't have to worry about contractors overrunning your costs unless staffing issues send you looking for more resources. When these dependent risks are present, your decisions are even more complicated as you take into account both the probability of the risks occurring and the financial impact of each outcome. On top of all this, you probably have to convince your executive team that you're proposing the right solution.

Decision trees are helpful tools for getting a handle on all your options and communicating your analysis to others. Calculating the probability of the different combined risks helps identify which risks to actively manage. Evaluating both probability and impact helps identify the least costly option, as the figure illustrates.

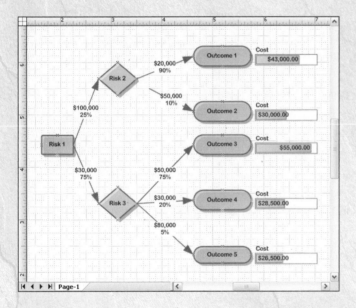

In Visio, you can take your pick from several templates as the foundation for a decision tree. Because a decision tree is merely shapes with connectors between them and a smattering of text, the primary factor in your choice is the shapes you want to use to indicate the decisions and outcomes. For example, the diagram in the example uses the Basic Flowchart template, but you could use the Block Diagram template, or even the Organization Chart template.

To create a decision tree, drag the shapes you want for the starting point, the decisions, and the outcomes onto the drawing page. Then, select the Connector tool and glue the shapes together. Finally, use the text blocks within the shapes as well as separate text block shapes to label the diagram.

Following the Flow of Data

As you might expect, data flow diagrams focus on how data flows during the execution of processes and where data is stored. Visio 2007 includes two templates for documenting the flow of data between processes and data stores:

- **Data Flow Diagram** — The Data Flow Diagram template offers basic shapes for representing data flow within a process. You can use it to analyze processes for potential improvements. Choose File ➪ New ➪ Business ➪ Data Flow Diagram. Visio opens the Data Flow Diagram Shapes stencil, which contains shapes for processes, entities, states, and data stores; and a few other stencils for connections and annotation.

- **Data Flow Model Diagram** — The Data Flow Model Diagram also models data flows, but is based on the Gane and Sarson symbology. Choose File ➪ New ➪ Software and Databases ➪ Data Flow Model Diagram. Visio opens only the Gane-Sarson stencil.

CROSS-REF To learn more about data flow diagrams and how to construct them, read the introduction to data flow diagrams at `http://en.wikipedia.org/wiki/Data_flow_diagram`. For a detailed explanation of the Gane-Sarson methodology, read *Structured Systems Analysis: Tools and Techniques* by C. Gane and T. Sarson (New York: IST, Inc., 1977).

Adding Data Flow Shapes

As it turns out, the shapes on the Data Flow Diagram Shapes stencil are more numerous than those for Gane-Sarson. Here are some of the shapes you use to document processes on the Data Flow Diagram Shapes stencil and the corresponding shapes from the Gane-Sarson stencil:

- **External Interactor** — An external input, such as a customer; a terminator, or state. The Gane-Sarson shape is called Interface.

- **Data Process or Oval Process** — A process that transforms data in some way, such as a validation process transforms a submitted application into an approved application. An Oval Process shape contains a control point for creating multiple data flows from the process. The Gane-Sarson shape is called Process.

- **State** — A state achieved during a process, such as submitted, approved, or rejected. The Gane-Sarson stencil doesn't have a shape to correspond to State.

- **Data Store** — A source or destination for data that is internal to the process, such as a database, a file, or a filing cabinet. The Gane-Sarson stencil includes a Data Store shape.

- **Entity** — An entity that performs a process. The Gane-Sarson stencil doesn't have a shape to correspond to Entity.

- **Entity Relationship** — A shape that indicates the relationship between entities, that is, a connector between entities. The Gane-Sarson shape is called Data Flow, but it's still basically a connector.

Showing Flow Between Shapes

You indicate flow on a data flow diagram with Center to Center shapes. To create a data flow, follow these steps:

1. Drag a Center to Center shape onto the page near the two shapes between which data flows.

2. To change the direction of the flow, choose Shape ➪ Rotate or Flip, and then choose Flip Horizontal or Flip Vertical.

3. Glue the end points of the Center to Center shape to the connection points at the center of each of the process shapes. Visio highlights the end points with red squares when the shapes are connected.

4. To change the curvature of the data flow arrow, drag the green selection handle in the middle of the arc to a new location. To change the location of arrows, drag one of the control handles.

Showing a Data Loop

To indicate a loop in the process, follow these steps:

1. Drag a Loop on Center shape onto the page until Visio displays a red square around the connection point on the process that loops.

2. To change the size or position of the loop, drag the end point. To change the location of the ends of the loop, drag the control handle and the selection handle.

Analyzing and Documenting Workflow

Work flow diagrams show the interactions and flow of control for business processes — in short, who does what and when, and perhaps, who to talk to when something is amiss. In Visio 2007, the Work Flow Diagram template has had a facelift. Instead of the Work Flow Diagram Shapes stencil that opened in Visio 2003 and included shapes for departments, the new and improved Work Flow Diagram stencil opens three new stencils with spiffed up shapes:

- **Department stencil** — Includes shapes that represent different departments and personnel. For example, Accounting, Information Services, and Shipping appear on the Department stencil as they still do on the Work Flow Diagram Shapes stencil (which is still available if you choose File ➪ Shapes ➪ Business ➪ Work Flow Diagram Shapes). The shapes on the Department stencil look like fancy clip art with several colors and a three-dimensional quality perfect for presentations.

- **Workflow Objects stencil** — Has a small selection of shapes for storage repositories such as Box and CD Rom, files, and generic employees, such as Person.

- **Workflow Steps stencil** — Includes shapes for actions and milestones in a process, such as Analyze, Meeting, and Reject.

The new workflow shapes are better-looking cousins to the earlier Work Flow Diagram Shapes. They don't contain any shape data and you still use basic Visio drag, drop, and glue techniques to build your workflow diagrams. In addition to the three-dimensional quality, these shapes are designed to work with the new Theme Color feature (see Chapter 7) to highlight parts of the shape with the theme color you choose while keeping other parts of the shape in their original colors, as illustrated in Figure 16-2.

TIP   Hidden in the ShapeSheets for these new workflow shapes is a property that changes the skin color of the people in the shapes. To create a diagram that shows organizational diversity, select a shape and then choose Window ⇨ Show ShapeSheet. In the User-defined Cells section, modify the formula in the User.SkinColor property.

FIGURE 16-2

New stencils for the Work Flow Diagram template have shapes that are more three-dimensional and work with the new Themes feature.

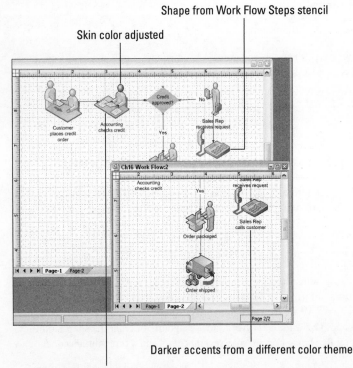

Shape from Work Flow Steps stencil

Skin color adjusted

Darker accents from a different color theme

Shape from Department stencil

Identifying Causes and Effects

Completed cause and effect diagrams, also called *fishbone diagrams* because they resemble a fish's skeleton, show both potential and existing factors that produce a result. In reality, the power of cause and effect diagrams is their ability to influence teams to explore the roots of problems instead of focusing immediately on solutions.

Fishbone diagrams start with the problem posed in the form of a question. For example, if the problem is stated as "Late Delivery," your teammates could easily launch into an energetic solution session for delivering earlier. But, by stating the problem as "Why is delivery late?" you get people thinking about the underlying cause.

Causes are grouped into major categories. You can choose your own categories or work with one of the standard category sets. For example, for service industries, the categories are typically equipment, policies, procedures, and people. Manufacturing industries often use manpower, methods, materials, and machinery. As you begin brainstorming causes, you can attach them to the branch for their corresponding category. For each cause, keep asking why it happens. Eventually, you'll uncover the underlying causes that produce the problem. At that point, the team can start brainstorming corrections to the true cause.

CROSS-REF Cause-and-effect diagrams are also called Ishikawa diagrams, after their creator, Dr. Kaori Ishikawa, who initiated a well-known approach to quality management in the Kawasaki shipyards. For more information about these diagrams, navigate to `www.isixsigma.com/library/content/t000827.asp` or `http://en.wikipedia.org/wiki/ishikawa_diagram`.

In quality management, you typically create a cause-and-effect diagram after investigating the problems associated with a product or service and ranking them in a *Pareto chart* (also known as an *80-20 chart*). You use the effect ranked highest in the Pareto chart as the starting point for your cause-and-effect diagram. For example, when you determine that the most frequent customer complaint is late delivery, you can construct a cause-and-effect diagram to explore the reasons for this.

When you create a new drawing with the Visio Cause and Effect Diagram template, Visio creates a new file with one drawing page .It automatically adds an Effect shape for the effect you are studying and four Category boxes to classify the causes. To create a cause-and-effect diagram, follow these steps:

1. Choose File ➪ New ➪ Business ➪ Cause and Effect Diagram.

2. To label the problem or effect you're studying, select the horizontal arrow on the page and type a question that describes the problem.

3. To label the categories that Visio adds automatically, select an angled line and type the category name. Visio places the text in the boxes at the end of the fishbones.

4. To add, delete, or move cause categories, use one of the following methods:

- **Add a category**—Drag a Category 1 or Category 2 shape onto the page and position it so the arrowhead touches the horizontal line of the Effect shape. When you release the mouse button, Visio automatically glues the Category shape to the Effect shape. With the shape selected, type the category name.

- **Delete a category**—Select a Category box or line and press the Delete key.

- **Move a category**—Drag a Category shape to the new location until it snaps to the Effect shape.

5. To show major causes within a category, drag Primary Cause shapes onto the page until the arrowheads snap to category lines, as shown in Figure 16-3.

FIGURE 16-3

Cause and Effect shapes snap anywhere on shape geometry.

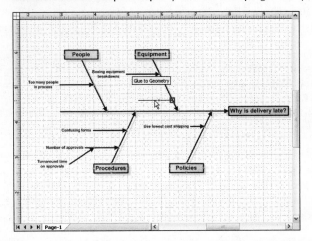

NOTE   The only differences between the two versions of primary and secondary cause shapes are the direction of the line and the location of the text relative to the line.

6. To illustrate secondary causes that contribute to primary causes, drag Secondary Cause shapes onto the page until the arrowheads snap to primary cause lines.

7. For each cause, select the shape, and type a description of the cause.

TIP   If text in Cause shapes overlaps on the page, click Align Left or Align Right on the Formatting toolbar to reposition the text within Cause shapes. You can also distribute the text over several lines by adding line breaks. To do this, select the Text tool, click the text, and press Ctrl+Enter.

Value Stream Mapping

Value stream mapping is an offshoot of the lean manufacturing process developed by Toyota. It documents the flow of information and materials in a process and can highlight delays and waste to eliminate. Although value stream mapping originated in manufacturing, all kinds of organizations use it to improve their business processes.

CROSS-REF To learn more about value stream mapping, search the Web for the term.

Visio Professional 2007 includes a Value Stream Map template, which opens the Value Stream Map Shapes stencil. The Process shape on this stencil includes several shape data properties for evaluating performance, such as Cycle Time and Batch Size, so you can use Data Graphics to present process statistics, as shown in Figure 16-4. The other shapes on the stencil are primarily drag-and-drop shapes that you label using their text blocks.

CROSS-REF See Chapter 10 to learn how to apply Data Graphics to shapes on a diagram.

FIGURE 16-4

Data Graphics present the shape data associated with the Value Stream Process shape to help evaluate process performance.

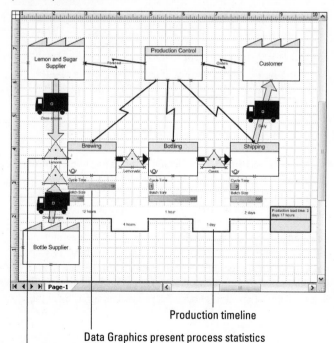

Production timeline

Data Graphics present process statistics

Process shape has shape data

An Overview of Other Business Process Templates

In addition to the templates described so far in this chapter, Visio 2007 provides several other templates in the Business category. These templates support specialized methodologies and diagrams, but they all respond to drag-and-drop, glue, and other standard Visio techniques. This section is an overview of the available templates, what the diagrams are used for, and a few hints on creating them in Visio.

ITIL

ITIL (Information Technology Infrastructure Library) is a set of processes that represent the best practices for managing IT services within organizations. Developed by the Office of Government Commerce in the United Kingdom, it has grown into a worldwide standard.

CROSS-REF   To read about ITIL from the source, navigate to the ITIL web site from the Office of Government Commerce, www.itil.co.uk/.

The stencils that open when you use the new Visio 2007 ITIL template should be familiar for the most part. From Basic Flowchart Shapes to Computers and Monitors to Cross-Functional Flowchart Shapes, many shapes used in ITIL diagrams are nothing new. However, the ITIL Shapes stencil is new and offers workflow shapes whose names conform to standard ITIL terminology.

TQM Diagrams

TQM is a structured approach to business process improvement that focuses on building quality into products and services from the beginning. All employees and departments participate, from top-level management on down. The TQM Diagram template includes shapes and connectors that conform to the symbology for Total Quality Management projects.

Audit Diagrams

Audit diagrams document processes, including accounting, financial management, fiscal information tracking, money management, decision flowcharts, financial inventories, and other types of financial transactions. For example, you can model the process for an online stock trade to ensure that checks and balances are in place to satisfy the Securities and Exchange Commission.

The Audit Diagram template opens several stencils, including the Audit Diagram Shapes stencil, with shapes for processes, operations, documents, and data repositories, as well as basic flowchart shapes such as On-page and Off-page references and connectors. To mark the process elements that you want to research further, use Tagged Process and Tagged Document shapes. After adding Tagged shapes to the page, right-click a Tagged shape and choose Tagged or Untagged from the shortcut menu to toggle the Tagged setting.

Six Sigma Templates

Six Sigma projects include phases for defining, measuring, analyzing, improving, and controlling processes. The Visio DMAIC Flowchart template, available on Microsoft Office Online, is designed to document requirements and steps for a Six Sigma project. When you create Visio diagrams or other documents as part of your project, you can link these documents directly to the DMAIC diagram to keep your project information organized.

The DMAIC Flowchart template includes an overview page and two additional pages for drilling down. Each page includes shapes to get you started. The quickest way to find the DMAIC template is to choose Help ➪ Microsoft Office Online. Select the Templates tab and then, in the Templates search box, type **Visio Six Sigma** and click Search. Click the link for the DMAIC flowchart with the units you want and then click the Download Now button.

Fault Tree Analysis Diagrams

Fault tree analysis studies the events that can lead to failure in an effort to prevent them from occurring. Fault tree analysis diagrams are frequently used in the Analyze phase of Six Sigma business process improvement projects. Fault tree diagrams use a top-down structure to represent the routes within a system that can lead to failures. You can use logical operators to interconnect events or conditions that contribute to a failure.

 Bell Telephone Laboratories developed fault tree analysis in 1962 for the U.S. Air Force so they could analyze the Minuteman missile system.

IDEF0 Diagrams

The IDEF0 communication methodology uses context diagrams, parent/child diagrams, and node trees to model business and organizational processes. Context diagrams are high-level diagrams that show activities and external interfaces. IDEF0 node trees show an entire decomposition in one diagram. Parent/child diagrams illustrate the relationships between processes.

CROSS-REF To learn about the IDEF0 methodology, navigate to `www.idef.com`.

Context diagrams show the relationships between activities. To create a context diagram, drag Activity box shapes onto the page and type the process name and process ID in their corresponding fields in the Shape Data dialog box. If the process is a decomposition, also type the ID for the decomposition diagram in the Sub-diagram ID field. One-legged connector shapes connect activities to external interfaces.

Parent/child diagrams also show activities and connections. On these diagrams, you can use the IDEF0 connector to create a variety of connections between processes. After adding parent and child processes, choose one of the following methods to connect the processes on your diagram:

- **Joined Arrows** — Drag one IDEF0 connector between connection points on two activity boxes. Drag a second IDEF0 connector onto the page and glue one end to a connection point on another activity box. Drag the free end of this connector so its arrowhead overlaps the first connector's arrow.

- **Forked Arrows** — Drag one IDEF0 connector between connection points on two activity boxes. Drag a second IDEF0 connector onto the page and overlap its start point with the start point of the first connector. Glue the other end of the second connector to another activity box.

- **Branching Arrows** — Drag one IDEF0 connector between connection points on two activity boxes. Select the connector, press and hold Ctrl, while you drag to create a copy of the connector to create a branch. Press F4 to create additional branches. Connect the end points of the branches to the appropriate activity boxes.

> **TIP** Aligning the branches requires that you connect the beginning points of all the branches. Drag the control handle on each branch to reposition its middle leg to the same beginning point as all the other branches.

Node trees show the decomposition of a context diagram. They are similar to work breakdown structures used by project managers to decompose work into smaller and smaller chunks.

> **TIP** Add a hyperlink between a node and the page that contains process details when you want to quickly view the details associated with a node.

Creating SDL Flowcharts

SDL flowcharts use shapes and connectors based on International Telecommunications Union standards to illustrate communications and telecommunications systems networks. These shapes are similar to other types of flowchart shapes, including procedures and decisions. Some include control points that you can use to modify the position of dividers within the shapes.

Event Process Chain (EPC) Diagrams

EPC diagrams are part of the SAP R/3 modeling methodology for business engineering. EPC diagrams illustrate business process workflows by showing the transfer of control in processes as a chain of events and functions.

The EPC Diagram Shapes stencil includes the following shapes:

- **Functions** — Represent processes or activities, such as creating forms or checking inventory.

- **Events** — Trigger functions or is the result of functions. In EPC modeling, events also represent process states, such as Order Form Created.

- **Organizational Units** — Indicate the part of the organization responsible for a process or activity.

- **Information/Material** — Represent data elements such as forms or data records.

- **Logical Operators** — Specify how events and functions interact. For example, the AND operator indicates that both events must occur to trigger a function, whereas the exclusive OR (XOR) operator specifies that the occurrence of only one of the events triggers the function. An OR operator indicates that the occurrence of any or all of the events triggers the function.

- **Connectors** — Show the relationships between the components on the diagram.

Summary

The Visio Business templates support numerous popular methodologies for analyzing, improving, and documenting business processes, such as TQM and ITIL. Although you need to know how to document business processes using the different types of diagrams, such as Cause and Effect or Fault Tree Analysis, you can use standard Visio techniques to produce and format your drawings.

Chapter 17

Scheduling Projects with Visio

*P*rojects are endeavors with a definite beginning and end — typically, special undertakings that are distinct from the routine day-to-day operations of an organization. Projects don't just happen on their own; they take conscientious planning, tracking, and management. Project managers monitor and control schedules, tasks, resources, and budgets to ensure that the projects they manage achieve their goals.

Typically, project managers use software such as Microsoft Project to build, track, and revise project plans. Although Visio is not a project management tool, it comes in handy for communicating the project plans and status built in other programs. If you don't have project management software, the Timeline, Gantt chart, and PERT (Program Evaluation and Review Technique) chart templates can step up to build and report on very simple projects. The Calendar template provides shapes to construct calendars that span one day to several years, which you can use for projects or tracking your kids' school activities.

In addition to building project management diagrams by dragging and dropping shapes, you can also import data from Microsoft Project or other applications into Visio diagrams. If you develop a prototype schedule in Visio, you can jumpstart projects by exporting your schedules to Microsoft Project.

In this chapter you learn how to work with the Visio Calendar, Timeline, Gantt chart, and PERT chart templates to represent your projects and other time-sensitive information.

Exploring the Project Scheduling Templates

In project management circles, the Gantt chart is the granddaddy of formats for presenting project schedules. However, other formats are useful in specific situations. For example, a timeline is great for summarizing a project with milestones. A PERT chart focuses on the interdependencies between tasks rather than the chronology. And, nothing beats a calendar for presenting information to audiences unfamiliar with project management techniques.

Visio covers the bases with four project-related templates that present project and date-related information in distinct ways. Each template automatically sets the page to a letter-size sheet with landscape orientation and uses inches drawn at a one-to-one scale (millimeters for metric templates). Use the template that corresponds with the type of information you want to convey:

- **Calendar** — The Calendar template displays reminders, meetings, special events, milestones, and more in a format that everyone is comfortable with. The Visio Calendar template works for projects but can display activities scheduled by day, week, month, or year for any purpose. The template includes the Calendar Shapes stencil and the Calendar menu.

- **Timeline** — As its name implies, the Timeline template is set up to show tasks and events along a horizontal or vertical timeline, for example, a one-line summary of the countdown of events to a space shuttle launch. The Timeline template includes the Timeline Shapes (with shapes for milestones and intervals), Background, and Borders and Titles stencils, along with the Timeline menu.

- **Gantt Chart** — Although you can drag shapes onto a page to create Gantt charts in Visio, you're more likely to build these charts by importing data from Microsoft Outlook or Microsoft Project. The Gantt Chart template includes the Gantt Chart Shapes, Background, and Borders and Titles stencils, along with the Gantt Chart menu and toolbar.

- **PERT Chart** — The PERT Chart template provides a pair of PERT chart boxes, or *nodes*, to show the interdependencies between tasks without taking timing into account. There are shapes for node connectors, callouts, and legends. The PERT Chart template includes the PERT Chart Shapes, Background, and Borders and Titles stencils.

To create a drawing using one of the project scheduling templates, choose File ➪ New ➪ Schedule and then choose Calendar, Gantt Chart, PERT Chart, or Timeline. To use one of the project scheduling stencils in a different template, choose File ➪ Shapes ➪ Schedule and then choose the stencil you want. Visio adds the appropriate menu to the Visio menu bar when you drag a shape from one of the template stencils onto the drawing page.

Constructing Calendars

The Calendar template is a tremendous help when you want to create calendars, whether you want a monthly calendar to track the athletic, music, and social events in your kids' lives or you plan to create a customized yearly calendar with candid photos of your goldfish. When you add appointments, events, or project tasks to calendar days, Visio automatically associates them with the corresponding calendar dates. Calendar art is nothing more than clip art for highlighting special dates, such as birthdays, conveniently included in the Calendar Shapes stencil.

Calendar shapes come with a configuration dialog box, so you can set up the calendar before the shape hits the page. For example, in a monthly calendar, you can specify the month and year and the first day of the week. By default, weekends are shaded, but for folks on a 7 × 24 schedule, you can show all days unshaded. Specifying a language for a calendar determines its date format and the language for days and months.

You can create stand-alone calendars by creating a new Visio drawing using the Calendar template. To add a calendar to another type of drawing, simply open the Calendar Shapes stencil and drag one of the calendar shapes onto your drawing.

CROSS-REF Since the 2003 edition of Visio, you can import appointments and other schedule information from your Microsoft Outlook Calendar (see the "Importing Outlook Calendar Data into Visio" section later in this chapter).

Date Formats and Language

When you add any of the calendar shapes to a drawing, the Configure dialog box for that shape appears and you're confronted with a choice of language, and possibly Date Format. The language you choose not only determines the names you see for days and months, but also sets the date format to one commonly used for that language. For example, choosing English (U.S.) sets the date format to <day_name, month_name dd, yyyy>. However, if you choose French (Canada), the date format changes to <day_name, dd month_name yyyy>; for example, "mercredi, 19 septembre, 2007."

If the Configure dialog box includes a separate drop-down list for date format, you can choose a date format to fit the size of the calendar or your preferred method of showing the date. For example, if the calendar is small, you might choose a date format that presents the date as follows: m-dd-yy. In the Date Format drop-down list, Visio illustrates the date formats using today's date.

Creating Daily Calendars

For a calendar consisting of a single day or multiple nonconsecutive days, you can drag Day shapes onto a drawing and then add Appointment shapes for the time slots that are booked. To create a calendar of selected days, follow these steps:

1. Drag the Day shape onto the drawing.

2. In the Configure dialog box that appears, modify the configuration settings if necessary, and then click OK to add the shape:

   ■ The date box is set to today's date by default. Click the drop-down arrow to choose the date from a calendar or type the date.

   ■ Optionally, choose the language to change the date format and the language for the days and months.

   ■ To change the date format separate from the language, in the Date Format drop-down list, choose the format you want.

3. Repeat steps 1 and 2 for each day in your calendar.

4. Resize and move the Day shapes into the size and position you want.

5. To change the date or date format in a Day shape after you've created it, right-click the shape and, from the shortcut menu, choose Configure.

Creating Weekly Calendars

Weekly calendars are great for planning out the next several days, whether for yourself, your spouse, or your project team. To create a weekly calendar, follow these steps:

1. Drag the Week shape onto the drawing.

2. In the Configure dialog box that appears, modify the configuration settings if necessary, and then click OK to add the shape:

- In the Start Date box, enter the date for the first day of the week you are adding. The start date doesn't have to be at the standard beginning of a week. For example, you can create a weekly calendar that runs from Tuesday through Saturday.

- In the End Date drop-down list, select the number of days for the week and the resulting end date, for example, if the start date is 3/5/2007, you might choose 5 days - 3/9/2007.

- Specify the language and date format, if necessary.

- To specify whether the weekend days are shaded, select either the Yes option or No option.

- To include a title for the week (for example, Week of Monday, March 05, 2007), check the Title check box.

3. To add additional weeks, repeat steps 1 and 2.

4. Resize and move the week shapes into the size and position you want.

> **TIP** Rather than add multiple individual weeks, you can create a multiweek calendar for several consecutive weeks, for example, to track events for the next six weeks. Drag the Multiple Week shape onto the drawing. In the Configure dialog box, specify the start and end dates and any other calendar formatting options, and then click OK.

Creating Monthly Calendars

Although the Month shape represents only a single month, you can create multi-month calendars by adding several Month shapes to separate pages within a drawing file. To create a monthly calendar, follow these steps:

1. Drag the Month shape onto the drawing.

2. In the Configure dialog box that appears, modify the configuration settings if necessary, and then click OK to add the shape:

- In the Month box, choose the name of the month.

- In the Year box, type the calendar year.

- In the Begin Week On drop-down list, select the day on which the weeks should begin.

- Specify the language and date format, if necessary.

- To specify whether the weekend days are shaded, select either the Yes option or No option.

- To include a title for the week (for example, Week of Monday, March 05, 2007), select the Title check box.

3. By default, Month shapes fill the page. If necessary, resize and move the Month shape into the size and position you want.

> **TIP** Thumbnails for the previous and next month are a helpful addition to any calendar. Drag the Thumbnail Month shape to a position near your Month calendar. In the Shape Data dialog box, enter the month and year and then click OK.

4. To add additional months, add a new page by choosing Insert ⇨ New Page and then clicking OK in the Page Setup dialog box. Repeat steps 1 through 3 to set up the month on the new page.

> **TIP** To show the phases of the moon, drag the Moon Phases shape onto a day of the month. Right-click the shape and choose New Moon, First Quarter, Last Quarter, or Full Moon.

Creating Yearly Calendars

With a few clicks on your part, Visio can lob an entire year's worth of calendar onto a single page. To create a yearly calendar, follow these steps:

1. Drag the Year shape onto the drawing.

2. In the Shape Data dialog box that appears, modify the configuration settings if necessary, and then click OK to add the shape:

   - In the Year box, type the calendar year.

   - In the Begin Week On drop-down list, select the day on which the weeks should begin.

   - Specify the language, if necessary.

3. If necessary, resize and move the Year shape into the size and position you want. By default, it fills the page.

4. To add another year to the drawing, add a new page by choosing Insert ⇨ New Page and then clicking OK. Repeat steps 1 through 3 to set up the year on the new page.

> **TIP** Although the Calendar template sets the page to landscape orientation, the Year shape is better suited to a portrait page orientation. Choose File ⇨ Page Setup and then select the Print Setup tab. Under Printer Paper, select Portrait and then click OK.

Modifying and Formatting Calendars

After you've added a calendar to a drawing page, you can modify it in various ways:

- **Change configuration settings** — To change any of the settings you specified when you first added the calendar, right-click the shape and, from the shortcut menu, choose Configure.

- **Specify colors with a theme** — To change a calendar's suite of color, choose a Visio theme, as described in Chapter 7.

- **Change the calendar title** — For the Week shape, the Month shape, and the Multiple Week shape, you can change the calendar title by double-clicking the calendar's title and then typing the title you want. When you're finished, press Esc. To change the title for a year calendar, right-click the calendar and then choose Configure from the shortcut menu. Enter a year in the Year text box and then click OK.

Adding Appointments and Events to Calendars

Most of the time, the whole point of a calendar is to remind you when tasks, events, or appointments occur. The Calendar Shapes stencil is but one stencil, but it's packed with shapes for adding appointments, such as meetings or classes, and events, such as conferences or birthdays, as shown in Figure 17-1. Appointments and events are associated with the calendar dates you specify, so they automatically move to the correct box in a calendar if you change the dates on which they occur.

- **Add an appointment** — The Appointment shape represents events or tasks that occur on only one day. To carve out the time for an appointment, drag the Appointment shape onto a calendar day. In the Configure dialog box, enter the start and end times for the appointment. In the Subject box, type the subject of the meeting, the name of a one-day project task, or the name for any other kind of short event. You can specify the time and date format displayed as well. Click OK to add the appointment to the day.

- **Add a multiple-day event** — Drag the Multi-Day Event shape onto your calendar. In the Configure dialog box, enter the subject and location of the event or project task as well as the start and end dates. Click OK.

- **Add text** — Double-click the day in which you want to add text, type the text you want and press Esc. The text box and day box are grouped so that you can move and resize them as one. You can use standard techniques to edit and delete text in calendars.

CAUTION   Text added to a day box is not associated with the actual day, in the way that appointments and events are. To move the text to another day, you can cut and paste it from one day box to another.

FIGURE 17-1

Add appointments, events, project tasks, and reminders to a calendar.

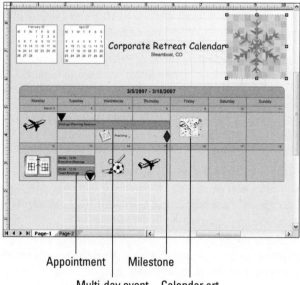

Appointment   Milestone

Multi-day event   Calendar art

- **Revise an appointment or event** — Right-click it and, from the shortcut menu, choose Configure.

- **Delete an appointment or event** — Right-click it and choose Cut.

Exploring Calendar Art Shapes

The Calendar Shapes stencil includes icon-like shapes, or *calendar art,* that can enhance the appearance of calendars or visually categorize appointments and events. Shapes such as Clock or Meeting are ideal for flagging scheduled appointments and events. You can add reminders to the calendar with shapes such as Important and To Do. Travel days stand out when you add shapes, such as Travel-Air or Travel-Car. You can make important project dates stand out with Milestone or Completion shapes. Don't forget the fun activities in a schedule. Add shapes such as Birthday, Sports, or Celebration to highlight the lighter moments.

Calendar art is associated with the day, but not the date, to which you add it. This means that changing the date in a calendar doesn't automatically move the art to the correct box. You must cut and paste calendar art to change dates.

Importing Outlook Calendar Data into Visio

If you already have appointments and events scheduled in Microsoft Outlook, the easiest way to create an appointment calendar in Visio is by importing your Outlook calendar into a Visio calendar shape. The Import Outlook Data Wizard works with Microsoft Outlook 2002 and later. To import Outlook calendar appointments into a Visio calendar, follow these steps:

1. Create a new calendar drawing by choosing File ⇨ New ⇨ Schedule ⇨ Calendar — or open an existing calendar drawing into which you want to import Outlook appointments.

2. Choose Calendar ⇨ Import Outlook Data Wizard.

NOTE If the Choose Profile dialog box appears, select the Outlook profile you want to use in the Profile Name drop-down list. Most people have only one Outlook e-mail profile, but you might create two to keep your work calendar and e-mail separate from your personal data.

3. In the first wizard page, select an option to specify whether you want to import Outlook appointments into a new or existing Visio calendar and then click Next. The Selected Visio Calendar option is available only if an existing Visio calendar is selected on the drawing page.

4. In the second wizard page, specify the range of dates and times that contain the appointments you want to import and then click Next. By default, Visio sets the Start Date at today's date and the End Date one week later.

TIP To import only some calendar items, click Filter to specify words in the appointment and event Subject fields. In the Filter Outlook Data dialog box, select the Subject Contains check box, type in the word or phrase for which you want to filter, and click OK.

5. In the third wizard page, which appears only if you import appointments into a new Visio calendar, specify the calendar type and properties as you would if you were dragging a calendar shape onto a drawing, and click Next.

6. In the last wizard page, review the import properties you specified. To change any of the import properties, click Back to open and edit the appropriate wizard page. When all import properties are set the way you want, click Finish.

NOTE If there are too many appointments to fit in a single day in your calendar, they stack on top of each other. To show all the appointments, resize the calendar.

Consolidating Schedules for Several Individuals

Suppose you want to combine the schedules of team members on a project into one Visio calendar. By routing the Visio calendar to each team member, they can import their calendars and you receive the calendar at the end with everyone's schedule in it. To combine appointments from different people, follow these steps:

1. Import your own appointments into a new or existing Visio calendar.

2. When you're finished, choose File ⇨ Send To ⇨ Routing Recipient.

3. In the Routing Slip dialog box, enter the e-mail addresses for the people whose appointments you want to import.

4. In the Message Text box, tell the recipients how to run the Import Outlook Data Wizard on the calendar.

5. Under Route to Recipients, select One After Another.

6. To ensure that the calendar comes back to you after everyone adds their items, select the Return When Done check box.

7. Click OK.

Each recipient in turn runs the Import Outlook Data Wizard on the calendar to add his or her appointments. When the last recipient finishes, the calendar returns to you, complete with everyone's imported appointments.

Documenting Project Timelines

The Timeline template shows phases, tasks, intervals, and milestones along a horizontal or vertical bar. A timeline drawing, as illustrated in Figure 17-2, can help you communicate dates and show progress toward a deadline.

Here are some of the more important features that the Timeline template provides:

- **Timeline** — The Timeline shapes stencil includes several variations of timeline shapes to highlight intervals, including divisions, ticks on a line, or marks on a ruler.

- **Milestones** — Highlight significant achievements and important dates in a schedule with one of several Milestone shapes. When you add a Synchronize Milestone shape to an expanded timeline, it links with a milestone in the main timeline and positions itself according to the main timeline milestone date.

- **Intervals** — Interval shapes help you demarcate and annotate the phases within a timeline and synchronize intervals across multiple timelines on a single page.

- **Markers** — Annotate the timeline to show the current day or elapsed time. The Today Marker automatically moves to the current date as set in your computer system clock. You can also add the Elapsed Time shape to show the duration of a project up to the current date.

- **Expanded Timeline** — An Expanded Timeline shape magnifies a segment of an overview timeline to show more detail. You designate the start and end date of the expanded timeline and Visio automatically draws lines to those dates on the overview.

- **Data exchange with Project** — Import data from Microsoft Project into your Visio Timeline or export Visio timelines to Microsoft Project.

FIGURE 17-2

Horizontal or vertical timelines show project intervals and milestones, whereas expanded timelines show the detail within intervals.

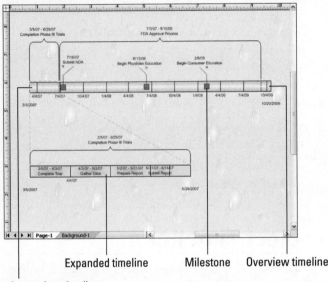

Interval on timeline   Expanded timeline   Milestone   Overview timeline

Creating Timelines

A timeline shows the activities during a period of time, including phases (intervals) and key dates (milestones). To create a timeline with the Timeline templates, follow these steps:

1. Choose File ➪ New ➪ Schedule ➪ Timeline.

2. From the Timeline shapes stencil, drag one of the Timeline shapes, such as Block Timeline or Cylindrical Timeline, onto the drawing.

3. In the Configure Timeline dialog box, under Time Period, specify the start and finish dates for the timeline. For long periods of time, the start and end times usually don't matter. However, if the timeline is for only a few days, specify the start and finish times as well.

4. Under Scale, in the Time Scale drop-down list, choose the time interval that you want Visio to annotate on the timeline, such as Months or Quarters. If you select Quarters in

the Time Scale drop-down list, you can also specify the Start Fiscal Year On date. If you select Weeks in the Time Scale drop-down list, you can specify the first day of the week.

For the micromanagers in the audience, the Time Scale drop-down list includes intervals as short as seconds, minutes, and hours.

5. Click OK to close the Configure dialog box and add the shape to the drawing.

6. Drag the new timeline into position on the drawing.

If you don't see dates or times on the timeline, right-click the Timeline shape and choose Configure Timeline. Select the Time Format tab. Select the check boxes to control whether start and finish date and interim time scales appear in the timeline. You can specify the date format for the dates and, for interim markings, whether the date appears or only tick marks.

Showing Intervals on a Timeline

Intervals show activities during a segment of the time covered by the timeline, such as the duration of the research phase of a project. To add an interval to a timeline, follow these steps:

1. From the Timeline shapes stencil, drag one of the Interval shapes, such as Block Interval or Cylindrical Interval, onto the timeline.

2. In the Configure Interval dialog box, specify the start and finish dates for the interval. If the time is also important, specify the start and finish times as well.

3. In the Description text box, type the text or title that you want to appear with the interval in the timeline.

4. Change the date format if necessary.

5. Click OK and the Interval shape spans the specified dates in the timeline. The start and finish dates and the description are labels within the Interval shape.

Highlighting Milestones

Milestones show an event on a particular date, such as the kickoff date of a phase or the due date of a major deliverable. To highlight milestones in a timeline, follow these steps:

1. From the Timeline shapes stencil, drag one of the Milestone shapes, such as Diamond Milestone or Pin Milestone, onto the timeline.

2. In the Configure Milestone dialog box, specify the date in the Milestone Date box. If the time is important, specify it in the Milestone Time box.

3. In the Description text box, type the text for the milestone.

4. To change the date format, select a format from the Date Format drop-down list.

5. Click OK. The Milestone shape moves to the specified date in the timeline, and the date and description appear above or below the timeline itself as the milestone's label.

If part of the Milestone shape is not visible, right-click any visible part of it and then choose Shape ⇨ Bring to Front from the shortcut menu.

Expanding Timelines to Show Detail

In a timeline that spans years, activities that occur over a few days or weeks are barely noticeable, but they might be essential to the success of the project. For example, the coding for a new software program goes on for months, but the acceptance test at the customer site might span a week, jam-packed with activity. Expanded Timeline shapes create a more detailed view of a portion of the total timeline. An overview timeline could show the development phases for the software project as a whole. Then you could expand one part of the timeline to show the detailed tasks during acceptance testing.

Visio timelines can expand to several levels — from an overview timeline to an expanded timeline, and finally a second or even third timeline for minute detail. One overview timeline can have several expanded timelines to place several key points in the project under a microscope. To expand the detail for a higher-level timeline, follow these steps:

1. Add and configure the overview timeline.

2. From the Timeline Shapes stencil, drag the Expanded Timeline shape onto the drawing.

3. In the Configure Timeline dialog box, specify the start and finish dates for the expanded timeline. If the time is also important, specify the start and finish times as well.

> **NOTE** The dates that you specify for the expanded timeline must be within the date range of the overview timeline.

4. In the Time Scale box, select the time intervals based on the length of time that the expanded timeline covers. For example, if the expanded timeline covers two months, you might choose Weeks.

5. Click OK to add the expanded timeline to the drawing. Visio takes care of drawing any milestones, intervals, and date markers in the expanded timeline that occur during the expanded time period in the overview timeline. As illustrated in Figure 17-2, gray, dashed lines correlate the start and finish dates on the expanded timeline and the overview timeline.

6. Draw any additional milestones or intervals on the expanded timeline.

> **NOTE** Milestones and intervals added to the expanded timeline do not appear on the overview timeline. Visio synchronizes changes to milestones or intervals, so elements that you add to the overview timeline within the date range of an expanded timeline appear on the expanded timeline as well.

You can use the mouse to change the expanded timeline in the following ways:

- **Move** — Drag the expanded timeline to the location you want on the drawing. The expanded timeline remains associated with the overview timeline even when you move it.

- **Resize** — Select the expanded timeline and drag any of the four selection handles to the size you want. This only resizes the timeline and does not change the dates.

- **Change the start or end date** — Right-click the expanded timeline, and, from the shortcut menu, choose Configure Interval. Type the new dates.

Synchronizing Milestones and Intervals

Visio takes care of synchronizing milestones and intervals from overview to expanded timelines, but you can synchronize milestones and intervals across any number of timelines on a page. You can synchronize existing milestones or intervals or you can create the synchronization using the Synchronized Milestone and Synchronized Interval shapes.

To synchronize a milestone or interval with another, follow these steps:

1. Use one of the following methods to select a milestone or interval:

   ▪ **Existing shapes** — Select the milestone or interval you want to synchronize.

   ▪ **New shapes** — From the Timeline Shapes stencil, drag the Synchronized Milestone or Synchronized Interval shape onto the timeline.

2. Choose Timeline ➪ Synchronize Milestone or Timeline ➪ Synchronize Interval.

3. In the Synchronize With drop-down list, select the milestone or interval with which the selected shape should be synchronized.

4. Select the date format if necessary and click OK. A gray, dotted line appears, showing the synchronization between the milestones or intervals. Now, if you modify one of the synchronized milestones or intervals, the other one changes to follow.

NOTE Deleting the gray, dotted line does not remove the association between synchronized milestones or intervals. To break the link between synchronized milestones or intervals, delete one of the synchronized shapes. You can then add it back as a regular unsynchronized shape.

Modifying Timelines

For existing timelines, you can modify dates and times as well as the overall look of a timeline. Simply select the timeline and choose Timeline ➪ Configure Timeline (or right-click it and, from the shortcut menu, choose Configure Timeline). The Configure Timeline dialog box includes fields to modify the following aspects of timelines:

■ **Start and end dates** — Select the Time Period tab if necessary. Under Time Period, change the start or finish dates.

■ **Time Scale** — This setting specifies the tick marks and interim dates that show on the timeline. Select the Time Period tab if necessary. In the Time Scale drop-down list, select the timescale you want, such as Months or Weeks.

■ **Date or time format** — Select the Time Format tab. Under Show Start and Finish Dates on Timeline, in the Date Format drop-down list, select the date format you want. To change the date format for interim dates, select the format under Show Interim Time Scale Markings on Timeline.

- **Timescale date and markings** — Select the Time Format tab. If you don't want any tick marks or dates between the start and finish dates, clear the Show Interim Time Scale Markings on Timeline check box. If you don't want interim dates to show, clear the Show Dates on Interim Time Scale Markings check box. Your start and finish dates, as well as any milestone and interval dates, still appear.

- **Revise dates automatically** — When you want dates to dynamically change as you move milestones or intervals along the timeline (this is the default), on the Time Format tab, select the Automatically Update Dates When Markers Are Moved check box. Clear this check box if you do not want dates to update automatically.

TIP You can edit all date and time formatting for different elements in a timeline. Select the timeline and then choose Timeline ⇨ Change Date/Time Formats. Change the date format for the elements you want, such as the start and finish dates or the milestone dates.

- **Change the timeline orientation** — Select the timeline and drag the selection handles to move the timeline to a vertical orientation.

- **Change the type of timeline** — Right-click the timeline and, from the shortcut menu, choose Set Timeline Type. In the Timeline Type drop-down list, select the type of timeline you want.

- **Show arrowheads** — Right-click the timeline and then choose Show Start Arrowhead or Show Finish Arrowhead.

NOTE To delete a timeline, select it and press Delete. All milestones, intervals, and any other shapes associated with the timeline are deleted as well.

Importing and Exporting Timeline Data

If Microsoft Project and Visio are both installed on your computer, you can transfer timeline data from one program to the other. For example, if you threw a timeline together in Visio to get an initial OK from the management team, the information from that timeline can be the foundation of a project schedule in Microsoft Project. Conversely, you can build a nice summary of tasks from a Microsoft Project schedule by importing them into a Visio timeline.

NOTE Data in text files or Microsoft Excel workbooks won't convert directly into Visio timelines, but they do import into Visio Gantt charts. If you export that data to Microsoft Project, the Import Timeline Data wizard can read the data from Microsoft Project data and build either a timeline or a Gantt chart in Visio.

Importing Microsoft Project Data into a Timeline

Bring information from Microsoft Project into Visio when you want to present or report project status. You can import all tasks in the project or just top-level tasks, summary tasks, milestones, or any combination of these. To import information from Microsoft Project into a Visio timeline, follow these steps:

1. In Visio, open an existing Timeline drawing or create a new one.

2. Choose Timeline ➪ Import Timeline Data.

> **NOTE** If Import Timeline Data doesn't appear on the Timeline menu, Microsoft Project is not installed on your computer.

3. In the wizard page, click Browse. Navigate to the Microsoft Project file with the information you want to import into Visio. Select the file, click Open, and click Next.

> **CAUTION** To successfully import project data, make sure that the Microsoft Project file that you're importing is not currently open.

4. Select the type of tasks you want to import from Microsoft Project to Visio. If you want to build a timeline of the entire project, choose All. But, for presentations, you're more likely to choose Top Level Tasks and Milestones. Other choices include Top Level Tasks Only, Milestones Only, and Summary Tasks Only. Click Next

5. Select the Timeline, Milestone, and Interval shapes you want to use on the timeline. Summary tasks in your project schedule turn into Interval shapes; milestones transfer directly to Milestone shapes, and the overall schedule is a Timeline shape. Click Next.

> **NOTE** Any Microsoft Project tasks with 0 duration import as milestones.

6. Review the import properties you specified. To change any of the import properties, click Back.

7. When all import properties are set the way you want, click Finish. Visio imports the selected tasks. If you imported into an existing Timeline drawing, Visio creates the imported timeline on a new page.

Exporting Visio Timelines to Microsoft Project

Suppose you built a simple project as a Visio Timeline, perhaps as a proposal or to gather input from team members. Now it's time to initiate the project, and you want to get a head start on the plan by transforming the Visio Timeline into a Microsoft Project schedule so you can take full advantage of scheduling, resource allocation, and budget tracking features. In Microsoft Project, intervals from a Visio timeline become tasks with start dates, finish dates, and durations. To export information from a Visio timeline to Microsoft Project, follow these steps:

1. In Visio, open and select the timeline.

2. Choose Timeline ➪ Export Timeline Data.

> **NOTE** If you select a timeline that's associated with expanded timelines, a prompt asks if you want to export markers on the expanded timelines. To export only the overview timeline, click No. To export all information, click Yes.

3. In the browser window that appears, navigate to the folder in which you want to save the exported project file.

4. In the File Name text box, type a name for the project file. Be sure that the Save As Type box is set to Microsoft Project File, and then click Save.

5. In the wizard page again, click Next. A message will appear saying that the project has been successfully exported. Click OK.

6. Open Microsoft Project and review the project file you just created with the exported timeline data.

Scheduling Projects Using Gantt Charts

The Gantt chart is one of the most popular diagrams for showing project task information. In a Gantt chart, task data appears in a table next to charts of task bars, milestone markers, and other symbols that indicate the timing and relationships between tasks along a horizontal timescale. If you're comfortable with Microsoft Project, you'll probably turn to the Visio Gantt Chart template only for presentations or status reports, such as the one shown in Figure 17-3. But the Gantt Chart template can be helpful for building basic project schedules if you don't have easy access to Project.

FIGURE 17-3

Create a Gantt chart in Visio to gain consensus on a proposed project plan or to report ongoing progress.

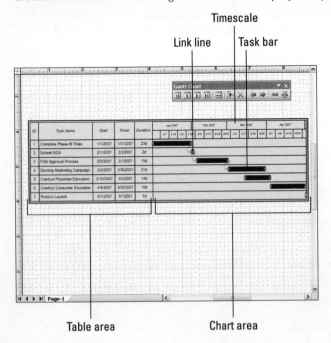

Timescale

Link line   Task bar

Table area   Chart area

With the Gantt Chart template, you can perform the following actions:

- **Draw a Gantt chart** — Use the Gantt Chart Frame, Column, and Row shapes to define the Gantt Chart drawing.

- **Draw task bars and milestones** — Indicate the scheduling of tasks with the Task Bar shape, and the major accomplishments with the Milestone shape.

- **Create a hierarchy of project tasks** — Indent and outdent tasks to create summary tasks and subtasks.

- **Link tasks** — Show dependencies between predecessor and successor tasks using the Link Lines shape.

- **Specify working time** — Indicate the normal working days and hours for your project team.

- **Explain the Gantt chart to others** — Use the Title, Legend, Text Block, and Horizontal Callout shapes to clarify the meaning of the Gantt chart.

- **Add columns** — Insert predefined or custom project fields, such as Resource Names or % Complete, in the table area of the Gantt chart.

- **Format the Gantt chart** — Specify the look of the task bars, milestones, summary task bars, and text.

- **Exchange project data** — Import and export project information with Microsoft Project.

NOTE A Visio Gantt chart is primarily a visual representation of a project. It uses SmartShapes technology and can perform basic calculations among start dates, finish dates, and durations. However, a Visio Gantt chart does not calculate resource allocation, budget estimates, and other invaluable project management information, which is the domain of a software tool such as Microsoft Project.

Creating Gantt Charts

Start a new Gantt chart drawing by setting up the overall parameters of the project. Next, specify the details for individual tasks. Add milestones, organize the tasks into a hierarchical outline or work breakdown structure, and link tasks together to show task dependencies.

To create a new Gantt chart in Visio, follow these steps:

1. Choose File ⇨ New ⇨ Schedule ⇨ Gantt Chart. A new drawing with the Gantt Chart Shapes stencil, the Gantt Chart toolbar, and the Gantt Chart menu appears, along with the Gantt Chart Options dialog box.

NOTE To add a Gantt chart to an existing drawing, choose File ⇨ Shapes ⇨ Schedule ⇨ Gantt Chart Shapes. From the Gantt Chart Shapes stencil, drag the Gantt Chart Frame shape onto the drawing.

   2. In the Gantt Chart Options dialog box, specify the following settings:

   ▓ The number of tasks you want to include in the Gantt chart.

   ▓ The major and minor timescale units. Under Time Units, don't make your timescale too detailed for the timescale date range you specify. For example, if your start and finish dates span a year, and you set the time units for days within months, your chart will balloon far beyond the boundaries of a standard page.

   ▓ The duration time units you prefer for tasks, such as Days for smaller projects and Weeks for multi-year projects.

   ▓ The anticipated start and finish date for the project.

   3. When you're finished, click OK to build and display the Gantt chart.

To enter task details in the table area of your Gantt chart, follow these steps:

   1. Zoom into the Gantt chart if necessary to see text in the columns.

   2. In the Task Name column, double-click Task 1 in the first row and change Task 1 to the first task name in your project. When you're finished, double-click Task 2 and repeat the process. Repeat this for each task in your project.

   3. In the Start and Finish columns, double-click a date and change it to the date for the corresponding task. When you're finished, click outside the field. Visio automatically calculates the Duration field from the start and finish dates you enter and displays the duration in the Duration units you chose.

 To create a milestone, double-click the Duration cell and type 0.

Although you can specify the start date, finish date, and duration for tasks by typing in their table columns, dragging task bars within the timescale of the Gantt chart can be faster, although sometimes less accurate. When you select a task bar, selection handles and control handles appear, which you can use to define task information. Specify task details in the chart area of the Gantt chart as follows:

   ■ **Change the start date** — Drag the left green selection handle to the left or right until it's under the start date you want. The date in the Start Date column for the task changes as well.

   ■ **Change the finish date** — Drag the right green selection handle to the left or right until it's under the finish date you want. The date in the Finish Date column for the task changes as well.

   ■ **Change the task duration** — Drag either green selection handle until the task bar spans the number of days you want. The amount in the Duration column for the task changes as well.

   ■ **Indicate progress in the task bar** — Select the task bar shape and then drag the left yellow control handle toward the right in the task bar to display a progress bar.

Navigating in Gantt Charts

Even high-level summaries of projects can contain dozens of tasks that span months, if not years, making it almost impossible to see the entire project on the screen at the same time. Tools on the Gantt Chart toolbar, illustrated in Figure 17-4, help you find the task bars or dates you want in a Gantt chart. Use the following tools to scroll the chart area within your Gantt chart.

FIGURE 17-4

Navigate to the tasks you're looking for with tools on the Gantt Chart toolbar.

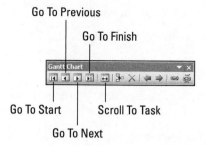

Go To Previous

Go To Finish

Go To Start

Scroll To Task

Go To Next

- **Go To Start** — Shows the first task bar in the project.
- **Go To Previous** — Shows the time period just before the period currently displayed.
- **Go To Next** — Shows the time period just after the first period currently displayed.
- **Go To Finish** — Shows the last task bar in the project.
- **Scroll To Task** — Shows the task bar for the currently selected task.

Adding Milestones

Milestones are a great way to show progress on a project, because a completed milestone is a much more quantifiable status than a task that's 40 percent complete. It's a good idea to add milestones that you can complete from reporting period to reporting period. To add a milestone to your Gantt chart, follow these steps:

1. From the Gantt Chart Shapes stencil, drag the Milestone shape onto the task above which you want to insert a new milestone. Visio inserts a new task above that point and the milestone marker appears in the chart area.

2. Double-click New Task in the Task Name column and then change the placeholder name to the milestone name.

3. Change the Start Date field to the milestone date. In Visio, milestones must have duration of 0, so as soon as you enter the start date, Visio updates the finish date to the same date. You can also drag the milestone marker in the chart area to the date you want. Visio updates the fields in the Start and Finish columns automatically.

> **NOTE** Leave the duration for the milestone task at 0. If you add a duration, the milestone marker changes to a Task Bar shape in the chart area. Likewise, if you change a task duration to 0, its task bar changes to a milestone marker.

Organizing Tasks

You can set up the tasks in your Gantt chart as an outline of summary tasks and subtasks. This is helpful for setting up individual tasks within phases or for subdividing tasks with larger scope into their component tasks, as in a work breakdown structure. To arrange tasks into a hierarchy of summary tasks and subtasks, follow these steps:

1. Add all the summary tasks and subtasks in the proper order for the hierarchy.

> **TIP** To move a task, select the entire task row by clicking the task's ID. Drag the row to the location you want.

2. Select the task you want to transform into a subtask of the task immediately above it.

3. On the Gantt Chart toolbar, click the Indent tool or choose Gantt Chart ➪ Indent, which makes the following changes:

   ■ Visio indents the selected task to show that it's a subtask of the task above it.

   ■ Visio bolds the font for the task above, indicating that it's now a summary task of the indented tasks below it.

   ■ The summary task information calculates rolled up values for all its subtasks.

   ■ The task bar for the summary task is marked with triangular end points, and represents the start, finish, and duration for all subtasks.

Setting the Project Working Time

By setting a project's working time, you can show the days and times in which work occurs on the project — for example, Monday through Friday 8:00 A.M. through 5:00 P.M. If a project is behind schedule and Saturdays turn into workdays, you can specify that as well.

To specify working days and times for the project, choose Gantt Chart ➪ Configure Working Time. Under Working Days, select the check boxes for the days of the week designated as workdays and clear the check boxes for days off. Under Working Time, enter the start and finish times for the working days. Working time is reflected with different colors in the chart area of the Gantt chart and can affect task finish dates.

Linking Tasks

Many tasks can't start until other tasks are completed, a condition which is known as a *task dependency,* or *task link.* For example, frosting a cake isn't advised until the cake is baked and cooled. To link tasks to show their dependencies, follow these steps:

1. Select the first task or the predecessor you want to link. Then Shift-click or Ctrl-click the second task, or successor. Click as many tasks as you want in the order that you want them linked.

 **Shift-clicking selects one task at a time, rather than a series of consecutive tasks.**

2. On the Gantt Chart toolbar, click the Link Tasks tool. Visio links the selected tasks in a finish-to-start relationship, as shown by link lines in the chart area of the Gantt chart. Start and finish dates are recalculated to reflect the scheduling changes caused by the new task links.

 **Tasks are linked in a finish-to-start relationship only. There is no way to represent start-to-start, finish-to-finish, or start-to-finish task links in a Visio Gantt Chart.**

Annotating Gantt Charts

You can polish up Gantt charts by adding annotations. From the Gantt Chart Shapes stencil, drag one or more of the following shapes to the Gantt chart drawing:

- Title
- Legend
- Text Block (8-point, 10-point, or 12-point text)
- Horizontal Callout or Right-Angle Horizontal (callout)

 **For information on annotation techniques, see Chapter 6.**

Modifying the Content in Gantt Charts

After you've built your Gantt chart, you can modify it to make adjustments or to add detail. For example, you can add or delete tasks to reflect changes in the project scope, or you can show or hide columns in the table area of the Gantt chart, depending on the information the audience needs. Use one or more of the following methods to modify the content of your Gantt chart:

- **Add a new task** — Click the task above which you want to add a new task. On the Gantt Chart toolbar, click New Task. You can also choose Gantt Chart ⇨ New Task or drag the Row shape from the Gantt Bar Shapes stencil onto the Gantt chart. Type the name of the new task along with the start date, finish date, duration, and any other task information.

- **Delete a task** — Select the task you want to delete. On the Gantt Chart toolbar, choose Delete Task, or choose Gantt Chart ⇨ Delete Task.

- **Rename a task** — Double-click the task name, edit the name, and press Esc.

- **Add a column to the table** — Click the column heading to the left of where you want the new column. Choose Gantt Chart ⇨ Insert Column or drag the Column shape from the Gantt Chart Shapes stencil onto the Gantt chart. In the Column Type drop-down list, select the field you want to add, such as % Complete or Actual Duration. To add a custom column, select one of the User Defined fields, such as User Defined Duration. Click OK.

- **Remove a column from the table** — Click anywhere in the column you want to delete. Choose Gantt Chart ⇨ Hide Column.

Formatting Gantt Charts

To clarify a Gantt chart, you can format how Visio displays time-based information. You can also change the appearance of chart area elements to make a Gantt chart more compelling. Format your Gantt chart in the following ways:

- **Change timescale dates and units** — Choose Gantt Chart ⇨ Options and then select the Date tab if necessary. Under Time Units, specify the major and minor time periods for the timescale in the chart area of the Gantt chart. Under Duration Options, specify the time unit for duration. Under Timescale Range, specify the start and finish dates for the project.

- **Change the look of task bars and milestones** — Choose Gantt Chart ⇨ Options and then select the Format tab. Specify the shapes on the ends of task bars and summary bars and the shape of milestones. You can also display text to the left and right of task bars as well as inside the bars by choosing field name in the drop-down lists for the Left Label, Right Label, and Inside Label fields.

- **Change text formatting in the table area** — Select the text you want to format and then choose Format ⇨ Text.

- **Change colors in the Gantt chart** — Apply a theme or select the item whose color you want to change and choose Format ⇨ Fill.

Importing and Exporting Gantt Chart Data

You can import and export Gantt chart data between Visio and Microsoft Project. When Microsoft Project is installed on your computer with Visio, the Import and Export commands appear on the Visio Gantt Chart menu.

Importing Project Data into Visio Gantt Charts

Bring information from Microsoft Project into Visio when you want to create a Gantt chart based on existing project management information. Although you can create summary reports in Microsoft Project, a Visio Gantt chart offers simple techniques for creating summaries of projects.

To import Microsoft Project data into a Visio Gantt chart, follow these steps:

1. In Visio, choose File ⇨ New ⇨ Schedule ⇨ Gantt Chart or open an existing Gantt chart drawing.

2. Choose Gantt Chart ⇨ Import.

3. In the first wizard page, select Information That's Already Stored in a File and then click Next.

4. Select Microsoft Office Project File and click Next.

> **TIP** In this wizard page, you can also choose to import project information from existing Excel spreadsheets (XLS files) or text files (TXT or CSV files).

5. Click Browse and navigate to the Microsoft Project file that contains the information you want to import as a Visio Gantt chart. Select the file and click Open. Click Next.

6. Specify the major and minor timescale units, as well as the duration time units to be used. Click Next.

7. Select the type of tasks you want to import from Microsoft Project to Visio. Click Next.

> **NOTE** Any Microsoft Project tasks with 0 duration are imported as milestones. Any tasks designated as milestones but containing a duration, such as one day, are imported as a milestone with a 0 duration.

8. Review the import properties you specified. To change any of the import properties, click Back. When all import properties are set the way you want, click Finish. The selected task information is built as a Gantt chart in Visio. If you are importing into an existing drawing, Visio creates the imported Gantt chart on a new page.

Entering Gantt Chart Data via an Excel or Text File

You might find it more efficient to enter large amounts of project data in an Excel (XLS) file or a text (TXT or CSV) file and then import that file into Visio. You can create these data files directly in Excel or an application that produces text files, or create the spreadsheets or text files within Visio. To import data from a spreadsheet or text file, follow these steps:

1. In Visio, choose File ⇨ New ⇨ Schedule ⇨ Gantt Chart or open an existing Gantt chart drawing.

2. Choose Gantt Chart ⇨ Import. The Import Project Data Wizard appears.

3. Select Information That I Enter Using the Wizard and then click Next.

4. Select Microsoft Excel or Delimited Text and then click Next. If you select Microsoft Excel, an Excel spreadsheet opens. If you select Delimited Text, a text file opens. Either option provides preset column headings and sample data in cells to help you enter project information for your Visio Gantt chart.

5. Under the New File Name box, click Browse. Navigate to the folder in which you want to store the new file. Type a name for the file in the File Name box. Make sure that Microsoft Office Excel Workbooks (*.xls) or Text Files (*.txt; *.csv) is selected in the Save As Type box and then click Save.

6. Click Next. Microsoft Excel or Notepad will appear, with a page containing project-related headings that contribute to your Gantt chart in Visio, such as Task Name, Duration, and Start Date. In a Notepad text file, the information for the different columns is separated by commas.

7. Replace the sample data under the headings with your own project data.

8. When you're finished, save and close the file. For best results, close Microsoft Excel or Notepad as well.

9. In Visio, continue to step through the remaining wizard pages to import the information from the file you just created.

10. Check the information in the final wizard page and then click Finish. Visio creates a Gantt chart using the selected task information from your data file.

Exporting Visio Gantt Charts to Microsoft Project

To export project information from your Visio Gantt chart into Microsoft Project, follow these steps:

1. In Visio, open and select the Gantt chart and then choose Gantt Chart ⇨ Export.

2. Select Microsoft Office Project File and then click Next.

3. In the next wizard window, click Browse, and, in the browser window that appears, navigate to the folder in which you want to save the exported project file.

4. In the File Name box, type a name for the new project. Be sure that Microsoft Project File is selected in the Save As Type box. Click Save.

5. In the wizard page again, click Next.

6. In the next wizard page, review the export properties you specified and click Back to change any properties. When all export properties are set the way you want, click Finish. A message indicates that the project has been successfully exported. Click OK.

7. Open Microsoft Project and the project file you just created with the exported Visio Gantt chart data.

 To learn more about importing and exporting information between Visio and other applications, see Chapter 9.

Building PERT Charts

Like Gantt charts, Program Evaluation and Review Technique (PERT) charts are also popular for displaying project task information in a network layout. Each task is represented by a box, or *node*, which is connected to other nodes in the PERT chart via their task links much like a flowchart. Each node contains task information such as task duration, start date, and finish date. Unlike Visio Gantt charts and timelines, the Visio PERT chart does not calculate start dates, finish dates, or durations, and is nothing more than a visual representation of your project tasks. In most cases, you're better off building a PERT chart in Microsoft Project so it can calculate values for you.

Creating PERT Charts

To create a new PERT chart in Visio, follow these steps:

1. Choose File ➪ New ➪ Schedule ➪ PERT Chart. A new drawing containing the PERT Chart Shapes stencil appears.

2. From the PERT Chart Shapes stencil, drag the PERT 1 or PERT 2 shape onto the drawing. The PERT 1 shape creates a task node containing the task name, with six boxes in which to enter task details. The PERT 2 shape contains four additional boxes for task details.

NOTE Both the PERT 1 and PERT 2 task node shapes contain placeholder project information, including duration, early start, slack, scheduled finish, and more. As you add PERT shapes to a drawing, be sure to replace the placeholders with any type or format of information you want.

3. Drag a PERT node shape onto the drawing for each project task you want to show. It's best to use the same type of PERT shape for all tasks.

4. To enter a task name, select a node and then type the task name. When you're finished, press Esc.

5. To enter other task information in the node, first select the node, select a text box, type the information, and then press Esc.

TIP To empty a text box, select it, press the spacebar, and then press Esc.

6. To show a task dependency from one node to another, click the Connector tool on the Standard toolbar, click the Line Connector, Line-curve Connector, or Dynamic Connector in the PERT Chart Shapes stencil, and then drag from the predecessor to the successor node.

7. To add a legend or callouts, drag the Legend shape onto the drawing and update the text to help others understand the information in your task nodes, or drag the Horizontal Callout or Right-Angle Horizontal shape onto the drawing.

CROSS-REF If you want to show specific information, you can create your own PERT task node shape. To learn more about customizing shapes, see Chapter 32.

Summarizing Projects on PERT Charts

The PERT chart's Summarization Structure shape is meant to create a high-level graphical overview of a project, like a work breakdown structure. However, to build this diagram, you must drag shapes onto the drawing and connect them to build the hierarchy. Unfortunately, the Summarization Structure shape doesn't include even the smart behaviors that help you build organization charts quickly. So, in most cases, it's much more effective to use a program such as Microsoft Project to summarize the structure of a project.

Summary

Visio provides templates for creating calendars, timelines, Gantt charts, and PERT charts to help plan and communicate project information. Calendars and PERT charts strictly reflect the information you provide. By contrast, timelines and Gantt charts calculate start dates, finish dates, and durations, to assist with basic project scheduling.

The Visio project scheduling tools are designed to exchange information easily with other programs. You can import appointments from your Microsoft Outlook calendar and exchange timeline information with Microsoft Project. You can import and export Gantt chart information with Microsoft Project, Microsoft Excel, and text files.

Chapter 18

Documenting Brainstorming Sessions

Brainstorming is an effective way to get people to think outside the box while generating new ideas or creative solutions to challenging problems. During brainstorming sessions, participants are free to express any idea that springs to mind without judgment from other team members. When anything goes and criticism is withheld, as-yet-unidentified options and innovative solutions often surface. Brainstorming can help flesh out ideas for any purpose, including business strategy, research, new applications for existing products — even the plot of a novel.

This chapter begins by describing how to create brainstorming diagrams to show the relationships between topics and numerous levels of subtopics. You'll learn how to transform the organized chaos of brainstorming diagrams you build during sessions into readable and meaningful diagrams. Finally, you'll learn how to add a legend to a diagram to identify the symbols used.

Exploring the Brainstorming Template

Brainstorming sessions tend to be fast-paced and a bit disorganized, so it's hard to capture the flood of information on a diagram. In the Visio Brainstorming template, the Brainstorming Shapes stencil contains only five shapes for documenting the relationships between main topics and other topics. You can quickly drag and drop topic shapes and connectors to reflect the ideas participants lob into the discussion.

Visio Brainstorming tools support fast capture of ideas as well as more meticulous analysis, organization, and refinement of results:

- The Brainstorming outlining tool works equally well for documenting existing ideas before brainstorming begins or refining the results of a brainstorming session.

- Add topics by choosing one of the topic commands on the Brainstorming menu or toolbar or by right-clicking the diagram or shape and choosing a command from the shortcut menu. Topic commands are faster than dragging and dropping because they create and connect Topic shapes at the same time.

- The Legend Shapes stencil provides shapes for annotating brainstorming diagrams so that they make sense after the session ends. Attach Legend shapes to topics to convey shorthand messages, such as order of priority or ideas that need more work.

- Whenever you have time during or after your session, create a legend, modify topic and connector styles, or optimize the layout of the topics with the Auto-Arrange command.

 **NOTE** The Brainstorming template replaced the Mind Mapping template provided by Visio 2002.

When you open or create a Brainstorming diagram using the Brainstorming template, Visio opens several stencils, floats the Brainstorming toolbar in the drawing window, adds the Brainstorming menu to the Visio menu bar, and opens the Outlining window (see Figure 18-1).

FIGURE 18-1

The Brainstorming template includes menus, a toolbar, commands for layout and adding brainstorming shapes, and stencils.

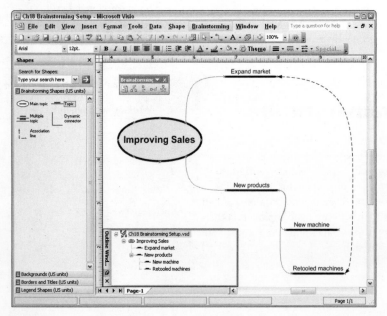

Accessing Visio Brainstorming Tools

Visio provides Brainstorming commands in several places, so that the tools you need are close at hand regardless where you've positioned the mouse or how you prefer to work.

- **The Brainstorming menu** on the Visio menu bar contains all the brainstorming commands that Visio offers.

- **The Brainstorming toolbar**, which can dock or float as you prefer, includes the most commonly used Brainstorming commands.

- **Shortcut menus** with context-sensitive commands appear when you right-click a Brainstorming diagram or shape.

NOTE If the Brainstorming toolbar doesn't appear when you open a Brainstorming diagram, choose View ➪ Toolbars ➪ Brainstorming. To display the Outline window when it is hidden, choose Brainstorming ➪ Outline Window.

Exploring the Visio Brainstorming Shapes

Brainstorming diagrams can contain a hierarchy of topics with one or more main topics connected to as many levels of subtopics as you want. Isolated ideas need only Topic shapes without connections to other topics.

The Brainstorming Shapes stencil contains three Topic shapes and two connectors:

- **Main Topic** represents a central theme or idea. You can add more than one Main Topic to a diagram.

- **Topic** represents ideas or topics. Although Visio provides only one Topic shape, you can use it to create peer-level or subordinate topics.

TIP For diagrams with several hierarchical levels, it's a good idea to modify Topic shapes to visually differentiate levels in the hierarchy. To do this, right-click the shape and, from the shortcut menu, choose Change Topic Shape. In the Change Topic dialog box, select the type of shape you want and click OK. Some brainstorming styles, such as Mosaic1 and Starburst (described in the section "Choosing Brainstorming Styles to Change Topic Shapes") display different shapes for each level of the hierarchy.

- **Multiple Topics** opens a dialog box in which you can type text for multiple topics. Visio adds separate Topic shapes for each line you type in the dialog box.

- **Dynamic Connector** connects topics when you glue each end to shapes representing topics on the drawing page.

- **Association Line** is a connector that shows an ancillary relationship between two topics.

The Legend Shapes stencil, which opens automatically when you use the Brainstorming template, contains the Legend shape itself as well as symbol shapes that you can glue to Topic shapes as shorthand reminders.

The Brainstorming template also opens the Backgrounds stencil and the Borders and Titles stencil.

Creating Brainstorming Diagrams

The easiest way to create a Brainstorming diagram is to create a new drawing using the Brainstorming template and then use Brainstorming commands to add topics to the page. To create a Brainstorming diagram, choose File ➪ New ➪ Business ➪ Brainstorming Diagram.

Adding and Connecting Topics

Visio doesn't automatically connect Topic shapes when you drag and drop them from the Brainstorming stencil, but connections between Topic shapes are important, particularly when you rely on the Outline window to rearrange topics and their subordinates. Fortunately, the commands on the Brainstorming menus create shapes and automatically connect them to other shapes.

TIP If you find relationships easier to see in an outline, right-click topics in the Outline Window to open a shortcut menu with commands for adding, deleting, modifying, and rearranging topics.

Adding Main Topics

A Brainstorming diagram can contain more than one main topic, although it's easier to maintain focus when you ration your main ideas. To create a main topic, choose your favorite method from the following:

- **Shortcut Menu** — Right-click the drawing page and, from the shortcut menu, choose Add Main Topic.
- **Outline Window** — Right-click the diagram file name in the Outline window and choose Add Main Topic.
- **Menu or Toolbar** — From the Brainstorming menu or the Brainstorming toolbar, choose Add Main Topic.
- **Drag and Drop** — Drag the Main Topic shape onto the drawing page.

Adding Subtopics

To create and automatically connect a topic as a subordinate to a topic at a higher level, use your favorite method from the following:

- **Shortcut Menu** — Right-click a topic on the drawing page and, from the shortcut menu, choose Add Subtopic.
- **Outline Window** — Right-click a topic in the Outline window and choose Add Subtopic.
- **Menu or Toolbar** — Select a Main Topic or Topic shape on the diagram and, from the Brainstorming menu or the Brainstorming toolbar, choose Add Subtopic.
- **Drag and Drop** — Drag the Topic shape onto the drawing page.

Linking Topics to Reference Documents

Suppose that an off-beat marketing strategy on your brainstorming diagram hinges on a consumer study. Adding hyperlinks to supporting documentation is an easy way to help readers understand the rationale behind brainstorming ideas — without cluttering the diagram itself.

To add a hyperlink, select a topic and choose Insert ⇨ Hyperlinks. In the Address box, click Browse, and then click Local File. If necessary, choose the type of file you want to link to in the Files of type list. Navigate to the file you want and click Open. To name the hyperlink, in the Description box in the Hyperlinks dialog box, type a name and click OK.

To view a hyperlinked document, right-click the topic and, from the shortcut menu, choose the hyperlink name.

Adding Topics at the Same Level

Peer topics are topics at the same level as other topics in the hierarchy. Create a peer topic connected to the same Main Topic shape using one of the following methods:

- **Shortcut Menu** — Right-click a topic on the drawing page and, from the shortcut menu, choose Add Peer Topic.

- **Menu or Toolbar** — Select a Topic shape at the same level as the topic you want to create and, from the Brainstorming menu or the Brainstorming toolbar, choose Add Peer Topic.

 The shortcut menu in the outline window doesn't offer a command for adding a topic at the same level. You must right-click a topic at the next level up and choose Add Subtopic.

Adding Multiple Topics at Once

When one topic spawns a host of subtopics, use the Add Multiple Subtopics command to add and connect several topics simultaneously to the same higher-level topic. To add multiple subtopics, follow these steps:

1. Right-click the Topic or Main Topic shape to which you want to add multiple subtopics and, from the shortcut menu, choose Add Multiple Subtopics.

 You can also initiate the Add Multiple Subtopics command by first selecting the higher-level topic and then, from the Brainstorming menu or toolbar, choosing Add Multiple Subtopics.

2. In the Add Multiple Topics dialog box, type text for each topic and press Enter to type the text for the next topic.

3. After you have added all the subtopics you want, click OK. Visio creates a separate Topic shape for each line of text arranged vertically and all connected to the original Topic or Main Topic shape, as illustrated in Figure 18-2.

FIGURE 18-2

Visio adds and connects several Topic shapes and positions them vertically, but you can rearrange them later.

Type one topic on each line

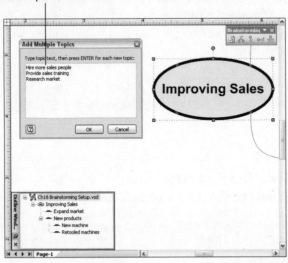

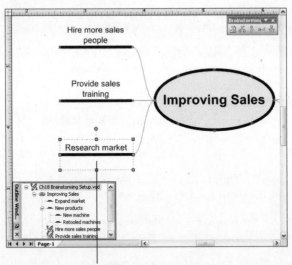

Visio creates and connects a Topic shape for each line

Building a Brainstorming Diagram from an Outline

The Outline window, true to its name, presents main topics and lower-level topics in a hierarchical view. Although an outline of brainstormed ideas might not have the panache of a free-form brainstorming diagram, it is particularly convenient when your diagram spans several pages. You can easily rearrange topics in the outline without affecting the appearance of your diagram and without navigating between pages. If you prefer organizing ideas in outline form, the Outline Window enables you to add, remove, and modify topics, while Visio makes the corresponding changes to the diagram itself.

Here are actions you can perform within the Outline Window:

- **Add main topics** — Right-click the name of the diagram at the top of the Outline Window and, from the shortcut menu, choose Add Main Topic.

- **Add topics** — Right-click any main topic or lower-level topic and, from the shortcut menu, choose Add Topic or Add Multiple Topics. The Outline Window doesn't offer the Add Peer Topic command.

- **Delete topics** — Right-click a topic and from the shortcut menu, choose Delete Topic.

- **Reorder topics** — Right-click a topic and, from the shortcut menu, choose Move Up or Move Down until the topic appears in the correct position in the outline.

- **Rename topics** — Right-click a topic and, from the shortcut menu, choose Rename. In the dialog box that appears, type the new text that you want to appear for the topic.

- **Find topics** — Finding shapes on a multi-page brainstorming diagram is easy when you right-click a topic in the outline and, from the shortcut menu, choose Select on Page. Visio pans to the section of the drawing that contains the topic you right-clicked and centers the topic in the drawing window.

- **Move topics** — If you decide to move one branch of ideas to another page, in the Outline Window, right-click the parent topic and, from the shortcut menu, choose Move Topic to New Page. In the Move Topics dialog box, choose the New Page or Existing Page option, specify the name of the page, and click OK.

> **TIP** The Outline window docks along the side of the drawing area or within the Shapes window. You can also float it within the Visio window, but sometimes the Outline Window gets in the way regardless of where you place it. To hide the Outline Window when you want to work on the diagram, right-click the title bar of the Outline Window and, from the shortcut menu, choose AutoHide. AutoHide rolls the window up to only its title bar when you position the mouse pointer over the drawing page. When you position the mouse pointer over the Outline Window title bar, it automatically expands to its full width.

Connecting Topics Manually

If you use the Brainstorming menus to add topics, connecting shapes manually is rarely necessary. However, connecting topics is required if you want to add ancillary relationships between topics, reconnect shapes that became disconnected inadvertently, or connect shapes that you added by dragging and dropping. Connecting topics on a brainstorming diagram is nothing more than creating a shape-to-shape connection, described by the following steps:

1. From the Brainstorming stencil, drag the Dynamic connector until one end is over a connection point on the first shape you want to connect. Visio shows that a connection point is selected by highlighting it with a red box.

2. Drag the other end point of the connector to the inside of the second topic shape. Visio indicates the pending shape-to-shape connection with a red box around the entire topic. After you release the mouse button, solid red squares at the connector's end points indicate that the shapes are connected correctly.

3. To change the curvature of a curved connector or Association connector, drag a green eccentricity handle. Drag a green diamond vertex to modify the angle of a straight connector.

NOTE   In Visio, dynamic connectors can connect a subordinate shape to only one higher-level shape. If an idea links to more than one strategy, *association lines* represent these additional connections to more than one higher-level topic, Association lines can also represent relationships between topics at the same level. The links you create with Association line connectors don't appear in the Outline Window or when you export the diagram.

Exporting and Importing Brainstorming Ideas

After generating a brainstorming diagram, chances are you'll want to export that valuable information into another format, such as a Microsoft Word document for a report to management, a Microsoft Excel workbook to track assignments, or an XML file for converting to yet another format. In addition, although most brainstorming diagrams are the result of brainstorming meetings, on occasion you might come to the meeting with initial ideas—the CEOs top-priority business initiatives or the pet projects from the sales department. This section describes how to export and import brainstorming ideas.

Exporting Brainstorming Topics

Visio exports brainstorming topics to Microsoft Word, Microsoft Excel, or XML, but, the files it produces in each case are all XML files. The difference is that each XML file uses a schema specific to the end format. As it turns out, you can open any of these XML files with Word, Excel, or an XML editor.

To export information from a brainstorming diagram, do the following:

1. Choose Brainstorming ➪ Export Data and, on the submenu, choose To Microsoft Office Word, To Microsoft Office Excel, or To XML.

2. In the File Save dialog box, navigate to the folder in which you want to save the file, in the File Name text box, type the name of the file, and click Save.

3. If you save to Word or Excel, Visio launches Word or Excel and opens the XML file in that program. If you save to XML, Visio simply saves the XML file.

Importing Brainstorming Topics

Visio imports only XML files and it has a particular schema that it expects. If you plan to import information into a Visio brainstorming diagram, the easiest way to obtain the correct schema is by exporting a small brainstorming diagram to XML format. Open the exported XML file in an XML editor and you can copy, paste, and edit topics to create the items you want to import.

To import information to a brainstorming diagram, do the following:

1. Choose Brainstorming ➪ Import Data.

2. In the File Open dialog box, navigate to the folder that contains the file you want to import and double-click the name of the file. Visio imports the items to the current drawing.

Cleaning Up Brainstorming Diagrams

Brainstorming diagrams are meant to capture and present ideas unearthed during brainstorming sessions. The initial arrangement of topics is often as exuberant as the discussion that brought the topics to light. After collecting everyone's thoughts, the next step is organizing the ideas for review and further refinement. In Visio, you can rearrange topics automatically or move individual topics to resolve specific layout issues. Formatting brainstorming diagrams makes them more attractive and helps communicate their content.

Moving and Reordering Topics

Whether you want to relocate topics within a page, change topic level in the diagram hierarchy, or move topics (with or without their subordinates) to other pages, Table 18-1 describes the movement methods that Visio offers.

TABLE 18-1

Methods for Moving and Reordering Topics

Method	Steps
Rearrange topics automatically	Choose Auto-Arrange Topics on the Brainstorming menu or toolbar. As long as the Topic shapes on your diagram are connected using shape-to-shape connections, Visio automatically rearranges the Topic shapes for you.
Move a topic on the same page	Select the Topic shape and drag it to a new position on the page. Visio moves any subordinate Topic shapes and automatically repositions their connectors.
Move a topic and its subordinates to another page	Select the top-level topic that you want to move on the drawing page or in the Outline window and then choose Brainstorming ⇨ Move Topic to New Page. In the Move Topics dialog box, choose the New Page or Existing Page option. To move the topic and its subordinates to a new page, type the name for the new page. To move them to an existing page, select the page name from the Existing Page list. Click OK to close the dialog box and move the topics.
Copy a topic without its subordinates to another page	Right-click the Topic shape you want to copy on the drawing page and choose Copy from the shortcut menu. Navigate to the page onto which you want to copy the shape, right-click the page, and, from the shortcut menu, choose Paste. To copy several Topic shapes without their subordinates, hold the Shift key and select all the topics you want. Then, right-click one of the selected shapes and, from the shortcut menu, choose Copy.
Change a topic's level in the hierarchy	In the Outline window, drag the topic whose level you want to change and drop it on top of the topic you want as its superior. Visio moves the topic and all its subordinates underneath the superior topic and changes connections on the diagram.

NOTE When you move a Topic shape and its subordinates, the top-level Topic shape remains on the original drawing page with an arrow symbol that indicates that the topic appears on another page. To navigate to the corresponding Topic shape on the other page, right-click the Topic shape and, from the shortcut menu, choose Go to Sub Page. To navigate back to the original page, right-click the moved Topic shape and choose Go to Page Containing Parent.

Changing the Layout Style

The *layout style* governs how Visio positions Topic shapes on the drawing page, for instance from left to right, top to bottom, or with a Main Topic shape in the center of the page with Topic shapes radiating outward. If your brainstorming ideas would make more sense with a different layout, follow these steps to change the layout style:

1. Choose Brainstorming ⇨ Layout.

2. In the Layout dialog box, choose the layout you want. Visio displays a preview of the layout style in the box on the right.

3. To specify the connector style, in the Connectors box, choose Curved or Straight. For example, if your diagram contains a lot of crossing connectors, choose Straight to make connections easier to follow.

4. To view the selected layout, click Apply. If the results are acceptable, click OK. Otherwise, select a different layout and click Apply until you like the results.

NOTE The layout you select applies to the current drawing page and to any Topic shapes you add to that page. The connector style applies only to the current drawing page.

Choosing Brainstorming Styles to Change Topic Shapes

The *brainstorming style* determines the shapes that Visio applies to topics at each level of the hierarchy. For example, out of the box, Visio uses the Simple style, which uses an oval for main topics and horizontal bars for all other topics. The Starburst style sports a starburst shape for main topics and different geometric shapes for the second, third, and fourth levels.

NOTE You should apply a brainstorming style only after you're sure the diagram is finished. If you add new Topic shapes or move Topic shapes to different levels in the hierarchy, you must reapply the brainstorming style.

Follow these steps to change the brainstorming style:

1. Choose Brainstorming ➪ Brainstorming Style.

2. In the Brainstorming Style dialog box, choose the style you want. Visio displays a preview of the layout style in the box on the right.

3. To view the selected style, click Apply. If the results are acceptable, click OK, otherwise select a different style until you like the results.

TIP Applying a Visio Theme to your brainstorming diagram changes shape colors to a predefined set of coordinated colors for all pages in your diagram. On the Formatting toolbar, click the Theme button. In the Theme - Colors task pane, click a theme to apply it.

Resizing the Page to Fit the Diagram

If your colleagues are on a roll hatching ideas, but you've run out of room on the drawing page, you can add Topic shapes beyond the borders of the page. Later, when you have time, resize the page to fit your diagram using one of the following methods:

■ **Dragging the page borders** — To display the page borders, on the Standard Toolbar, click the drop-down arrow next to the Zoom box, and then choose Page. Press the Ctrl key and position the pointer on a page border. When the pointer changes to a two-headed horizontal arrow, drag the pointer to resize the page. The status bar displays the current height and width as you drag the border.

 To change either the height or width, position the pointer near the middle of a side. To change both height and width, position the pointer near but not on the corner of the page. The pointer changes to a two-headed diagonal arrow. Drag to change both dimensions. If you position the pointer directly on a corner, the Rotation tool appears, enabling you to rotate the entire page.

- **Using Page Setup** — Choose File ⇨ Page Setup and select the Page Size tab. Select the Size To Fit Drawing Contents option and click OK.

Working with Legends

Symbol shapes, such as Attention, Note, or Priority 1, add reminders and flags to your brainstorming diagrams. For example, identify the priority of ideas by associating Priority 1 shapes with your top priority Topic shapes, or indicate Topic shapes that include additional information or need additional work with Note or To Do shapes. When you move a Topic shape, any attached symbol shapes move with it.

A Legend shape displays the symbol shapes included on a page, the symbol descriptions, and the number of times each one appears on that page. The Legend shape on the page updates automatically as you add symbol shapes to the page, so you can choose whether to add the Legend shape or symbol shapes first.

TIP Although the Legend Shapes stencil opens automatically for only the Brainstorming, Network Diagram, and Building Plan templates, you can use it with any type of Visio drawing.

Adding Symbols to Topics

Each instance of a symbol shape connects to only one Topic shape. To add a symbol shape to a Topic shape, follow these steps:

1. If the Legend Shapes stencil isn't open, choose File ⇨ Shapes ⇨ Business ⇨ Brainstorming ⇨ Legend Shapes.

2. To display the Legend Shapes stencil, in the Shapes window, click the Legend Shapes stencil title bar.

3. Drag a symbol shape from the Legend Shapes stencil and drop it near the Topic shape with which you want to associate it.

NOTE As soon as you drop a symbol shape on a drawing page that contains a Legend shape, the symbol appears in the legend with its description and an updated count of the number of occurrences on the page.

4. To connect the symbol to a topic, drag the yellow control handle from the symbol shape to a blue connection point on the Topic shape. The control handle turns red when it's attached.

To detach a symbol from a topic, select the symbol shape and drag the red square away from the Topic shape. To delete a symbol, select it and press Delete.

TIP Symbols are simple shapes with little room for text in their text blocks and no shape data for storing information. You can attach a Note shape to a Topic shape, but you can't include the note text in that shape. To add a reference to associated information to a Note shape, select the shape, type an alphanumeric ID, such as A1, and then add a text block to the drawing with the Text tool to add the identifier and note text.

Creating Legends and New Symbol Shapes

If the symbols on the Legend Shapes stencil aren't sufficient, you can transform any Visio shape into a symbol — as long as your drawing page contains a Legend shape. For example, instead of using the Note shape from the Legend Shapes stencil, you might want to use a customized Note shape with shape data for the note text.

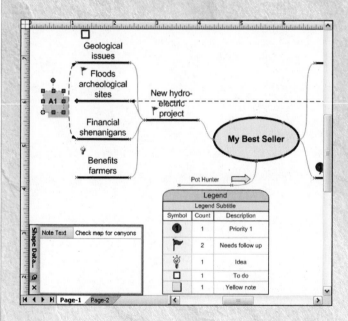

Chapter 6 explains how to perform the following tasks with legends and symbol shapes:

- Add a Legend shape to a drawing page
- Specify what appears in legend columns
- Turning other Visio shapes into symbols

Summary

Documenting the abundance of brainstorming ideas seems daunting, but the small selection of Visio brainstorming shapes and the brainstorming template tools keep the process simple. The commands on the Brainstorming menu or toolbar automatically add and connect Topic shapes at the same time. Alternatively, you can export brainstorming results to Microsoft Word, Microsoft Excel, or XML for inclusion in reports or other business documents. If you have topics already documented, you can convert them to XML and import them into a Visio brainstorming diagram.

Cleaning up a first draft might entail selecting a different layout or moving individual Topic shapes. Applying styles and themes put the final touches on a professional-looking brainstorming diagram. Symbol shapes act as shortcuts for annotating your diagram, whether you want to prioritize your ideas or identify topics that need more refinement. The Legend Shapes stencil includes a number of predefined symbols, but any Visio shape can become a Legend symbol.

Chapter 19

Analyzing Results with PivotDiagrams

V isio 2007 Professional includes a new PivotDiagram template. If you've never used pivot tables in Excel, you might wonder what the big deal is. And, people who can't live without Excel pivot tables wonder why they would need pivot diagrams in Visio, too.

When you have oodles of data to analyze, the sheer magnitude of information can make it almost impossible to make sense of the numbers or spot trends. In many situations, you want to look at the information from different perspectives, similar to rotating a painting to determine whether it's an impressionist landscape or just blotches of paint. For example, depending on what your management team is asking about, you might want to see sales totals by region, by product line, or by type of customer. Serious sessions in Excel with the SUMIF and COUNTIF functions can give you this kind of information, but pivot tables in Excel, and now pivot diagrams in Visio, provide tools to do just that with fewer headaches. With pivot features, you choose the category to use to slice and dice your data, specify the subtotals you're interested in, and the programs shuffle the data around to show totals by category.

Pivot diagrams in Visio perform the same kinds of transformations as pivot tables in Excel. In fact, an Excel workbook is one source for data you can use to build pivot diagrams in Visio. The main difference is that Visio pivot diagrams show the results visually, so you can communicate what you've discovered to your audience more effectively. By applying Data Graphics to a pivot diagram, you can make crucial information stand out in ways that Excel only dreams about.

This chapter begins with a quick introduction to the Visio PivotDiagram template. The chapter describes how to create pivot diagrams and link them to data. You'll also learn how to add categories and totals to a pivot diagram to analyze data in different ways. Finally, the chapter explains how to fine-tune a pivot diagram by filtering the content.

IN THIS CHAPTER

Understanding Visio pivot diagrams

Creating pivot diagrams

Copying pivot diagrams

Categorizing data

Totaling values

Showing combined categories

Formatting pivot diagrams

An Introduction to Visio PivotDiagrams

Unlike most of the other built-in templates in Visio, the PivotDiagram template seems to be less about shapes and all about built-in data-related functionality. When you create a new drawing from the PivotDiagram template, Visio launches the Data Selector wizard (see Chapter 10) so you can select the data source you want to use to fuel your pivot diagram. Because Visio pivot diagrams take advantage of the new Data Link features for data access, pivot diagrams can connect to the data sources that Data Link supports: Excel, Access, SQL Server, SQL Server Analysis Services, OLE/DB or ODBC data sources, and SharePoint lists.

When the data source is in place, the resulting Visio pivot diagram contains one shape, the Pivot Node, which initially displays the total of all of the rows in your data. Likewise, the PivotDiagram Shapes stencil contains only one master: Pivot Node. If you drag that master onto a drawing page and complete the steps in the Data Selector, all the nifty PivotDiagram features appear in your Visio drawing, as shown in Figure 19-1.

FIGURE 19-1

The PivotDiagram template includes only one stencil with one master, but the built-in features enable you to analyze data from different perspectives.

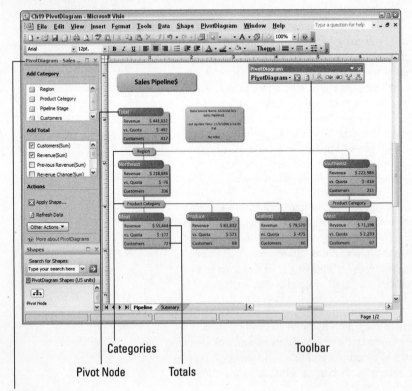

Categories

Toolbar

Pivot Node

Totals

PivotDiagram window

When you've created a Visio pivot diagram and connected it to a data source, the PivotDiagram tools appear in a couple of places:

- The PivotDiagram window, which includes access to many of the PivotDiagram tools, joins the Shapes window on the side of the main Visio window.

- The PivotDiagram menu, which also includes most of the PivotDiagram tools, appears in the Visio menu bar.

- The PivotDiagram toolbar floats within the drawing area. The toolbar has a drop-down menu with several PivotDiagram commands, but the icons are primarily for layout.

The Add Category and Add Total sections appear prominently at the top of the PivotDiagram window. These two lists are the source for the slicing and dicing you do on your data:

- **Categories** — In a pivot diagram, categories represent fields in your data source that you can use to summarize results. For example, for a sales-oriented pivot diagram, your Excel workbook data source might include columns for sales region, product lines, customers, sales reps, and so on. As you'll learn later in this chapter, you can choose the category you want for the top-level of analysis, such as region, and then choose subcategories to further evaluate results, for instance to see the performance of product lines in each region.

NOTE Because categories summarize information, the fields or columns that you use for categories must include duplicate values. That is, your data must include several records or rows with the same sales region, product line, customer, and so on. The pivot diagram counts the occurrences or uses those occurrences to identify which values to sum. The fields in the data source that you total, such as revenue or the number of customer support calls, may or may not have unique values.

- **Totals** — In the PivotDiagram window, the Add Total section includes totals you can calculate, whether they are sums of the values, such as revenue, or counts, such as the number of customers in a region.

- **Combining categories** — On the PivotDiagram menu are the Promote and Merge commands. You can use these to display combinations of categories. By selecting two nodes within a category level and choosing PivotDiagram ⇨ Merge, the pivot diagram shows the combined values for those nodes in a new combined node. For example, if you want to see the revenue for ice cream and cake in one node, while keeping candy separate, simply apply the Merge command to ice cream and cake. Alternatively, you can use the Promote command to combine a category with a subcategory, for example, to summarize data for desserts that are ice cream and chocolate flavored.

Creating Pivot Diagrams

Whether you choose the PivotDiagram template to create a new drawing or drag the Pivot Node shape from the Pivot Diagram Shapes stencil onto a drawing page, the Data Selector wizard starts. And, when you complete all the steps in the wizard, you have what looks like a single shape on the page. However, this Pivot Node is connected to the data source so you can use the PivotDiagram categories and totals to analyze that data in any way you want, which subsequently appears on the Visio drawing page. This section describes the different methods you can use to create a Visio pivot diagram.

Creating a New Pivot Diagram

The most common method for creating a pivot diagram is to create a new drawing using the PivotDiagram template:

1. Choose File ⇨ New ⇨ Business ⇨ PivotDiagram. The Data Selector wizard, which is the same one that Data Link uses, appears.

2. On the first screen of the Data Selector wizard, choose the type of data source that contains your data. The options are the same as those described in detail in the section "Linking a Drawing to a Data Source" in Chapter 10:

    - Microsoft Office Excel 2007 workbook
    - Microsoft Office Access 2007 database
    - Microsoft Office SharePoint Services list
    - Microsoft SQL Server database
    - Microsoft SQL Server Analysis Services
    - Other OLEDB or ODBC data sources

3. As you do for Data Link, you must choose the file or web site for the data source, and depending on the type of data source, the worksheet or table that contains the data.

4. On the Connect To Data screen, select the columns and rows to include.

5. Click Next and then click Finish. A progress message appears while Visio imports the data. When the import is complete, the drawing page contains three shapes, illustrated in Figure 19-2:

    - The Pivot Node, which contains the data from the data source
    - A text box for the name of the PivotDiagram
    - A data legend with information about the data source, such as the file name and the date and time of the last update

The Pivot Node totals one of the columns or fields in the data source. To learn how to analyze the data in different ways, continue to the section "Categorizing and Totaling Results," later in this chapter.

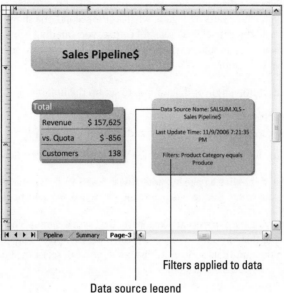

FIGURE 19-2

The legend identifies the data source and any filters applied to the data.

Filters applied to data

Data source legend

Creating a Pivot Diagram in Another Type of Visio Drawing

Suppose you're working on another type of drawing, such as a workflow diagram for a process you're trying to streamline. You can include a pivot diagram as part of the workflow diagram to highlight the steps that jam up. To create a pivot diagram in another type of drawing, open the drawing and then choose Data ⇨ Insert PivotDiagram. The Data Selector wizard appears, and the rest of the steps are exactly the same as the ones you perform when you create a pivot diagram using the PivotDiagram template.

NOTE Visio pivot diagrams use layout settings just like many other templates. If you insert a pivot diagram into another type of drawing, the pivot diagram blends in by assuming the layout settings for the drawing page it's on. If you want the pivot diagram to use its typical layout, choose Shape ⇨ Configure Layout. In the Configure Layout dialog box, in the Style drop-down list, choose Hierarchy.

Creating Multiple Copies for Different Viewpoints

If you want to be prepared for any question the executives might ask, you probably want to create more than one version of a pivot diagram, each with its unique take on business results. For example, the first pivot diagram might show sales by region, whereas the second one shows sales by product line, and yet another, for sales by type of customer. Of course, you can drag a Pivot Node onto a drawing page for each pivot diagram, but why repeat the Data Selector wizard steps to connect to the same data source time and again?

The easiest way to create additional pivot diagrams is to set up the first one and then copy it to another drawing page:

1. Select the Pivot Node and press Ctrl+C.
2. Click the tab for the drawing page on which you want to paste the new pivot diagram.

 You can also right-click a page tab and choose Insert Page. For that matter, you can make the window for another Visio drawing file active and select a page in that file.

3. On the new drawing page, press Ctrl+V to paste the pivot diagram. The page contains the same three shapes that appeared when you created the first pivot diagram, and the diagram is connected to the same data source. At this point, you can select the Pivot Node or any other node to recategorize the diagram.

TIP If you want to drill down in more detail in only one part of the pivot diagram, you can select a node lower in the hierarchy, for example, the ice cream product line in the southern region. When you copy that node to another drawing page, the new pivot diagram starts with that combination of categories. The copy includes a legend, which shows that the data is filtered for both ice cream and the southern region.

Categorizing and Totaling Results

The whole point of pivot diagrams is to look at your data in different ways. As soon as you connect a pivot diagram to a data source, Visio evaluates the field or columns in the data source and adds the ones you can summarize to the Add Category list. It also adds the totals you can calculate in the Add Total list. To slice your data in some way, pick a node on the diagram and then pick the category.

Categorizing Data

To break your data up into chunks based on a category, do the following:

1. If you just created a pivot diagram, select the Pivot Node on the drawing page. For an existing diagram that you've already broken down to one or more levels, select the node you want to break down further.

2. In the PivotDiagram window, in the Add Category list, click the category you want to use to categorize the data. Visio adds nodes underneath the selected node for each value in the category, as shown in Figure 19-3.

3. To break down the data into additional subcategories, do the following:

  a. In the PivotDiagram window, right-click the category you just added and, from the shortcut menu, choose Select All to select the lowest level of nodes.

  b. In the Add Category list, choose the next category you want to apply. Visio adds another level of nodes to the diagram broken down by that category.

4. If you change your mind and want to pivot the diagram a different way, select the node whose subnodes you want to recategorize and then, in the PivotDiagram window, in the Add Category list, click the new category for the break down. For example, to switch a sales by region pivot diagram to a sales by product line diagram, select the top Pivot Node and click the product line category.

FIGURE 19-3

Choosing a category adds another level of nodes to the diagram.

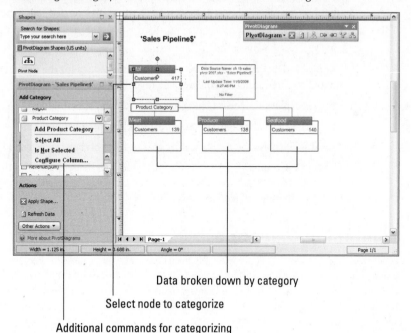

Data broken down by category

Select node to categorize

Additional commands for categorizing

NOTE The shortcut menu that appears for categories has other entries as well. Add Category Name adds a node as if you clicked the category name. It's faster to just click the category name. If you want to filter the data that you show in the diagram, choose Configure Column. In the Configure Column dialog box you can specify criteria to choose the rows of data to show. To remove the filter, in the Show Data Where Product Category box, choose (Select Operation).

Totaling or Summarizing Data

In most cases, the information you're looking for is a total of some kind — the revenue the company has reaped, the number of commercial customers, or the total difference between the estimated and actual values. By default, nodes on a pivot diagram include one total, which is for the first field or column in the data source that can be totaled. To add other calculations to your nodes, do the following:

1. In the PivotDiagram window, in the Add Total list, click the calculation you want to display in the nodes. In the example in Figure 19-3, to show the total revenue for each node, you would click the Revenue(Sum) entry. Visio displays a green check mark for each calculation you turn on.

2. If you don't see the calculation you want, in the PivotDiagram window, in the Add Total list, hover the pointer over the field you want to recalculate and click the arrow that appears. On the shortcut menu, choose the calculation you want from the following choices:

   ■ **Sum** — Adds the numeric values for the field from each subnode, for example, the total revenue for the subnodes or the total number of customers for the subnodes.

   ■ **Average** — Displays the average of the numeric values for the field from each subnode. You might calculate the average revenue per customer.

   ■ **Min** — Displays the lowest value for the field from all subnodes.

   ■ **Max** — Displays the highest value for the field from all subnodes.

   ■ **Count** — Displays the number of records or rows of data represented by the subnodes.

   ■ **Configure Column** — If you want to filter the data that you show in the diagram, choose Configure Column. In the Configure Column dialog box you can specify criteria to choose the rows of data to show.

Combining Categories

Sometimes, the categories that are available aren't exactly what you need. For example, you want to analyze sales revenue for a few different product lines that are ripe for cross-marketing purposes. To combine two or more of the breakdowns within a category, select the nodes for those values and choose PivotDiagram ➪ Merge. The title bar for the new node shows both categories, as shown in Figure 19-4.

NOTE When you merge categories in one branch of the pivot diagram, those categories remain separate in the other branches, as illustrated in Figure 19-4.

FIGURE 19-4

You can combine the results for two categories with the Merge command.

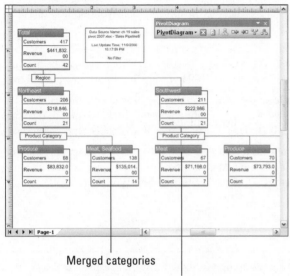

Merged categories

Category not merged in other branch

As its name implies, the Promote command promotes a subcategory to combine it with a category to build a filter for your data. Instead of combining two categories at the same level like Merge, Promote filters data with a category and one or more of its subcategories, such as a parent category ice cream with the flavor subcategories, chocolate and vanilla. To build a filter, select the subnode on the drawing page that you want to combine and then choose PivotDiagram ⇨ Promote.

Formatting Pivot Diagrams

The techniques that you use to format Visio pivot diagrams are probably familiar. Pivot diagrams use the standard Visio layout tools, so you can adjust the arrangement and spacing as you would a flowchart or organization chart. In addition, the nodes on a pivot diagram accept data graphics, so you can show the results of your analysis in eye-catching ways. Here is an overview of the formatting you can apply to pivot diagrams:

■ **Data graphics** — By default, Visio adds calculations to nodes as text items in a data graphic. To change the callouts on your pivot diagram, select a node and then choose PivotDiagram ⇨ Edit Data Graphic. Visio applies the data graphics to all the nodes on the diagram. See Chapter 10 for detailed instructions on adding or editing callouts for data graphics.

■ **Themes** — You can further spruce up a pivot diagram by applying a theme, as described in Chapter 7.

■ **Layout** — Change the layout spacing or arrangement of shapes by choosing Shape ➪ Configure Layout. See Chapter 4 for instructions on applying the layout options to a drawing page.

■ **Page size** — Because pivot diagrams change size as you add or remove categories, Visio automatically resizes the page to fit your diagram. To stop Visio from trying to be helpful, choose Shape ➪ Configure Layout and clear the Enlarge Page To Fit Drawing check box.

■ **Hiding the pivot diagram data legend** — Although the data legend contains valuable information, it could get in the way of a more important message. If you want to turn the legend off, choose PivotDiagram ➪ Options. Uncheck the Show Data Legend check box.

■ **Renaming the diagram** — If you want to add your own title to a pivot diagram, you can simply delete the title that Visio adds and add your own. However, if you want to rename the diagram, choose PivotDiagram ➪ Options. In the Title box, type the new title.

Summary

Pivot diagrams in Visio 2007 Professional transform data in the same ways that pivot tables do in Excel — breaking data down by category and summarizing results in different ways. The PivotDiagram template contains only the Pivot Node shape, but it offers commands for categorizing and totaling data. The template also takes advantage of other new Visio 2007 data features, including Data Link for connecting to a data source, and Data Graphics for emphasizing results.

This chapter describes how to create pivot diagrams, link them to data, and analyze results with categories and totals. Formatting pivot diagrams is similar to formatting other types of diagrams, although the PivotDiagrams options include a few additional settings.

Part IV

Using Visio in Information Technology

Chapter 20

Modeling and Documenting Databases

Whether you are designing a database from scratch or trying to figure out how to modify an order database to support online sales, an accurate and up-to-date model of the database schema helps everyone see what they're working with. Most dedicated database folks use data modeling tools such as ERwin or Oracle Designer. If you don't have that luxury, you can use the Database Model Diagram template in Visio Professional to document database models for both relational and object-relational databases.

With a working knowledge of database concepts and database management practices, you can use the Database Model Diagram template to build a diagram from scratch, import a model from another application, or reverse-engineer an existing database. When you have a database model to work with, you can add to or modify objects, including tables, columns, parent-child relationships, indexes, and code.

In addition, Visio provides templates for Object Role Modeling diagrams and Express-G diagrams, two other methodologies for modeling database. However, these templates produce only diagrams, not models.

In this chapter, you learn how to create database models from scratch, as well as how to import models from other applications and reverse engineer models from existing databases. You'll learn how to set database options and preferences in Visio and work with elements in Visio database model diagrams.

NOTE The database templates are available only in Visio Professional and Visio for Enterprise Architects. Moreover, if you want to generate SQL code from a Visio database model, you must use Visio for Enterprise Architects, which is available with Visual Studio Professional.

Exploring the Database Model Templates

Visio Professional includes three templates to help you produce database documentation, but only one builds a model of your database.

■ **The Database Model Diagram** template helps design and document logical and physical database models. The results go beyond simple diagrams of database schema — physical data modeling encompasses tables, views, relationships, stored procedures, and other elements, using either relational or IDEF1X notation. The Database Model Diagram template adds the Database menu to the Visio menu bar and provides several specialized windows for viewing and modifying database properties. In this template, the Reverse Engineer Wizard takes an existing database and builds a Visio database model for you.

CAUTION   If Visio encounters an unexpected error from either an internal or external source, your template-specific menus, such as the Database menu, can disappear. If Visio shuts down unexpectedly, restart it and then choose Tools ⇨ Options. Select the Advanced tab, check the Enable Automation Events check box, and click OK. Save any open drawings, exit and restart Visio, and then reopen your database model diagram.

■ **ORM** diagrams show database models at a conceptual level, in a way that humble humans as well as the people who hold the purse strings can understand. ORM diagrams show database objects, the relationships between them, the roles they play, and constraints.

■ **Express-G** diagrams use a notation for modeling data graphically developed as part of the ISO standard, STEP (Standard for the Exchange of Product Model Data), which in turn, is used to define industry foundation classes within the AEC/FM industries. Express-G diagrams help database designers visualize large information models by showing relationships between objects and other components within a data model.

The Express-G and ORM templates are straight diagramming templates. They don't include the Database menu or special database modeling features. You use basic Visio techniques, including dragging shapes onto the page, connecting them, and adding shape data to document and annotate your diagrams.

NOTE   In earlier versions of Visio Professional, you could create a physical database from a Visio model. Since, Visio 2003, this capability is part of Visio for Enterprise Architects, which is part of Visual Studio Professional, *not* Visio Professional. Confusingly, Visio for Enterprise Architects 2003 came out before Visio 2003, so it actually contains Visio 2002. Visio for Enterprise Architects 2005 contains Visio 2003.

You can import logical database models from Visio Professional into Visio for Enterprise Architects, and then use that tool to transform them into physical database schemas or DDL scripts. To learn more about Visio for Enterprise Architects, search `http://msdn.microsoft.com/vstudio` for Visio-based database modeling.

Boning Up on Database Modeling

If you're fairly new to data modeling and database design, trying to learn what you need to do might be a bit overwhelming. In reality, you probably want to learn more about data and databases before you can decide whether Visio Professional has the tools you need. With this knowledge, you can more easily choose the right Visio template and build the diagram you want. The following are some educational resources to help you learn more about data modeling and database models:

- *Data and Databases: Concepts in Practice* by J. Celko (San Francisco: Morgan Kaufmann, 1999) is a good introductory book to database technology.

- *An Introduction to Database Systems, Seventh Edition,* by C. J. Date (Boston: Addison-Wesley, Inc., 2003) describes the fundamentals of data theory.

- Just because documentation is helpful, you don't necessarily need a lot of it. To learn how to produce just the right amount of database documentation, read *Agile Modeling* by Scott Ambler (Indianapolis: Wiley, 2002).

- *Agile Database Techniques* (Indianapolis: Wiley, 2003) by Scott Ambler and one of Scott Ambler's web sites, www.ambysoft.com/agileDatabaseTechniques.html, both introduce the process of data modeling and also discuss database refactoring.

- To learn more about what you need to model physical databases, see www.agiledata.org/essays/umlDataModelingProfile.html.

- *Information Modeling and Relational Databases: From Conceptual Analysis to Logical Design* by T. A. Halpin (San Francisco: Morgan Kaufmann, 2001) is the book about ORM diagramming.

Exploring Database Model Shapes

Although relational and object-relational notations both use entities, columns, views, and relationships, the shapes on the Entity Relationship stencil and Object Relational stencil appear and behave according to the rules of their respective database modeling methods. The Entity Relationship stencil includes an Entity shape to represent tables, a View shape to show combinations of columns assembled from other tables, and a Relationship connector to shows parent-child relationships. In addition, you can use the Category shape to relate multiple child tables to a parent table. Parent to Category and Category to Child connectors link tables to categories and create foreign keys within parent tables.

The Object Relational stencil includes all of the shapes from the Entity Relationship stencil plus a few specific to object-relational modeling. The Table Inheritance and Type Inheritance connectors configure child tables or types to inherit the attributes of a parent automatically. You can nest object-relational tables in a model by using the Type shape to define a type and then assigning it as the data type for a column in another table.

Keeping Visio Database Shapes Up-to-Date

In Visio releases, the Database Model Diagram template implements new shape behaviors frequently — and automatically. When you open a database model diagram, Visio opens the Update Shapes dialog box if it finds older versions of Database Model shapes. Microsoft recommends that you keep shapes up-to-date, but you don't have to do so. Shapes left as they are continue to behave as they did in the earlier version of Visio and include the same shortcut menus as they had in the earlier version.

NOTE   If you don't update shapes when you open a diagram, but change your mind later, don't worry. You can update shapes at any time by choosing Tools ⇨ Add-Ons ⇨ Visio Extras ⇨ Update Shapes.

Creating Database Models

With the Database Model Diagram template, you can choose from three methods for creating database models:

- **Existing database** — If you want to document an existing database, the Reverse Engineer Wizard extracts information from the database and uses it to build a model.

- **Databases documented in other applications** — The Database menu includes an Import command, which brings database models created in other applications into Visio.

- **New database** — If you're starting from scratch, you can build a model by dragging and dropping shapes and connectors onto the drawing page.

Reverse Engineering Existing Databases

The easiest way to build a model of an existing database is with the Reverse Engineer Wizard in Visio Professional. Of course, that's only because the hard work on developing the database is already done. The Reverse Engineer Wizard enables options when they match features provided by the target database management system and then walks you through the steps to extract information from your database. While it's reverse engineering your database, the wizard analyzes the database schema and tells you about any problems it finds in its Output window.

Setting Up Data Sources

The Reverse Engineer Wizard can step you through every aspect of setting up the data source you want to reverse-engineer, but the end-to-end process can tax even the longest attention span. To keep the reverse engineering process as simple as possible, set up database drivers ahead of time from the Database menu and define data sources using the Data Sources (ODBC) administrative tool in the Windows Control Panel.

NOTE   To define data sources with Windows Control Panel tools, choose Start ⇨ Settings ⇨ Control Panel ⇨ Administrative Tools ⇨ Data Sources (ODBC).

Visio Professional provides several default database drivers that work specifically with the Database Model Diagram template. You can use the following drivers or combine one of the generic drivers with an ODBC driver provided by your database vendor:

- Generic OLE DB Provider
- IBM DB2 Universal Database
- Microsoft Access
- Microsoft SQL Server
- ODBC Generic Driver
- Oracle Server

NOTE   In Visio Professional 2007, the INFORMIX Online/SE Server and Sybase Adaptive Service Enterprise drivers are no longer available.

To set up a default Visio driver to work with your database management system, follow these steps:

1. Choose Database ➪ Options ➪ Drivers and select the default driver you want to use.

2. If you want to associate a vendor's ODBC driver with the selected Visio driver, click Setup and do the following:

   a. Select the check box for the ODBC driver you want to use.

   b. To specify other settings, such as the comment style or semicolons at the end of SQL statements, select the Preferred Settings tab.

   c. Click OK when you're done.

3. To configure the properties for the data types in the database, select the Default Mapping tab. For each category in the Category list, choose attributes such as type, size, length, and scale. For example, you can set Text data types to Fixed Length, Variable Length, and so on.

4. On the Default Mapping tab, specify the category to use by default for new columns in tables. In the Default Category Type for Column Creation list, select a category in the drop-down list.

5. Click OK.

Using the Reverse Engineer Wizard

With the Reverse Engineer Wizard, you can specify the information to extract from your database and whether you want Visio to create a diagram for you. To reverse engineer a database, follow these steps:

1. To create a new database model diagram file and present the database modeling tools, choose File ➪ New ➪ Software and Database ➪ Database Model Diagram.

2. Choose Database ➪ Reverse Engineer.

3. On the first Reverse Engineer Wizard screen, shown in Figure 20-1, use one of the following methods to set up or connect to a data source and then click Next.

FIGURE 20-1

When you set up the driver and data source ahead of time, you can simply select them when you run the Reverse Engineer Wizard.

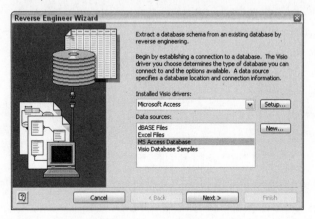

- **Set up a database driver and data source** — Click Setup if you haven't yet configured a database driver for your database management system. If the data source you want doesn't exist, click New. When you create a new data source, you specify whether the data source is a File Data Source, which means you can move it to another computer, or a User Data Source or System Data Source, which means it applies to only your computer. Then, you must select the database file that contains your database.

- **Use an existing driver and data source** — Select the Visio driver appropriate for the database you want to reverse engineer. For example, to reverse engineer a Microsoft Access database, in the Installed Visio Drivers list, select Microsoft Access. In the Data Sources list, select the data source you want to use.

4. If a driver-specific dialog box appears, such as the Connect Data Source dialog box, follow its instructions. Typically, this dialog box appears so you can type a user name and password to access the database.

5. In the Select Object Types to Reverse Engineer screen, shown in Figure 20-2, clear the check boxes for any objects you don't want to extract from the database. By default, all the check boxes are selected to extract everything. Click Next.

FIGURE 20-2

Choose the types of objects you want to extract from the database.

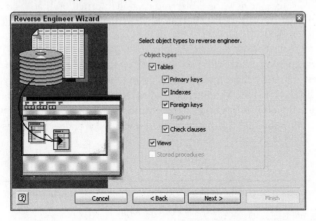

6. The next screen shows the tables and views in the database. Select the check boxes for the tables and views you want to extract. To extract everything, click Select All. Click Next.

7. If you selected the Stored Procedures check box in step 5, select the check boxes for the procedures that you want to extract, or click Select All to extract them all. Click Next.

8. To automatically create a diagram of the reverse-engineered database, on the next screen, shown in Figure 20-3, select Yes, Add The Shapes To The Current Page option and click Next.

FIGURE 20-3

Creating a diagram of the database model is as simple as selecting an option.

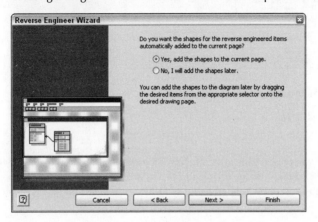

 If you select No, I Will Add the Shapes Later, you can create the diagram by dragging objects from the Tables and Views window onto the drawing page.

9. On the final screen, review the tables and catalog information that Visio will reverse engineer. To change any selections, click Back. To extract the information shown, click Finish.

After the database model is reverse-engineered into Visio, the Tables and Views window, as its name suggests, shows the tables and views extracted from the database, shown in Figure 20-4. Every table and view in the database model shows up in this window, even if they don't appear on the drawing page. If you didn't tell Visio to add shapes to the drawing during the reverse-engineering process, you can drag objects from the Tables and Views window onto the drawing page. And, if you selected the option to build the diagram, the drawing page shows the tables and views.

FIGURE 20-4

You can see the results of reverse engineering in the various Database Model Diagram windows.

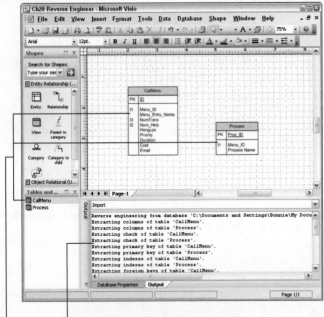

Reverse engineering actions and results

Reverse engineered tables and views

 After you drag a table onto the drawing page, you can display tables related to it by right-clicking it on the drawing page and, from the shortcut menu, choosing Show Related Tables. Visio adds all the related tables to the drawing page and makes the connections between them.

Meanwhile, the Output window summarizes the tasks accomplished during the reverse engineering process. You can select the objects that the process extracted, including tables, attributes, and code. As Visio validates the model, it displays any errors or problems, such as conflicting names, in the Output window. If you plan to correct the identified problems, you can copy the messages in the Output window to another application for reference. Right-click the Output window and choose Copy Message or Copy All Messages.

Updating Reverse-Engineered Database Models

Just because you reverse-engineered a database, it doesn't mean that the database stays the same. As business needs change, so do the databases that support them. At the same time, you have better things to do than to update your Visio database model diagram by hand. Fortunately, Visio can review the physical database and update the Visio model with changes. The Refresh Model command starts a wizard that compares your diagram to the current physical schema. If it finds any changes, you can decide whether to update your diagram.

 **NEW FEATURE** In Visio 2007, the Refresh Model Wizard now recognizes tables added to the database.

To refresh a model, choose Database ➪ Refresh Model. The Refresh Model Wizard displays the additions and changes since you created or updated the model. By default, the Resolution option is set to No Change, so you need to take action only when you want to update the model. To update the model, do one of the following:

- **Update one difference** — Select the item that you want to update and select the Refresh Model option.

- **Update a category of changes** — Select an entire category by selecting the category in the tree structure. Then select the Refresh Model option.

- **Update all differences** — Select the highest level entries and select the Refresh Model option.

TIP You can also mix and match updating and ignoring differences. For example, if you want to update all but one item in a category, select the category and select the Refresh Model option. Then, select the item you want to ignore and select the No Change option.

After you select what you want to update or ignore, click Next to view the changes that the wizard is about to make to your model. If you notice any issues, click Back and correct them. Otherwise, click Finish. Although new tables appear in the Tables and Views window, you have to drag them onto the drawing page to see them in the diagram.

NOTE The Update Database option under the Resolution heading provides compatibility with Visual Studio .NET Enterprise Edition and is always disabled in the Refresh Model Wizard.

Building Database Models from Scratch

The only reason you might turn to Visio to build a database model is to prototype a database design. You don't have an existing database to reverse engineer and you might not want to dive into the details of using a full-blown database design tool. If that's the case, you can create a database model in a new Visio drawing file and, during the process, specify the modeling options you want to use. Here are the basic steps to follow:

1. Choose File ⇨ New ⇨ Software and Database ⇨ Database Model Diagram. Visio creates a new drawing file with a blank page, adds the Database menu to the Visio menu bar, and opens the database-related windows, such as Tables and Views.

2. To specify the notation options you want to use, choose Database ⇨ Options ⇨ Document.

3. To specify the settings for how you want to show the elements of your model, in the Database Document Options dialog box, select the General tab, if necessary.

> **TIP**   If you aren't sure what settings you want, keep the Visio default settings. You can choose Database ⇨ Options ⇨ Document at any time to change the settings when you identify an option you want.

4. Select the symbol set you want to use (IDEF1X or Relational).

5. In the Names Visible On Diagram section, select the names you want to see on the diagram.

6. Select the Table tab and then choose the check boxes for the attributes you want to display for tables. Choose options to specify the order in which keys appear and which data types to show.

7. Select the Relationship tab and specify the notation you want to use and how to display relationship names. By default, Visio shows relationship, but you can add Crow's feet, cardinality, and referential actions.

8. Click OK to apply the settings.

> **TIP**   After you specify settings, you can save them as the default for all new database models by clicking Defaults and then choosing Set As. If you want to restore the settings to their original values, choose Restore Original. Restore reverts settings to those set last for the document.

Importing Database Models from Other Applications

If you modeled a database using ERwin or Visio Modeler, there's not much reason to bring those files into Visio. However, importing ERwin ERX or Visio Modeler IMD files might come in handy, if you don't have access to the software that created the models.

To import an ERwin or Visio Modeler file, create a new database model diagram and then choose Database ⇨ Import. On the submenu, choose either Import ERwin ERX File or Import Visio Modeler .IMD File. In the dialog box that appears, type the path and file name for the file you want to import, or click Browse and navigate to the file, and then click Open. Clicking OK in the Import dialog box begins the import process and displays progress in the Output window. After the import is complete, Visio shows the imported tables in the Tables and Views window.

Working with Database Models

Whether you have a reverse-engineered model or one you built from scratch, you can create and edit the objects in your Visio database model. The tables, columns, views, and relationships are all editable and you can modify a host of attributes such as data type, keys, cardinality, referential integrity, indexes, and extended attributes. Before you launch into a wild session of editing, it helps to understand the various windows that the Database Model Diagram template contains. From then on, you can flit from window to window based on the work you want to do.

 If you change the schema of a database in Visio, you also must update your database.

Working in the Database Windows

The Database Model Diagram template includes several windows that make it easy to access any table or view in your database model and to view or modify database objects. Although some of these windows appear automatically when you use the Reverse Engineer Wizard, choosing Database ⇨ View enables you to open any window you want: Tables and Views, Types, Code, and Output.

By default, Visio anchors the Tables and Views window, Types window, and Code window at the bottom of the Shapes window, which increases the screen area available for your drawing, as shown in Figure 20-5. When more than one of these windows is open at a time, Visio simply adds tabs for each window so you can switch the visible window simply by clicking its tab. For hard-core sessions with a window, floating an individual window in the drawing area might be preferable. To do this, right-click the window's tab and, from the shortcut menu, choose Float Window.

 To float the entire window containing the Tables and Views, Types, and Code tabbed windows, drag the window title bar to a new location.

FIGURE 20-5

Database windows are more numerous than in other templates, but they work the same way.

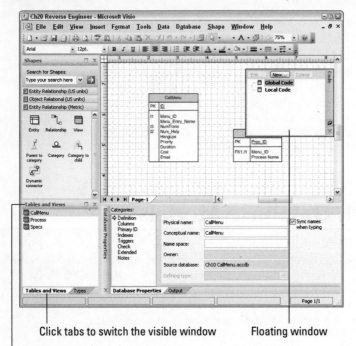

Click tabs to switch the visible window   Floating window

Window docked within the Shapes window to save screen area

The Output window docks in the same window as the Database Properties window by default. You can float either of these windows by right-clicking their tabs and, from the shortcut menu, choosing Float Window.

As you work on database model diagrams, the Database Properties window is likely to be your favorite. This is the window that provides access to the properties associated with objects in your database model. Whether you are just looking or want to modify properties, when you select an object, such as a table, the categories of properties for that object appear in the Database Properties window, as shown in Figure 20-6.

To access the properties for an object, choose one of the following methods:

- **Database Properties window closed** — Double-click the shape on the drawing page.

- **Database Properties window open** — Select the object in the model or on the drawing page.

- **Database Properties window hidden** — Select the object in the model or on the drawing page and then move the pointer over the Database Properties title bar.

FIGURE 20-6

The Database Properties window provides access to all the properties for the selected object.

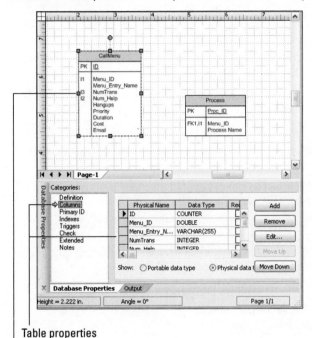

Table properties

Selected table

> **TIP**   When the Database Properties window is floating and you want it to remain open, in the title bar, click the Push Pin to turn off AutoHide.

The Database Properties window displays different categories depending on the object you select in the Tables and Views window or on the drawing page. Here are the properties that appear for different objects:

- **Table categories** — The following categories appear when you select a table in the Tables and Views window or an Entity shape on the drawing page:

    - **Definition** — Specify the physical and conceptual names for the object and whether you want to synchronize the two as you type. You can use name spaces to distinguish similarly named tables in a model. Owner and Source Database specify the owner and file for the source database. You can create typed tables using the composite data types in the Defining Type field.

    - **Columns** — Add, remove, edit, or change the order of columns in a table. You can specify column data types, primary and foreign keys, required fields, and whether to show physical or portable data types.

- **Primary ID** — Add, edit, or remove primary keys in a table. You can specify whether to create an index using the primary key.

- **Indexes** — Create, edit, rename, or delete indexes. You can also specify the type of index you want to create, or, if your database management system supports them, extended attributes.

- **Triggers** — Add, edit, or remove triggers associated with a table. The Code Editor window opens when you click Add or Edit.

- **Check** — Add, edit, or remove check clauses associated with a table. The Code Editor window opens when you click Add or Edit.

- **Extended** — If your database supports extended attributes, set them in this tab.

- **Notes** — You can use this property to add notes about the object or reminders of changes you want to make.

- **View categories** — View objects include Definition, Columns, Extended, and Notes, which are the same as for tables. The following categories are unique to views and appear when you select a view in the Tables and Views window or a View shape on the drawing page:

  - **Join Criteria** — Add or edit the columns that join to create the view and any criteria for the join.

  - **SQL** — Create or edit the SQL statements that create the view.

- **Relationship categories** — The following categories appear when you select a Relationship connector on the drawing page:

  - **Definition** — Specify the parent, child, and foreign key for a relationship.

  - **Name** — Specify the verb phrases to use to describe the relationship, the physical name of the foreign key, and any notes.

  - **Miscellaneous** — Specify or modify the cardinality, relationship type, and whether a child table must have a parent.

  - **Referential Action** — Specify the action to take to check referential integrity when a parent is updated or deleted.

Working with Tables and Columns

Tables in a Visio model appear as Entity shapes on a database model diagram. The appearance of Entity shapes on a drawing page depends on the stencil from which they came and the database options you've chosen. For example, Relational notation shows tables as rectangles with a shaded area at the top for the conceptual table name, whereas IDEF1X notation shows them with rounded corners.

Adding Tables to a Model

Adding a table to your model is easy in Visio. However, you must remember that adding a table represents a change to the database schema. If the model is of an existing database, you must add the table to your database as well. To add a table, follow these steps:

1. Drag an Entity shape from the Entity Relationship or Object Relational stencil onto the drawing page.

2. In the Database Properties window, in the Physical Name box, type the table name. If the Sync Names When Typing check box is selected, Visio updates the Conceptual Name automatically.

Adding Columns to Tables

You can add columns to any table in your model, whether or not it appears on the drawing page or has been reverse-engineered from an existing database. When you use *relational notation,* Visio shows *keys,* which specify uniqueness in tables (see Figure 20-7).

FIGURE 20-7

With relational notation, keys and indexes appear in tables on the drawing page.

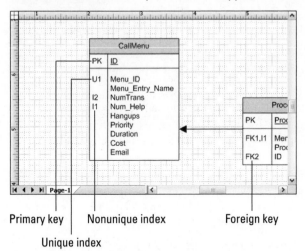

Primary key   Nonunique index   Foreign key

Unique index

To add columns to a table, follow these steps:

1. Select the table in the Tables and Views window (or if it's easy to spot, select the Entity shape for the table on the drawing page).

2. In the Database Properties window, in the Categories list, click Columns and then use one of the following methods to add the column:

   ▪ Click a blank Physical Name field (use the scrollbar to find a blank line, if necessary) and type the name for the column. To add additional columns, press Enter or the Down Arrow key and type the next column name.

   ▪ Click Add. Visio generates a default name for the column using the naming conventions you specified in the Database Modeling Preferences dialog box (described in Setting Modeling Preferences later in this chapter). To change the name, double-click the physical name and type a new one.

3. Select the Portable Data Type or Physical Data Type option to specify the data types you want to use. Portable Data Types are generic data types you can use in any database. Physical Data Types correspond to the data types supported by the database management system you're using.

4. To designate a column as a primary key for the table, select the PK check box for the column you want to act as the primary key.

NOTE   To display primary keys at the top of the table, which makes them easier to spot, choose Database ⇨ Options ⇨ Document, select the Table tab, and then select the Primary Keys At Top option. To display a separator line between the primary keys and other columns, select the Draw Separator Line check box.

5. To specify or change the data type for a column, click the column's Data Type field and select the data type from the list.

NOTE   You can also change the data type when you click Edit to access other column properties.

6. To require values for the column, so the field can't be left blank in a record, select its Req'd check box.

7. To edit other properties of the column, click Edit and select the following tabs to specify column properties:

   ▪ **Definition** — Specify physical and conceptual names and whether you want to synchronize the two as you type. You can specify a default value and whether the default value is a literal or an expression. For optional columns, check the Allow NULL Values check box.

   ▪ **Data Type** — Specify the data type you want in the Data Type box and choose whether you want to use portable or physical data types.

   ▪ **Collection** — Specify whether the column contains a single value, a set of values, an ordered list of values, or values that can include duplicates.

   ▪ **Check** — Specify check clauses for the column.

   ▪ **Extended** — If your database supports extended attributes, set them in this tab.

   ▪ **Notes** — Add notes about the column.

Understanding Keys

A *key* is one or more data attributes that help identify an entity and find it in the database, such as an Employee ID identifies the people that work for your company. Entities should have unique keys to increase the performance of the database, but additional duplicate keys often exist. A key that is made up of more than one attribute is called a *composite key*. A key defined by attributes that already exist in the real world is called a *natural key*. For example, U.S. citizens receive a Social Security Number (SSN) that is unique to them. For a project limited to the United States, the SSN could be used as a natural key for a Person entity, assuming privacy laws allow it and you're comfortable that your security procedures protect against identify theft.

In a logical data model, an entity type has zero or more *candidate keys*, also referred to as *unique identifiers*. For example, for American citizens, the SSN is one candidate key for the Person data entity, and a unique combination of name and phone number is potentially a second candidate key. In a physical data model, candidate keys can act as the *primary key* or an *alternate key* (also known as a *secondary key*), or not act as a key at all. A primary key is the preferred key for an entity type.

There are two strategies for assigning keys to tables. The first is to simply use a natural key, one or more existing data attributes that are unique to a business concept. For example, in a Customer table with two candidate keys, such as `CustomerNumber` and `SocialSecurityNumber`, you could use either candidate key. The second strategy is to introduce a new column to act as a key, which is called a *surrogate key* because it has no business meaning. For example, an AddressID column in an Address table is useful as a surrogate key, because addresses don't have an easy natural key. You would need all of the columns of the Address table to form a key for itself, so a surrogate key is a much better option. The primary advantage of natural keys is that they exist already, so you don't need to introduce a new "unnatural" value to your data schema. However, because they have business meaning, they might change if your business requirements change. If you decide to use surrogate keys, you can use one of the following strategies:

- **Key values assigned by the database** — Most leading database vendors, such as Oracle, Sybase, and Informix, implement a surrogate key strategy called *incremental keys*. Although each strategy uses a similar concept, some assign values uniquely across all tables, whereas others assign values that are unique only within a single table.

- **MAX() + 1** — A common strategy is to use an integer column, starting the first record at 1, and then using the SQL MAX function to set the value for a new row to the maximum value in this column plus one.

- **Universally unique identifiers (UUIDs)** — UUIDs are 128-bit values that are created from a hash of the ID of your Ethernet card or an equivalent software representation, and the current date and time of your computer system. The algorithm for doing this is defined by the Open Software Foundation (`www.opengroup.org`).

- **Globally unique identifiers (GUIDs)** — GUIDs are a Microsoft standard that extend UUIDs, if an Ethernet card exists. If one doesn't, GUIDs hash a software ID and the current date and time to produce a value that is guaranteed to be unique to the machine that creates it.

- **High-low strategy** — With this approach, your key value, often called a *persistent object identifier* (*POID*) or an object identified (*OID*), comprises a unique HIGH value that you obtain from a defined source and an *n*-digit LOW value that your application assigns itself. Each time that the application obtains a HIGH value, the LOW value is set to zero and begins to increment. An implementation of a HIGH-LOW generator can be found at www.theserverside.com.

Reordering Columns

Sometimes, columns in a table don't appear in the right order. For example, after you define the primary key and index columns, you can reorder the columns so that the primary keys appear first. Or, if you create several new columns, you can move them to their place in line. To reorder columns in a table, in the Tables and Views window, select the table, and then, in the Database Properties window, click Columns. Click the column you want to move and then click Move Up or Move Down.

 You can move several columns at once by Shift-clicking the first and last column in a group of contiguous columns and then dragging the columns to a new location in the column list.

Removing Tables

To remove a table from your drawing, follow these steps:

1. Select the table you want to remove on the drawing page and then press Delete.

2. In the Delete Object dialog box, click Yes to remove the table from the model. Click No to remove the table only from the drawing page.

NOTE If the Delete Object dialog box doesn't appear, choose Database ➪ Options ➪ Modeling and select the Logical Diagram tab. Below the When Removing an Object from the Diagram heading, select the Ask User What to Do option.

Reusing Table Attributes by Categorizing Tables

In Visio, categories simplify the creation of several tables of the same type or tables that share the same attributes. Categories contain common columns, the primary key, and the discriminator, which is the column that Visio uses to determine the category to which a table belongs. For example, a Resource table can contain all the columns common to every resource on a project, and a discriminator that uses the `Resource_Type` column. Category tables such as employee and contractor include columns specific to those types of resources, as illustrated in Figure 20-8.

A parent table contains all the common columns and attributes for a category and must include a column for the discriminator. The Category shape links the parent table to the child tables that inherit the columns and attributes. Child tables automatically inherit the primary key from the parent. *Complete categories* include all possible subtypes and are indicated by double lines in Category shapes. Categories that don't include all possible subtypes use single lines.

FIGURE 20-8

A category stores the common columns and other attributes, while child tables inherit columns and attributes from the category.

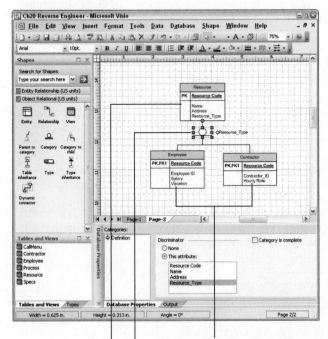

Child tables with columns unique to that type

Single line on Category indicates incomplete category

Parent table with common columns

To set up a category and the child tables that use it, follow these steps:

1. Drag a Category shape onto the drawing page.

2. To link a parent table (the table with the common attributes) to a category, on the Standard toolbar, click the Connector tool.

3. In either database stencil, click the Parent to Category shape.

4. On the drawing page, drag from the parent Entity shape to the Category shape.

5. To link the category to the child tables, with the Connector tool still selected, in a stencil, click the Category to Child connector and drag from the Category shape to a child Entity shape.

6. Repeat step 5 for each child table that uses that category.

7. To specify the properties for the Category, double-click the Category shape on the drawing page. In the Database Properties window, select the Category Is Complete check box if the category represents all subtypes. To specify the column that acts as the discriminator for child tables, select the This Attribute option and choose the column in the list.

Creating Additional Data Types

Each column has an associated data type that determines the kind of information you can store in that column. In the Database Properties window, the Portable Data Type option specifies data types that are independent of the specific database management system you're using. However, if you are building a diagram for a specific type of database or have reverse-engineered a database, you can choose the Physical Data Type option to use the data types that apply to that database management system.

 The Types window lists built-in logical data types for your target database, composite types, and portable data types you create within the User Defined Data Types dialog box.

You can define your own data types, including composite data types, which contain sets of several data types. Although they appear as distinct elements in a diagram, you can optimize the performance of your database by storing several values as one element. You can also use composite data types to create typed tables and views.

To create a user-defined data type, follow these steps:

1. Choose Database ⇨ User Defined Types

2. In the User Defined Types dialog box, click Add.

3. In the Add New User Defined Type dialog box, type the name for the new data type. If you want to base the new type on an existing user-defined data type, check the Copy From check box and then select the data type in the list.

4. Click OK to close the Add New User Defined Type dialog box.

5. Back in the User Defined Types dialog box, specify the data type category, data type, size, length, and scale as necessary. If you want to add a description of the data type, type text in the Description box.

6. Click OK when you're done.

To create a composite data type, follow these steps:

1. From the Object Relational stencil, drag a Type shape onto the drawing page.

2. Double-click the Type shape and then click Definition in the Categories list in the Database Properties window.

3. Type the name you want for the composite data type in the Name text box and choose one of the Composite Type options.

4. To specify other properties for the composite type, click a category in the Database Properties window and specify the properties and settings you want.

Defining Database Views

Database views provide more than a pretty picture of your data. By using views, you can assemble information from several tables without modifying the structure of your underlying database. Views are also helpful for manipulating data and controlling access to information, or encapsulating access to database tables, as described in *Agile Database Techniques* by Scott Ambler (Indianapolis: Wiley, 2003). In Visio, you can create views by dragging View shapes from the Entity Relationship or Object-Relational stencils onto the drawing page. When you do this, Visio automatically creates the SQL code to define the view.

To create a view, follow these steps:

1. Drag a View shape onto the drawing page.
2. In the Database Properties window, in the Physical Name box, type the view name.
3. In the Categories list, click Columns and then use one of the following methods to add columns to the view:
   - Click the Physical Name field in a blank line and type the name you want for the column. To add additional columns, press Enter or the Down Arrow key and type the next column name.
   - Click Add. Visio generates a default name for the column using the naming conventions you specified in the Database Modeling Preferences dialog box. If you want to change the name, double-click the physical name and type a new name.
4. To specify the source for the column, click the name of the column and then click Edit.
5. In the View Column Properties dialog box, select the Source tab, click the Known Column in Another Table or View option, and then click Change.
6. In the Pick a Column dialog box, select the column from the list of tables and views in your database model and click OK.

 You can also choose the Derivation Rule option and then specify how to derive the information you want to display in the column.

7. Modify any of the other settings you want in the View Column Properties dialog box and then click OK. Repeat steps 4 through 6 to edit each column added in step 3.

Creating Relationships Between Tables

In Visio, you use the Relationship connector to create parent-child relationships. When you create a relationship between two tables, the child table receives the foreign key attributes of the parent. The properties for a Relationship connector define the parent and child in the relationship, the key you want to use to join the tables (in case the primary key isn't what you want), the referential integrity rules, and the optionality and cardinality of the relationship.

To add a relationship between two tables and specify the relationship's properties, follow these steps:

1. On the Standard toolbar, click the Connector tool.

2. Position the Connector tool over the parent Entity shape. When Visio outlines the parent Entity shape in red, drag to the center of the child Entity shape and, when the child Entity shape is outlined in red, release the mouse button. Visio changes the Relationship connector end points to red and displays the primary keys for the parent Entity shape as foreign keys in the child Entity shape.

3. To specify the properties for the relationship, double-click the Relationship connector on the drawing page.

4. To specify the cardinality of the relationship, in the Database Properties window, in the Categories list, click Miscellaneous. Choose one of the cardinality options.

5. To specify the referential integrity actions, in the Categories list, click Referential Action and then select options to specify the actions to perform when a parent is updated or when a parent is deleted. You can choose from No Action, Cascade, Set NULL, Set Default, and Do Not Enforce.

Creating and Editing Indexes

Database management systems use indexes to speed up searching and sorting the records in databases. You can enhance the performance of a database by defining indexes for columns you plan to search frequently. In the Database Properties window, you can create and edit indexes for columns or modify indexes extracted during the reverse engineering process.

NOTE When Visio extracts indexes during reverse engineering, it automatically applies uniqueness constraints to primary keys, unique indexes to alternate keys in an IDEF1X model, and non-unique indexes for inversion entries in an IDEF1X model.

To create a new index for a table, follow these steps:

1. In either the Tables and Views window or on the drawing page, select the table that you want to index.

2. In the Database Properties window, in the Categories list, click Indexes and then click New.

3. In the Create Index dialog box, type a name for the index and click OK.

4. To specify the type of index, in the Index Type drop-down list select a type, such as Unique Index Only.

5. In the Available Columns list, select each column you want to include in the index and click Add.

NOTE Shift+click or Ctrl+click to select multiple columns for the index.

6. In the Indexed Columns list, uncheck the ASC check box if you want the index to use descending sort order. The Disp. Name field represents the index notation that Visio displays on the database model diagram.

 If you want to specify extended attributes for a database management system that supports them, click Options.

Editing Database Code

The Code window lists the code that is associated with your model, including code extracted during reverse engineering of a database. From the Code window, you can view the code for your model, write new code, and edit or delete existing code. In the Database Properties window, when you click the Check and Triggers categories, respectively, check clauses and triggers for tables are also available.

Chances are, if you know enough to write code for a database, you won't do it in Visio. However, if the mood strikes you while you are looking at a database model in Visio, it's a good idea to use a mirror file to store your code. A *mirror file* is a separate file that you can save, access outside of Visio, and manage using your source code control application. To specify a mirror file for code, in the Code Editor window, select the Properties tab and, in the Mirror File File Name box, type the path and file name.

Code listed in the Global Code category includes stored procedures, functions, and other platform-specific types of data definition language code. Local code includes triggers and check clauses specific to a table or column in a model. Entries in the Code window include the name of the table or view that uses the code. You can work with the following types of database code:

- **Check clauses** — To define check clauses for a table, select the Entity shape that represents the table and then, in the Database Properties window, click the Check category. Click Add to open the Code Editor. You can specify the check clause name on the Properties tab and type the SQL statements in the box on the Body tab. Check clauses appear in the Local Code list in the Code window.

- **Stored procedures** — To create a stored procedure, in the Code window, click Global Code and then click New to open the Code Editor dialog box. On the Properties tab, type the name of the stored procedure. Select one of the Stored Proc, Function, or Raw DDL options. Type the SQL statements on the Body tab.

 If the Code window isn't open, choose Database ➪ View ➪ Code.

- **Triggers** — To create a trigger, select the Entity shape that represents the table to which you want to apply a trigger and then, in the Database Properties window, click the Triggers category. Click Add to open the Code Editor. You can specify the trigger name on the Properties tab and type the SQL statements in the box on the Body tab. Triggers appear in the Local Code list in the Code window.

- **View SQL Code** — Select the View shape whose SQL code you want to edit and, in the Database Properties window, click the SQL category. Clear the Auto-Generated check box and edit the code.

Table 20-1 lists several tools and shortcuts that the Code Editor offers. To specify settings for editing code, in the icon bar within the Code Editor window, click the Window Properties icon (which looks like a tab) to open the Window Properties dialog box.

TABLE 20-1

Code Editing Tools in the Code Editor Window

Editing Task	Visio Method
Insert code skeleton	To insert a skeleton for the type of code you are editing, on the Code Editor toolbar, click the Insert Code Skeleton icon, which looks like a magic wand.
Print the code for an item	On the Code Editor toolbar, click the Print icon.
Change code keyword colors	To highlight code keywords in different colors, in the Code Editor toolbar, click the Window Properties icon, select the Color/Font tab, select Keywords in the Item list, and select the color you want to use. You can specify colors and font styles for several other code elements, such as line numbers and comments.
Automatically indent lines	To specify how the Code Editor indents lines, in the Code Editor toolbar, click the Window Properties icon, select the Language/Tabs tab. Select the Follow Language Scoping option and select the language to which the scoping rules should be applied. Select Copy from Previous Line to use the same indentation as the previous line.
Assign keyboard shortcuts	To define keyboard shortcuts for frequently used commands, in the Code Editor toolbar, click the Window Properties icon, select the Keyboard tab, select a command, and assign a keyboard shortcut.
Number lines automatically	In the Code Editor toolbar, click the Window Properties icon, select the Misc tab, select the Numbering style you want to use in the Style drop-down list, and type the starting line number in the Start At box.

Setting Database Options and Preferences

The Database Model Diagram template includes a passel of settings and preferences that control how database shapes look and behave on a drawing as well as within the database model. For example, database document options control the type of notation you use and whether one-to-many relationships appear as crow's feet. For the database management systems that support them, extended attributes enable you to fine-tune object definitions in your model. Visio applies default options to new diagrams you create, which make it easy to get started with your diagram. However, when you identify specific settings you want, you can change the options and save them for future Visio database models.

Setting Modeling Preferences

Database modeling preferences control shape behavior and naming. Visio uses the settings you choose for each new database model diagram you create until you change the settings again. To specify modeling preferences, choose Database ➪ Options ➪ Modeling. In the Database Modeling Preferences dialog box, select the Logical Diagram tab to specify the following preferences:

- **When Removing an Object from the Diagram** — The options under this heading enable you to remove objects from the model as well as the drawing page, remove objects from the drawing page only, or, possibly the safest option, ask the user what to do.

- **Show Relationships After Adding Table to Diagram** — Select this check box to show relationships for new tables you add to a diagram.

- **Show Relationships After Adding Type to Diagram** — Select this check box to show relationships for new types you add to a diagram.

- **Sync Conceptual and Physical Name in New Tables and Columns When Typing** — Select this check box if you want changes you make to a name in one name field in the Database Properties window to propagate automatically to the other field.

Select the Logical Misc tab to specify the following preferences:

- **FK Propagation** — Select Propagate on Add if you want Visio to create a foreign key relationship between parent and child tables when you connect them with a Relationship connector. Check Propagate on Delete if you want Visio to remove a foreign key relationship when you delete a Relationship connector.

- **Name Conflict Resolution** — Select the action you want Visio to take when you add a foreign key that uses the same name as a column in the child table. For example, Auto-Rename automatically assigns a new name to the foreign key, whereas Ask prompts the user about what to do. You can also allow duplicate names or disallow them completely.

- **Default Name Prefixes** — Specify the prefix you want to use in the default conceptual name for objects in a database model. Visio initially sets these prefixes to abbreviations of the object type, but you can set any prefix you want.

- **Default Name Suffixes** — Specify the suffix you want to add to the default prefix of the default conceptual name for objects in a database model.

- **FK Name Generation Option** — Specify the objects to use when automatically generating foreign key names. In the FK Name Generation Option list entries, Suffix represents the default suffix for foreign keys that you specify in the Default Name Suffixes section.

Specifying Notation and Other Display Options

The appearance of shapes in a database model diagram is set in part by the database notation type. For example, IDEF1X notation shows a table as a rectangle with rounded corners with the conceptual name above the rectangle. Relational notation shows the conceptual name in a shaded section at the top of the table rectangle. Visio database options include other display settings, such as the level of detail you want to show and whether you use crow's feet to show relationships. Choose Database ➪ Options ➪ Document and then use one of the following methods to specify notation and display options:

- **Notation** — To specify IDEF1X or relational notation, select the General tab, under the Symbol Set heading, select either IDEF1X or Relational. The preview pane on the right shows a sample of the notation.

- **Names Visible on Diagram** — Choose an option to specify the names that appear on the diagram. The option Based On Symbol Set shows names based on what is in the shapes you use. However, you can show conceptual names, physical names, or, for diagrams with lots of whitespace, both types of names.

 You can switch between IDEF1X and relational notation whenever you want. If you collaborate with someone who prefers a different notation, you can change the notation each time you receive a copy of the file.

- **Crow's Feet** — To specify how relationships are shown, select the Relationship tab. To use crow's feet to show one-to-many relationships, select the Crow's Feet check box. With the box cleared, the relationship connector is a single line.

- **Cardinality** — When the Crow's Feet check box is cleared, you can select this check box to display cardinality notation on Relationship connectors.

- **Referential Integrity** — Select the Relationship tab and choose this check box to display symbols on Relationship connectors to indicate referential integrity constraints.

- **Level of Detail** — Click the Table tab and select or clear check boxes to specify the information you want to display on your diagram. For example, you can show keys, indexes, annotations, data types, and IDEF1X optionality.

Creating Express-G and ORM Diagrams

Express-G and ORM are notations designed to convey special types of information. For example, Express-G notation is specifically designed to document product data so it can be interpreted and exchanged via computer. Object Role Modeling (ORM) captures business rules and describes them in terms of real-world objects and the roles they play in processes. ORM diagrams provide documentation for these rules so you can design databases to support them.

NOTE You can build ORM models that you can engineer into databases using Visio for Enterprise Architects.

Using Express-G to Create Entity-Level and Schema-Level Diagrams

The Express-G stencil contains shapes and connectors to represent entities and relationships for Express-G diagrams. Because this template builds diagrams not models, constructing an Express-G diagram is a matter of dragging shapes from the Express-G stencil onto the drawing page. To build an Express-G diagram, choose File ➪ New ➪ Software and Database ➪ Express-G. Visio creates a new letter-size drawing and opens the Express-G stencil.

The shapes on the Express-G stencil include shape data for attributes. The Shape Data dialog box appears whenever you drop one of these shapes onto the drawing page. For example, the Base Types shape includes a Data Type shape data property that sets the type to Binary, Boolean, Integer, and so on. In addition, once the shapes are on the drawing, you can right-click them and, from the shortcut menu, choose Set Data Type to change the value.

> **NOTE** If your diagram is larger than one page, the To-page Reference and From-page Reference shapes are perfect for navigating to the continuation on another drawing page. You can connect these reference shapes to Entity shapes and data shapes and add hyperlinks to jump between pages.

In Express-G notation, foreign schemas are represented by USED Entity or REFERENCED Entity shapes. To further specify the schema details, right-click the shape and, from the shortcut menu, choose Set Schema Details.

As with other database diagrams, connectors represent relationships. You can specify or change the type of relationship by right-clicking the connector, from the shortcut menu, choosing Set Attributes, and typing or selecting values in the shape data fields.

> **NOTE** If you want to view only entities and data shapes on an Express-G diagram, without the spaghetti of relationships, you can assign your Express-G entities and relationships to different layers and then turn off the visibility of the relationship layer. To assign shapes to a layer, select the shapes you want, choose Format ➪ Layer, select the layer to which you want to assign the shapes, and then click OK.

Creating Object Role Model Diagrams

As with most methodologies and notations, you can't fake your way through them. It's easy to create a Visio drawing file for an ORM diagram, but connecting the shapes in the right way requires specialized knowledge. To learn more about the uses and benefits of ORM diagrams, see the books and online resources provided in the "Exploring the Database Model Templates" section earlier in this chapter. Then, armed with your newfound expertise, you can create an ORM diagram by choosing File ➪ New ➪ Software and Database ➪ ORM Diagram.

Many of the shapes on the ORM Diagram stencil are quite basic. In many cases, they don't even have shape data. Here are a few of the special behaviors in these shapes.

- **Subtypes** — To indicate that one entity type is a subtype of another, glue a Subtype connector to the two Entity shapes with the arrowhead on the connector pointing to the subtype Entity shape.

- **Relationships, roles, or facts** — In ORM, predicates containing one or more roles indicate relationships between entity types or between entity types and value types. To show relationships and roles, drag a Predicate shape with enough role boxes to relate all the associated entity types on the drawing and then glue Role connectors between each Entity shape and a role box on the Predicate shape:

  - **Unary** — Includes only one role.

  - **Binary** — Indicates relationships or roles between two entity types or between an entity and a value. This is the most commonly used predicate.

  - **Vertical Binary** — Shows a binary relationship but is oriented vertically.

  - **Ternary** — Indicates relationships between three entities.

  - **Quarternary** — Indicates relationships between four entities.

> **NOTE**   To add the names for the roles that an entity type plays, double-click the Predicate shape, place the insertion point between the ellipsis in the text block, and type the name for the role.

- **Nested entities** — To indicate nested entities in an ORM diagram, use the Rectangle tool to draw a rectangle around the Predicate shape you want to designate as an objectified predicate. Select the rectangle, choose Format ➪ Corner Rounding, click the third rounding option in the top row, and click OK.

- **Mandatory roles** — Mandatory roles mean that every member of an entity type must play that role, so null values are invalid in the relationship. To indicate that a role is mandatory, glue the Mandatory Role connector to the Entity shape and the Predicate shape.

- **Uniqueness** — To show that a role is unique, drag the Uniqueness Constraint shape onto the drawing page and place it directly above or below the Predicate shape you want to constrain. The Uniqueness shape is sized so that it can be glued to the connection points above a role in a Predicate shape.

- **Frequency** — To show that each instance of a role occurs a specific number of times, drag the Frequency Constraint shape onto the drawing page and place it near the Predicate shape you want to constrain. With the Frequency Constraint shape selected, type the number of times the role occurs.

- **Subset or Equality constraints** — To indicate a subset or equality constraint, drag the Subset Constraint or Equality Constraint shape onto the drawing page and place it between the Predicate shapes whose roles you want to constrain. Glue one end point of the Constraint shape to a role box on one Predicate shape. Glue the other end point of the Constraint shape to the corresponding role box on the other Predicate shape.

- **Ring constraints** — To show that an entity type plays two roles in a predicate, drag a Ring Constraint shape onto the drawing page near the Predicate shape with the two roles you want to constrain. Depending on whether the Predicate shape includes two or more roles, use one of the following methods to indicate a ring constraint:

  - **Two roles** — For a binary predicate, you show only the constraint type with no line between the roles. Right-click the Ring Constraint shape and choose Format ⇨ Line from the shortcut menu. Click None in the Pattern drop-down list and click OK.

  - **Three or more roles** — Glue the end points of the Ring Constraint shape to the roles you want to constrain in a Predicate shape.

NOTE To designate the type of ring constraint, double-click the Ring Constraint shape and type a two-letter abbreviation. Use *ir* to indicate an irreflexive constraint or *as* to indicate an asymmetric constraint.

- **External constraints** — To show an external constraint, drag one of the External Constraint shapes (Ext. Freq., Ext. P, Ext. Mand., or Ext. Uniq.) onto the drawing page near the Predicate shape whose roles you want to constrain. Glue a Constraint connector to the role box on the Predicate shape and a connection point on the External Constraint shape.

Summary

Database templates are available only in Visio Professional. You can draw Express-G and ORM database diagrams using basic Visio techniques. If you use the Database Model Diagram template, you can create both a diagram and model for your database. You can also reverse engineer existing databases into Visio models. In a Visio model, you can create additional objects, modify the properties of objects, and update the physical database with those changes.

Chapter 21

Building UML Models

eveloping a software architecture and design before you start writing code is an important step toward meeting the software system's requirements and making customers and end users happy. With upfront planning, software is much easier to develop and maintain in the long run. In modeling a software system, you progressively construct the details of that system, alternately decomposing high-level objectives and broad requirements into manageable pieces, and then assembling software components into packages and eventually a complete run-time system. Models and diagrams help you visualize both high-level architecture and low-level components, so you can make the most of design opportunities or spot potential problems early on.

The Unified Modeling Language (UML) is a popular and comprehensive approach to modeling software systems through each phase of the software development life cycle. The Object Management Group (www.omg.org), which is a not-for-profit consortium, produces and maintains specifications for interoperable enterprise applications including UML. If you discover that the Visio UML tools fall short of your expectations, OMG publishes a list of available UML tools at www.uml.org/#Links-UML2Tools.

The UML Model Diagram template in Visio Professional includes tools and stencils for building UML 1.3-compliant diagrams. (The most recent specification is UML 2.0.) Although the Visio UML shapes come with significant built-in sets of shape data properties for designating characteristics, the Visio UML template has its limitations. As with many complex methodologies, you have to understand UML to build meaningful UML diagrams. Visio is a

THIS CHAPTER

Choosing the right UML diagram

Working with the Model Explorer and other UML windows

Organizing models with packages

Working with shapes and model elements

Creating UML models

Creating different types of diagrams in UML models

Reverse-engineering source code into UML models

421

diagramming tool, not a tutorial on UML. Furthermore, although the UML template is awash with shape data, properties, and options, getting your UML information back out of a Visio UML model isn't always possible.

 The UML Model Diagram template is available only in Visio Professional and Visio Studio .NET Enterprise Architect.

This chapter introduces the UML Model Diagram template and the tools and stencils it offers. You learn how to use the UML modeling tools that Visio provides. In addition, you will learn how to work with each type of UML diagram in the template, and examine the differences between the Visio UML shapes and terminology and those used in the current UML 2.0 specification.

Learning More About UML

If you are new to UML and don't know where to start or which Visio diagram you should use, consider the following resources to increase your knowledge:

- *The Object Primer, 3rd Edition: Agile Model Driven Development with UML 2.0* by Scott Ambler (New York: Cambridge University Press, 2001). This book is a distillation of software development practices and provides a comprehensive description of all 13 UML diagrams in addition to other critical models.

- *UML Distilled, Third Edition,* by M. Fowler (Boston: Addison Wesley Longman, 2003). This book provides a very good overview of UML, along with helpful descriptions of the most common UML diagrams.

- *Applying UML and Patterns: An Introduction to Object-Oriented Analysis and Design and the Unified Process* by C. Larman (Upper Saddle River, N.J.: Prentice-Hall, 2001). This book uses UML in great detail to describe object-oriented analysis and design.

- www.agilemodeling.com/essays/umlDiagrams.htm provides a good overview of the UML 2.0 diagrams. Think of it as a free, online version of UML Distilled.

- *Elements of UML Style* by S. Ambler (Cambridge University Press, 2002). This book provides UML style guidelines that help you develop readable, high-quality diagrams.

CROSS-REF As the industry-standard language for specifying, constructing, and documenting software systems, UML covers a lot of ground. To learn more about this methodology and its notation, search for UML at http://msdn.microsoft.com or go to the Object Management Group's web site, www.omg.org. For more information about UML diagrams, see www.agilemodeling.com/essays/umlDiagrams.htm.

Exploring the UML Model Diagram Template

Large or complex software systems demand a coordinated development approach in which teams collaborate on each phase of a system's life cycle. The UML methodology and the UML Model Diagram template in Visio Professional provide tools to model and document the phases of system design and development. In addition to constructing models for new software systems, the UML Model Diagram template can help you model existing systems by reverse-engineering projects created in several Microsoft programming languages and generating UML static structure models for those projects.

The UML Model Diagram template includes stencils for several types of UML diagrams that document different aspects and phases within development projects. Although this chapter focuses on the basic approach for specifying properties, the shapes on the UML stencils include dozens of properties for the different attributes and conditions that UML supports. You can double-click any shape on a drawing page or an element in the Model Explorer window to open its UML Properties dialog box. If you're not sure what a property does, click the Help button in the lower-left corner of the dialog box.

One Visio drawing file can contain all the models and diagrams for a software system, whether you build and work on models in the hierarchical view of the UML Model Explorer window or through shapes on drawing pages. The UML Model Explorer window is a handy tool for navigating to the diagram or component you want.

CAUTION Visio is far from a perfect solution for producing UML models. Its UML templates do not fully support the entire UML notation and it isn't compliant with the current UML specification. The Visio UML template doesn't support Composite Structure diagrams, Interaction Overview diagrams, Object diagrams, and Timing diagrams for the UML 2 specification. Depending on your system, these discrepancies could be inconsequential or the death knell for Visio as your UML tool.

TIP Sandrila (www.sandrila.co.uk/), a company based in the United Kingdom, sells a Visio add-in that supports UML 2.0.

Choosing the Right UML Diagram

Within the Software and Database template category, Visio Professional provides a lone template for UML: the UML Model Diagram template. When you create a new drawing file with this template, Visio opens a bevy of stencils — each one for a different type of UML diagram. In the Shapes window, the stencils appear in alphabetical order. Table 21-1 identifies the stencils in the order you typically use them as you proceed through phases in your project.

TABLE 21-1

UML Diagrams and Visio UML Stencils

UML Diagram	Stencil	Description
Use Case Diagram	UML Use Case	Use cases identify how external actors interact with your system to analyze system usage requirements. The stencil shapes represent use cases, actors, and their interrelationships.
Class Diagram (the name of the Visio stencil doesn't match the UML diagram)	UML Static Structure	This diagram shows the classes of a system, including their operations, attributes, and inter-relationships. It's typically used for several purposes, including the exploration of domain concepts in the form of a domain model, the analysis of requirements, or the presentation of the detailed design of object-oriented software.
Package Diagram	UML Static Structure UML Use Case	Shows how related elements in a software system are grouped into packages. Commonly used to group classes or use cases.
Activity Diagram	UML Activity	Similar to flowcharts and data flow diagrams in structured development, these diagrams show high-level business processes, including data flow, or help model complex logic within a system.
Statechart Diagram (called a State Machine Diagram in UML 2)	UML Statechart	This diagram shows the dynamic behavior of entities in response to events based on their current state and help explore complex behavior of classes, actors, subsystems, or components. Also used for modeling hardware and real-time systems.
Sequence Diagram	UML Sequence	Shows the classifiers, such as classes, objects, components, and use cases that participate in an interaction, and the sequence and timing of events that they generate. The diagram helps explore potential usage of a system, to determine the complexity of classes, and to detect potential bottlenecks in an object-oriented system.
Collaboration Diagram (called a Communication Diagram in UML 2)	UML Collaboration	This diagram typically depicts how objects communicate. It shows instances of classes, their interrelationships, and messages exchanged during an interaction.

UML Diagram	Stencil	Description
Component Diagram	UML Component	This diagram describes the components within an application, system, or enterprise, as well as component interrelationships, interactions, and public interfaces. Also used as a high-level architecture model.
Deployment Diagram	UML Deployment	Shows the structure of deployed systems and the configuration and deployment of hardware and software components. Also used as a high-level architecture model.

TIP If you would like to see examples of each of the UML diagrams, you can download a sample file from Microsoft Office Online. Choose Help ↪ Microsoft Office Online. On the Microsoft Office Online web site, click the Downloads tab. In the Download search box, type **Visio sample** and click Search. Click the link for Visio 2003 Sample: 20 Sample Diagrams and follow the instructions for downloading and installing the file on your computer. After you extract the sample VSD files, you can open them as you would any Visio drawing file. Visio 2007 updates the diagram to incorporate changes to the UML shapes.

Exploring the UML Menu

When you use the UML Model Diagram template to create a new Visio drawing file, the UML menu appears on the Visio menu bar. Unlike the Database menu, which appears when you use the Visio Database Model Diagram template, the UML menu doesn't include any wizards. It provides commands for defining software system elements such as models, packages, stereotypes, and events; displaying the UML template windows; and specifying options for the appearance and behavior of elements in your models and diagrams. Choose UML on the menu bar and then choose the following commands on the UML menu:

- **Models** — Opens the UML Models dialog box, in which you can create or modify models for your system.

- **Stereotypes** — Opens the UML Stereotypes dialog box, in which you can create or modify stereotypes to extend UML functionality.

- **Packages** — Opens the UML Packages dialog box, in which you can create or modify packages for your system. This dialog box lists the packages for the different data types within your system, not packages that you create within models.

- **Events** — Opens the UML Events dialog box, in which you can create or modify events within any of the packages in your system. This dialog box lists all packages in your system in the Packages list box. To add an event, select the package to which you want to add the event and then click New.

425

- **View** — From the submenu, choose Model Explorer, Properties, or Documentation to toggle between hiding and displaying those windows.

- **Options** — Opens the Options dialog box, in which you can specify options for default behavior for shapes, packages, object names, and more in the UML Model Diagram template.

> **CAUTION**   If Visio encounters an unexpected error from either an internal or external source, your template-specific menus, such as the UML menu, can disappear. If Visio shuts down unexpectedly, restart it and then choose Tools ➪ Options. Select the Advanced tab, check the Enable Automation Events check box, and click OK. Save any open drawings, exit and restart Visio, and then reopen your UML model drawing file.

Updating UML Shapes

When you open a UML model drawing file, the Update Shapes dialog box appears if Visio finds older versions of UML shapes on the drawing. Microsoft recommends that you keep shapes up-to-date, but you don't have to. If you choose not to update shapes, they continue to behave as they did in the earlier version of Visio and include the same shortcut menus they had in the earlier version.

> **NOTE**   If you don't update shapes when you open a diagram, you can update them later by choosing Tools ➪ Add-Ons ➪ Visio Extras ➪ Update Shapes.

Working with UML Models

UML models act like blueprints for the software system you want to develop. Whether you work in the Model Explorer window or with shapes on drawing pages, you can use UML models and diagrams to develop your system from requirements through deployment.

Working with the Model Explorer

One Visio drawing file is all it takes to represent an entire software system, although that drawing file is typically chock-full with multiple models and dozens of diagrams. The Model Explorer window displays the elements of a system as a hierarchy with folders and subfolders for the models and packages that comprise the software system as well as elements for components and diagrams that present different views of the system. In addition to viewing and navigating a system, the UML Model Explorer includes commands for creating diagrams, adding elements, and applying properties to those elements.

Visio adds several elements to a new model by default. As you work, you can create additional elements and organize them within the model hierarchy, as shown in Figure 21-1.

FIGURE 21-1

The hierarchy of the Model Explorer Window simplifies finding and working on model elements.

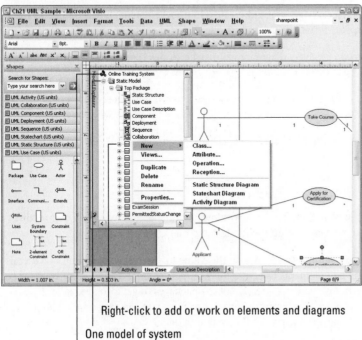

Right-click to add or work on elements and diagrams

One model of system

System being modeled

Here is a guide to the elements you find in the Model Explorer hierarchy:

- **System** — The top node of the hierarchy represents the system you're designing or documenting.

- **Static Model** — By default, Visio creates a static model, which contains all the packages, elements, and diagrams for a single software system model.

- **Top Package** — By default, for every static model in the system, Visio creates a top package, which contains all the elements and diagrams for that model. Right-clicking the top package displays a shortcut menu with commands for adding new elements to the package, renaming the package, or specifying its properties.

- **Package** — Organize model elements by creating additional packages within a model, including subpackages nested underneath higher-level packages. The shortcut menu for packages includes commands for adding new elements to the package, duplicating, renaming, or deleting the package, or specifying its properties.

- **Static Structure-n** — By default, for each model in the system, Visio creates a Static Structure diagram within the top package. This diagram corresponds to a drawing page of the same name in the Visio drawing file.

- **Elements** — Within a package, you can create additional elements, including subsystems, classes, interfaces, data types, actors, and use cases.

 The shortcut menu for every diagram in a model includes commands to open the diagram, rename it, or delete it from the model.

- **Data Types** — By default, Visio includes packages for common data types, including C#, C++, IDL, and Visual Basic. You can also create packages for your own data types.

 You can't delete the built-in Data Type packages, because the UML Add-on tool uses them.

Using the UML Model Windows

The UML Model Diagram template includes three windows that make it easy to view and modify UML model elements and to navigate within the models and diagrams for your system:

- **The Model Explorer window** — Shows all the elements in a software system in a hierarchical tree. You can add elements or access their properties from this window. In the Model Explorer Window, you can collapse all the packages to get a good idea of the overall architecture of your system. From the other direction, you can expand the hierarchy of packages to look at one specific part of the system in detail.

 To navigate to the drawing page for a diagram, in the Model Explorer window, double-click the diagram name. If you want to display the drawing page that contains an element such as an actor, right-click the element, from the shortcut menu, choose Views, select the diagram in the UML Diagram list, and then click OK.

- **The Properties window** — Shows the main properties associated with the selected element for reference only. To edit element properties, right-click the element in the Model Explorer window and, from the shortcut menu, choose Properties. Although this window does not appear by default, open it at any time by choosing UML ⇨ View ⇨ Properties.

- **The Documentation window** — Displays the documentation tagged value of the element you select on the drawing page or in the Model Explorer window. You can add documentation to the selected element by typing text in this window. Although this window does not appear by default, open it at any time by choosing UML ⇨ View ⇨ Documentation.

When you open a UML drawing file, the Model Explorer window appears by default and docks on the left side of the screen. You can hide or show each of the UML windows by choosing UML ⇨ View and then choosing the name of the window you want to hide or display. Visio anchors the Properties and Documentation windows within the Model Explorer window when you first display them and adds tabs for each window so you can switch among them.

You can also float each window individually by right-clicking its tab and, from the shortcut menu, choosing Float Window. If you want to float the entire window with all the UML windows in it, drag the window title bar to a new location.

Organizing Models with Packages

Similar to the work breakdowns that project managers use to get a handle on the work in a large project, each model you create decomposes into a hierarchy of packages. For large or complex software models — or for people who have trouble juggling one tennis ball, packages organize models and diagrams into more manageable pieces.

By default, each model in Visio includes a top package, which is like a cargo ship that holds all the elements, packages, and diagrams you create within that model. Additional lower-level packages organize elements into smaller groups, as a cargo ship holds the containers that trucks eventually transport and the containers hold shipping crates and boxes.

CROSS-REF For more information about when to use Package diagrams and how to make the most of the Package diagrams you develop, see `www.agilemodeling.com/artifacts/ packageDiagram.htm`.

Each element in a model belongs to only one package. However, a package can hold any kind of element, including other packages, so you can partition the elements in your model any way you want. Packages work as organizers even for diagrams when they begin to multiply. Use the following methods to package the contents of your model:

- **Create a package** — In the Model Explorer window, right-click a package and, from the shortcut menu, choose New ➪ Package. Visio adds a package icon underneath the package you right-clicked.

- **Add a package to a diagram** — Display the drawing page you want to package. From the UML Static Structure, UML Use Case, UML Component, or UML Deployment stencil, drag the Package shape onto the drawing page. Visio adds a package icon to the model with the diagram within the package.

- **Create a diagram from a package** — To automatically create a new diagram whenever you create a package in your model, choose UML ➪ Options and select the UML Add-on tab. Make sure the Create a Diagram Page When a Package or Subsystem Shape Is Added to a Document check box is selected.

- **Show package contents in a diagram** — In the Model Explorer window, right-click the package whose contents you want to show on a diagram and, from the shortcut menu, choose the type of diagram. Visio displays the stencil for that type of diagram and opens a blank drawing page. Drag shapes for the elements in the package onto the drawing page.

- **Partition a diagram** — In the Model Explorer window, create multiple diagrams within one package. You can drag the elements you want to include onto as many of the diagrams in the package as you want. Each instance refers to the same element in the UML model. To view all the references to an element on diagrams, in the Model Explorer window, right-click the element and, from the shortcut menu, choose Views.

Specifying UML Options

The UML Model Diagram template comes with several options for controlling the behavior of the elements and drawings in your model. For example, you can choose whether to use the lollipop or class-like version of an Interface shape, or whether to prompt before deleting a model element when you delete a shape on a drawing page. To specify the options for the UML Model Diagram template, choose UML ⇨ Options. In the UML Options dialog box, select the UML Add-on tab and then specify one or more of the following options:

- **Shape Ctrl-Drag Behavior** — To duplicate the UML element the shape represents, select Copy Object in the drop-down list. To create a view of the element the shape represents so that you can drag the shape to another package or diagram, choose Copy Object View in the drop-down list.

- **Create a Diagram Page When a Package or Subsystem Shape Is Added to a Document** — Select this check box to create a diagram page for every package or subsystem shape you add to a model.

- **Create Watermark on Drawing Page** — Select this check box to display a watermark that identifies the type of diagram the drawing page represents.

- **Prompt for Model Element(s) Delete on Delete of Shape(s)** — Select this check box if you want Visio to ask you if you want to delete the model element that a shape represents when you delete the shape on a drawing page.

- **Delete Connectors When Deleting Shapes** — Select this check box to delete the connectors glued to a shape when you delete the shape.

- **On Drop of an Interface from the Model Explorer** — Select an option to specify the default style you want to use for Interface shapes. If you want to show attributes on an Interface shape, select the Class-Like Interface Shape option.

- **Auto Assign Name to Newly Created UML Model Element** — Select this check box if you want Visio to generate names for elements you create. You can rename the elements after you create them.

Working with Shapes and Model Elements

You can work on a UML model through shapes on drawing pages or directly in the Model Explorer window. In addition to adding and removing elements, you can modify their properties to match the characteristics of your system. UML diagrams can display element properties within the shapes for those elements on drawing pages.

Adding and Removing Elements in a Model

The Model Explorer window is a great place to add, modify, and delete elements, but it also reflects the work you do on drawings. When you drag shapes from stencils onto a drawing page, Visio adds the elements that those shapes represent to the UML model. Conversely, because diagrams are merely views of a model, not everything that appears in the Model Explorer window has to appear on a drawing page. You can add elements to a model without adding them to a diagram. As well, you can drag an element in the Model Explorer window onto several drawing pages, creating multiple views of the same element.

Choose your favorite of the following two methods for adding elements to a model:

- **Model Explorer window** — Right-click a package or subsystem, from the shortcut menu, choose New, and then choose the type of element you want to add to the model.

- **Drawing page** — Drag a shape that represents the element you want to add onto a drawing page.

NOTE   Because elements added in the Model Explorer window do not appear on diagrams by default, after you add an element in the Model Explorer window, you can add it to a diagram by dragging its icon from the Model Explorer window onto a drawing page.

If you want to remove an element from a model, right-click it in the Model Explorer window and then, from the shortcut menu, choose Delete. Visio deletes the element from the model as well as all UML diagrams.

CAUTION   If you delete a shape from a diagram, Visio deletes only the view of the element that the shape represents. The element remains in the model in the Model Explorer window.

Displaying Information in UML Shapes

The shapes on the UML stencils represent each element in the UML 1.3 specification, and the behaviors for these shapes conform to UML 1.3 rules. However, you can change the values that appear on shapes to show the information you want.

To specify the values that appear in UML shapes, right-click any shape in a UML diagram and, from the shortcut menu, choose Shape Display Options. The rather monumental UML Shape Display Options dialog box opens with check boxes for showing or hiding information. If the data doesn't apply to the selected shape, the check boxes are grayed out. Depending on the types of shape you select, the following categories might appear:

- **General options** — These are basic properties, such as the name, stereotype, operation parameters, and properties.

- **Attribute** — For connectors, you can show attribute types, the initial value for attributes, and attribute multiplicity.

- **End Options** — For connectors, you can specify whether to show end names, multiplicity, navigability, and visibility.

- **Suppress** — When you want to prevent attributes, operations, and template parameters from displaying, select the corresponding check boxes in this section. For connectors, you can suppress information at each end of the connector.

- **Applying options to other shapes** — The last section in the dialog box is not labeled, but it's powerful nonetheless. To apply the shape display options you selected to other shapes of the same type on the current drawing page, select the Apply to the Same Selected UML Shapes in the Current Drawing Window Page check box. To also apply the options to new shapes of that type dropped on the page, select the Apply to Subsequently Dropped UML Shapes of the Same Type in the Current Drawing Window Page check box.

Specifying Element Properties

The elements you add to a UML model include numerous built-in properties that mirror UML notation and behavior. To configure an element in a model, double-click the element in the Model Explorer window or the shape that represents it on a drawing page. Visio opens a properties dialog box for that element, such as the UML Class Properties dialog box for a class or the UML Use Case Properties dialog box for a use case. In the dialog box, you can select the category of properties you want to edit and then specify the properties you want, as shown in Figure 21-2.

FIGURE 21-2

UML property dialog boxes contain categories of properties, each with numerous attributes to define the elements in a UML model.

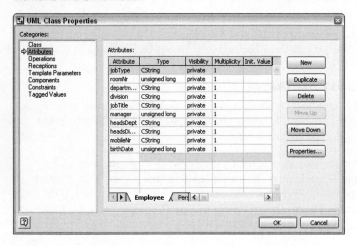

NOTE   Stereotypes, constraints, and tagged values in a UML *<element>* Properties dialog box extend the behavior of their corresponding elements in a system. To specify a stereotype for an element, select Class in the Categories list and then select the stereotype you want in the Stereotype drop-down list. To specify constraints, select Constraints in the Category list and then define the constraints you want. To add tagged values to an element, select Tagged Values in the Category list, select the tag you want, and then type the value in the Tag Value box.

Creating UML Models

The UML Model Diagram template opens the stencils for each type of UML diagram, docks the Model Explorer window on the left side of the screen, and adds the UML menu to the Visio menu bar. In addition, the UML Model Diagram template creates several default model elements, including the Static Structure-1 diagram and a corresponding drawing page by the same name. UML diagrams use a letter-size drawing page with portrait orientation and no drawing scale. To create a new UML drawing file, choose File ➪ New ➪ Software and Database ➪ UML Model Diagram.

Software systems often have more than one model. To create a new model within the Visio drawing file, follow these steps:

1. Choose UML ➪ Models.

2. In the UML Models dialog box, click a blank cell in the model column and type the name of the model.

3. To specify the properties for the model, click Properties in the UML Models dialog box. Select a category and specify the properties within that category. Click OK when you're done.

> **NOTE** The properties for a model are also available by right-clicking the model in the Model Explorer window and, from the shortcut menu, choosing Properties.

4. To add a diagram to the model, right-click a package in the Model Explorer window, from the shortcut menu, choose New, and then choose the type of UML diagram you want to add. You can also right-click an empty area on a drawing page and, from the shortcut menu, choose Insert UML diagram. Visio performs the following actions:

   ▪ Creates a blank page named *<diagram>-n*, where *<diagram>* is the name of the type of diagram you are creating and *n* is the next number in the sequence of that type of diagram in the drawing file

   ▪ Displays the new page in the drawing window

   ▪ Expands the UML stencil for that type of diagram in the Shapes window

   ▪ Adds an icon for the diagram to the element to which you added it in the Model Explorer window

5. To add an element to the model, in the Model Explorer window, right-click a package, from the shortcut menu, choose New, and then choose the element you want to add. You can also add subclasses to an existing class by right-clicking the class and, from the shortcut menu, choosing New ➪ Class.

> **NOTE** If you right-click a class or use case in the Model Explorer window and choose New from the shortcut menu, you can create diagrams appropriate for the selected element. You can choose from Static Structure Diagram, Activity Diagram, and Statechart Diagram for classes; and Activity Diagram and Statechart Diagram for use cases.

Working with Static Structure Diagrams (UML 2 Class Diagrams)

Early in the development life cycle, conceptual class diagrams show the real-world objects represented by your system and the relationships between them, such as Applicants registering for courses. These diagrams help clarify the terminology used within the context of the system, and the classes, including their operations, attributes, and interrelationships. Although class diagrams start at the domain level, they can also show the detailed design of object-oriented software. As you progress in the development cycle, class diagrams show the software classes that the system implements and how they relate to each other, as shown in Figure 21-3.

FIGURE 21-3

Visio static structure diagrams are the equivalent of UML Class diagrams.

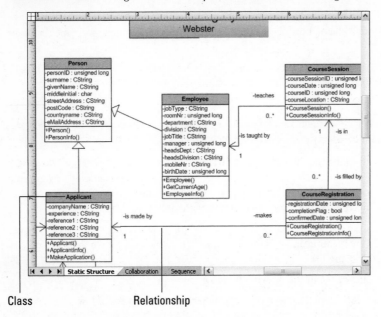

Class   Relationship

Creating Static Structure Diagrams

To create a static structure diagram in an existing model, follow these steps:

1. In the Model Explorer window, right-click the package to which you want to add the static structure diagram and, from the shortcut menu, choose New ⇨ Static Structure Diagram.

2. Add elements that represent real-world objects or the software classes that implement them to the model (or drag corresponding shapes onto the diagram drawing page).

3. Specify the properties for the object or class, as described in the section "Specifying Element Properties" earlier in this chapter.

NOTE When you select the Tagged Values category in the UML Object Properties or UML Class Properties dialog box, Visio displays default tags for the selected element. To add a value to a tag, select the tag you want to modify in the Tags list and type text in the Tag *<tag>* Value box. You can also add additional tags by clicking New.

4. To change the display for the shape, set the Shape Display Options as described in the section "Displaying Information in UML Shapes" earlier in this chapter.

5. To show relationships between objects and classes, drag Association, Link, Dependency, Generalization, or Composition connectors onto the drawing page and glue them to the related objects or classes.

6. To specify properties for a relationship, double-click its relationship shape on the drawing page to open the UML *<relationship>* Properties dialog box, specify the properties you want, and click OK.

NOTE Use a generalization relationship when you want to show that one object is a kind of another, such as a ferry is a kind of transportation. On the other hand, association relationships use the verb *knows*. For example, for an object to call a method in another object, it has to know about the object with the method, and therefore, the two must be related with an association. Composition relationships represent one object being part of another, such as your living room is a part of your house.

Troubleshooting UML Relationships

Relationship shapes have a few idiosyncrasies that might make them behave in ways you don't expect or don't want. For example, unlike other model elements, relationships appear only on drawing pages, not in the Model Explorer window. You must work on the drawing page to view, add, modify, or delete relationship shapes. Other behaviors that you might observe include the following:

- **Qualifier associations do not appear on diagrams** — The UML specification indicates that qualifier associations appear in UML diagrams, with attributes listed inside a box shape. In Visio, attributes of qualifiers are stored in the properties pages for association ends. To view qualifier attributes, follow these steps:

  1. Double-click an Association shape with a qualified association, which Visio indicates with a small box at one end.

  2. In the UML Association Properties dialog box, under Association Ends, select the End Name that has the attributes you want to view and click Properties.

  3. In the UML Association End Properties dialog box, under Categories, select Qualifier Attributes to view the details of the qualifier association.

- **Relationships persist after you delete them** — If you delete an Association shape on one drawing page, the relationship it represents persists if shapes representing the associated elements exist on other pages. To remove a relationship from a model, you must delete all views of an Association shape.

TIP You can find all the views of an association in the Model Explorer window by right-clicking the Association shape and, from the shortcut menu, choosing Views. To display a view, in the UML Diagram dialog box, select a view in the list and then click OK.

- **Association shape labels don't appear near the right shape** — When several lines connect to the same shape, you might not be able to distinguish the shape to which a label applies. However, you can unlock and move text labels for Association shapes to make diagrams more legible by following these steps:

  1. Select the Association shape whose label you want to move.

  2. Choose Window ➪ Show ShapeSheet.

  3. In the Protection section of the ShapeSheet, in the LockTextEdit cell, change the value from 1 to 0 and press Enter.

  4. To move the label for the Association shape, click the Text Block tool on the Standard toolbar, select the label you want to move, and then drag it to a new location.

- **Relationship lines appear for every instance of the same class** — If you create more than one instance of a class, Visio displays relationship lines between every instance of the class and the other classes to which it relates, which can quickly make your diagrams difficult to decipher. You can turn off the automatic display of additional relationship lines and then use the Show Relationships command from a shape's shortcut menu to view its relationship lines. To turn off the automatic display of additional relationships lines, follow these steps:

  1. Use the registry editor to access HKEY_CURRENT_USER\Software\ Microsoft\Office\12.0\Visio\Solution\UML Solution.

  2. Right-click the Automatic Instance UML Relationships registry entry, from the shortcut menu, choose Modify, and change the value in the Value Data box from 1 to 0.

Creating Use Case Diagrams

Use case diagrams show how users (called *actors*) interact with and generate events in a system. Use case diagrams begin by showing the use cases in context within the system, highlighting the interaction of processes, rather than individual steps. As your analysis progresses, you can refine the diagrams to show more detail.

Creating a use case diagram isn't any different than creating other types of UML diagrams. In the Model Explorer window, right-click the package or subsystem to which you want to add the use case diagram and, from the shortcut menu, choose New ➪ Use Case Diagram. Then, drag shapes from the UML Use Case stencil onto the drawing page or add the use cases within the Model Explorer window. Finally, define any properties for the use cases. Use the following shapes to show the actors, use cases, and interactions between them:

- **System Boundary** — Although few people use this symbol, you can drag this shape onto the drawing page to indicate the boundary of the system.

 You can move a system boundary and all the use cases it contains by dragging a selection rectangle around the System Boundary shape and then dragging it to a new location.

- **Use Case** — Clearly named, this shape represents use cases in the system.

- **Actor** — For each actor in the system, drag this shape onto the drawing page and place it outside the system boundary.

 **Because an actor represents a role played by an external object, one physical object might be represented by more than one actor, and vice versa.**

- **Communicates** — This connector indicates a relationship between an actor and a use case. On the Standard toolbar, click the Connector tool, select the Communicates connector in the UML Use Case stencil, and then drag from the Actor shape to the Use Case shape.

TIP **To show the actor who initiates an interaction as the primary actor in a use case, double-click the Communicates shape. In the UML Association Properties dialog box in the Association category, select the IsNavigable check box for the end to which you want to add an arrow and click OK.**

- **Extends** — To extend the behavior of one use case to another, connect them with the Extends connector in the UML Use Case stencil.

- **Uses** — The Uses connector indicates that one use case uses the behavior of another. Visio draws an arrowhead at the end of the connector glued to the Use Case shape that uses the behavior.

NOTE **In the UML 2 specification, the term *uses* has been replaced with *includes*.**

Creating Activity Diagrams

Similar in many ways to data flow diagrams used in structured development, UML activity diagrams model business and software processes and can depict logic for complex business rules and operations. They represent flows triggered by internally generated actions, whereas statechart diagrams show flow in response to external events. Because activity diagrams emphasize parallel and concurrent activities, they work well for modeling workflow, analyzing use cases, and making sure that multithreaded applications perform properly.

To create an activity diagram in an existing UML model, follow these steps:

1. In the Model Explorer window, right-click the subsystem, package, class, use case, or operation to which you want to add the activity diagram and choose New ⇨ Activity Diagram from the shortcut menu.

2. To indicate responsibility for activities, drag Swimlane shapes from the UML Activity stencil onto the drawing page for each class, person, or organizational unit you want to represent. You can double-click Swimlane shapes to add names and other property values or drag side selection handles to resize the lanes.

3. Drag Action State or State shapes onto the drawing page for each state you want to represent. Use the Initial State and Final State shapes for the first and last states.

4. To show the flow of control as one state changes to another, connect Control Flow shapes to Action State or State shapes. Click the Connector tool on the Standard toolbar, click the Control Flow shape in the UML Activity stencil, and then drag from the shape that represents the source state to the shape representing the state to which it changes.

NOTE If you want to further define the transition between states, double-click the Control Flow shape on the drawing page and specify events, guard conditions, action expressions, and other information in the UML Transition Properties dialog box.

5. Double-click any shape to open its UML *<element>* Properties dialog box, specify the properties you want, and click OK.

Showing Complex Transitions

When one state forks into multiple parallel states, or several states synchronize into one state, you can use Transition shapes with Transition (Fork) and Transition (Join) shapes to show the transition. To represent a complex transition, follow these steps:

1. Drag a Transition (Fork) or Transition (Join) shape from the UML Activity stencil onto the drawing page.

2. Drag a Transition shape from the UML Statechart stencil onto the drawing page and glue it to the source State shape and the Transition (Fork) or Transition (Join) shape. When several states synchronize into one, repeat this step to add transitions from each of the original State shapes into the Transition (Join) shape.

3. Drag a Transition shape from the UML Statechart stencil onto the drawing page and glue it to the Transition (Fork) or Transition (Join) shape. When one state forks into multiple parallel states, repeat this step to add transitions from the Transition (Fork) shapes to each forked State shape.

4. To use signal icons instead of transition strings with signal icons, drag Signal (from the UML Static Structure stencil), Signal Send and Signal Receipt shapes (from the UML Activity stencil) onto the drawing page to represent the signals. Glue the control handles on these shapes to the source and destination Action State shapes.

5. To associate the Signal Send or Signal Receipt shapes to a Signal, right-click the Signal Send or Signal Receipt shape and, from the shortcut menu, choose Properties. In the Categories list, select Actions and then click Properties. In the UML Send Action Properties dialog, in the Signal drop-down list, choose the signal you want to associate to the action. The Signal Send or Signal Receipt shape then displays the name of its associated signal.

Creating State Machine Diagrams

A *state machine* shows how an object responds to events, showing the event responses that occur depending on the event and the object's current state, as illustrated in Figure 21-4. For example, a dog that's sleeping might respond to a pat on the head by biting your hand, whereas a dog that's awake might respond by sitting on your foot. State machines are typically connected to classes or use cases. In Visio Professional, a statechart diagram represents a state machine, whereas flow driven by internally generated actions, rather than external events, appear instead on an activity diagram.

FIGURE 21-4

Visio statecharts show events and an object's responses to them.

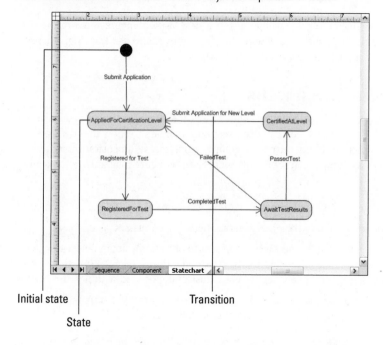

Create a statechart diagram as you would other diagrams. In the Model Explorer window, right-click the class or use case that you want to document in the statechart diagram and, from the shortcut menu, choose New ⇨ Statechart Diagram.

Here is an overview of the shapes you can use to build elements in a state machine:

- **State** — Drag the following shapes onto the drawing page:
  - **State** — Represents the sequence of states through which an object passes
  - **Composite State** — Represents concurrent, mutually exclusive, or nested substates

> **NOTE**   Visio creates a new statechart drawing page for a composite state. To show the substates within the composite state, in the drawing window, click the tab for the new statechart diagram and then drag State, Transition, Shallow History, or Deep History indicators and other shapes onto the drawing page.

- **Transition** — To show transitions from one state to another in response to an event, on the Standard toolbar, click the Connector tool, select the Transition shape in the UML Statechart stencil, and then drag from the first State shape to the next. If an object remains in the same state in response to an event, use the arc-shaped Transition shape and glue both ends to the same State shape.

> **TIP**   You can show a state forking into multiple states or several states synchronizing into one by connecting Transition shapes to Transition (Fork) and Transition (Join) shapes, as described in the "Showing Complex Transactions" sidebar earlier in this chapter.

- **History** — To show that an object resumes a state it held last within a region, drag a Shallow History or Deep History shape onto the drawing page and use Transition shapes to connect it to the source and destination State shapes.

Creating Sequence Diagrams

Sequence diagrams help model the dynamic logic within classes and show the time sequence of events generated by actors. For example, you can use a sequence diagram to show the messages generated in a real-time transaction. In a sequence diagram, the horizontal dimension shows the actors or objects and the vertical dimension represents time. UML sequence diagrams are typically used to do the following:

- Expand and validate the logic for potential usage of a system.
- Walk through the invocation of the operations defined by your classes.
- Detect bottlenecks generated by messages being sent to objects within an object-oriented design. By examining the messages sent to objects and the time it takes to run an invoked method, you can identify design changes to better distribute the load within your system.
- Identify the classes in an application that are going to be complex, which therefore might benefit from the development of state machine (statechart) diagrams.

Create a sequence diagram in the Model Explorer window by right-clicking the package or subsystem to which you want to add a sequence diagram and, from the shortcut menu, choose New ➪ Sequence Diagram.

Here is an overview of the shapes and configuration you can apply to a sequence diagram (illustrated in Figure 21-5:

FIGURE 21-5

Sequence diagrams show the lifespan of actors, when they perform actions, and the communication that occurs between them.

Object Lifeline for actor

Activation represents when the actor acts

Message

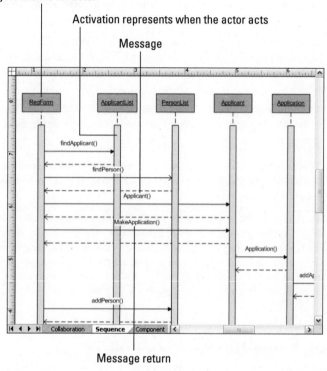

Message return

- **Actor's lifespan** — The Object Lifeline shape indicates the length of an actor's life in an interaction. Use one or both of the following methods to define an actor's lifespan:

  - Drag an Object Lifeline control handle to define the length of the actor's life in the interaction.

  - To indicate that the actor or object is destroyed during the interaction, right-click the Object Lifeline shape, from the shortcut menu, choose Shape Display Options, and select Destruction Marker. To apply this change only to the selected actor, in the UML Shape Display Options dialog box, make sure that the Apply to the Same Selected UML shapes in the Current Drawing Window Page check box is cleared. Visio adds a black X at the end of the object's lifeline.

■ **Classify an Object Lifeline shape** — Double-click an Object Lifeline shape to open the UML Classifier Role Properties dialog box. In the Classifier Role category, select the classifier that the actor represents in the Classifier drop-down list. Visio changes the appearance of the Object LifeLine shape to reflect the classifier you choose.

 You can define your own classifiers by clicking New, specifying the properties for the class, and clicking OK.

■ **Timing** — To indicate when an actor performs an action, drag an Activation shape onto the drawing page and glue it to the actor's Object Lifeline shape. Drag the end points of the Activation shape to correspond to the period during which the actor performs the action.

■ **Communication** — To indicate communication between actors, drag a Message shape onto the drawing page and glue it from the Actor shape sending the message to the Actor shape receiving the message.

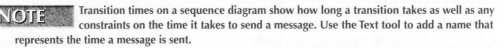 Transition times on a sequence diagram show how long a transition takes as well as any constraints on the time it takes to send a message. Use the Text tool to add a name that represents the time a message is sent.

The convention is to show the name, usually a letter, in the left margin, aligned with the message to which it applies. If a message does not arrive instantaneously, slant the message line and name each end of the line. For the message received name, you can use the message sent name with a prime appended to it. For example, if the time a message is sent is a, the message receipt time is a'. You can indicate a constraint for the message time, such as $b - a < 1$ sec by dragging a Constraint shape onto the drawing page near the message name, double-clicking the Constraint, and adding text in the Body text box.

Creating Collaboration Diagrams

Collaboration diagrams are now called *communication diagrams* in the UML 2 specification. Like sequence diagrams, these diagrams show the exchange of messages between objects in response to system events, and imply the relationships between classes. The difference between sequence diagrams and collaboration diagrams is that collaboration diagrams do not show the sequence of messages by their position on the diagram. Instead, numbers associated with messages indicate sequence. You can use communication diagrams to accomplish the following:

■ Provide an overview of collaborating objects, particularly within a real-time environment.

■ Allocate functionality to classes based on the behaviors within a system.

■ Model the logic of complex operations, particularly those that interact with many other objects.

■ Analyze the roles that objects play within a system, and the relationships associated with those roles.

Create a collaboration diagram in the Model Explorer window by right-clicking the package to which you want to add a collaboration diagram and, from the shortcut menu, choose New ⇨ Collaboration Diagram.

Here is an overview of the shapes and configuration you can apply to a collaboration diagram:

- **Add roles** — For each actor or object role that collaborates in the interaction, drag a Classifier Role shape from the UML Collaboration stencil onto the drawing page. To name the classifier role or specify other properties, double-click a Classifier Role shape. In the UML Classifier Role Properties dialog box, type the name in the Name box or select other properties for the role and then click OK.

- **Add set of objects** — To represent a set of objects, drag a Multi-Object shape onto the drawing page.

NOTE A multi-object shows operations that affect an entire set of objects as a unit, rather than a single object within the set. To indicate an operation on each object in the set, use a single message with an iteration and include a many indicator (*) in the target role name.

- **Links objects** — Glue Association Role shapes to the Classifier Role or Multi-Object shapes. Double-click an Association Role shape to specify its name, message flow, message label, multiplicity, and other property values.

Creating Component Diagrams

UML component diagrams are typically used as architecture-level artifacts, to model the business and the technical software architectures for systems. You can enhance your architectural work by iterating between component diagrams for business and technical software architecture and UML deployment diagrams or network diagrams for the physical architecture.

Component diagrams help distribute development across large teams. Initially, your architectural modeling focuses on modeling the high-level software components and the interfaces to those components. After the team defines and agrees to the interfaces for the system, it's much easier to assign development among smaller teams. As work progresses and you identify new requirements or needed changes, you can negotiate changes between teams and implement the modified interfaces. Component diagrams can also help partition a system into components and show dependencies between components.

Create a component diagram in the Model Explorer window by right-clicking the package or subsystem to which you want to add a component diagram and, from the shortcut menu, choose New ⇨ Component Diagram.

Here is an overview of the shapes and configuration you can apply to a collaboration diagram:

- **Component** — Drag a Component shape from the UML Component stencil onto the drawing page.

- **Interface** — To add an interface to a component, drag an Interface shape from the UML Component stencil onto the drawing page and glue the end point without the circle to a connection point on the corresponding Component shape.

> **TIP** To list the operations for an interface, right-click the Interface shape and, from the shortcut menu, choose Show As Class-like Interface. The operations you specify in the Operations category in the UML Interface Properties dialog box appear in the lower half of the rectangular Interface shape.

- **Relationships** — To show relationships between components or between a component and another component's interface, on the Standard toolbar, click the Connector tool, select the Dependency shape on the UML Component stencil, and then drag from one Component shape to the Component shape that depends on it.

Creating Deployment Diagrams

Deployment diagrams, like component diagrams, provide views of your system implementation. These diagrams illustrate the structure of run-time systems as well as the configuration and deployment of hardware and software. In deployment diagrams, nodes represent processing resources such as computers, mechanical processing devices, or human resources who perform processing activities. Components represent physical modules of code or business documents. You can use deployment diagrams to accomplish the following:

- Identify installation issues for your system
- Identify the dependencies between your system and other current or planned production systems
- Document the deployment configuration of a business application
- Configure the hardware and software for an embedded system
- Represent the network and hardware infrastructure of an organization

Create a deployment diagram in the Model Explorer window by right-clicking the package or subsystem to which you want to add a deployment diagram and, from the shortcut menu, choose New ⇨ Deployment Diagram.

Here is an overview of the shapes and configuration you can apply to a deployment diagram:

- **Node** — For resources in a system, drag a Node shape from the UML Deployment stencil onto the drawing page.
- **Component or object** — Drag Component and Object shapes into the Node shape to which they belong.
- **Interface** — To add an interface to a component, drag an Interface shape from the UML Deployment stencil onto the drawing page and glue the end point without the circle to a connection point on the Component shape.
- **Relationships between nodes** — On the Standard toolbar, click the Connector tool, click the Communicates shape on the UML Deployment stencil, and then drag from one Node shape to another.

- **Relationships between components** — To show relationships between components or between a component and another component's interface, on the Standard toolbar, click the Connector tool, click the Dependency shape on the UML Deployment stencil, and then drag from one Component shape to the Component shape that depends on it.

Reverse-Engineering Code into UML Models

If you develop projects using Microsoft Visual C++, Microsoft Visual Basic, or Microsoft Visual C#, you can use Visual Studio to reverse-engineer your source code into UML and generate a UML diagram in Visio from your project's class definitions. For example, if you're maintaining a legacy system, you can build a UML model to better understand the system and make software maintenance easier.

Visio includes Visio UML Add-Ins for Visual Basic and Visual C++, which provide toolbars you can use to reverse-engineer source code to create a UML static structure model in Visio. The reverse-engineered code elements appear in the Visio Model Explorer window. Then, to create a UML static structure diagram, you can drag the elements from the model onto the drawing page.

NOTE   You can't use the Visio UML reverse engineering add-ins for Visual Basic 6.0 and Visual C++ 6.0 simultaneously on the same machine.

Visual Studio .NET can reverse-engineer code written in the following languages. Depending on the language you use, you can reverse-engineer different aspects of a project. Visio reverse-engineers the following elements:

- **Visual C++ 6.0** — Classes, user-defined types, enumerated types, member functions, member variables, and method parameters

- **Visual C++ 7.0** — Namespaces, classes, enums, structs, unions, member operations, member variables, method parameters, typedefs, template definitions, Inline function specifier, cv qualifier, conversion-function ID, and operator-function ID

- **Visual Basic 6.0** — Classes, modules, and forms, functions and subroutines, parameters, constants, member variables, properties, events, and user-defined types

- **Visual Basic .NET and Visual Studio .NET** — Namespaces, classes, interfaces, enumerated types, structures, properties, delegates, member operations, member variables, method parameters, events, and constants

- **Visual C#** — Namespaces, classes, interfaces, enumerated types, structs, properties, delegates, member operations, member variables, method parameters, and constants

NOTE   To prepare to use the Visio add-ins, shut down all instances of Visio and Visual Studio that are running and then run and close Visio once.

Reverse-Engineering Visual C++ Code

Before you can reverse-engineer a Visual C++ project, you must customize Visual C++ with the Visio UML Add-In, and you must generate a Browse Information file, which the UML Add-In uses to generate a UML model from the source code in your project.

To customize Visual C++ so you can reverse-engineer code, follow these steps:

1. In Visual C++, choose Tools ➪ Customize.

2. In the Customize dialog box, select the Add-Ins and Macro Files tab, select the Visio UML Add-In in the Add-Ins and Macro files list, and click Close. The Visio UML Add-In toolbar appears.

3. To generate a Browse Information file in Visual C++, open the project you want to reverse-engineer and choose Project ➪ Settings.

4. In the Project Settings dialog box, choose the type of build configuration you want, select the C/C++ tab, and then click Generate Browse Info.

5. Select the Browse Info tab, specify the name and location of the Browse Information file, click Build Browse Info File, and then click OK.

To reverse-engineer a Visual C++ project, follow these steps:

1. Build the project in Visual C++.

NOTE If you modify a Visual C++ project after you reverse-engineer it, you must rebuild the project and reverse-engineer it into Visio again to see the changes.

2. In Visual C++, click the Reverse Engineer UML Model button on the Visio UML Add-In toolbar. After the reverse-engineering process is complete, Visio opens a blank static structure diagram drawing page. The Model Explorer includes elements for the class definitions from your project's source code.

NOTE If more than one project exists in the Visual C++ workspace, in the Select Project dialog box, select the project you want to reverse engineer and click OK. If more than one Browse Information file exists in the project hierarchy, in the Select Browse File dialog box, select the file you want and then click OK.

3. To create a static structure diagram, drag elements from the Model Explorer window onto the drawing page.

CAUTION The Browse Information file API contains a bug that sometimes corrupts class names and class method names. When you reverse-engineer a project with corrupted names, the elements with corrupted names are usually not added to the UML model. However, if the reverse-engineering process doesn't detect the name corruption, the corrupted names are added to the model. Visio creates a log file that lists the errors detected during the reverse-engineering process. By default, the file is written to C:\Temp\project.txt.

Reverse-Engineering Visual Basic Code

Before you can reverse-engineer a Visual Basic project, you must customize Visual Basic with the Visio UML Add-In.

To reverse-engineer Visual Basic code, follow these steps:

1. To customize Visual Basic, in Visual Basic, choose Add-Ins ⇨ Add-In Manager. In the Add-In Manager dialog box, select Visio UML Add-In. For Load Behavior, select Loaded/Unloaded and Load on Startup and click OK. The Visio UML Add-In toolbar appears.

2. Open the project that you want to reverse-engineer.

3. In Visual Basic, click the Reverse Engineer UML Model button on the Visio UML Add-In toolbar. Visio opens a blank static structure diagram drawing page and populates the Model Explorer window with the elements that represent the class definitions in the source code.

4. To create a static structure diagram, drag elements from the Model Explorer window onto the drawing page.

Summary

The UML Model Diagram template is available only in Visio Professional and Visio Studio .NET Enterprise Architect. To model a software system, you can create a single Visio drawing file and create the multiple models and diagrams of your system within that file. You can work on the model of your system in the Model Explorer window or by modifying shapes on drawing pages. However, the shapes on drawing pages represent views of the elements in your model. You can add elements to a model without displaying them on diagrams or you can add views of elements to more than one diagram. You can also reverse-engineer existing source code into Visio models.

Chapter 22

Building Software Development Diagrams

Whether or not you use the Unified Modeling Language (UML) to model a software system, other types of software diagrams come in handy for designing and documenting software architecture, program structure, and memory management. In addition to the UML Model Diagram template, Visio Professional includes several additional software templates for visualizing your system based on other software development methodologies.

The software templates described in this chapter help you document your software systems with diagrams — but they don't produce a model like the one that the UML Model Diagram template offers. The good news is that basic Visio techniques are all you need to construct these software diagrams: dragging and dropping, gluing with connectors, dragging control handles, applying shape data, and using the occasional configuration command from shortcut menus. The bad news is that, like many other Visio templates, these templates don't check whether your software diagrams conform to the rules and notations of the software methodologies they represent. Nor do they offer an opinion on whether your diagram makes any sense software-wise.

This chapter describes the different software templates that Visio Professional offers and briefly introduces how to put them to use. For prototyping user interfaces, you will learn how to develop diagrams for application windows, wizards, dialog boxes, menus, and toolbars.

Choosing the Right Software Template

The Visio template you should choose depends on what you want to document, as well as the methodology you use, and the phase of development you're in. In addition to the UML Model Diagram template, Visio provides seven other software templates. Many of the shapes on software stencils fit the standard software modeling notations. However, it's up to you to build diagrams that conform to the rules and syntax of the associated software methodology.

These software templates don't include specialized menus or toolbars. New diagrams based on these templates open with a new letter-size drawing page using portrait orientation and no drawing scale.

In Visio 2007, the software templates are in the Software and Database category along with the templates for database and web site documentation. Here are the additional software diagram templates that Visio Professional 2007 provides:

- **COM and OLE** — These diagrams show the structure of Component Object Model (COM) and Object Linking and Embedding (OLE) components for an application and the interfaces between them.

- **Data Flow Model Diagram** — The Gane-Sarson notation helps you design software by modeling the flow and transformation of data across processes, interfaces, data flows, and data stores.

NOTE   To learn about data flow diagrams (DFDs) and structured analysis, read *Structured Systems Analysis: Tools and Techniques* by Chris Gane and T. Sarson (Prentice-Hall, 1977) or go to www.agilemodeling.com/artifacts/dataFlowDiagram.htm.

- **Enterprise Application** — When you're building large enterprise-wide applications, this template helps show the logical and physical architecture of systems. Logical diagrams document processes, components, interfaces, and boundaries. The physical architecture of an application includes mainframes, servers, workstations, laptops, interfaces, and communication links.

- **Jackson** — The Jackson software design method focuses on the system activities that affect data. You can develop data structure diagrams, system network diagrams, and program structure diagrams using the Jackson notation, as described in *Problem Frames: Analyzing and Structuring Software Development Problems* by Michael Jackson (Addison-Wesley, 2000).

- **Program Structure** — These diagrams show program architecture, structure, and memory management with shapes that represent language elements such as functions and subroutines; and memory objects, such as stacks, pointers, and bytes.

- **ROOM** — ROOM diagrams are specifically suited to documenting real-time systems using object-oriented concepts and real-time software techniques. They show the structure of system components and the system's response to events.

- **Windows XP User Interface** — Create prototypes of Windows XP interfaces — from individual buttons and message boxes to tabbed dialog boxes and application windows.

NOTE To learn more about other types of software diagrams, read *Agile Modeling* by Scott Ambler (Indianapolis: Wiley, 2002) and *The Object Primer, 3rd Edition: Agile Model Driven Development with UML 2.0* by Scott Ambler (New York: Cambridge University Press, 2001).

Constructing COM and OLE Diagrams

COM is a Microsoft standard, in which software components can be written in more than one language and communicate through object interfaces. OLE is a subset of COM functionality. With COM and OLE diagrams, you can show the software components associated with processes and how those software components relate to each other through interfaces.

Understanding the Elements of COM

COM controls the identification, structure, and interaction of software components. COM specifications regulate the following:

- The structure of component interfaces
- Communication between components, including communication across process and network boundaries
- Shared memory management
- The dynamic loading of components
- Error and state reporting
- Unique identification of components and interfaces

Client applications and component objects interact through interfaces that define the behavior and responsibilities of the component objects. *Interfaces* are collections of functions that component objects make available to client components or applications. All component objects must implement the IUnknown interface, which counts references to determine component lifetime, and enables clients to determine whether an object supports a required interface and then to connect pointers to object interfaces.

Each component object and interface receives a globally unique identifier, or GUID, which is a 128-bit integer that is unique across space and time. By uniquely identifying component objects and interfaces, you can prevent component objects from connecting to the wrong components or interfaces. For computers with Ethernet cards, GUID integers are based on the computer used to create the component or interface and the date and time at which the component or interface was created. If a computer doesn't have an Ethernet card, the GUID is only guaranteed to be unique on that computer.

By using Vtables, you can write component objects in any language that uses pointers to call functions, such as C, C++, or Microsoft Visual Basic. COM specifies the layout of Vtables in computer memory and a standard method for calling functions through Vtables.

Working with Visio COM and OLE Diagrams

The COM and OLE template opens only the COM and OLE stencil. To create a COM and OLE diagram, choose File ➪ New ➪ Software and Database ➪ COM and OLE.

Here are a few special features built into COM and OLE shapes:

- **IUnknown Interface** — COM objects include the IUnknown interface by default. Drag the control handle for the IUnknown interface to reposition it.

- **Add an interface to a COM Object shape** — To add your own interfaces to a COM Object shape, drag the control handle in the center of the shape in the direction you want the interface to point. To name the new interface, subselect it, type the name for the interface, and then press Esc.

- **Change the appearance of a COM Object shape** — Visio provides two formats for COM objects. To choose the style, right-click the shape and, from the shortcut menu, choose the COM Style command, which toggles between COM Style 1 and COM Style 2.

- **Add Vtables** — Vtable shapes include a shape data property that specifies the number of cells in the table. When you drag a Vtable shape onto the drawing page, the Shape Data dialog box appears, so you can type the value for the number of cells. You can also set the number of cells in the Shape Data Window. To add text to table cells, subselect a cell and type the text you want. You can drag Vtables inside COM Object shapes.

 To specify the number of cells in a Vtable at a later time, right-click the table and choose Set Number of Cells from the shortcut menu.

- **Create relationships between COM objects, Vtables, and interfaces** — Defining the relationships between elements comprises dragging Reference or Weak Reference connectors onto the drawing page and gluing them to shapes or interfaces.

NOTE To change the angle of the bend in a Reference connector, drag its control handle to a new location.

Working with Data Flow Model Diagrams

The Gane-Sarson methodology represents software systems and business processes with data flow diagrams (DFDs) that show the data stores that hold data, the processes that transform data, and the data flows that processes generate. This approach to designing software begins by defining top-level processes and then decomposes those processes into lower-level processes, as shown in Figure 22-1.

To create a Data Flow Model diagram, choose File ➪ New ➪ Software and Database ➪ Data Flow Model Diagram. Visio creates a new drawing file containing the Top Process drawing page and opens the Gane-Sarson stencil.

FIGURE 22-1

Data Flow Model diagrams use the Gane-Sarson notation and you can use standard Visio techniques to show high-level flows on one page and details on another.

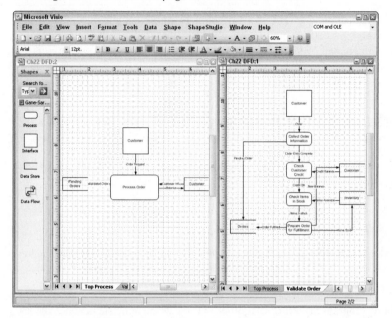

NOTE   Beginning in Visio 2003, the Model Explorer and DFD menu are no longer available.

Here are the shapes on the Gane-Sarson stencil and what they represent:

- **Process** shapes represent the business processes that operate on data.
- **Interface** shapes represent interfaces to entities, such as customers.
- **Data Store** shapes show data repositories for your system.
- **Data Flow** connectors indicate the flow of data between processes or to and from data stores. Use the Data Flow connector as you would any other connector by clicking the Connector tool on the Standard toolbar, selecting the Data Flow connector in the Gane-Sarson stencil, and then dragging from the source to the destination of the data flow.

NOTE   To define a lower-level process, create a new drawing page for the subprocess. You can create additional drawing pages to define additional levels of subprocesses in your system until you have described the entire data flow model in sufficient detail.

Creating Software Diagrams with Jackson Notation

The Jackson software design methodology encompasses the software system life cycle from analysis to physical design. With this methodology, you analyze the effect of system actions on input and output data streams, not functional tasks. The Jackson design method uses system network diagrams to show the interactions between processes at the top-level of a software system. Each data stream shown on the system network diagram is described in detail by a data structure diagram. Finally, each program is described by a program structure diagram.

> **NOTE** Jackson methodology has been overshadowed by UML-based object techniques in recent years, although it is still a common technique within traditional development circles.

- **System Network Diagrams** — The Jackson template doesn't offer any shapes or tools for network diagrams. You use Visio drawing tools to create circles that represent data streams, rectangles for programs, and arrows for the flow of data streams in and out of programs. A Jackson system network diagram is a network of arrows connecting data stream circles to program rectangles.

> **CAUTION** In the Jackson methodology, you should not connect two circles or two rectangles. In addition, each circle can have only one arrow pointing into the data stream and one arrow flowing out.

- **Jackson Data or Program Structure Diagrams** — The Jackson template includes shapes for creating tree diagrams to document data structure or program structure (choose File ⇨ New ⇨ Software and Database ⇨ Jackson). Tree structure diagrams comprise four basic components: sequence, elementary, selection, and iteration with elementary components the lowest components in a tree structure. To build tree structure diagrams, use the following shapes and methods:

  - **Process, Procedure, or Procedure 2 shape** — Drag these from the Jackson stencil onto the drawing page to show programs and procedures. To specify the component type, right-click the shape and, from the shortcut menu, choose No Symbol for a sequence or lowest-level component, Show Asterisk to denote an iteration (loops of various types), or Show Circle for a selection (similar to option buttons).

  - **Tree Connector shape** — This connector works like the tree connectors on the Connectors stencil. You can glue the trunk to a Process, Procedure, or Procedure 2 shape or drag control handles to glue branches to other shapes.

Modeling Large-Scale Application Architectures

Enterprise architecture diagrams present either logical or physical views of the architecture of a large-scale or enterprise system. During analysis, you can use these diagrams to show the system architecture at a conceptual level with Process, Object, and Datastore shapes. During design and development, you can add physical details to the architecture diagram, showing the mainframe computers, servers, workstations, and laptops that make up the equipment infrastructure. Later in the system life cycle, you can use these diagrams to design test platforms and plan delivery and support services.

The shapes on the Enterprise Architecture stencil don't have any extra bells and whistles. You simply drag and drop them onto the drawing page, glue them together with connectors, and use other standard Visio techniques to format and annotate them. To create an enterprise architecture diagram, choose File ➪ New ➪ Software and Database ➪ Enterprise Application.

Creating Program Structure Diagrams

The Program Structure template tackles a different part of software documentation from the program structure diagrams you create with the Jackson template — it focuses on memory management and language-level functions within structured or procedural computer programs. To document program structure or memory management using non-Jackson symbology, choose File ➪ New ➪ Software and Database ➪ Program Structure and then use shapes from the following stencils:

- **Language Level Shapes** — Function/Subroutine, Function w/Invocation, Switch, and other shapes show the structure of functions within higher-level programs.

- **Memory Objects** — This stencil includes shapes that represent memory, such as Array, Pointer, Byte, or Variable. To connect pointers to memory cells, drag end points or control handles from Pointer shapes and glue them to connection points on shapes representing memory objects, such as Array.

Modeling Real-Time Systems with ROOM Diagrams

The Real-Time Object-Oriented Modeling (ROOM) language combines real-time and object-oriented software techniques that you can use to show both the structure of system components and a system's response to events. To create a new ROOM diagram, choose File ➪ New ➪ Software and Database ➪ ROOM.

To learn more about using ROOM to model software, read *Real-Time Object-Oriented Modeling* by Bran Selic (Wiley, 1994).

The shapes on the Visio ROOM stencil support ROOM structure and behavior diagrams. To build ROOM Structure diagrams, use Actor Class shapes for the software components. Drag Actor Reference or Modified Actor Ref. shapes inside Actor Class shapes and specify the object type by right-clicking the shape, and then choosing Select Actor Reference Type. Port shapes connect the Actor Classes and Binding shapes represent the communication paths between ports.

For ROOM Behavior diagrams, use State Context and State shapes to indicate the states that can exist. The different types of transition shapes show how states progress from one to the other.

To create multiple paths for a transition, drag a Choicepoint shape onto the drawing page and glue the beginning end points of each diverging Transition shape to it.

Prototyping User Interfaces

Prototyping user interfaces before writing code is a tremendous timesaver and results in more usable interfaces. You can sketch out the menus, toolbars, dialog boxes, and controls for your application and walk through them with end users to obtain feedback on their usability before casting them in concrete code. Developers often use specialized prototyping tools or Visual Basic to develop user interface prototypes. However, if you don't have access to tools such as these and want to experiment with interface elements, you can use the Visio Professional Windows XP User Interface template.

Exploring the Windows XP User Interface Template

The Windows XP User Interface template doesn't contain specialized menus, toolbars, or commands, but it does open several stencils for different types of interface elements. Many of the interface shapes include shortcut commands for configuring the shapes. For example, you can show the interface element as enabled or disabled or specify the type of button. To create a new user interface drawing, choose File ➪ New ➪ Software and Database ➪ Windows XP User Interface. Visio opens the following stencils:

- **Windows and Dialogs** — Contains shapes for forms, panels, tab controls, group boxes, status bars, buttons, and icons
- **Wizards** — Contains shapes for simple and advanced wizard windows, including welcome screens, completion screens, and interior screens for all the steps in between
- **Toolbars and Menus** — Contains shapes for menu bars, top-level menu items, dropdown menu items, toolbars, toolbar buttons, and other toolbar icons
- **Icons** — Includes commonly used interface icons such as the Recycle Bin, My Documents, Help, and Folder
- **Common Controls** — Includes shapes for frequently used controls, such as command buttons, option buttons, check boxes, scroll bars, list boxes, combo boxes, tree nodes, and sliders

Creating an Application Window

You can prototype the entire user interface for an application by starting with an application window. You begin with a blank form and add the user interface elements you want, as shown in Figure 22-2. To create a mock-up of an application interface, follow these steps:

FIGURE 22-2

You can mock up a Windows XP user interface in Visio before writing code.

1. From the Windows and Dialogs stencil, drag a Blank Form shape onto the drawing page. With the shape selected, type the title for the application window.

2. To add an icon to the left of the text in the title bar, such as the Visio icon that appears in the Visio title bar, right-click the Blank Form shape and, on the shortcut menu, check the Room for Icon check box.

> **NOTE**   The Blank Form shortcut menu also includes commands for choosing the background color for the form. The easiest is to choose White Background or Gray Background. If you must have just the right color of green, choose Custom Background and, in the Fill dialog box, choose the fill color you want. To change the appearance of the Blank Form shape to that of an inactive window, clear the Active window on the shortcut menu.

3. To add buttons to the title bar of the form, such as Close, drag a Windows Buttons shape onto the right end of the title bar area of the Blank Form shape and select the type of button you want. The typical arrangement of buttons for a form is the Minimize button, the Maximize button, and the Close button from left to right, respectively.

4. Drag a Status Bar shape from the Windows and Dialogs stencil and glue it to the bottom edge of the Blank Form shape. To add a divider to the status bar, drag a Status Bar Divider shape onto the page and glue it to the control handle for the Status Bar shape or another Status Bar Divider shape. To add text to the Status Bar or Status Bar Divider, select the shape and type the text you want.

> **NOTE** If the Status Bar Divider shape disappears when you glue it to a Status Bar shape, right-click the Status Bar shape and, from the shortcut menu, choose Shape ⊏ Send to Back.

5. To indicate that the window is resizable, drag a Window Resize shape from the Windows and Dialogs stencil and glue it to the connection point at the bottom right of the Status Bar shape.

6. Drag Scroll Bar shapes (horizontal and vertical) from the Common Controls stencil onto the bottom and right edges of the Blank Form shape, respectively. To specify how far the window scrolls, right-click the Scroll Bar shapes and, from the shortcut menu, choose Set Thumb Size.

> **NOTE** To indicate that only a small portion of the document is visible, make the thumb size small. To indicate that most of the document is visible, make the thumb size large.

7. Drag shapes from the other Windows XP User Interface stencils to add menus, toolbars, icons, and common controls to the form.

Prototyping Wizards

The Wizards stencil includes shapes for the different types of screens you see in wizards. You can use the Simple-Single Page shape to create one-page wizards. For wizards with two or three pages, use the Simple Wizard shapes. Use the Advanced Wizard shapes for wizards with more than three pages.

Creating Wizards with One to Three Pages

To create a wizard with only a few pages, do the following:

1. Drag a Simple-Single Page or Simple-Welcome shape from the Wizards stencil onto the drawing page. With the Wizard shape still selected, type the name of the wizard.

2. To add a graphic to the panel on the right side of the screen, select the box within the right panel and then choose Insert ⊏ Picture to specify the graphic file to use.

3. To add the wizard instructions, select the Wizard Text text block in the main panel of the Wizard shape and type the instructions for your wizard.

4. To construct the user interface elements of the wizard, drag shapes from the Common Control stencil onto the main panel of the Wizard shape.

> **NOTE** Because each Wizard shape includes the navigational buttons appropriate for that wizard page, you can't edit them.

5. For each remaining wizard page, create a new drawing page and repeat steps 1 through 4.

Creating Wizards with More Than Three Pages

Wizards with more than three pages include a welcome page, a completion page, and two or more interior pages. You define the contents of each page as you would for a simpler wizard by dragging shapes from the Common Controls stencil onto the corresponding Wizard shape. To create the pages for an advanced wizard, use the following shapes:

- **Welcome** — Drag an Advanced-Welcome shape from the Wizards stencil onto the drawing page.

- **Interior pages** — Create new drawing pages and then drag an Advanced-Interior shape onto each drawing page. It's common practice to include a graphic related to the one on the Welcome page at the right edge of the Wizard Text box.

- **Completion** — Create a new drawing page and then drag an Advanced-Completion shape onto the drawing page.

Simulating Wizard Navigation

Simulating the navigation through a wizard enables you to test its usability before you write the code to implement the wizard. To simulate wizard navigation, you must place the shapes for each wizard page on a separate drawing page and save the drawing file. To configure wizard diagrams so that you can simulate their behavior, follow these steps:

1. Navigate to the drawing page that contains the first page for the wizard.

2. To modify the drawing size to fit the wizard page exactly, choose File ⇨ Page Setup and select the Page Size tab. Under Page Size, select the Size to Fit Drawing Contents option.

3. To add navigation to the Navigation buttons on the Wizard shape, follow these steps:

   a. Draw a rectangle around each button using the Rectangle tool on the Drawing toolbar.

   b. Make the rectangle invisible by right-clicking the rectangle, choosing Format ⇨ Line, and in the Pattern drop-down list, selecting 00: None. Click OK. Remove the fill in the rectangle by right-clicking the rectangle, choosing Format ⇨ Fill, and in the Pattern drop-down list, selecting 00: None. Click OK.

   d. With the now-invisible rectangle selected, choose Insert ⇨ Hyperlinks. Click the Browse button to the right of the Address box and choose Local File.

   e. In the Link to File dialog box, double-click the drawing you are working on.

   f. Click the Browse button to the right of the Sub-address box.

   g. In the Hyperlink dialog box, in the Page box, select the Visio drawing page that contains the wizard page to which the button should take you and click OK. In the Hyperlinks dialog box, click OK.

4. To run the simulation, choose View ⇨ Full Screen and click the Next or Back buttons to navigate from page to page.

Building Menus and Toolbars

With the Windows XP User Interface template, you can mock up menu bars, drop-down menus, and toolbars. Menus and toolbars can reside on their own on drawing pages or appear in a form you prototype.

Creating Menu Bars

To create a menu bar, follow these steps:

1. From the Toolbars and Menus stencil, drag a Menu Bar shape onto the drawing page. To add the menu bar to a blank form, glue the Menu Bar shape to the connection points on the bottom edge of the title bar area in the Blank Form shape.

2. To display the gripper dots at the left edge of the menu bar that indicate that you can move the menu bar around, right-click the Menu Bar shape and clear the Lock Menu Bar check box.

NOTE You can add an icon at the beginning of a menu bar, similar to the menu bar in Visio and other Microsoft Office programs. Drag an Icon shape from the Icons stencil and glue it to the left edge of the Menu Bar shape. To learn how to import a graphic file for the icon into a Visio shape, refer to Chapter 9.

3. To add a menu item to the menu bar, drag a Top-level Menu Item shape from the stencil and glue it to the connection point at the top-left corner of the Menu Bar shape. With the shape selected, type the menu name.

NOTE If the menu item has an associated keyboard shortcut, underline the appropriate letter in the menu name.

4. Repeat step 3 to glue additional Top-level Menu Item shapes to the right edge of the previous Top-level Menu Item.

Creating Drop-Down Menus

In most cases, you create drop-down menus associated with the top-level menu items in a menu bar. However, you can also build standalone versions. To create a drop-down menu, follow these steps:

1. With a Top-level Menu Item in position, glue a Drop-down Menu Item to the bottom of the Top-level Menu Item shape. Type the name of the drop-down menu item.

2. To add additional drop-down menu items, drag the Drop-down Menu Item shape and glue it to the bottom of the previous Drop-down Menu Item shape.

3. To resize all the drop-down menu items to the width of the widest menu entry, drag from the vertical ruler and glue the blue guide line to the connection point at the bottom right of the widest entry. Then, select each Drop-down Menu Item shape and glue its green end point to the guide line.

4. To specify options for drop-down menu items, right-click each shape and, from the short-cut menu, choose Menu Item Properties. For example, you can change the menu item to Checked, Radio, Cascading, or Separator. The Menu Item Position option sets the justification and width of the menu item, depending on whether it is the top, bottom, or a middle entry in the menu.

Building Toolbars

You can build toolbars within a blank form or on their own on the drawing page. To create a toolbar, follow these steps:

1. Drag a Toolbar shape from the Toolbars and Menus stencil onto the drawing page. To add the toolbar to a blank form, glue the Toolbar shape to connection points for shapes within the blank form, such as the connection points on the bottom of a Menu Bar shape.

2. To display the gripper dots at the left edge of the toolbar that indicate that you can float the toolbar, right-click the Toolbar shape and clear the Lock Toolbar check box.

3. To add buttons to the toolbar, drag a Toolbar Buttons shape or an XP Toolbar Buttons shape from the Toolbars and Menus stencil and glue it to the left edge of the Toolbar shape. In the Shape Data dialog box, choose the type of button you want, such as File or Save.

NOTE To change the type of button later, right-click the button and, from the shortcut menu, choose Set Button Type.

4. To add a separator between buttons, glue a Toolbar Separator shape to the right edge of the previous Toolbar Buttons shape.

5. Repeat steps 3 and 4 to add buttons and separators to the toolbar.

NOTE You can also add shapes to indicate a toolbar drop-down menu or the Overflow Chevron.

Designing Dialog Boxes

The Windows and Dialogs stencil includes shapes for prototyping dialog boxes. After you create a dialog box, you can use shapes from the Icons and Common Controls stencils to add interface elements to the dialog box.

Creating a basic dialog box is similar to creating a form. In fact, you begin with the Blank Form shape and add Windows Button shapes and Command buttons, as you would for a form. Add other interface elements by dragging shapes from the Common Controls stencil. To make it easy to align user interface shapes, drag guide lines from the horizontal and vertical rulers onto the drawing page and glue shapes to the guide lines. To assemble interface elements into logical groups, drag Group Box or Group Line shapes onto the Blank Form shape. Group Box shapes include boundary line and heading text to identify the group. Group Line shapes include a horizontal line and heading text.

 If you add text blocks to the Blank Form shape with the Text tool, use Tahoma 8-point text to match the text in the user interface shapes.

For dialog boxes with tab controls, follow these steps:

1. Drag a Tab Control (Body) shape from the Windows and Dialogs stencil onto the Blank Form shape. Drag the Tab Control (Body) selection handles to resize the shape to fit inside the Blank Form shape with room along the top for tabs.

2. Drag a Tab Control (Tabs) shape from the Windows and Dialogs stencil and glue it to the top edge of the Tab Control (Body) shape. With the Tab Control (Tabs) shape selected, type the tab name.

3. Repeat step 2 for each tab, gluing the new tab to the connection point at the bottom right of the previous tab.

4. To display a tab as the front tab, right-click the tab you want in front and, from the short-cut menu, choose Foreground Tab.

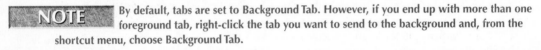 By default, tabs are set to Background Tab. However, if you end up with more than one foreground tab, right-click the tab you want to send to the background and, from the shortcut menu, choose Background Tab.

Summary

Unlike the UML template, which provides a modeling environment, other Visio software templates provide only diagramming tools. They don't validate your diagrams to make sure you correctly apply the rules of the selected methodology. You can use basic Visio techniques such as drag and drop, formatting, control handles, and shape data to build your diagrams.

Chapter 23

Mapping Web Sites

Whether web sites are a few pages and a handful of links or a complex spider's web of pages and related web sites, they require careful planning during the development stage. After sites are up and running, they require conscientious maintenance to ensure that they serve the needs of their organizations and, in turn, those organizations' customers. Visio 2007 provides powerful tools for documenting and managing both stages of web site work.

The Conceptual Web Site template helps plan the structure and navigation of web sites as well as proposed content before you build anything. The Web Site Map template is an invaluable aid for maintaining and refurbishing existing web sites. It includes tools that build models of existing web sites whether they reside on servers or local hard drives and then develop web site map diagrams from those models.

A site map makes it easier to maintain a web site because the current content and organization is plain to see. Furthermore, a Visio site map detects broken links. This chapter explains how to use Visio to create and fine-tune conceptual Web diagrams, as well as how to generate and work with web site maps for existing web sites.

Planning Web Sites

Internet and intranet web sites have become crucial components of business strategy. How will customers find your company if it isn't on the Web? More importantly, why would they buy from you if they can't find out whether your products are what they need? Smart web site development begins like any project—with a set of business objectives and a laundry list of items that should be published on the web site.

Think of the Conceptual Web Site diagram as a storyboard or flow diagram, showing major web site features and the relationships among them. This type of diagram can help web designers:

- Develop and examine the user interactions in your proposed web site
- Create and refine the navigation flow throughout the site
- Determine whether any elements or connections are missing or unnecessary
- Determine whether the web site is too complex and redesign accordingly

The Conceptual Web Site Template

The Conceptual Web Site template helps Web designers visualize the organization of web site elements in a web site, as shown in Figure 23-1. As new ideas or proposed design changes arise, it's much easier to change the conceptual diagram than prototype Web programming. Then, when the concept diagram is complete and officially approved, the transformation of that conceptual web design into a living, breathing web site that serves its target audience is faster, less costly, and less error-prone—and the site is more likely to satisfy the needs of the people who visit it.

NOTE   Although a well-thought-out layout is essential for an effective web site, providing content that visitors are interested in is equally important. Web designers and information architects often analyze business objectives, customer support calls, search terms used, and survey results to identify the information that people want. The Conceptual Web Site template won't help identify this information, but you can add hyperlinks between the reference documents and the shapes on the diagram that represent corresponding web site components.

The Conceptual Web Site template is much like many other Visio templates. It includes stencils with shapes for depicting the plan for and navigation of the components of a web site—the web pages themselves, elements on web pages, popups, and forms that visitors fill in. The template includes the Conceptual Web Site Shapes stencil, Web Site Map Shapes stencil, and Callouts, Backgrounds, and Borders and Titles stencils. Because it is a web site layout tool, it doesn't offer any specialized commands or menus.

FIGURE 23-1

Use the Conceptual Web Site template to plan the content, organization, and interactions of a new web site.

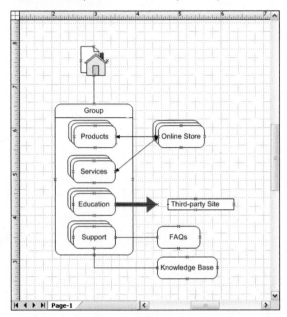

Creating Conceptual Web Diagrams

Because of the hierarchical nature of most web sites, you might find it useful to create an overview page for your conceptual web diagram that shows all the major groups of pages. Then, you can create additional pages to show detail for each of the major groups. For detailed planning, it might make sense to break it down to another level of detail showing the elements on each web page. For simple web sites, an overview of the overall web site structure with additional diagram pages for the details of each individual web page is often sufficient. Either way, a high-level overview page sets up the big picture of a web site or web application before tackling the details of individual pages.

To create a new conceptual web diagram, follow these steps:

1. Choose File ➪ New ➪ Software and Database ➪ Conceptual Web Site.
2. From the Conceptual Web Site Shapes stencil, drag the first shape, such as the Home page shape, onto the drawing.
3. To label the shape, select it, if necessary, and then type its label.
4. Repeat steps 2 and 3 for each web site component you want to document.

5. Drag connector shapes—for example, the 2-Way Data Connection connector or Dynamic connector—onto the drawing. Position the connector(s) to show links among the different shapes on the drawing.

> **TIP** Shapes from the Web Site Map Shapes stencil come in handy when you're creating conceptual Web diagrams that include detail about individual web pages. For example, page-specific shapes, such as Database, FTP, Newsgroup, and Search can identify the purpose of each page.

> **CROSS-REF** Because conceptual web diagrams are composed of shapes and connectors, you can use familiar Visio techniques to fine-tune their appearance. Refer to the following chapters if you need to learn how to format, annotate, or layout Visio diagrams:
>
> Chapter 7 covers formatting diagrams, shapes, text, and other components.
>
> Chapter 4 describes how to layout, align, and distribute shapes.
>
> Chapter 6 discusses text and methods of annotating diagrams.
>
> Chapter 8 provides instructions for adding hyperlinks to pages to navigate from the overview page to detailed pages.

Working with Web Site Maps

A Visio web site map diagram is a versatile tool for maintaining or revamping an existing web site. Not only does the Visio web site map discover and show the structure and elements contained in the web site, but it can also find changes since the last time you generated the site map, including any broken links. Use the web site map as the starting point for reorganizing existing content, adding new elements, or merging or deleting duplicated pages. By keeping an eye on the overall web site structure, you can ensure that the web site is meeting both the organization's needs and those of the target audience.

> **NOTE** The generated web site map is a visual representation of the elements in a specified web site. Although you can add and remove shapes representing web page elements and repair broken links on the drawing, these actions don't change the web site itself. For best results, keep track of the changes you're making on the site map. When you're finished modeling your changes in the Visio web site map, make those changes to the web site itself.

The Web Site Map Template

The Web Site Map template helps web developers analyze and maintain existing web sites. The Generate Site Map command is the core of the template, because it sifts through every web page and link on an existing web site to discover every web page, published document, graphic file (pictures and icons), media file (videos and audio recordings), and other information along with all the links between those elements. The Generate Site Map command then builds a model of those elements to show the hierarchy and relationships between them. Because the Generate Web Site Map command follows all the links to elements, it also finds any broken links on the site without tedious interactive testing.

The Web Site Map Shapes stencil contains shapes for many of the components found on web sites: HTML or XML content, scripts, graphics, audio, and video. Content-related shapes represent documents created with Microsoft Word, Excel, or PowerPoint. Shapes for protocols such as FTP, Mailto, Newsgroup, and Telnet are also included. The stencil provides two types of connectors.

The Web Site Map menu provides commands for alternative ways to view the information in the web site model:

- The Filter Window presents the contents of the web site by category, such as Graphic, HTML, Script, and Text. You can filter the view to show or hide an entire category by selecting or clearing the category check box. For example, if you want to focus on the organization of HTML pages, you might turn off the display of graphics. By expanding a category, you can select or clear individual components.

- The List Window shows the same components as the Filter Window, but in an outline view. In this window, you can display or hide components by expanding or collapsing branches of the outline.

- On the Web Site Map menu, the Reports command opens a dialog box with reports you can run for the web site, such as the Web Site Map Links with Errors report, which shows broken links.

Generating Maps of Existing Web Sites

When you use the Web Site Map template to create a drawing, Visio dutifully creates the drawing, but more importantly, launches the Generate Site Map command. After you specify the URL for the web site for which you want to generate a map, Visio sifts through the site and builds a model of its contents. Visio uses the information in the model to build the web site map diagram on the drawing page.

 Depending on the size and nature of the web site it might be necessary to run the web site map after hours or during slow periods of activity.

Generating a Web Site Map

To generate a web site map, follow these steps:

1. Choose File ➪ New ➪ Software and Database ➪ Web Site Map. The Generate Site Map dialog box appears.

2. In the Address box, type the address for the web site you want to map. Enter the full path, including the protocol and the name of the individual page — for example, `www.bonniebiafore.com/index.htm`.

TIP   Generate Site Map can also discover the contents of an HTML file, ASP page, or Internet shortcuts stored on your local computer. For example, you can generate a site map for your as-yet-unpublished web site from the development files on your computer. In the Generate Site Map dialog box, click Browse. Navigate to the folder that contains your web site files and select the file for the home page or another web page, for example, `..\My Documents\WebSite\index.htm`. Click Open.

3. Click Settings to review the discovery and diagramming settings and modify them if necessary. For example, if the first pass at mapping your site doesn't show all the links you want, you can increase the number of levels to discover. To learn about specifying the settings for discovery, see the section "Configuring What Visio Adds to the Web Site Map Model and Diagram," later in this chapter.

4. To generate the site map, click OK.

   ▨ Visio finds and scans the web site or web documents, verifies their links, searches the specified levels of the hierarchy, and finally displays the web structure and elements as a site map on the drawing page. The time Visio takes to do all this depends on the number of levels and links you specified for mapping, the speed of your computer, and the speed of your Internet connection. The site map appears as a series of shapes linked in a hierarchy, as illustrated in Figure 23-2.

   ▨ The shapes in a web site map include indicators, such as the double downward pointing chevrons, which indicate a link that you can expand, or the red X, which indicates a broken link.

Keeping Your Web Site Up-to-Date

Use the Generate Site Map command to regularly check your web site and keep it up-to-date. Broken links are not only annoying to site visitors, they also give an unfavorable impression of your organization. An effective method for reviewing your web site is to work iteratively, as follows:

1. Generate the site map of the web site using the Visio Web Site Map template.

2. Review the map to detect any broken links and other problems or potential problems with the site.

3. Model the repairs to the broken links and other issues in the web site map. By tracking markup in the web site map, as described in Chapter 11, you can create a clear guide to the changes needed in the actual web pages.

4. Open the source files for the web pages that need updating. Use your changes in the Visio web site map as your punch list for modifying the source files.

5. In Visio, generate the web site map again to confirm that your changes work the way you expect.

Repeat this process as often as the nature and expected usage of your web site demands. For generally static web sites, a monthly check might be sufficient. For sites whose content is constantly changing, a daily check might be required.

FIGURE 23-2

Visio shapes represent each site map element; connectors show the hierarchical relationship between the levels.

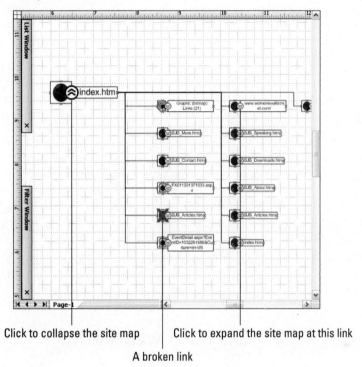

Click to collapse the site map   Click to expand the site map at this link

A broken link

Mapping Protected Areas on Web Sites

If the web site map contains areas that are not filled in or that are covered with a red X, the culprit might be protected areas that require you to click a link or enter a password to continue the site map discovery process. In other words, you need to interact with the web site itself for the site map discovery process to move forward.

To allow mapping of protected areas in a web site, follow these steps:

1. On the drawing page, find the shape representing a link to the web page that requires you to click a link or enter a password.

2. Right-click the link and, from the shortcut menu, choose Interactive Hyperlink Selection. The Interactive Discovery dialog box appears and loads the selected web page.

3. Work with the web page as needed and then click Close. The links you navigated to are added to your site map.

Viewing Elements on Web Site Map Diagrams

Each shape in a web site map diagram represents a page, a link, or other element in the web site. The connectors between shapes indicate the relationships between elements on the site. The different shapes also delineate the type of page or element. For example, by comparing the shapes on the drawing with the shapes in the Web Site Map Shapes stencil, you can learn more about each element — determining whether an element is an HTML page, a graphic image, a style sheet, or JavaScript.

Even the smallest web site can contain an astounding number of elements, when you take into account the graphics files for buttons and icons, and files for visitors to download. To thoroughly review a web site map, you might want to expand or collapse different areas of the site. This section describes several methods for doing so.

- **Expand links** — A freshly generated web site map might only display a single large shape, which shows the name of the web site you mapped. There's no need to panic. To reveal the levels or child links beneath this page, double-click the shape. You can also right-click a shape and then, from the shortcut menu, choose Expand Hyperlink.

- **Expand all links** — To see all child links beneath a shape, right-click the shape and then, from the shortcut menu, choose Select All Hyperlinks Beneath.

- **Collapse links** — Hide child links of a shape by double-clicking an expanded shape, or by right-clicking the shape and, from the shortcut menu, choosing Collapse Hyperlink.

Finding Shapes in Web Site Maps

Suppose you're searching for a particular shape or link in your web site map — for example, a plug-in on the second level of the hierarchy. You can find shapes by searching for shape names, text, shape data, and so on. To find a shape in your web site map, follow these steps:

1. Choose Edit ⇨ Find. The Find dialog box appears.

2. In the Find What box, type the text associated with the shape you're looking for, such as the name of a web page you plan to remove from the site.

3. Under Search In, select an option to specify whether you want Visio to search the current selection, the current page, or all pages on the drawing.

4. Under Search In, select the check boxes that indicate the element you want to search — the element that contains the text you entered in the Find What box. For example, because web page names appear in shape text, select the Shape Text check box to find a web page name. You can check multiple check boxes.

5. Under Options, select the check boxes for any additional search parameters you want to use.

6. Click Find Next. Visio finds and highlights the first shape that meets your search criteria. Click Find Next to continue finding the next shape(s) that meets the search criteria.

Working with the Web Site Map Model

The model that Visio creates of a web site stores every iota of information about that web site and becomes an integral part of the web site map drawing file. This model contains information about every Web element that the Generate Site Map command discovered and how they relate to each other. Initially, Visio uses the model to lay out the web site map diagram, but the true power of the model is that it enables you to look at what you want when you want.

The List Window and the Filter Window offer views of the web site map model, as illustrated in Figure 23-3. The List window lists every web page element in alphabetical order. The Filter window lists the web page elements by file type. These windows open as part of the Web Site Map template and list every element in the web site map, whether or not those elements actually appear on the drawing. You can easily add the element back into the drawing by dragging it from either of these windows onto a drawing page.

FIGURE 23-3

The List window and Filter window display all elements in the memory model, which in turn builds the web site map.

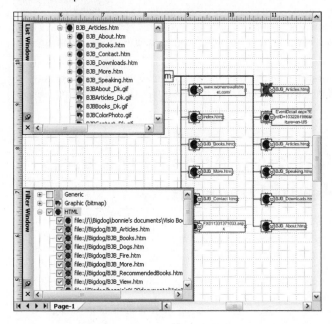

After the initial generation of the web site model and associated diagram, the model retains control over the elements of the web site map. You can delete elements from the web site diagram to focus on key elements, but the deleted elements remain in the model. After you've digested and decided how to renovate the major elements of the web site, you can add deleted elements back to the diagram by dragging them from the model. However, if you delete elements within the List Window or the Filter Window, Visio deletes those elements from both the diagram and the model.

NOTE Remember that any changes made to the web site map and the model affect only the Visio diagram, not the actual web site. You must make changes to the web site files separately using a web site development tool or text editor.

From either the List Window or the Filter Window, you can perform the following actions on model elements:

- **Expand links** — Click the plus sign to the left of an element to expand it. The plus sign changes to a minus sign when an element is expanded.

- **Collapse links** — Click the minus sign to the left of an element to collapse it.

- **Regenerate portions of a web site** — To tell Visio to scan a portion of the web site for changes, right-click an element and, from the shortcut menu, choose Refresh. Visio navigates to the element you right-clicked and finds all its child links.

- **Find a shape on the diagram** — To center the drawing page on an element in the model, right-click the element in the List or Filter Window, and, from the shortcut menu, choose Show On Page.

- **Add hyperlinks to an element** — To add hyperlinks to a model element, right-click the element in the List or Filter Window, and, from the shortcut menu, choose Configure Hyperlinks.

- **Delete elements** — To delete an element from the web site map model and diagram, right-click the element in the List or Filter Window, and, from the shortcut menu, choose Delete.

Configuring What Visio Adds to the Web Site Map Model and Diagram

When you run the Generate Site Map command, the Generate Site Map dialog box includes a Settings button, which enables you to specify how Visio discovers and generates the web site model, as well as how it configures the resulting web site map diagram.

To change the settings for the model discovery and the diagram configuration, follow these steps:

1. Choose Web Site Map ➪ Generate Site Map to open the Generate Site Map dialog box.

2. In the dialog box, in the Address box, type the URL for the web site you want to map. You can also click Browse and navigate to the folder that contains the web site files.

3. Click Settings.

4. In the Web Site Map Setting dialog box, on the Layout tab, which appears by default, adjust the following settings to specify how much detail the command should discover and how Visio should configure the web site diagram:

- **Discover** — Enter the number of levels you want Visio to discover. Although the default is 3 levels, you can specify from 1 to 12 levels. Also enter the number of elements or links you want Visio to discover. The default is 300 links, but you can specify up to 5,000. If you enter a number of levels or links greater than those contained in the web site, Visio will discover all levels and links.

- **Layout Style** — Select the placement and routing styles Visio uses on the site map. To change these styles, click Modify Layout. In the Configure Layout dialog box that appears, specify the styles you want under the Placement and Connectors labels.

- **Shape Text** — Specify the text to include in the shape, such as Relative URL (the default), Absolute URL, File Name Only, HTML Title, and No Text. For example, to obtain full disclosure, choose Absolute URL. To keep the text on the diagram short and sweet, choose Filename Only, which displays only the file name and not the path for the folder in which the file resides.

 You can also change shape text after a site map is generated by choosing Web Site Map ➪ Modify Shape Text.

- **Shape Size** — To control the size of shapes at each level, which provides an easy-to-see cue of the level you're reviewing on the drawing, select the Shape Size Varies With Level check box. By default, the root level is displayed at 200 percent, level 1 at 100 percent, level 2 at 75 percent, and further levels at 50 percent.

5. If necessary, change settings on the Extensions, Protocols, and Attributes tabs to specify the types of elements that Visio can discover and represent with a shape in your web site map. By default, Visio selects all the listed extensions, protocols, and attributes available. Clear the check box for any extension, protocols, and attributes that you want to exclude from the site map. If there are other extensions, protocols, and attributes you want that are not in the list, click Add. The following list describes the options available on the tabs on the Web Site Map Settings dialog box:

- **Extensions tab** — Specify the types of programs, files, and scripts that you want to display or hide when Visio generates the web site map. The Shape column shows the shape Visio uses to represent the file type on the site map.

- **Protocols tab** — Review or change the Internet protocols that Visio is able to discover when generating a web site map. Examples include FTP, File, Mailto, Newsgroups, Search, and so on. The Shape column shows the shape Visio uses to represent the protocol on the site map.

- **Attributes tab** — Indicate the HTML attributes that Visio should include in its discovery and generation of web site maps. Examples include Code, SRC, HREF, Background, and Action.

6. To fine-tune the criteria that Visio uses in generating a web site map, click the Advanced tab. The following list describes the options available:

■ **Analyze all files discovered** — Choose this option to discover every file related in some way to the web site regardless where the linked files are located.

■ **Analyze File Within Specified Domain** — Discover files only on the specified domain. For example, this option discovers files only on the servers in the domain, not servers that host hyperlinked files.

■ **Analyze Files Within Specified Direcotry** — This option discovers files within a directory on a hard drive.

■ **Include Links to Files Outside of Search Criteria** — Select this check box to discover links to files beyond those specified by the Search option you chose. For example, you might choose to search for files within a specified domain, but want to see links to files on other web sites.

■ **Display Duplicate Links As Expandable** — Select this check box to view all links, whether they repeat links on other pages or within the same page.

■ **HTTP Authentication** — If the web site you are going to generate requires login, in the Name and Password boxes, type a valid combination of user name and password. Visio uses this user name and password during the current session, but you'll have to re-enter the user name and password the next time you open the drawing file.

7. After specifying site map settings, in the Web Site Map Settings, click OK. In the Generate Site Map dialog box, click OK to generate the site map applying your new parameters.

8. Regenerate the site map to see the results of the setting changes.

> **NOTE**   If an element that you know exists on a web site doesn't appear in the Visio model, the likely culprit is an omission on the Extensions, Protocols, or Attributes tab. For example, if the Program check box is cleared, Visio won't discover programs with .exe and .bin extensions. Review the Web Site Map settings and then re-run the Generate Site Map command to recapture the web site elements.

Formatting Web Site Maps

You can modify the look of a web site map in a few ways without having to run the Generate Site Map command. For example, you can switch between a hierarchical view and a page-centric view, or move portions of the site map to other pages with page connectors between them.

Displaying Links to and from a Single Shape

By default, Visio displays a web site map in a hierarchical view, which displays all the shapes in the web site map in a top-down scheme according to the layout pattern you've chosen — for example, Compact Tree, Flowchart, or Circular. The hierarchical view provides a complete big picture of the web site to the level of detail you have specified.

However, you can also create the page-centric view, which is great for focusing on the relationships of one particular element in a web site and is shown in Figure 23-4. In the page-centric view, the shape you select becomes the center of the page. Links *to* the shape appear to the left of the central shape. Links from the shape appear to the right of that shape.

FIGURE 23-4

The page-centric view shows the links to and from a single selected shape.

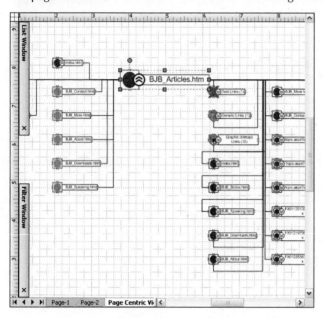

To create a page-centric view, choose Web Site Map ⇨ View ⇨ Page Centric. Visio adds only one page, named Page Centric View, to the drawing file for the page-centric view. If you select another shape for the page-centric view, the new shape replaces the previous shape in the one page-centric viewing page.

Displaying Site Maps Across Multiple Pages

Sometimes, dividing a web site map across multiple pages helps separate categories and might help others understand the structure of the web site. Here are commands for distributing a web site map across several pages and adding navigational aids to each page:

- **Move a portion of a site map to a separate page** — Right-click the shape with the child links you want to move and, from the shortcut menu, choose Make Subpage. The Make Subpage command creates a new page, moves the child links to the new page, and creates an Off-Page Connector shape on both pages.

- **Navigate between pages** — Double-click the Off-Page Connector shape. You can also right-click the Off-Page Connector shape and choose the name of the hyperlink from the shortcut menu.

TIP Many web sites have duplicate links throughout the site or even on the same page. For example, a web page might contain a link to contact information in the midst of a paragraph of content, in a navigation pane, and in a standard menu at the bottom of the page. In Visio, duplicate links are grayed out in your web site map by default. To hide duplicate links in your web site map, when you generate your site map, click Settings, and, on the Advanced tab of the Web Site Map Settings dialog box, clear the Display Duplicate Links As Expandable check box.

Identifying Web Site Problems

Web site map diagrams indicate problem elements — a link or element that is either broken or that Visio could not discover — with a red X. Finding these broken links can help you determine whether you have a problem with the web site itself, or just with the process of generating the site map. Finding broken links is the step that Visio can take, but you must determine the reasons for the breakage and then resolve the problems. You can also run reports that summarize problems with the web site map.

Finding Broken Links

If you're a Webmaster responsible for maintaining a web site, it's a good idea to periodically generate the web site map specifically to find broken links. You can then go straight to the problems in the web site itself and, with a little luck, fix them before your audience finds them.

Broken links are easy to spot, either on the drawing page or in the List window or Filter window. If you see a broken link in one of the windows and you don't see the element in your drawing, right-click the link in the window and choose Show on Page from the shortcut menu.

Getting information about the broken link is your first step toward fixing the problem. Position your mouse pointer over the broken link on the drawing to show the shape tag button. Click that button and read the information it provides. Broken links can be caused by the following errors:

- File Not Found (incorrect file name, incorrect file location, or missing file)
- Site Not Found
- Error <number> (for example, 504: Gateway not found)
- Access Denied
- Password Required
- Site Timed Out

Try these strategies to fix a broken link in your web site map:

- **Check the web address** — Make sure you entered every part of the address — from the protocol (http:// or ftp://, for example) through to the page (home.htm or default.asp, for example), and that there are no typographical errors. If you entered a file path to an HTML file, make sure the path is typed accurately, and that the file is actually there.

- **Refresh the link** — If a broken link is caused by a timeout error, refreshing the link or the parent of the link should resolve the error. To do so, right-click the link and choose either Refresh or Refresh Parent of Hyperlink.

- **Enter required information** — If a link connects to an area of the web site that requires user entry, such as a password or a click, use the Interactive Discovery dialog box. To summon the dialog box, right-click the broken link and, from the shortcut menu, choose Interactive Hyperlink Selection. The Web page appears in the dialog box. Make the required entry and then click Close.

- **Fix the link on the web site** — If the link is still displayed as broken, fix the link on your web site or report the broken link to your Webmaster. After you fix the link, update the shape representing the link on your drawing. Right-click the link and then choose Refresh Parent of Hyperlink. Visio regenerates the parent HTML or ASP page and the first level of links from that page.

Running Web Site Map Reports

You can generate reports of site links, including broken links. Visio provides three built-in reports. You can modify these reports to suit your requirements or create entirely new reports. To generate a web site map report, follow these steps:

1. Choose Web Site Map ⇨ Reports. The Reports dialog box appears, listing the three built-in reports: Inventory, Web Site Map All Links, and Web Site Map Links with Errors.

2. Select the report you want and then click Run.

3. In the Run Report dialog box, select the report format — for example, Excel, HTML or XML. Visio Shape is the default, which displays the report as a shape in the current drawing page.

4. If Visio Shape is selected in step 3, then specify whether to save the report with a copy of the report definition or a link to the report definition and then click OK.

5. For the other format choices, select where to save the report, and Visio generates the report according to the specifications.

Summary

You can use the web diagramming tools in Visio 2007 to assist with web site development and maintenance efforts. The Conceptual Web Site template provides resources for brainstorming, organizing, and prototyping a new web site or application interface. The Web Site Map template includes sophisticated functionality for generating a web site map, including all elements and links, and representing them as shapes in a drawing. It's easy to examine the map for broken links or other problems and then make the changes on the actual web site using a separate web site editor. The web site map makes it easy to diagram future improvements to web sites without committing the actual changes.

Chapter 24

Creating Network Diagrams

With people collaborating throughout organizations as well as over the Internet, computer networks have become the backbone of organizations large and small. Just like buildings and software, networks need to be designed well to satisfy requirements and adjust to changing needs. In addition, they must be regularly maintained to keep things humming. Whether you're just starting to design a network or document a network that has evolved, Visio network diagrams are helpful for representing the equipment in small to medium-sized networks and how that equipment interconnects.

Visio Standard doesn't offer much in the information technology arena, but it does offer a template to build basic network diagrams. Visio Professional includes templates for producing detailed logical or physical network diagrams, documenting directory services, or laying out network equipment in racks.

You can integrate Visio network diagrams with other applications to store network information in databases, track equipment in spreadsheets, or present network designs in PowerPoint presentations. Shape data associated with some network shapes and the Visio Data Link feature enable you to update your diagrams based on the information stored in a database. In addition, you can produce reports, such as inventories, for your network equipment.

Every version of Visio seems to introduce changes to the network templates. Several networking features went away in Visio 2003 and, in Visio 2007, the Novell Directory Services template has disappeared as well due to contractual reasons. This chapter identifies the changes to network diagrams in Visio 2003 and Visio 2007. Then, you'll learn how to use the templates that remain to create logical and physical network diagrams as well as diagrams that drill down to lower levels of detail. You will also learn how to use Visio templates to create directory services diagrams and diagrams of equipment in racks.

Exploring Network Templates

The Network template category in Visio Professional includes templates for documenting high-level network designs, detailed logical and physical network designs, and the arrangement of network equipment in equipment racks. In addition, you can use network diagrams as architecture-level artifacts and as an alternative to UML deployment diagrams. Because network tools changed significantly in Visio 2003, this section identifies both the additions and subtractions from network features in Visio 2003 and Visio 2007.

What's New in the Network Templates?

Here are the additions to networking features in Visio 2003 and Visio 2007.

Visio 2003 Network Enhancements

The enhancements to networking in Visio 2003 aren't a substitute for the features that were lost, but they have their benefits. Here are the improvements in Visio 2003 Standard and Professional.

- **Enhanced Basic Network Diagram template** — Available in both Visio Standard and Visio Professional, the shapes in the Basic Network Diagram template look more professional and behave more consistently.

- **Detailed Network Diagram template** — This template for documenting logical and physical network topology replaces the Logical Network Diagram template in Visio 2002.

- **Rack Diagram template** — New in Visio 2003 Professional, this template provides shapes in standard industry sizes for determining rack space requirements for new equipment. The shapes in this template fit together precisely so you can stack equipment and accurately estimate the rack space you need.

- **Consistent set of network shape data** — In Visio 2003 and previous versions, shape data are called custom properties, but, by either name, Visio 2003 networking shapes possess a set of shape data that work with the three predefined reports (Network Device, Network Equipment, and PC Report) introduced in Visio 2003.

Visio 2007 Network Enhancements

Visio 2007 Professional introduces a few new networking shapes:

- **Rack Mounted Servers stencil** — The Detailed Network Diagram template opens this new stencil, which contains 16 rack servers drawn in the isometric perspective. These masters represent the same types of servers available on the Servers stencil except that they are drawn as rack-mounted equipment.

- **XML Web Service** — This one lonely new shape on the Detailed Network Diagram stencil represents an XML web service.

What's Missing?

Visio 2003 dropped the following networking features:

- Visio's AutoDiscovery features.

- Using SNMP to discover network resources in a local or wide area network and create a diagram of your existing network.

- Importing directory services information, such as an existing Active Directory structure.

- The Directory Navigator in the Directory Services Diagram templates.

- The Visio Network Equipment Sampler. However, you can still download network shapes from equipment manufacturers' web sites.

Mapping your network is still possible with third-party products, such as the Optiview Console and LAN MapShot products from Fluke Networks. Both use Visio to generate diagrams of discovered network devices. An article about Fluke Networks' network tools is available at `www.microsoft.com/casestudies/casestudy.aspx?casestudyid=50463`.

The only casualty in Visio Professional 2007 is the Novell Directory Services template and its stencils, which are no longer available due to contractual reasons.

Choosing the Right Template

The Network templates available in Visio 2007 do not include network-specific menus, toolbars, or add-ons. When you create a new network diagram, Visio creates a letter-size page using portrait orientation and no drawing scale, and opens stencils appropriate for the type of network diagram. The Basic Network Diagram template is the only template available in both Visio Standard and Visio Professional. The other templates are available only in Visio Professional. Table 24-1 lists all the network templates available in Visio 2007.

Using Visio with Large Networks

If you work with networks with thousands of nodes, you can use Visio Professional to produce high-level network diagrams, although you have better things to do than draw diagrams that show every end node. For example, you might produce diagrams that show all the routers on your network or the service providers that support your network in different geographical areas.

When you manage large networks, you can use products such as HP OpenView Network Node Manager, MicroMuse Precision, or What's Up Gold to automatically generate maps of your network topology. Even with advanced discovery and mapping tools such as these, the maps they generate might require editing to represent your network the way you want.

TABLE 24-1

Visio 2007 Network Templates

Network Template	Description
Basic Network Diagram	Design and document simple networks or produce presentation graphics for high-level networks.
Detailed Network Diagram	Design and document logical network connections or the layout and physical connections for network equipment at specific sites. Produce drill-down diagrams showing progressive levels of network detail.
Active Directory	Design and document directory services for sites using Microsoft Active Directory.
LDAP Directory	Design and document directory services for sites using Lightweight Directory Access Protocol (LDAP).
Rack Diagram	Optimize the use of space in equipment racks and document the configuration of equipment in racks.

Creating Logical and Physical Network Diagrams

In Visio, network diagrams respond to all the standard Visio drawing techniques — dragging and dropping from network stencils to add equipment, representing network topology by gluing connectors from Ring Network or Ethernet shapes to network equipment shapes, and using text, annotation shapes, and shape data to add information.

Setting Up Network Diagrams

Regardless whether you are creating a basic network diagram or one with levels of detail, the same techniques apply. To create a network diagram, choose File ➪ New ➪ Network and then choose either Basic Network Diagram (in Visio Standard and Visio Professional) or Detailed Network Diagram (in Visio Professional only). Visio creates a new drawing file containing a single letter-sized drawing page.

If your network is larger than a few servers or you plan to include details, change the drawing size by choosing File ➪ Page Setup, selecting the Page Size tab, and specifying the page size you want. The default orientation is Landscape, which more often than not is just what you want.

CROSS-REF If you want to spiff up a diagram of a wide area network (WAN), you can add a graphic image of a map to a background page (see Chapters 3 and 8) or drag a shape from the Backgrounds stencil.

Adding Nodes and Network Topology

Whether you're trying to sort out the jumble of your small business' network or documenting the wide area network for your organization, topology and equipment shapes are the backbone of your diagram. Topology shapes, such as the Ethernet shape, have control handles that you drag and glue to the shapes that represent nodes and devices in your network.

If you've created other types of Visio diagrams, adding topology and nodes is straightforward as the following steps show:

1. Add topology shapes from the Network and Peripherals stencil by dragging a Ring Network shape or Ethernet shape onto the drawing page. Resize the shape if you want by dragging its selection handles.

NOTE The Cloud shape on the Network Locations stencil is perfect for showing stratospheri-cally high-level connectivity. For satellite or microwave links, glue a Comm-Link shape, which looks like a lightning bolt, to shapes on a network diagram.

2. From stencils such as Computers and Monitors, Network and Peripherals, or Detailed Network Diagram, drag shapes that represent network devices onto the drawing page.

3. To connect the equipment to the Visio topology shapes, on the drawing page, select a topology shape, such as an Ethernet shape. Visio displays yellow control handles designed to connect to shapes that represent network devices. Drag a control handle and glue it to a connection point on a device shape, illustrated in Figure 24-1. When the device shape is glued to the topology shape, its connection point turns red.

4. If you use all the connectors that the topology shape shows by default, add additional connectors by dragging one of the control handles within the topology shape, as shown in Figure 24-1.

FIGURE 24-1

Add more connectors to a topology shape by dragging control handles.

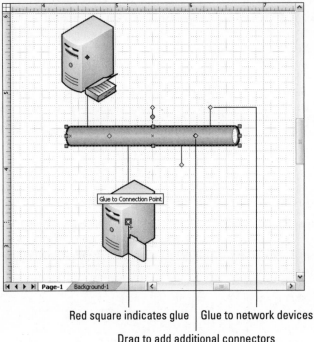

Glue to Connection Point

Page-1   Background-1

Red square indicates glue | Glue to network devices

Drag to add additional connectors

TIP Unused connectors hanging in the space of a drawing page are as unsightly as cables hanging from ceilings in an office. To hide unused connectors, drag their control handles back to the interior of their Ethernet or Ring Network shape.

5. Annotate and format the shapes as described in Chapters 6 and 7.

Creating Drill-Down Diagrams

Large or complex networks typically don't fit on a single page without looking as twisted as the wires in a telephone closet. *Drill-down diagrams* show the high-level view of your network on one page, with detail on additional pages. For example, you can use Cloud and Building shapes from the Network Locations stencil to represent the campus buildings that your network supports. Then, the network equipment within each building can appear on additional drawing pages, as shown in Figure 24-2. If necessary, you can create additional drawing pages for details such as the network equipment within a computer room in one of the buildings.

CROSS-REF To navigate more easily between these drawing pages, add hyperlinks from the high-level drawings pages to the detailed drawing pages. Chapter 8 describes how to add hyperlinks and navigation shapes to drawings.

FIGURE 24-2

Drill-down diagrams show additional detail on other drawing pages.

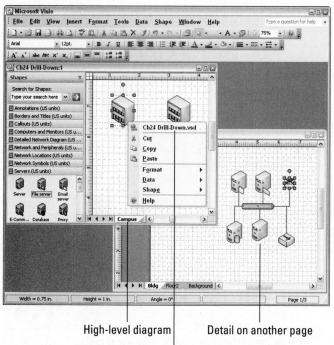

High-level diagram | Detail on another page

Use hyperlinks to navigate between pages

Documenting Directory Services

Visio Professional provides templates for documenting Active Directory Services or LDAP Directory Services. Directory services diagrams come in handy for designing new directories, redesigning existing directories, or planning the migration of your current network directory. You can also use the Visio Directory Services templates to plan network resources and set network policies.

To create a directory services diagram, choose File ➪ New ➪ Network and then choose the template for the type of directory services diagram you want to create. Filling in the diagram is nothing more than dragging shapes from the directory service stencil onto the drawing page. Defining the parent-child relationships between directory services objects is connecting the parent and child with the Directory connector. (On the Standard toolbar, click the Connector tool, click the Directory connector in one of the stencils [such as LDAP Objects], and then, on the drawing page, drag from the shape that represents the parent object to the shape that represents the child object.)

Showing Relationships in Active Directory Diagrams

The shapes on the Active Directory Sites and Services stencil can do double duty by helping to plan networks and directories or to show how information is distributed and replicated to servers in your network. The following shapes show relationships between network domains, sites, and services:

- **Site shapes** — Represent regions for network connectivity, whether by geography or function, and can show one or more LANs and their interconnections.

- **Domain shapes** — Define security and administrative boundaries in a network. These shapes exist within a site or sites and move along with the shape that represents a site. A domain might have one or more domain controllers. Sites can also have one or more domain controllers.

- **Site link shapes** — Represent the transport links that communicate information between sites. The Site-Link Bridge 3D shape delineates a set of site links joined to form a larger link or bridge.

- **Replication connection shape** — Represents intersite replication between two domain controllers.

Laying Out Equipment Racks

When rack space is at a premium or your equipment rooms are simply packed to overflowing, rack diagrams can help the people installing equipment find the right location and make the right connections. The shapes in the stencils for rack diagrams conform to industry-standard sizes, so they fit together precisely. Connection points are positioned on equipment shapes, so they snap to racks and other rack-mounted equipment. Rack-mounted equipment glues together so your rack configurations stay connected even when you move them.

When you create a rack diagram in Visio Professional, the following stencils open along with your new drawing:

- **Annotations** — Add labels, notes, and reference symbols to your rack layouts.
- **Callouts** — Add labels and notes to your rack layouts.
- **Free-standing Rack Equipment** — Include equipment that doesn't attach directly to racks, such as monitors, printers, and laptops.

- **Network Room Elements** — Include shapes for elevation views of doors, windows, chairs, and tables.

- **Rack-mounted Equipment** — These shapes are useful for showing network equipment that attaches directly to racks, such as routers, switches, patch panels, shelves, and servers.

To create a rack diagram, follow these steps:

1. Choose File ➪ New ➪ Network ➪ Rack Diagram.

2. From the Rack-mounted Equipment stencil, drag the Rack shape or Cabinet shape onto the drawing page. The Rack shape has open sides, whereas the Cabinet shape has sides to enclose the equipment.

3. To change the height of the rack or cabinet, choose one of the following methods:

   ▦ Select the Rack or Cabinet shape and then drag a selection handle at the top or bottom of the shape up or down.

   ▦ Right-click the shape, from the shortcut menu, choose Properties. In the Shape Data window, in the Height in U field, type a new number of units. One unit represents the spacing between two bolting slots in a rack.

NOTE The U height is the number of units a rack holds or a piece of equipment uses — a unit is a set of holes for bolting equipment into a rack. By default, the U height appears above Rack and Cabinet shapes and to the left of Rack-mounted equipment shapes. To hide the U height, right-click a shape and choose Hide U Sizes from the shortcut menu. To show U height, right-click a shape and choose Show U Sizes.

4. To change the width between rack holes, right-click the Rack or Cabinet shape, from the shortcut menu, choose Properties, and in the Width Between Holes field, type the number of inches or centimeters for metric.

5. To add equipment to a rack or cabinet, drag equipment shapes from the Rack-mounted Equipment stencil and glue them to the Rack or Cabinet shape. Connection points at the lower corners of the equipment shape glue to the connection points on the Rack or Cabinet shape, as shown in Figure 24-3, and turn red to indicate that they are glued.

FIGURE 24-3

Equipment shapes glue to the connection points on Rack and Cabinet shapes.

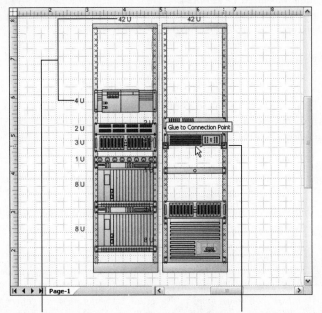

Units for a rack and equipment   Gluing rack-mounted equipment to a rack

Enhancing Network Diagrams

Network diagrams might relate to high technology, but they can benefit from formatting and good annotation. Many of the standard Visio techniques apply to network diagrams, from color-coding with data graphics and numbering shapes to adding shape data and producing reports. This section describes some of the techniques you can use and where to find more detailed instructions elsewhere in this book.

Formatting and Annotating Network Diagrams

Use one or more of the following Visio features to enhance the appearance of your network diagrams:

- **Text and annotation shapes** — As described in Chapter 6, you can add text shapes or annotation shapes from the Annotations stencil to document equipment and connections.

- **Number network shapes** — With dozens if not hundreds of network devices on a diagram, numbers can help identify the equipment in question. The Number Shapes add-on, described in Chapter 6, can number shapes in sequence as you add them to your diagram or after all the shapes are in place.

■ **Format shapes** — Network shapes respond to standard Visio formatting commands, described in Chapter 7.

■ **Label or visualize network data** — Network diagrams can contain a lot of information, much of which might be stored in shape data. For example, you might want to color-code equipment based on the LAN to which it belongs. Or, you might simply want to display shape data values in or around the network shapes. The new Data Graphics feature, described in Chapter 10, enables you to show data in various ways: color-coding, text, status bars, and icon sets.

Storing and Reporting Network Information

The shapes in the network diagram templates include a consistent set of shape data for the network devices they represent, such as asset numbers, serial numbers, manufacturer, part number, and IP address. If you fill in shape data for network device shapes, you can view that information in the Shape Data window or produce reports.

Here are the data-related tasks you can use with network shapes and where to find detailed instructions elsewhere in this book:

■ **Add shape data values** — See Chapter 10 for instructions on adding values to shape data.

■ **Create new shape data properties** — Although the predefined shape data is rather robust, you can add your own shape data properties, as described in Chapter 33.

■ **Link shapes to fields in a database** — See Chapter 10 for instructions on using the new Data Link feature to connect database fields to shape data in shapes on Visio drawings.

■ **Generate equipment reports** — Chapter 10 describes how to run reports and customize them. Visio includes built-in reports, including the Inventory report, Network Device, Network Equipment, and the PC Report.

Summary

The Network templates are primarily drag-and-drop tools. In Visio Standard, you can create basic network diagrams. With Visio Professional, you can create basic and detailed network diagrams for logical and physical networks, directory services diagrams, and rack layouts. In addition to dragging, dropping, and connecting network shapes, you can use shape data to enhance the appearance of diagrams and report on the devices in your network.

Part V

Using Visio for Architecture and Engineering

Chapter 25

Working with Scaled Drawings

If you've ever experimented with a new office layout by shoving your office furniture around, you already appreciate the value of working with scaled drawings. The objects you place in a design on paper are much easier to move around than the real-world building components, furniture, and equipment you deal with during construction. By scaling real-world objects up or down, you can work on them at a manageable size, manipulate them into the results you want, and easily share them with colleagues. In addition, by drawing real-world objects at different scales, you can show more or less detail without consuming entire forests of trees. For example, the layout of a factory floor might be scaled to $\frac{1}{8}$" = 1' 0", whereas the connection details between a steel column and a floor joist could be $\frac{1}{2}$" = 1' 0".

For technical drawings such as architectural plans, accuracy is essential or the construction crews in the field will wield their hammers getting pieces to fit together — usually with results that don't make the client or the architect happy. To make field assembly go smoothly, the shapes on drawings must be placed precisely *and* accurately.

Although Visio isn't meant to replace a computer-aided design (CAD) application, it has tools to produce precise plans. Visio includes shapes that are designed to work on scaled drawings and adjust to the scale you're using. Many of architectural and engineering shapes in Visio include behaviors that help you lay out components on your plan, although Visio shapes can't deliver the equivalent of a bachelor's degree in architecture or engineering.

In this chapter, you'll learn which Visio templates produce scaled drawings, how to use scale and units to your advantage, and how to indicate dimensions on your scaled drawings. Following chapters cover the specifics of creating different types of scaled drawings, such as office layouts, building plans, and different types of engineering plans.

CROSS-REF To learn about methods for positioning shapes precisely, see Chapter 4.

Exploring Scaled Drawing Templates

Visio provides several drawing templates specifically designed to produce scaled drawings. Visio automatically sets the units and drawing scale to those common for the type of drawing. Scaled shapes on the associated stencils resize to the scale of the drawing as long as the drawing and shape scales aren't too disparate. Besides, you can specify whatever drawing scale you want on any Visio drawing and create shapes to work at that scale.

Choosing the Right Scaled Drawing Template

If you use Visio Standard, the Office Layout template works not only for office layouts but for building plans as well. In fact, it's your only choice because the Office Layout template is the only Visio Standard template with shapes for walls.

Visio Professional offers plan templates suited to different types of plans. For example, although the Office Layout template is available, the Floor Plan template is better because it creates a standard architectural size page and opens stencils with more shapes for walls, doors, windows, and other common building components. Most of the Maps and Floor Plan templates use an architectural page size of 36 inches by 24 inches and a drawing scale of ¼" = 1' or an A4 page using a drawing scale of 1:50 and millimeters for metric units. In the Engineering category, the Parts and Assembly Drawing template uses engineering page size and scale.

CROSS-REF For a complete list of the stencils that open with templates, see Chapter 42.

Most scaled drawing templates belong to the Maps and Floor Plans category in Visio Professional. The Parts and Assembly Drawing template is part of Visio Professional and is the only scaled drawing template in the Engineering category. Here are the Visio templates available for scaled drawings:

- **Office Layout** — The Office Layout template is the only scaled drawing template available in both Visio Standard and Visio Professional. It's useful for laying out seating plans, tracking inventories of furniture and other office equipment, planning office space and cubicle placement, or planning for office moves. The template sets the page to a letter-sized landscape page and an architectural scale of ½" = 1' 0". The metric template sets the paper and scale to A4 and 1:25.

- **Floor Plan** — The Floor Plan template is like the Office Layout template on steroids. Its stencils include more shapes for building components, so you can design buildings, facilities, and space layouts. Still, you don't want to try to design a skyscraper with it. The template sets the page to an ANSI Architectural size 36" × 24" and an Architectural scale of ¼" = 1' 0". Metric uses A4 at 1:50.

■ **Home Plan** — The Home Plan template provides shapes for laying out spaces and rooms in houses as well as designing the interior layout of each room with furniture and fixtures. The template sets the page to an ANSI Architectural page of 36" × 24" and an Architectural scale of ¼" = 1' 0". Metric uses A4 at 1:50.

■ **Plant Layout** — At the other end of the spectrum from home and office layouts, the Plant Layout template is designed to help lay out large equipment on factory floors or the spaces required for storing, receiving, and distributing inventory. The template sets the page to an ANSI Architectural page of 36" × 24" and an Architectural scale of ¼" = 1' 0". Metric uses A4 at 1:50.

■ **Electrical and Telecom Plan** — As its name implies, this scaled template provides shapes for adding electrical and telecommunications services to floor plans. The template sets the page to an ANSI Architectural page of 36" × 24" and an Architectural scale of ¼" = 1' 0". Metric uses A4 at 1:50.

■ **Plumbing and Piping Plan** — This scaled template provides shapes for laying out plumbing fixtures and pipes on floor plans. The template sets the page to an ANSI Architectural page of 36" × 24" and an Architectural scale of ¼" = 1' 0". Metric uses A4 at 1:50.

■ **Reflected Ceiling Plan** — If you've ever stared up at a ceiling in a commercial building, you know that they are teeming with lights, sprinklers, and other assorted services. This scaled template provides shapes for laying out the reflected ceiling grids and tiles as well as the sprinklers, diffusers, and lights. The template sets the page to an ANSI Architectural page of 36" × 24" and an Architectural scale of ¼" = 1' 0". Metric uses A4 at 1:50.

■ **HVAC Plan** — The HVAC Plan template helps lay out the HVAC ductwork and devices that keep spaces comfortable regardless of the temperature outside. The template sets the page to an ANSI Architectural page of 36" × 24" and an Architectural scale of ¼" = 1' 0". Metric uses A4 at 1:50.

■ **HVAC Control Logic Diagram Plan** — This template shows the controls for the HVAC systems in a building. The template sets the page to an ANSI Architectural page of 36" × 24". Because this plan doesn't require precision placement, the scale is set to 1 to 1. Metric uses A4 at 1:1.

■ **Security and Access Plan** — This template includes shapes for laying out devices for access to building, such as doors and turnstiles, and security features, such as keypads, card readers, and cameras. The template sets the page to an ANSI Architectural page of 36" × 24" and an Architectural scale of ¼" = 1' 0". Metric uses A4 at 1:50.

■ **Site Plan** — The Site Plan template includes shapes for landscaping, from the layout of your backyard to the plans for a park or commercial development. Because site plans cover a lot of ground, the template sets the page to an ANSI Architectural page of 36" × 24" and an Engineering scale of 1" = 10' 0". Metric uses A4 at 1:200.

- **Space Plan**—Space planning is the task of designing the overall layout and relationships between spaces, such as operating rooms, supply closets, and the admitting area. The Space Plan template includes shapes for arranging spaces without the details inside them. Because space plans cover larger spaces, the template sets the page to a letter-sized landscape page but uses an Architecture scale of ⅛" = 1' 0". Metric uses A4 at 1:100.

- **Parts and Assembly Drawing**—The Parts and Assembly template, which is the only scaled template in the Engineering category, provides shapes for drawing schematics or detailed mechanical drawings of parts and how they go together. The template sets the page to an n ANSI Engineering B-size page of 17" × 11" and an engineering scale of ¼:1. Metric uses A4 at 1:10.

Working with U.S. and Metric Templates

All Visio templates are available in both U.S. and metric units. When you install the English language version of Visio, the installation procedure checks the settings on your computer and installs the templates that match. If you use both types of units in your work, you can install both sets of templates on your computer. With both sets of templates installed, you simply choose the template with the units you want to use when you create a new drawing.

To install both sets of templates after Visio is already installed, follow these steps:

1. If Visio is running, save your work and then choose File ➪ Exit.
2. Click the Start button and choose Control Panel.
3. In the Control Panel window, double-click Add or Remove Programs.
4. In the Currently Installed Programs list, select Microsoft Office Visio and then click Change.
5. On the first wizard page, select the Add or Remove Features option and then click Continue.
6. On the Installation Options tab, click the plus sign next to Microsoft Office Visio. If options that you want to use are preceded by a red X, such as Solutions (US units), Solutions (Metric units), Add-ons (US units), and Add-ons (Metric units), click the arrow to the right of the red X and then, from the shortcut menu, choose Run from My Computer.
7. Click Continue.

When both U.S. and metric unit templates are installed, Visio displays two options for each built-in template. When you choose File ➪ New and point to the template category, you'll see one template with (US units) after the template name (Visio doesn't use the periods for U.S.), and the other followed by (Metric). Choose the template that uses the units you want, and Visio sets up the drawing with the appropriate measurement units and drawing scale.

CAUTION When you work with both types of units, make sure when you open additional stencils that you pick the stencils whose units match the template you're working with. Mixing U.S. units and metric units in the same drawing can make shapes line up improperly.

Setting Drawing Scale and Units

Visio scaled drawing templates automatically set up your drawings with appropriate units and scale to show real-world objects at a manageable size. As you add shapes from scaled drawing stencils to the drawing page, the shapes resize to match the drawing scale you're using. In addition, the dynamic grid, rulers, and other Visio drawing aids obligingly adjust to the current units and scale to help you place shapes precisely. Perhaps the best aspect of working with Visio scaled drawings is that you can glue dimensioning shapes to the plan shapes on a scaled drawing. The dimensioning shapes automatically measure the distances they span and display them based on the units and drawing scale.

NOTE A shape won't resize if its scale is more than eight times larger or smaller than the scale of the drawing page. If shapes don't resize, make sure that you are using scaled shapes from stencils designed to work with the type of drawing you're using. Visio compares the scale of the drawing on which the master resides to the scale of the drawing on which you drop shapes. You can create masters that work at a specific scale by setting the scale on your master drawing before you create the master shapes.

Scaling Shapes Up or Down

A drawing scale represents how a distance on a piece of paper corresponds to a distance in the real world. Whether you use architectural or engineering formats, the scale makes full-size objects shrink to fit on the drawing page. For example, the typical U.S. architectural scale of $\frac{1}{4}$" = 1' 0" means that $\frac{1}{4}$ inch on a piece of paper is the equivalent of 1 foot in the real world, which works for most building plans. However, a large building might require $\frac{1}{8}$" = 1' 0". Conversely, you might use a scale of 1" = 1' 0" to show welds and connection details for a steel column. Site plans usually cover so much ground to capture the configuration of buildings, roads, parking lots, and more, that a scale of 1" = 10' 0" is needed.

NOTE Sometimes, scales appear in shorthand that shows only the ratio between paper size and real-world size. For example, the metric scale of $\frac{1}{8}$:1 means that the drawing on paper is one-eighth of actual size. In Visio, metric scales are represented as ratios, such as 1:50, which indicates that 1 meter on paper represents 50 meters actual size.

The smaller the drawing scale, the more you can show on the same size piece of paper. Table 25-1 shows the real-world distances you can show on a 36" × 24" architectural page at different scales.

CAUTION Because the drawing scale affects the size at which shapes appear on the drawing page, be sure to set the drawing scale *before* you add shapes to the drawing page. In addition, if you change the drawing scale after you've added shapes to the page, they might not resize properly. For example, the text blocks in title block shapes might not fit properly in their designated boxes.

TABLE 25-1

Distances You Can Represent on Scaled Drawings

Drawing Scale	Real-World Dimensions
1" = 10' 0"	360 feet × 240 feet
1/8" = 1' 0"	288 feet × 192 feet
1/4" = 1' 0"	144 feet × 96 feet
1" = 1' 0"	36 feet × 24 feet
1:50	45.72 meters × 30.48 meters

The shapes on scaled drawing stencils resize to match the drawing scale, as shown in Figure 25-1.

FIGURE 25-1

The ruler, grid, and scaled shapes all resize to match the scale of your drawing.

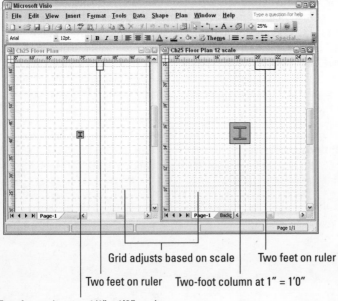

Grid adjusts based on scale   Two feet on ruler

Two feet on ruler   Two-foot column at 1" = 1'0"

Two-foot column at 1/4" = 1'0" scale

Setting Drawing Scale

Each drawing page in a drawing file can use a different drawing scale, which is handy when you want to lay out a floor on one page but need a larger scale drawing of a construction detail on another page. To specify the drawing scale for a drawing page, follow these steps:

1. Display the page whose drawing scale you want to set.

2. Choose File ⇨ Page Setup and select the Drawing Scale tab.

3. To specify one of the scales predefined in Visio, select the Pre-defined Scale option and select one of the following types of predefined scales:

   - **Architectural** — Relates a number of inches or a fraction of an inch on paper to 1 foot in the real world

   - **Civil Engineering** — Relates 1 inch on paper to a number of feet in the real world

   - **Metric** — Relates meters on paper to a number of meters in the real world

   - **Mechanical Engineering** — Represents a ratio to scale objects up or down. For example, a mechanical engineering scale of ⅛:1 relates a fraction of a unit to one unit in the real world in order to scale objects down to fit on the page. A scale of 8:1 relates multiple units on paper to one unit in the real world in order to scale objects up so they're legible on paper.

NOTE You can also create your own drawing scale by selecting the Custom Scale option and specifying the paper distance and its corresponding real-world distance.

4. In the Scale drop-down list, choose the predefined scale you want. The values in the Page Size boxes change to indicate how many measurement units fit on the page at the scale you've selected.

5. Click Apply to save the drawing scale with the drawing page. Although the shapes on the drawing resize to match the new drawing scale, and the distances shown in the rulers adjust to the new scale, the real-world dimensions of the shapes on the drawing page remain the same.

6. If you use background pages with your scaled drawings, display the background page and then repeat steps 2 through 4 to apply the same drawing scale to it.

Showing Scale on Drawings

When you work with scaled drawings, it's a good idea to indicate the drawing scale somewhere on the drawing page, so that someone viewing a hard copy of the drawing knows what the scale is and can measure objects on it correctly. Visio provides several shapes that automatically display the drawing scale for you. Table 25-2 lists some of the shapes you can use to show drawing scale.

To use one of these shapes, simply open the stencil on which the master is located and drag it onto the drawing page. By default, each shape shows the drawing scale differently, as outlined in Table 25-2 and shown in Figure 25-2. However, if you use the Drawing Scale shape from the Annotations stencil, you can change the scale type by right-clicking the shape and then, from the shortcut menu, choosing one of the scale styles.

TABLE 25-2

Shapes That Show Drawing Scale

Shapes	Stencil	Scale Style
Drawing Scale	Annotations	Mechanical Engineering $\frac{1}{48}$:1
Scale Symbol	Annotations	Graphical display of scaled distances
Scale	Title Blocks	Decimal format 1:48
Title Block Large	Title Blocks	Maintains drawing scale format specified
Title Block Small	Title Blocks	Maintains drawing scale format specified

FIGURE 25-2

You can configure scale format in some shapes, whereas others have fixed scale formats.

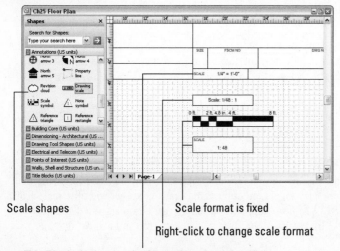

Scale shapes

Scale format is fixed

Right-click to change scale format

Title Block shape uses the format for selected scale

Specifying the Units Used to Dimension Shapes

When you work with scaled drawings, two types of units are important: page units and measurement units. Page units represent the distances or units on the printer page or piece of paper you print, such as inches or millimeters. Measurement units represent real-world distances or units for the actual sizes of the objects you're drawing, and these come into play mainly when you add dimensions to a plan or when you measure a distance. For example, in the architectural scale of ¼" = 1', the page units are inches and the measurement units are typically feet and inches.

If you use one of the scaled drawing templates, Visio automatically sets both the drawing scale and measurement units for you. You can switch between U.S. and Metric units by choosing a U.S. units template or a Metric template when you create a new drawing. In Visio, scale and units are inextricably linked. Drawing scales specify the relationship between page distances and real-world distances, so choosing a drawing scale in Visio also sets the measurement units and page units.

Setting Default Units

Although Visio templates set units for you, it's wise to set default units to use in case you decide to create a drawing without a template. To specify either U.S. or metric units, do the following:

1. Choose Tools ➪ Options and select the Units tab.

2. Under Default Units, select the Always Offer 'Metric' and 'US Units' for New Blank Drawings and Stencils check box.

3. If you want to change the units for the current page, click Change and then choose the new units in the Measurement Units drop-down list.

> **TIP** The list of measurement units includes units such as days and weeks. You can choose these units if you want to produce schedules in which 1 inch represents one week or some other length of time.

Specifying Measurement Units for a Page

Similar to drawing scale, measurement units can differ for each drawing page. For example, you can specify whether the rulers and drawing grid use inches, meters, or even miles, for different drawing pages in the same drawing file. To guarantee that a plan you're drawing fits on the page, you can specify the page size in measurements units. For example, if you want to draw a building that is 60 feet long and 30 feet wide, you can set your drawing page to 70 feet by 40 feet in measurement units. To specify measurement units, use one of the following methods:

- **Specify measurement units** — Choose File ➪ Page Setup and select the Page Properties tab. From the Measurement Units drop-down list, choose the units you want and then click Apply. Visio changes the distances you see on the rulers and adjusts the grid to match the new units.

- **Specify the page size in measurement units** — Choose File ➪ Page Setup and select the Drawing Scale tab. In the Page Size (In Measurement Units) boxes, type the distances you want to represent on the page. For example, to create a page that represents 70 feet

501

by 40 feet, type **70 ft.** in the first box and type **40 ft.** in the second box. Click Apply to change the page size. Visio shows the size of the drawing page and the printer paper in the preview pane, as shown in Figure 25-3.

FIGURE 25-3

You can change the drawing scale, the drawing page size, or the printer paper to make your plan print on the page.

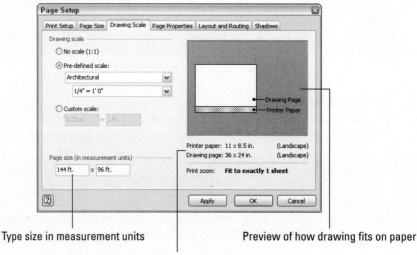

Type size in measurement units   Preview of how drawing fits on paper

Physical size of printer paper and drawing page

Dimensioning Scaled Drawings

Measuring a hard copy of a scaled drawing to determine the actual sizes of scaled objects isn't always possible. Only the most hard-core architects and engineers walk around with scales in their pockets. Yet, knowing the dimensions of objects on scaled drawings is crucial or the furniture you pick out at the store might not make it through the door at its destination. Typically, scaled drawings include dimensions to show sizes, offsets, and distances from reference points. Visio Professional provides stencils with shapes for dimensioning linear, radial, and angular distances.

NOTE   In Visio Standard, the only way you can add dimensions to a scaled drawing is with the Room Measurement shape and the Controller Dimension shape on the Walls, Doors, and Windows stencil.

Visio Professional provides two stencils with shapes specifically designed to glue to scaled shapes and show their dimensions. Although the shapes on each of these stencils share the same names and work the same way, they display dimensions in different formats. Depending on the type of drawing you are creating, you can open a dimensioning stencil by choosing File ➪ Shapes ➪ Visio Extras and then choosing either of the following stencils:

- **Dimensioning–Architectural** — For linear dimensions, architectural dimension shapes display the dimension value above the dimension line and use slashes at the ends of the dimension line.

- **Dimensioning–Engineering** — For linear dimensions, engineering dimension shapes display the dimension value in the middle of the dimension line and use arrowheads at the ends of the dimension line.

Adding Dimensions

Some scaled shapes, such as Room and Wall shapes, display dimensions automatically when you select them, but the dimensions disappear as soon as the shape is no longer selected. To annotate your drawings with dimensions that stay put, use dimension shapes instead. The Dimensioning stencils include a wide range of dimension shapes, but they all behave similarly. You drag a dimension shape onto the drawing page and glue its dimension lines to the shapes you want to measure. The dimension shape displays the dimension and recalculates the dimension automatically if you resize the shape.

Dimensioning Linear Distances

Dimension shapes include control handles you drag to define the distance to measure as well as the location of the dimension lines. The control handles that appear depend on the dimension shape you choose. For example, you can add linear dimensions from a vertical baseline by following these steps:

1. Drag the Vertical Baseline shape onto the page and position it at the bottom and to the left of the distances you want to dimension.

2. Drag the lower green end point and glue it to a geometry point that defines the baseline for all your dimensions, such as the corner of an exterior wall.

3. Drag the other green end point and glue it to a geometry point that defines the end of the first distance you want to dimension, such as the closest jamb of a window.

4. To reposition the text and vertical dimension line for the first dimension, drag the yellow control handle on the first dimension line to the left or right.

5. To define the next dimension, drag the yellow control handle between the dimension shape's selection handles to a position above the first dimension. Another yellow control handle appears at the end of the horizontal reference line. Drag this control handle and glue it to a point that defines the second distance you want to dimension, as illustrated in Figure 25-4.

FIGURE 25-4

Drag control handles to define and modify multiple dimensions.

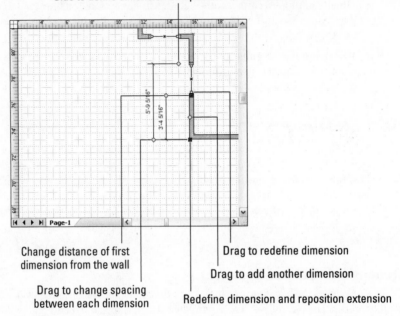

Glue to define dimension and position of extension line

Change distance of first
dimension from the wall

Drag to redefine dimension

Drag to add another dimension

Drag to change spacing
between each dimension

Redefine dimension and reposition extension

6. Repeat step 5 until you have added the dimensions you want.

7. To change the spacing between the vertical dimension lines, drag the yellow control handle on the baseline of the dimension shape to the left or right.

8. To change a dimension, drag a control handle or selection handle at the end of the horizontal reference lines up or down.

Dimensioning Radii

Shapes for dimensioning angles include selection and control handles you can drag to configure the angular dimension. For example, you can dimension a radius with the Radius shape by following these steps:

1. Drag the Radius shape onto the drawing and glue it to a point at the center of the radius you want to dimension.

2. Drag the yellow control handle onto the arc you want to dimension.

3. To position the radial dimension text, drag the green selection handle to a new location.

 To find out what a control handle does, position the pointer over the control handle to display a screen tip.

Dimensioning Angles

To dimension an angle, follow these steps:

1. Drag the Angle Even shape onto the drawing page and glue it to the origin of the angle you want to dimension, for example, one side of a door swing.

2. To change the angle of the line from which the angular dimension is measured, drag the selection handle on the Angle Even shape to a new location, for instance, to change the starting line from horizontal to vertical.

3. To specify the angle that you want to measure, drag the yellow control handle at the top of the Angle Even shape to the angled line, as shown in Figure 25-5.

FIGURE 25-5

Drag control handles to define an angular dimension.

Drag to define end of angle

Drag to position dimension text

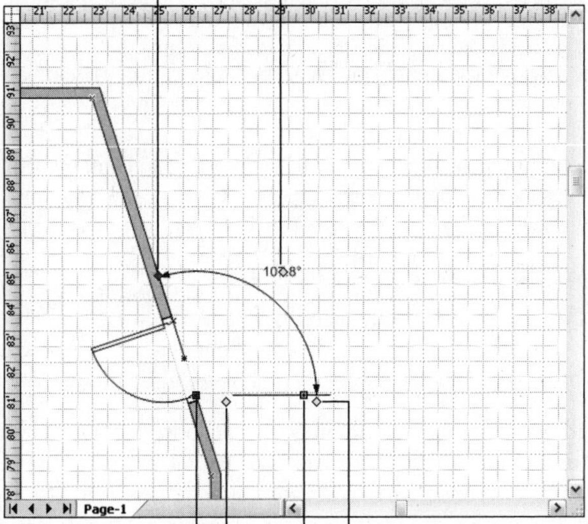

Drag to position angular dimension line

Drag to define start of angle

Drag to position extension

Drag to define origin of angle

NOTE You can also drag control handles on any of the Angle shapes to change the length of the extension line, the position of the angular dimension line, and the position of the dimension text.

Specifying Precision and Units for Dimensions

The dimensions that you add to drawings show distances based on the measurement units you've chosen for that drawing page. Sometimes, you need more or less precision for a few dimensions on the drawing page. To change precision and units for a dimension, follow these steps:

1. Right-click a dimension shape, such as Vertical, Radius Outside, or Angle Center, and, from the shortcut menu, choose Precision & Units. The Shape Data window appears.

2. To specify the number of decimal points of precision for the dimension shape, in the Precision drop-down list, select the entry with the right number of zeroes to the right of the decimal point.

3. To specify the units you want to use, from the Units drop-down list, select the units you want. For example, a floor plan might use Feet-Inch, but for dimensions that are less than 1 foot in length, you can choose Inches.

4. In the Units Display list, select an entry to specify whether or not to show the units in dimensions.

5. To change the angle of the dimension, type an angle in the Angle box.

6. Click OK.

CROSS-REF You can also show a shape's dimensions by inserting a geometry field in the shape's text block. To learn how to do this, see Chapter 33.

Calculating Area and Perimeter

Visio Professional also includes tools that automatically measure the area and perimeter of any closed shape. For example, you can calculate the area within the floor of a building to determine the number of sprinkler heads you need for fire protection, or the perimeter of a parking lot to order fencing.

To measure the area and perimeter of one or more shapes, choose Tools ➪ Add-Ons ➪ Visio Extras ➪ Shape Area and Perimeter. The Shape Area and Perimeter dialog box opens, displaying the area and perimeter in the set units, such as square feet and linear feet. You can keep the Shape Area and Perimeter dialog box open as you issue other commands. If no shapes are selected, the Total Area and Total Perimeter boxes display the words No Selection. As you select shapes, their area and perimeter values appear in the boxes.

If you select more than one shape, the Total Area and Total Perimeter values reflect the values of all the individual shapes combined. For example, to measure the square footage of several separate rooms, Shift-click each room shape. The Shape Area and Perimeter values reflect the total area and total perimeter of all the rooms combined.

To calculate the area and perimeter of the boundary for several shapes, such as the footprint of a building, use the Pencil or Line tool to trace the boundary you want to measure. To calculate the area and perimeter of the boundary, select the boundary you drew.

Measuring Areas with Holes

In many situations, you want to calculate the area for a space but want to ignore some space within it. For example, to calculate the rentable space within a building floor, you might calculate the area of the building boundary without the area for the building core, which contains stairs and elevators. You can use a Visio shape operation to help perform this calculation for you by following these steps:

1. On the Drawing toolbar, click the Line tool or Pencil tool.

2. Draw a shape around the building perimeter and then draw another shape around the building core.

3. With no other shapes selected, Shift+click the two shapes you just drew and then choose Shape ⇨ Operations ⇨ Combine. The Combine command creates a hole in the floor using the shape you drew around the building core.

4. Select the combined shape and choose Tools ⇨ Add-Ons ⇨ Visio Extras ⇨ Area Shape and Perimeter. The Total Area and Total Perimeter represent the values for the entire floor minus the values for the hole.

Summary

Scaled drawings make it easy to create plans in which accuracy and precision are important. In Visio, you can specify drawing scales and measurement units so your plan fits on the drawing page. Visio includes shapes designed to work with scaled drawings — they resize based on the scale you've set for the drawing page. Each drawing page can use a different drawing scale and measurement unit, so you can show a site plan on one drawing page, a floor plan on another, and a detail of a structural connection on yet another page.

Because accuracy is important, you can add dimensions to your scaled drawings to show the real-world sizes of the objects. Visio provides two stencils for dimensions, which display dimensions in either architectural or engineering formats. Dimension shapes include control handles and selection handles you can drag to define and configure dimensions.

Chapter 26

Creating and Managing Scaled Drawings

R ather than a competitor to CAD programs, Visio makes a perfect complement to the one you use. Visio building plan templates are ideal for fast prototyping — dragging and dropping Visio shapes is a quick way to experiment with different layouts. When you settle on a few candidate plans, the Visio drawings and their shapes export easily into your CAD application. Visio is also helpful when you want to include drawings created in CAD applications in broader presentations. After inserting CAD drawings into Visio drawings, you can put Visio tools to use adding presentation annotation and graphics that CAD doesn't handle as easily, if at all.

For folks who don't have access to a CAD program, Visio is an adequate substitute for reviewing CAD drawings or for producing smaller plan drawings from scratch. Visio Professional stencils offer numerous shapes for a variety of building plans.

Like CAD programs, Visio provides a feature called layers to organize drawing contents. Although Visio layers differ from their CAD cousins, they control shape behavior, such as whether shapes are visible when printed or whether they are editable. For example, reference lines that you use to build a plan might be visible on the screen but invisible at printing. Or, the standard notes applied to each drawing can be locked against editing, while the layers with the building plan remain editable.

Each shape can belong to multiple layers so you can manage shapes to suit your needs. With masters on stencils already assigned to layers, Visio automatically associates shapes with the proper layers as you drag them onto your drawings. In addition, dropping shapes with layer assignments onto a page automatically creates that layer for the page.

This chapter shows how to create plan drawings by using Visio plan templates, starting with an existing CAD drawing, or by using an existing Visio scaled drawing. You will also learn ways to use layers to control the behavior of shapes, and how to assign shapes to layers as easily as possible.

Creating Scaled Drawings

More often than not, a plan is kicking around that you can use to start a new drawing. But, every once in a while, you have to create a scaled drawing from scratch. For example, you might have a CAD drawing of a basic floor plan that you want to use as the backdrop for different furniture layouts. Similarly, you could have a CAD drawing produced by someone else and want to use Visio to review and comment on it. In these cases, the solution is to insert CAD drawings into Visio drawings as backgrounds and then drag and drop Visio shapes to add the remaining information.

Conversely, if you have a Visio drawing with some plan information, such as a building shell and core, you can use that as a basis for additional plans, such as building services, by copying and pasting just the shapes you want or the entire drawing into a new plan drawing. Even better, pasting the existing Visio floor plan onto a background page means you can display it in every foreground page you create. Then, to make sure that the underlying floor plan doesn't change, you can lock its layers so that the shapes on them can't be edited.

Creating New Scaled Drawings

Whether you're going to create a scaled drawing from scratch or set one up to hold an existing plan, follow these steps to prepare your Visio drawing file:

1. Create a drawing file set up for scaled plans by choosing File ➪ New ➪ Maps and Floor Plans and then choosing the template you want. Either way, the best approach is to choose a template that sets the scale to the one you plan to work with.

2. If necessary, change the default drawing scale or units, for example, to match the scale or page size of an underlying CAD drawing. See Chapter 25 for instructions on changing drawing scale, page size, and printer paper size.

NOTE In the Page Setup dialog box, the Printer Paper list includes paper sizes for only the current printer. To choose paper sizes for a different printer or plotter, choose File ➪ Print and then select the printer or plotter you want to use. In the Print dialog box, click Close without printing. Reopen the Page Setup dialog box to select a paper size for the new printer.

Referencing Existing CAD Floor Plans

If you have an existing CAD drawing, there's no reason to recreate it in Visio. It can act as a backdrop to additional plans constructed with Visio shapes. To insert the CAD drawing into a Visio drawing file, do the following:

1. Open the Visio drawing page into which you want to insert the CAD drawing.

2. Choose Insert ➪ CAD Drawing. By default, Visio sets the entry in the Files of Type box to AutoCAD Drawing.

3. Navigate to the folder that holds the CAD drawing you want to use, select the CAD file, and click Open. Visio does the following to set up the CAD drawing:

   ▪ Opens the CAD Drawing Properties dialog box and fills in the settings with CAD drawing units and a custom drawing scale that fits the drawing to the page. In most cases, the settings that Visio chooses make the CAD drawing and Visio drawing play well together, so they're usually best left as they are.

   ▪ Checks the Lock Size and Position check box, Lock Against Deletion check box, and View Extents check box so that the CAD drawing can't be moved, resized, or deleted in Visio.

4. Click OK to insert the CAD drawing on the Visio drawing page.

CROSS-REF To learn more about options and methods for importing CAD drawings into Visio, see Chapter 29.

5. If you want to use the inserted CAD drawing as a background for Visio drawing pages, choose File ➪ Page Setup and then select the Page Properties tab. Select the Background option, type the name you want for the background page, and click OK.

Starting from an Existing Visio Scaled Drawing

Sometimes, several plans share information, such as the basic building shell for multiple building service plans or the same basic floor plan elements for several floors in a high-rise building. If you already have these shared elements in a Visio drawing, you can copy and paste them into other Visio drawing pages so you can reuse the common shapes.

CAUTION When you paste shapes from a scaled drawing, Visio resizes the shapes using the drawing scale for the destination drawing page. If the drawing scales in the source and destination drawing pages are more than a factor of eight apart — for example, 1:12 and 1:200 — the pasted shapes might look very large or too small. To correct these overly swollen or shrunken shapes, change the scale of the destination drawing page to match the scale of the source page.

To copy an existing Visio plan into another drawing, follow these steps:

1. Open both the existing Visio scaled plan (the source) and the Visio drawing file into which you want to paste the reusable elements (the destination).

2. To display both drawing windows, choose Window ➪ Cascade or Window ➪ Tile.

3. Click the title bar for the source drawing window, select the shapes you want to copy, and then press Ctrl+C. If you want to copy the entire drawing page, choose Edit ➪ Copy Drawing or Ctrl+A.

4. Click the title bar for the destination drawing window and press Ctrl+V to paste the copied shapes onto the drawing page. Visio creates copies of all the selected shapes in the destination drawing.

Storing Scaled Drawing Content in One or More Drawing Files

The work on plan drawings is usually a collaborative effort, so the best approach for storing information in drawing files depends on how your team works and what you want to do. For example, if everyone wants to work on different building service plans simultaneously, you can place the data for each building service in a different Visio drawing file and link those drawings to Visio background pages in one drawing file that assembles the entire compilation of plans. (For information about linking and embedding files, see Chapter 8.)

Conversely, if you wear all the hats in the planning department, you can add all your shapes to the same drawing page and use layers to specify which shapes you see and whether they are editable. If you want more flexibility in your solo environment, you can place information on separate drawing pages, using some as background pages so they are available to several different foregrounds.

Although these techniques are more typical in a large office that uses CAD, it's worthwhile to plan your Visio drawing files, drawing pages, and layers before you create your plan drawings.

Managing Drawing Content with Layers

If you're familiar with CAD programs, you know that layers help you organize and manage the information on your drawings. In Visio, layers help you accomplish the following:

- View-specific categories of objects or shapes
- Print-specific categories of objects or shapes
- Display categories of objects or shapes in different colors
- Lock categories of objects or shapes to prevent editing
- Control whether you can snap or glue to shapes on a layer

For example, in a building plan, structural components could be assigned to one layer; walls, doors, and windows to another layer; furniture to a third layer; and electrical outlets to a fourth. When you work on the furniture layout, locking the other layers prevents you from moving building components inadvertently. If you want to evaluate whether the electrical outlets are sufficient, you might turn off the display of the structural and HVAC layers to focus on electrical components and their location in relation to cubicles, as illustrated in Figure 26-1.

FIGURE 26-1

Turning layer properties on and off specifies whether the shapes on a layer are visible, printable, editable, or available as reference points for snapping and gluing.

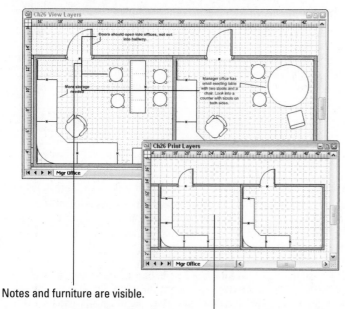

Notes and furniture are visible.

In this layer scheme, the notes and movable furniture are invisible.

By assigning review comments to their own layer, it's easy to print the plan with or without those comments. Turning off snap and glue for specific layers helps you attach new shapes only to appropriate shapes. For example, for adding electrical outlets to a plan, you can turn off snapping and gluing to all layers except the ones for walls. Each shape in Visio can belong to one layer, multiple layers, or no layers at all. For example, you could create separate layers for different building and cubicle panels to simplify the construction of your basic building shell and office layout. But then, you could introduce another layer that shows all those shapes on one layer, so you can focus on furniture, cabling, and electrical outlets.

Many shapes in Visio stencils already contain layer assignments, so you don't have to assign the shapes to layers or think up the kinds of layers you might want and what to name them. For example, as soon as you drop a Room or Wall shape onto a floor plan, Visio automatically creates a Wall layer, (if it doesn't already exist) and assigns the Room or Wall shape to it. If you have your own vision for layering, you can create customized versions of those shape masters with the layer assignments you want.

CAD Layers Versus Visio Layers

Although Visio layers and CAD layers share many characteristics, they also differ in several key ways. Visio layers don't determine the order in which shapes appear on the drawing. To specify whether a shape appears in front or in back of other shapes (the z-order), right-click the shape and, from the shortcut menu, choose Shape ⇨ Bring to Front or Shape ⇨ Send to Back.

In Visio, shapes can belong to no layer at all or multiple layers. In addition, each Visio drawing can have no layers or multiple layers, and the layers for each drawing page can be different. Finally, you can't group Visio layers.

Creating Layers

If you use shapes with layer assignments, you don't have to create layers at all. Visio automatically creates layers for the drawing page when you drop or copy a shape with a layer assignment onto the page. If the page already contains a layer with the same name, Visio adds the shape to the existing layer. However, you can also build a collection of layers manually to make sure that you have all the layers you want.

CAUTION If you use shape layer assignments to create your layers, copying the wrong type of shapes onto a page might also create layers you don't want. If a drawing page contains layers you don't want, clear the visibility of all layers except the ones that you question. The shapes that are still visible are the culprits. Then, you can reassign those shapes to other layers by following the instructions in the "Removing Layers" section later in this chapter.

You must create the layers you want for each drawing page separately, because new layers are added only to the current page. New drawing pages that you add to a drawing file don't inherit the layers associated with other existing pages in the drawing file. The one bright spot in this behavior is that every page in a drawing file can have a different set of layers.

TIP If you use the same set of layers over and over, a little preparation ahead of time can produce a reusable collection of layers. The prep work begins with creating a drawing file specifically to hold your layer collection—a template for layers. On the drawing page in the drawing file, create the layers you want by whichever method you prefer: create the layers manually or add a shape for each layer assignment you want. (When you've finished creating the layers, delete the shapes on the page.) Save the drawing file as a template, as described in Chapter 31.

Whenever you want to reuse that layer collection, you can create a new drawing file from the template or you can copy the drawing from the template into a new drawing. To copy the drawing, choose Edit ⇨ Copy Drawing. Switch to a new drawing page into which you want to copy the layers and press Ctrl+V. No shapes copy over, but the layers do.

To create a layer for a drawing page manually, follow these steps:

1. Choose View ➪ Layer Properties and then, in the Layer Properties dialog box, click New.

2. In the New Layer dialog box, type a name for the layer and then click OK. Visio creates the layer for the current page.

3. Back in the Layer Properties dialog box, in the row for the new layer, click the boxes for each property you want to apply to the layer, if they are not already selected, as shown in Figure 26-2.

FIGURE 26-2

In the Layer Properties dialog box, you select options to specify the behavior of each layer on a drawing page.

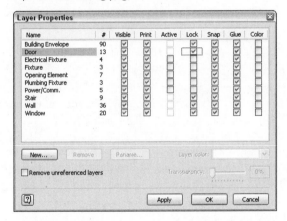

Renaming Layers

Suppose you want to give layers more meaningful or perhaps more concise names. You can rename existing layers for the current drawing page. The shapes on the layer remain the same. To rename a layer, choose View ➪ Layer Properties, select the layer you want to rename, and click Rename. In the Rename Layer dialog box, type a new name and click OK.

NOTE Rename only the layers that you create. Although you can rename the predefined layers assigned to shapes on Visio built-in stencils, Visio creates a new layer with the original layer name as soon as you add another shape with that layer assignment. For example, if you rename the Building Envelope layer to Building Outline and then add another Exterior Wall shape, you'll end up with a Building Envelope layer and a Building Outline layer.

Activating Layers

When you drag a shape that doesn't have a predefined layer assignment onto the page, Visio assigns it to the active layer. If no layers are active, Visio adds the shapes to the drawing page without assigning it to a layer at all. Although you can select shapes on a drawing page and assign them to a layer, it's much easier to assign shapes to a layer as you add them. To do this, activate the layer to which you want to assign the shapes by choosing View ⇨ Layer Properties and then, in the row for the layer you want to make active, select its box in the Active column. Visio makes that layer active for the current page and assigns shapes without predefined layer assignments to it.

> **NOTE** Making more than one layer active is a simple way to assign shapes to multiple layers. After setting several layers as active, simply add shapes to the page and Visio automatically assigns them to all of the active layers.

> **CAUTION** Make sure you reset the active layer as you work so that unassigned shapes you add are assigned to the correct layer. As long as you use built-in Visio shapes, an active layer isn't as important, because the shapes come with the proper layer assignments built in.

Removing Layers

Removing layers associated with a drawing page is easy. Perhaps, too easy, because removing a layer also deletes any shapes assigned to it. Before you delete a layer, be sure to reassign any shapes you want to keep to other layers first. To remove a layer, follow these steps:

1. Choose View ⇨ Layer Properties.

2. Verify that the layer you want to delete shows 0 in the # column, indicating that no shapes are assigned to the layer.

3. If the layer you want to remove does have shapes assigned to it, reassign them to a different layer by following these steps:

   a. To clearly see the shapes you want to reassign, clear the Visible cells for every layer *except* the one you want to remove.

   b. Click OK to close the Layer Properties dialog box.

   c. Select the shapes you want to reassign and choose Format ⇨ Layer.

   d. In the Layer dialog box, clear any layers that are selected and select only the layer to which you want to assign the shapes. Click OK. If all other layers are invisible, the reassigned shapes disappear from the screen.

4. Choose View ⇨ Layer Properties and, in the Layer Properties dialog box, select the layer you want to delete, and then click Remove.

5. Check the Visible cell for every layer you want to see on the page.

> **TIP** If your layers have gotten out of hand, it's easy to delete all unused layers. To do this, in the Layer Properties dialog box, select the Remove Unreferenced Layers check box and then click OK. Visio deletes all layers that contain no assigned shapes.

Controlling Shape Behavior with Layers

Layer properties control the behavior of shapes as a group — whether you can see shapes on the screen or when you print the drawing, whether you can edit shapes or snap and glue to them, and the color in which shapes appear. In the Layer Properties dialog box, you can specify the properties for any layer in the current drawing page. If you want the same properties for layers on other pages, you must apply those properties to each of those pages.

TIP Although you can't share layers between foreground drawing pages, background pages with layers achieve the same result. The layer properties you specify on the background page show up on every foreground page that uses that background. For example, you can put the basic building plan on a background page with different types of building components assigned to different layers. To hide the furniture shapes for all the foreground drawing pages in your file, you have to hide the furniture layer only on the background page.

To specify properties, choose View ➪ Layer Properties. In the Layer Properties dialog box, in the row for the layer you want to configure, select a property box to apply it. Clearing a box deactivates the property. You can control shape behavior with layers in the following ways:

- **Hide or show shapes on the screen** — Select the Visible box so that shapes assigned to the layer appear on the screen. To temporarily hide shapes on a layer, for example, to focus on specific shapes on the drawing, clear layers that aren't important for the task at hand.

- **Hide or print shapes on the printed output** — Select the Print box to include the shapes on the layer on printouts. To prevent shapes from printing, such as construction lines when you're printing final drawings, clear this property.

- **Assign shapes to layers automatically** — Select the Active box to set layers to active so that Visio assigns shapes to the layers automatically when you drop them on the page.

- **Prevent shapes from being edited** — When the Lock box is selected, you can't select, move, edit, or add shapes to the layer. In addition, you can't make the layer active.

- **Use shapes as reference for snapping** — Select the Snap box if you want to snap to shapes assigned to the layer. For example, if you are adding doors to a plan, you want to be able to snap to walls, but not to electrical outlets in the walls. When you clear a Snap box, you can't snap to shapes on the layer. However, if you move shapes on the layer, they can snap to shapes on snappable layers.

- **Glue to shapes** — Check the Glue box if you want to glue to shapes assigned to the layer. When you clear this property, you can't glue to shapes on the layer, although the shapes on the layer can glue to shapes on gluable layers.

NOTE If shapes are assigned to multiple layers and you don't want to snap or glue to them, you must clear the Snap or Glue properties for every layer to which the shapes are assigned.

- Color—Select the Color box to assign a color to the shapes on the layer. Each layer can use a different color, which overrides any color associated with graphic components of shapes on the layer, such as fill color or line color.

 Although you can't select, move, edit, or add shapes to a locked layer, you can change the color of shapes on a locked layer by setting the color in the Color column of the Layer Properties dialog box.

Assigning Color to a Layer

Assigning different colors to each layer makes it easier to identify different components in a plan, such as movable furnishings and non-movable furnishings, or new walls from existing walls. Colors can be opaque or transparent. For example, if you use filled rectangles to identify departments on a plan, you can make the layer for those rectangles transparent so that you can still see the furniture and building components. To assign a color for a layer on the current page, follow these steps:

1. Choose View ⇨ Layer Properties.
2. In the Color column, select the box for the layer you want to color.
3. Click the arrow next to the Layer Color box and then select a color in the Layer Color list.
4. To change the transparency for the color, drag the Transparency slider to the value you want: 100 percent makes the layer totally invisible; 0 percent makes the layer completely opaque. To make a color visible but transparent, choose a value between 0 and 100.

 Shapes assigned to multiple layers appear in their original colors.

Selecting Shapes Using Layers

Layers provide a convenient way to select groups of shapes, particularly when shapes are close together on the drawing page. To select shapes based on the layers to which they are assigned, choose Edit ⇨ Select by Type, and then use one of the following methods:

- **Select shapes on a specific layer**—Select the Layer option and then select the check box for the layer that contains the shapes you want. To select more than one layer, Ctrl-click each layer you want to select.
- **Select shapes without layer assignments**—Select the Layer option and then select the {No Layer} check box.

Assigning Shapes to Layers

The simplest way to assign shapes to layers is to use shapes with built-in layer assignments. When you drop a shape with a layer assignment onto a drawing page, Visio creates the appropriate layer if it doesn't already exist and assigns the shape to that layer. If you build your own shapes, you can assign layers to the masters, and Visio assigns them to layers automatically as well. If a shape doesn't have a predefined layer assignment, Visio assigns it to the active layer. The main issue with this technique is that you must remember to change the active layer to ensure that new shapes are assigned to the appropriate layer.

Inevitably, you end up with shapes with no assignment or incorrect assignments. For these situations, you can assign shapes to layers after you've added them to a drawing or assigned them to layers.

> **TIP** Assigning shapes to more than one layer makes drawings more flexible. For example, you can assign office furniture to both the Furniture layer as well as the Office Equipment layer. Then, the shapes for office furniture appear whenever either of those layers is visible.

Assigning Selected Shapes to Layers

To assign shapes that you've added to a drawing page to a layer, follow these steps:

1. Select the shape or shapes you want to assign and choose Format ⇨ Layer.

2. In the Layer dialog box, select the check box for the layer to which you want to assign the selected shapes. To select more than one layer, Ctrl-click each check box, as illustrated in Figure 26-3.

FIGURE 26-3

Select layer check boxes to assign shapes to one or more layers.

Assigning Masters to Layers

It's more effective to use masters with predefined layer assignments, because Visio adds them to the correct layer automatically. You can add layer assignments to shapes you create or to built-in masters. However, because Visio built-in stencils are copyrighted as well as read-only, you should make copies of the masters you want to change in a custom stencil or your Favorites stencil and edit them there.

NOTE If you edit the layer assignment for a master and then drag it onto the drawing page, the new shape you create uses the new layer assignment. However, any shapes you added prior to the layer change still use the previous layer assignment. You can change the assignment for those shapes by selecting them and choosing Format ⇨ Layer.

If you want to assign a Visio master to a layer or change its current layer assignment, copy the Visio master to your Favorites stencil or another custom stencil. To do this, right-click the master in the Visio stencil and then, from the shortcut menu, choose Add to My Shapes. Choose one of the custom stencils on the submenu, or choose Add to New Stencil or Add to Existing Stencil.

To assign a master to a layer or to change its layer assignment, open a Visio drawing so that the Shapes window appears, and then follow these steps:

1. Open the stencil that contains the master you want to assign to a layer. If the stencil is read-only, right-click the stencil title bar and, from the shortcut menu, choose Edit Stencil.

2. Right-click the master and, from the shortcut menu, choose Edit Master ⇨ Edit Master Shape.

3. In the master drawing window, select the master.

4. Choose Format ⇨ Layer.

5. Use one of the following methods to create a layer assignment:

   ▪ To change the layer assignment, clear a layer's check box to remove an assignment and check an existing layer box to assign the master to that layer.

   ▪ To assign a master to a new layer, click New. In the New Layer dialog box, in the Layer Name box, type the name of the new layer and click OK. Visio automatically assigns the master to the new layer. You can assign the master to additional layers by selecting other layer boxes or by clicking New and typing the name for the next layer.

6. To close the master drawing window, click the Close button for the master drawing window. When Visio prompts you to update the master, click Yes.

7. To save your changes, right-click the stencil's title bar and click Save.

8. To change the stencil to read-only, right-click the stencil title bar and choose Edit Stencil.

Assigning Groups to Layers

Groups of shapes can also have layer assignments. By default, when you select a group and choose Format ⇨ Layer to assign the group to a layer, all of the shapes in the group become members of the new layer, losing their previous layer assignments. However, you can change the group assignment while retaining the shapes' current layer assignments by selecting the Preserve Group Member Layers check box in the Layer dialog box. For example, if you build groups of shapes to represent standard office configurations that include office furniture as well as computer equipment, you can assign the shapes for furniture to the Furniture layer and shapes for computer equipment to the Electronics layer before grouping them. If you assign the group to the Office Equipment layer, the furniture and computer equipment could retain their previous layer assignments while also including assignments to the Office Equipment layer.

CAUTION If you see shapes on the drawing page, but can't select them, the layer might be locked. However, if you open the Layer Properties dialog box and find that the layer isn't locked, group layer assignments could be to blame. The problem arises when you assign individual shapes to one layer, and the group to which they belong to another layer. If the visibility of the group's layer is turned off but the visibility of the individual shape's layer is turned on, you can see the shapes because of their layer assignment, but you won't be able to select or edit them.

Summary

You can create blank plan drawings from Visio templates, CAD drawings, or existing Visio plan drawings. Layers help you manage the content of plan drawings by categorizing the shapes on your drawings. By specifying layer properties, you can control whether layers are visible on the screen, appear when you print your drawing, and are active or locked, and you can control both the appearance of shapes and access to shapes.

Chapter 27

Laying Out Architectural and Engineering Plans

Whether you're planning the remodel of a house, reorganizing office cubicles, or designing a major manufacturing facility, Visio floor plan templates get you started. Although Visio Standard provides only a few shapes for laying out office space, Visio Professional includes templates, stencils, and shapes for a variety of architectural and engineering plans.

Visio Professional plan templates start with the building shell — walls, doors, and windows — and continue with furnishings, electrical service, plumbing, HVAC, and other building services. The template list moves outdoors to include site and landscaping plans, security and access plans, and maps.

Unlike CAD programs, which come with oodles of drafting commands for drawing plan contents, Visio architectural and engineering capabilities stem from the shapes on the many architectural and engineering stencils. If you're familiar with basic Visio techniques including dragging control handles and setting shape data values to implement special behaviors, you already know most of what you need. Although Visio plan templates provide the tools you need to draw plans, they don't confer the specialized skills you need to determine what plans should contain or what constitutes an effective layout. You'll have to read other books to learn about that.

NOTE Many of the techniques described in this chapter work equally well for the basic office plans that Visio Standard supports and the more specialized architectural and engineering plans available only in Visio Professional. This chapter talks about content specific to Visio Professional.

CROSS-REF To learn about the different types of Visio plan templates and how to work with Visio plan drawings, see Chapters 25 and 26.

Working with Walls

In most cases, the first step in constructing a plan is placing walls to define the outside of a building to contain the stuff on the inside. Visio wall shapes represent different types of walls — structural exterior walls, interior walls, curtain walls that merely fill in the space between structural elements, and window walls. The Walls, Shell, and Structure stencil and the Walls, Doors, and Windows stencil both include built-in wall shapes with special behaviors for easily connecting walls and shape data for specifying attributes such as wall thickness and fire rating. You can also create custom wall shapes, for example, to show walls to be demolished with a hatched fill pattern.

If you began your design by creating a space plan in Visio, the easiest way to build walls is by converting Space shapes into Wall shapes. Otherwise, all the standard techniques for working with shapes (described in Chapters 4 and 5) apply to Wall shapes, too: dragging, dropping, snapping, gluing, and using the Connector tool. This section describes how to convert Space shapes into Wall shapes and provides tips for connecting walls using familiar techniques.

Converting Spaces into Walls

The Convert to Walls command makes quick work of turning the spaces in a space plan into walls that enclose those spaces. The command enables you to specify the type of wall you want to use, whether to display dimension lines, and whether to add guides to the walls it creates. In addition, you can delete the original Space shapes or keep them. For example, Space shapes come in handy when you want to show the square footage of spaces on the drawing or for tracking space in the building.

 To learn more about Space shapes and other methods for laying out spaces and rooms, see Chapter 28.

To convert Space shapes into walls, follow these steps:

1. Initiate the Convert to Walls command using one of these methods:

    ■ **Convert a single shape** — Right-click the shape and, from the shortcut menu, choose Convert to Walls.

    ■ **Convert several shapes** — Select the shapes you want to convert using any selection method and then choose Plan ⇨ Convert to Walls.

2. In the Convert to Walls dialog box, in the Wall Shape list, select the type of Wall shape you want to use.

> **TIP** The Wall Shape list might seem sparse, because it includes only the Wall shapes available on open stencils. If you don't see the Wall shape you want, click Cancel, open the stencil that contains the Wall shape you want, and begin again with step 1.

3. To automatically add dimensions to each segment of a wall that is created, select the Add Dimensions check box.

> **NOTE** Visio can be inconsistent with its automatic dimensioning, adding redundant dimensions in some places and leaving some wall segments undimensioned. If you don't like the dimensions that Visio adds, press Ctrl+Z to undo the results of Convert to Walls, and then rerun it with the Add Dimensions check box cleared.

4. To glue guides to each vertical and horizontal wall segment, select the Add Guides check box. Guides reduce frustration and time when you reposition wall segments. You can drag the guides to move wall segments while maintaining their connection to other walls.

5. To keep the Space shapes along with the converted Wall shapes, select the Retain option, which is ideal when you want to use spaces to track billable space.

> **NOTE** Although the Convert to Space Shape option might sound like it turns walls back into spaces, it applies only when you convert shapes other than Space shapes (for example, lines and closed shapes you draw with the Visio drawing tools) into walls. By choosing this option, the Convert to Walls command converts the selected shapes into Wall shapes and also creates a Space shape on the interior surface of the converted walls. It also deletes the original shapes, because you now have both Wall and Space shapes.

6. Click OK to convert the Space shapes into Wall shapes and add any additional elements you specified, as illustrated in Figure 27-1. Visio creates a separate Wall shape for each wall segment in the building. Because the Wall shapes are glued together, the intersections between Wall shapes are cleaned up.

7. If you added guides to the converted walls, you can reposition a Wall shape by dragging the guide glued to it. The advantage to this method is that Wall shapes glued to the guide resize without detaching from adjacent walls. One failing is that the original Space shapes don't resize with the walls. Fortunately, resizing a Space shape to match the new wall configuration is as simple as right-clicking the Space shape and, from the shortcut menu, choosing Auto Size.

FIGURE 27-1

The Convert to Walls command converts Space shapes or drawn geometry into Wall shapes, with optional dimensions and guides.

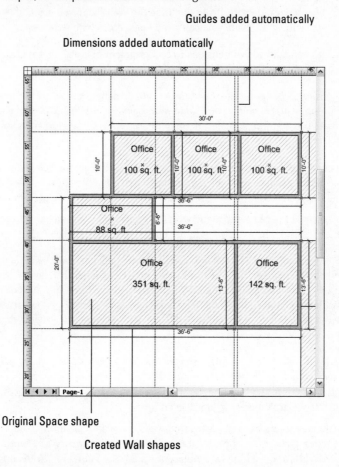

Guides added automatically

Dimensions added automatically

Original Space shape

Created Wall shapes

Creating Walls

When you add walls from scratch, standard techniques such as dragging, dropping, connecting, and gluing are about the only tools you need, and you can choose the easiest technique for the job at hand. For example, using the Connector tool with the Exterior Wall shape is the fastest way to create the exterior wall segments for a building shell. On the other hand, with short runs of interior walls, dragging and dropping Wall shapes might be more effective. As you add Wall shapes to the page and glue them together, Visio automatically cleans up the intersections so lines appear only for wall surfaces, as shown in Figure 27-2.

FIGURE 27-2

Although wall segments are rectangles, Visio cleans up wall intersections when you glue Wall shapes together.

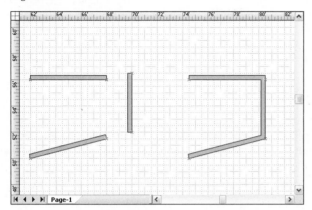

NOTE For Wall shapes to connect properly and clean up after themselves, snap and glue must both be turned on. Choose Tools ⟳ Snap & Glue. In the Snap & Glue dialog box, under the Currently Active heading, make sure that both the Snap check box and the Glue check box are selected. In addition, under both the Snap To and Glue To headings, make sure that the Connection Points and Shape Geometry check boxes are selected.

To create new walls on a drawing, use one of the following methods:

- **Drag and drop Wall shapes** — To add walls one at a time, drag the Wall shape you want onto the drawing page. To glue the new Wall shape to Wall shapes already on the page, glue its end points to connection points or shape geometry on existing Wall shapes. For example, you can create a wall divider by gluing a vertical Wall shape anywhere along the length of a horizontal wall. When you drag an end point to a connection point or shape geometry, Visio indicates the pending glue with a red square.

- **Use the Connector tool** — To add a sequence of connected walls, follow these steps:

  a. On the Standard toolbar, click the Connector tool.

  b. In a stencil, click the wall you want to add.

  c. For the first wall, drag between two points to define the beginning and end of the Wall shape. As soon as you complete this Wall shape, the pointer changes to the four-headed arrow, indicating that you can move the current point to a new location.

  d. To add another Wall shape that starts where the first Wall shape ends, move the pointer away from the end point and then move it back, but not quite over the end point, until the pointer changes to the Connector icon (a plus sign with a small connector next to it).

e. Drag from the current point to the end of the next Wall shape.

f. Repeat steps d and e to create additional wall segments.

> **NOTE** When you add Exterior Wall shapes to a plan, you want the selection handles on the interior surface of the building wall, so they are lined up at the inside corners and, thus, are easy to find when you're ready to glue other Wall shapes. If an Exterior Wall shape's selection handles are on the edge that represents the exterior of the building, right-click the shape and, from the shortcut menu, choose Flip Wall on Reference Line.

Connecting and Resizing Walls

Visio cleans up corners and other intersections when walls are glued together, but this convenience goes only so far. If you drag a Wall shape to another position, it separates from its friends and the corners fill in again. Most of the time, what you want Visio to do when you drag walls is to keep the walls glued together and lengthen or shorten wall segments as necessary. Guides attached to Wall shapes do just that. It's easy to glue Wall shapes to guides as you construct your plan using one of the following methods:

■ **Gluing to existing guides** — In many cases, you begin a floor plan by dragging guides onto the drawing page as reference lines. If you drag a Wall shape onto a page and drag its end points to guides, Visio glues the shape to the guide automatically.

■ **Creating guides with the Convert to Walls command** — If you convert Space shapes to Wall shapes as discussed in the section "Converting Spaces into Walls," the Convert to Walls command can automatically create and glue guides to the Wall shapes created during the conversion.

■ **Right-clicking a wall shape** — To add a guide to an individual Wall shape, right-click the shape and then, from the shortcut menu, choose Add a Guide.

When you drag a guide that is glued to a Wall shape, the Wall shape moves with the guide, and any Wall shapes that adjoin that Wall shape stretch or contract.

Modifying Wall Properties and Appearance

Length and thickness are obvious wall properties because they are visible on a drawing. But, shape data associated with Wall shapes controls every wall dimension: length, thickness, and height. Wall shapes include other shape data fields, as well as options for changing the appearance of walls on a plan.

Changing Wall Thickness and Other Properties

Some shape data fields for Wall shapes modify the configuration of the wall itself, whereas others store data for reference or reports. Here are the shape data fields for walls and what they do:

■ **Wall Length** — Changes the length of the wall segment, regardless of the angle the segment is rotated.

■ **Wall Thickness** — Changes the thickness of the wall segment, that is, from one wall surface to the other.

- **Wall Height** — Sets the height of the wall, which doesn't change the visible outline of the shape. However, you can produce legends, quantity take-offs, or bills of material based on these values.

- **Wall Justification** — Controls the alignment of the Wall shape. Centered draws the wall thickness equally on either side of the points you use to create the wall. Edge aligns the wall to the selection handles.

- **Wall Segment** — By default, this field is set to Straight for a perfectly straight segment. If you change this field to Curved, Visio adds a control point to the wall so you can modify the curvature of the segment.

- **Base Elevation** — Defines the elevation of the bottom of the wall.

- **Fire Rating** — Specifies the fire rating for the wall, which is useful in materials reports.

To change shape data, right-click a Wall shape and, from the shortcut menu, choose Properties. In the Shape Data dialog box, type or select values for the fields and click OK when you're done.

> **NOTE** If you modify shape data frequently, it's easier to dock the Shape Data window by choosing View ⇨ Shape Data Window. When you select a shape, its values appear in the fields. To change or enter a value, select the field you want to edit and type or select a value.

Changing the Way Walls Appear

By default, Visio shows walls as double lines — one line for each wall surface — although you might have to zoom in to see them. In Visio Professional, you can display walls in different ways. For example, to streamline a crowded drawing, you can display walls as single lines, or show walls as double lines with a reference so that the centerline of the wall is easy to spot. To change how Visio displays walls, right-click any Wall shape on a drawing page and, from the shortcut menu, choose Set Display Options. Make sure the Walls tab is selected and then select Double Line, Double Line And Reference Line, or Single Line.

> **NOTE** Changing the display options for walls affects all the Wall shapes on the current drawing page but not Wall shapes on other pages in the file.

Changing Wall Color and Line Style

In addition to the number of lines that represent walls, changing colors and line styles for Wall shapes is a common adjustment. For example, architects typically use different colors or fill patterns to identify existing walls, new walls, or walls to be demolished. If you simply want to make walls stand out, use the new Theme feature (on the Formatting toolbar, click Theme).

But, to show walls with specific colors that don't change with the selected theme colors, modify the styles associated with the wall. Wall shapes use the Wall line and fill styles, so you can change color, line style, and fill pattern, by modifying the Wall styles. In fact, to show different types of walls, you'll probably want to create a few new styles, for example, for walls targeted for demolition. To learn how to modify and create styles, see the section "Formatting with Styles" in Chapter 7.

Adding Doors, Windows, and Other Openings

Even jail cells have doors to enter or exit, so doors, windows, and other wall openings are essential. Glue and shape behaviors make it easy to add openings to walls. For example, when you add a door or window to a Visio Wall shape, the door or window automatically performs the following feats:

- Rotates to match the angle of the Wall shape
- Glues itself to the Wall shape
- Adjusts its width to that of the Wall shape
- Cleans up the Wall shape where the opening is located

Changing the configuration of a shape for openings is just as easy, whether you want a door to open to the inside or outside or swing to the left or right.

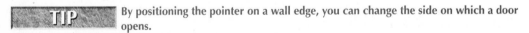

NOTE These configuration features are available whether you use shapes from the Walls, Doors, and Windows stencil in Visio Standard and Professional or the Walls, Shell, and Structure stencil only in Visio Professional.

Adding Openings to Walls

To insert a door, window, or other type of opening into a wall, simply drop an opening shape from a stencil onto a Wall shape. The shape rotates into position in the Wall shape, glues itself to the Wall shape, and changes its thickness to match the Wall shape, as shown in Figure 27-3.

TIP By positioning the pointer on a wall edge, you can change the side on which a door opens.

NOTE If an opening doesn't rotate to match the direction of the Wall shape, the shapes aren't glued together. Drag the shape over the Wall shape until you see the red square indicating pending glue, at which point the shape should be rotated to match the Wall shape. Release the mouse button to add the opening to the wall.

FIGURE 27-3

Shapes for openings automatically adjust to match the wall thickness and orientation. Control handles help you change shape configuration.

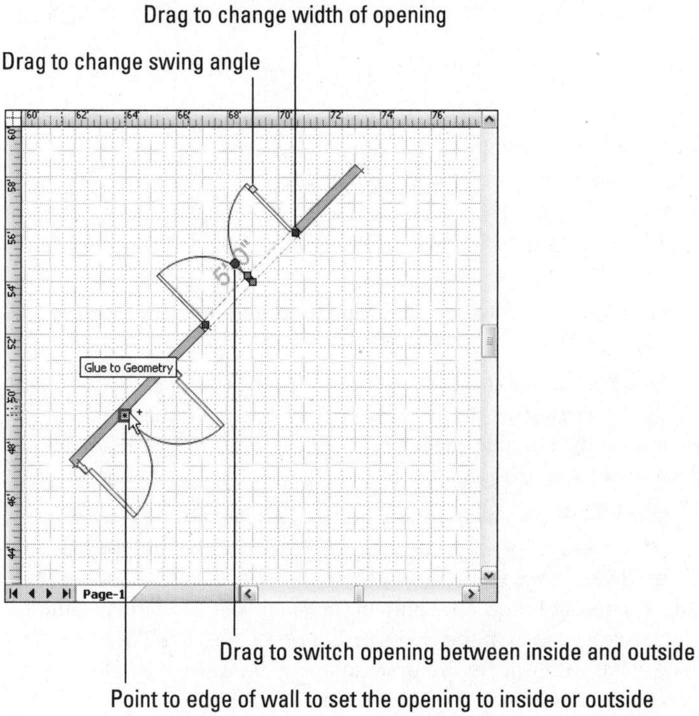

Drag to change width of opening

Drag to change swing angle

Glue to Geometry

Drag to switch opening between inside and outside

Point to edge of wall to set the opening to inside or outside

Modifying Doors, Windows, and Openings

Shapes for doors, windows, and other openings are teeming with shape data and shortcut menu commands for configuration. For example, the direction that a door swings (in or out) or the widths of openings are editable as is the appearance of doors, windows, and openings on each drawing page. Use one or more of the following methods to change door, window, and opening shapes:

■ **Reverse direction** — Right-click a Door, Window, or Opening shape, and then, from the shortcut menu, choose Reverse In/Out Opening. Or, drag the red circle control handle from one side of the wall to the other.

■ **Reverse swing** — Right-click a Door, Window, or Opening shape, and then choose Reverse Left/Right Opening.

■ **Reposition opening** — Drag the shape to a new position in the Wall shape. Visio heals the opening in the Wall shape at the original position and cleans up the Wall shape at the new location.

■ **Modify dimensions and other attributes** — Open the Shape Data window (choose View ➪ Shape Data Window) and then select the shape whose data you want to change. In the Shape Data Window, type or select new field values. For example, with the Double-Door shape, you can specify the width of the door, its height, the type of door, the percentage that the door is open on the page, its number, its fire rating, and its base elevation.

CROSS-REF   If you track building components in a database, you can import that data into shape data, as described in Chapter 10.

■ **Change the door and window components that appear** — Depending on the density of components, you might want to show every last detail of doors and windows or just the general outline of openings. For example, Visio displays the window frame and sash by default, but you can also show the header and sill. To change the components you see, do the following:

   a. Right-click any Door or Window shape on a drawing page and, from the shortcut menu, choose Set Display Options.

   b. On the Doors or Windows tab, select the check boxes for the components you want to see and then click OK. Changing the display options for Door or Window shapes affects the shapes only on the current drawing page, so you must repeat this step for each drawing page you want to change.

■ **Set default configurations** — When you use standard sizes, you can specify default properties for doors and windows. For example, you can set the width of the door frame to three feet so that every Door shape you add uses those dimensions. To do so, right-click a Door or Window shape and, from the shortcut menu, choose Set Display Options (available only in Visio Professional). On the Doors or Windows tab, click Properties. In the Set Door Component Properties or Set Window Component Properties dialog box, specify the default properties you want and then click OK. These settings affect only the current drawing page, so you must redefine these defaults for every page to which you want to apply them.

Creating Door and Window Schedules

For construction projects, door and window schedules identify each door and window in a set of construction plans and specify their dimensions and other attributes, so that the right components are installed in the right places. In Visio Professional, you can use reports to create door and window schedules that automatically collate information about the Door and Window shapes on your drawing. For example, the default Door Schedule report shows door number, door size, door type, and thickness. The default Window Schedule report includes window number, size, and type.

The easiest way to produce a door or window schedule is to drag a Door Schedule or Window Schedule shape from the Walls, Shell and Structure stencil onto the drawing page. These tabular shapes use the existing schedule report definition. If you want to show other shape data or customize the report, you can modify the report settings, as described in Chapter 10.

Adding Cubicles and Furniture

In Visio Standard, the Office Layout template is the only plan template available, but it includes the most frequently used shapes, including basic building components, office furniture, office equipment, office accessories, and cubicles. If you use Visio Professional, you can create more detailed floor plans with the Floor Plan template and choose from a broader range of shapes from the Cubicles, Office Accessories, Office Equipment, and Office Furniture stencils. There are two ways to add cubicles and furniture to plans: using predefined shapes that include the cubicles and their contents or adding each component individually. This section describes both methods.

Laying Out Office Cubicles

The Cubicles stencil has shapes that contain preassembled sets of cubicle walls, furniture, and equipment. As long as these ready-made Workstation shapes have the components you want, it's easy to lay out an office simply by dragging shapes onto the drawing page. You can build custom workstation shapes to conform to your organization's cubicle and office standards.

CROSS-REF See Chapters 32 and 33 to learn how to customize shapes and store them on custom stencils.

If the built-in shapes aren't what you want and you're not ready to customize shapes and stencils, you can resort to building cubicles piece by piece with Panel shapes, Panel Post shapes, and shapes for work surfaces and storage units from the Cubicles stencil. Similarly, free-standing pieces such as Round Table and Stool shapes from the Office Furniture stencil can join a collection of shapes for a cubicle or workstation. When you have the cubicles configured the way you want, the best approach is to create a group of the shapes and copy the group to lay out the rest of the office.

To create a cubicle from components, follow these steps:

1. From the Cubicles stencil, drag Panel or Curved Panel shapes onto the drawing page. To resize panels or change their orientation, drag the ends of the shapes.

2. To connect panels, follow these steps:

    a. Drag a Panel Post shape onto the page and glue it to one end of a Panel shape. Visio highlights the connection point in red to indicate that the two shapes are glued.

    b. Drag the end of the adjacent Panel shape and glue it to a connection point (a blue asterisk) on the Panel Post shape. The Panel shape rotates into position based on the connection point you choose. For example, if you drag an end of a vertically oriented Panel shape over the connection point on the right side of a Panel Post shape, the Panel shape switches to horizontal.

3. To add furniture and equipment to a cubicle, drag one or more of the following shapes into the cubicle:

    ▪ **Modular work surfaces** — From the Cubicles stencil, drag shapes such as Work Surface or Corner Surface and position them next to Panel shapes. To orient the shapes within the cubicle, use the shapes' rotation handles or the Rotate or Flip commands on the Shape menu.

- **Modular storage units** — From the Cubicles stencil, drag shapes such as Storage Unit and position them within the area defined by Panel shapes.

- **Suspended shelves and lateral files** — From the Cubicles stencil, drag shapes such as Susp Open Shelf or Suspended Lateral File on top of shapes for modular work surfaces. These shapes must align with the Panels in the cubicle because, in real life, they hang from the cubicle panels.

- **Chairs and other free-standing furniture** — From the Office Furniture and Office Accessories stencils, drag shapes into the cubicle area.

- **Computers and other equipment** — From the Office Equipment stencil, drag shapes into the cubicle area.

NOTE   You can also enhance cubicles by adding any shapes from the Cubicles stencil, Office Furniture stencil, Office Equipment stencil, or Office Accessories stencil to Visio built-in Workstation shapes or your customized workstations.

Connecting Modular Furniture

The shapes on the Office Furniture stencil are easy to connect because of the inward/outward connection points they contain (see Chapter 33 for an explanation of the different types of connection points). Modular furniture shapes glue and rotate to align with each other, similar to the Panel and Panel Post shapes for cubicles.

Labeling and Numbering Plans

Building plans are like other types of drawings when it comes to annotation. Text blocks, Callout shapes, and the text blocks within shapes themselves all have their place for showing information, notes, and comments. Because building plans are scaled drawings, you'll usually add Dimension shapes to show the distances represented on the plan.

In addition to these techniques, the Label Shapes add-on was developed specifically for building plans, although it's just as handy for labeling shapes on other types of drawings. With the Label Shapes add-on, you can display up to four shape data fields in labels on the shapes. For example, for a cubicle, you might display the person who sits there, the telephone number, the cubicle number, and department. You can also import data into the labels from other data sources.

The Number Shapes add-on is another tool you can't do without when working on plans, such as numbering the columns in a structural plan. This add-on numbers shapes in sequence or in the order you choose and uses the numbering scheme you specify, such as A-1 or Off-200.

To learn how to use the Number Shapes add-ons, see Chapter 6. The Label Shapes add-on is described in Chapter 28. And, of course, you can also label shapes with the new Data Graphics feature, described in Chapter 10.

To connect modular furniture, follow these steps:

1. From the Office Furniture stencil, drag a modular furniture shape, such as 45 Deg Table, onto the drawing page.

2. Drag a connection point on a second modular Office Furniture shape and position the pointer over a connection point on the first shape. When Visio highlights the connection points with a red square and rotates the second shape into the proper position, release the mouse button.

Modifying Cubicles

Predefined cubicles take care of adding cubicle walls, work surfaces, office equipment, and other furniture in one step. But, the built-in Visio Workstation shapes might not be set up the way you want. If a Workstation is close to what you need, you can modify one Workstation shape and duplicate it to populate your plan. For more significant changes or to create several versions of workstations, it's more effective to create a custom stencil of Workstation shapes, as described in Chapters 32 and 33.

To modify a single cubicle or workstation, follow these steps:

1. From the Cubicles stencil, drag one of the Workstation shapes, such as an L workstation, onto the drawing page. The Workstation is a group of shapes with group selection handles.

2. To change the size of the cubicle, drag the Workstation shape's selection handles. Shapes such as Panels and Work Surfaces resize automatically to match the new cubicle size. Shapes with fixed dimensions, such as chairs and suspended shelves do not change size, but might move within the boundary of the cubicle.

3. To move, delete, or format an individual component within the cubicle, select the Workstation shape and then subselect the individual shape inside it.

   ▪ Drag the shape to reposition it.

   ▪ Press Delete to delete it.

   ▪ Choose standard formatting commands to reformat it.

4. To add furniture or equipment to the cubicle, drag a shape, such as Telephone, from the Office Equipment stencil, onto the Workstation shape. With the Workstation shape selected, Shift-click the new shape, and then choose Shape ➪ Grouping ➪ Add to Group.

5. To copy the Workstation shape, press Ctrl+D to create a duplicate and then drag it into position.

TIP If the office layout is laid out on a grid, you can also use the Array Shapes command to create Workstation shapes at regular intervals both vertically and horizontally. For information on using Array shapes, see Chapter 4.

Laying Out Plant Floors

In addition to laying out smaller areas, Visio includes the Plant Layout template (Visio Professional only) for planning production lines and equipment layouts for manufacturing plants and distribution centers. As with other types of floor plans, you can start by creating a drawing with the Plant Layout template, or by laying out shapes over an existing Visio or CAD plan drawing. The Plant Layout template opens some stencils you'll see with other plan templates, but it also opens a few specific to plant layout, including the following:

- **Shop Floor–Machines and Equipment** — Includes shapes for machines such as lathes, saws, drill presses, and so on

- **Shop Floor–Storage and Distribution** — Includes shapes for equipment such as forklifts, cranes, shelves, and racks

- **Vehicles** — Includes shapes for cars, trucks, buses, and shapes that also show turning radii for vehicles

- **Warehouse–Shipping and Receiving** — Includes shapes for shipping doors, containers, cranes, and so on

The shapes on these stencils include the selection and rotation handles you know from other types of shapes. A few also include control handles for repositioning parts of the shape. However, these shapes are meant to stand alone and, therefore, don't include connection points for gluing the shapes together. Many of the shapes include shape data for specifying the size of equipment, the department to which it is assigned, or asset number.

Creating Building Services Plans

The equipment and services required to keep a building running are a lot more than just building walls, openings, and furnishings. If you've ever been stuck at work when the power goes out, you already know how important electrical service is to getting any work done. Visio Professional includes templates for each building service, including electrical, plumbing, HVAC, and security systems. Most of the building services shapes include shortcut commands and shape data for selecting the type of component, specifying component dimensions, or configuring the shapes in numerous ways. In addition, these shapes come with layer assignments so it's easy to manage building services plans with layers.

CROSS-REF   To learn more about using background pages, see Chapter 2. Chapter 26 describes how to use layers to control shapes. Chapter 29 discusses the use of CAD drawings within Visio drawings.

Adding HVAC Services

A few hours working in an office building when the air conditioning is turned off is all it takes to make HVAC an essential building service. Visio Professional offers two types of HVAC plans. An HVAC Plan shows the ductwork, registers, and diffusers that deliver and exhaust air. An HVAC Control Logic diagram represents the sensors and control equipment that determine the quantity and temperature of air that's delivered.

Drawing HVAC Plans

The HVAC Plan template in Visio Professional automatically opens the stencils with HVAC equipment and ductwork. To add HVAC shapes to an existing drawing, open the same stencils by choosing File ⇨ Shapes ⇨ Maps and Floor Plans and then selecting the following stencils:

- **HVAC Equipment** — Includes pumps, condensers, fans, and other types of HVAC equipment

- **HVAC Ductwork** — Includes ducts, junctions, ductwork connections, and transitions

- **Registers, Grills and Diffusers** — Includes shapes for openings that deliver, remove, or diffuse air

The HVAC Plan also opens the Drawing Tool Shapes stencil, which is also available when you choose File ⇨ Shapes ⇨ Extras ⇨ Drawing Tool Shapes. This stencil includes shapes that construct geometry useful in building ductwork.

Separating Building Services from the Basic Building

Building services plans use the exterior walls, structural elements, building core, interior walls, and, in some cases, cubicles to show where components such as sprinkler heads and electrical outlets go. In Visio, placing the basic building on a background page means that each building services plan can display the same basic plan for reference. If you build each building service plan on a different page in the same drawing file, you can associate the building plan background page with each building service plan foreground page. Using a background page also prevents you from inadvertently modifying the basic building plan as you work on building services.

Another approach is to add all the building service plan shapes to the same drawing page. In this case, layers on the drawing page act as separators, so you can turn off the layers you don't want to see for a specific plan. For example, you could turn off the HVAC layer when you're documenting the electrical service plan. With layers, you can protect the basic building components from editing by locking layers, such as Building Envelope, Wall, and Stair.

If you've imported a CAD drawing as a backdrop, you also lock that drawing against editing. To do so, right-click the drawing and then, from the shortcut menu, choose CAD Drawing Object ⇨ Properties. In the CAD Drawing Properties dialog box, on the General tab, make sure both the Lock Size and Position check box and the Lock Against Deletion check box are checked.

To lay out HVAC ductwork and equipment, follow these steps:

1. Create a new drawing with the HVAC Plan template. You can also open an existing floor plan page, a page with an imported CAD drawing, or insert a new page in a floor plan drawing file.

2. To keep shape data easily accessible, open the Shape Data window by choosing View ⇨ Shape Data Window.

3. Drag shapes that represent ductwork from the HVAC Ductwork stencil onto the drawing page.

4. To change the dimensions of a Ductwork shape, such as Straight Duct or Y Junction, use one of the following methods:

   ▪ **Selection handles** — Drag selection handles to change the length or width of ducts.

   ▪ **Control handles** — Drag control handles to change the angle of branches on ducts.

   ▪ **Enter dimensions** — If the Shape Data window is open, select the Ductwork shape and then, in the Duct Length or Duct Width box, type the new length or width, respectively.

5. To connect Ductwork shapes on the drawing page, drag a connection point on one Ductwork shape to a connection point on the next Ductwork shape. For example, drag from a Branch Duct shape to the shape that represents the main duct. Visio highlights the connection points with a red square to indicate they're glued and rotates the second Ductwork shape to match the orientation of the first.

> **NOTE** If you drag a Ductwork shape from a stencil, connect it to a Ductwork shape on the page by gluing one of its connection points to a connection point on the existing shape.

6. To change attributes for some ductwork shapes, right-click a shape and, from the shortcut menu, choose a command. For example, you can choose between Rectangular Duct and Circular Duct or specify whether the ends are open or closed.

7. To label ducts, use one of the following methods:

   ▪ **Show duct size** — Right-click a Ductwork shape and, from the shortcut menu, choose Show Duct Size. Shapes that represent rectangular ducts show the duct width and depth. Shapes that represent circular ducts show the diameter followed by the diameter symbol.

   ▪ **Add a text label** — As with other shapes, you can select a shape and type the label text you want. See Chapter 6 for instructions on working with shape text labels.

8. To add registers, grills, diffusers, and other types of HVAC equipment, drag shapes from either the Registers, Grills, and Diffusers stencil or the HVAC Equipment stencil and drop them on top of Ductwork shapes. These shapes don't glue to Ductwork shapes or rotate to match the orientation of Ductwork shapes. However, dragging the rotation handles on these shapes change their orientation.

Documenting HVAC Control Logic

HVAC Control Logic diagrams show the sensors, equipment, and wiring that control the HVAC air flow system drawn on an HVAC plan. Unlike the HVAC Plan template, the HVAC Control Logic Diagram template creates unscaled schematic drawings by default. These schematics are single-line or double-line drawings that represent ducts, sensors, and mechanical equipment to control the HVAC system. The built-in shapes for sensors, equipment, and ductwork are all the same width, so the controls align nicely with ductwork as shown in Figure 27-4.

FIGURE 27-4

Ducts and sensor shapes fit together so control diagrams are literally a snap to build.

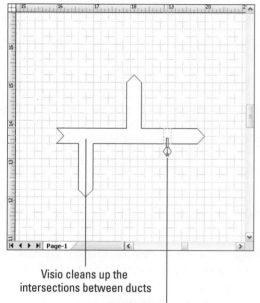

Visio cleans up the
intersections between ducts

The alignment boxes for devices match
duct width so they snap into place

Many of the HVAC control shapes use shape data to configure controls more precisely. The values in drop-down lists vary from control shape to control shape, but the shape data fields are as follows:

- **Control** — Choose Control if the shape represents equipment that controls what the HVAC system does. Choose Sensor if the shape represents a sensor that measures some aspect of the HVAC system, such as temperature or humidity.

- **Type** — This field includes values for different variations of the device. For example, the Light shape includes Not Assigned, Solar Radiation, Ambient Space, Occupancy, and Other.

- **Limit Switch** — This field specifies whether the device has a shut-off switch that activates when the environment hits a high or low value. The choices include None, LowLimit, HighLimit, and Both.

- **Reset Switch** — Choose either None or Manual depending on whether the device includes a switch to reset it.

- **B-O-M Tag** — If you use the Visio diagram to build a bill of materials for the HVAC controls, in this field, type the ID for the device.

- **Part Number** — To track part numbers, for example, to find devices quickly in case of a recall, type the part number in this field.

Start a new HVAC Control Logic diagram by choosing File ➪ New ➪ Maps and Floor Plans ➪ HVAC Control Logic Diagram. If you want to include the control diagram in an existing drawing file, create a new drawing page and open the HVAC Control and HVAC Control Equipment stencils.

Use one or more of the following techniques to construct the diagram itself:

- **Add reference lines to the diagram** — If you're building the diagram from scratch, drag guides from the horizontal and vertical rulers to provide reference points for adding ducts to the diagram.

- **Add ducts** — From the HVAC Controls Equipment stencil, drag Duct shapes, such as Duct, Return Duct, or Supply Duct, onto the drawing page and glue them to guides, the dynamic grid, or to other duct shapes. When ducts appear as double lines, add ducts to the diagram by positioning an end point of one Duct shape at the centerline of another Duct shape. If the ducts are perpendicular, the intersection between the ducts doesn't show any overlapping lines. If the intersection between ducts doesn't show an opening, as shown in Figure 27-4, right-click the Duct shape you are adding and choose Shape ➪ Bring to Front.

TIP   To show only the centerlines of ducts, right-click any Duct shape and, from the shortcut menu, choose Single Line Ducts.

- **Add equipment** — From the HVAC Controls Equipment stencil, drag Equipment shapes, such as Centrifugal Fan or Humidifier, onto the drawing page and place them on top of Duct shapes.

- **Add controls** — From the HVAC Controls stencil, drag Sensor shapes, such as Timer or Light, onto Duct shapes on the drawing page. The alignment boxes for these shapes match the width of Duct shapes so they snap into place, as illustrated in Figure 27-4.

- **Configure equipment and sensors** — In the Shape Data window, change the values for the shape data fields. If the Shape Data window isn't open, you can also right-click a shape and choose Properties from the shortcut menu.

- **Resize ducts** — Drag duct end points. If you want to constrain the Duct shape to its current rotation, press Shift as you drag its end point.

- **Move ducts and controls** — If a Duct shape is glued to a guide, dragging Equipment and Sensor shapes onto the Duct shape also glues those shapes to the guide. To reposition the Duct shape and all of its associated Equipment and Sensor shapes, drag the guide to a new location. Glue horizontal Duct shapes to horizontal guides, and vertical Duct shapes to vertical guides.

Creating Reflected Ceiling Plans

The Reflected Ceiling Plan template doesn't include any tools specifically for creating reflected ceiling plans, so you can actually start with any plan template you want. For the ceiling grid itself, drawing tools, guides, and the Array Shapes command do the trick. After the ceiling grid is constructed, open the Electrical and Telecom stencil and the Registers, Grills, and Diffusers stencil (if they aren't open already) and drag shapes for light fixtures, air diffusers, and smoke detectors onto the plan.

Ceiling grids often repeat the same pattern of tiles and devices throughout most of a building. Because of this, a shortcut for producing an entire reflected ceiling plan is to draw a portion of the grid including ceiling tiles and ceiling-mounted equipment. Then, you can use the Array Shapes command to repeat that pattern of tiles and equipment throughout the rest of the building. Follow these steps to reproduce a ceiling grid pattern:

1. Drag guides onto the page to define reference points for one or more ceiling tiles. For example, position horizontal and vertical guides along the bottom and left edges of a corner ceiling tile.

2. Because Visio doesn't include a ceiling tile shape, use the Rectangle tool to create a ceiling tile.

3. Copy the ceiling tile to create all the tiles for the part of the ceiling grid that repeats. For example, if an array of four ceiling tiles represents the repeating pattern of tiles and ceiling-mounted equipment, use the Array Shapes command, Ctrl+D (Duplicate), or a combination of Ctrl+C and Ctrl+V (Copy and Paste) to create the additional ceiling tiles.

4. If the ceiling-mounted equipment is located at regular intervals in the grid, from the Electrical and Telecom stencil, drag lighting shapes onto the drawing. From the Registers, Grills, and Diffuser stencil, drag Diffuser shapes onto the drawing.

NOTE If the ceiling-mounted equipment is not located at regular intervals or some rooms have special equipment, drag that equipment onto the drawing after creating the array of repeating equipment.

5. Select the shapes for the ceiling tiles and ceiling-mounted equipment that you want to repeat.

6. Choose Tools ⇨ Add-Ons ⇨ Visio Extras ⇨ Array Shapes.

7. In the Array Shapes dialog box, set the Spacing for Rows and Columns to zero, select the Between Shape Edges option, and then click OK. The Between Shape Edges option creates the next set of shapes in the array at the rightmost and topmost edge of the first set.

Adding Electrical and Telecom Services

The Electrical and Telecomm stencil includes shapes for lighting fixtures, electrical switches, outlets, and other electrical devices for a home, building, or manufacturing plant. Start a diagram from scratch by choosing File ⇨ New ⇨ Maps and Floor Plans ⇨ Electrical and Telecom. Because the Electrical and Telecomm template doesn't include any specialized commands, it's often easier to open the Electrical and Telecom stencil (File ➪ Shapes ➪ Maps and Floor Plans ➪ Building Plan ➪ Electrical and Telecom) while you work on an existing floor plan.

Most of the shapes on the Electrical and Telecom stencil include shape data for specifying different types of electrical devices. For example, the Switch Type property for the Switches shape can change the switch between Single Pole, 3 Way, 4 Way, Timer, and Weatherproof switches. Although the basic shape might stay the same, changing the type adds other graphics that indicate the specific type of component. A few shapes include shortcut menu commands for configuration, such as Flip Orientation, which flips the Switches shape about the horizontal or vertical access, depending on whether you want to attach the switch to a wall on the left or right side of a room.

Although switches and outlets are associated with walls in floor plans, Visio shapes for switches and outlets don't rotate into position as Door and Window shapes do. To correct this omission, drag the rotation handle on the shape to rotate the shape into position. You can also rotate a shape to the left or right by 90 degrees by selecting the shape and then pressing Ctrl+L or Ctrl+R.

Visio also doesn't offer a tool to draw wiring between electrical devices. You can use the Freeform tool or Pencil tool to draw connections between devices. To add wiring between two shapes, select the Freeform tool, drag from a connection point on the first shape, and drag slowly to the connection point where the wire ends.

CROSS-REF   To learn about the Freeform tool, see Chapter 4.

Adding Plumbing

The Plumbing and Piping Plan template creates a new scaled drawing and opens stencils with shapes you can use to show pipes, valves, and fixtures for water supply and wastewater disposal systems. You can start a diagram from scratch by choosing File ➪ New ➪ Maps and Floor Plans ➪ Plumbing and Piping Plan, but you can also open the following stencils by choosing File ➪ Shapes ➪ Maps and Floor Plans ➪ Building Plan while you work on an existing floor plan:

- **Pipes and Valves–Pipes 1 and 2** — These two stencils include dozens of linear shapes that represent pipelines and pipeline devices. Drag shapes for pipelines onto the page and glue each end to connection points on fixtures or plumbing equipment. You can resize pipeline shapes by dragging their end points.

- **Pipes and Valves–Valves 1 and 2** — These two stencils include all kinds of valves you can glue to the ends of pipeline shapes. For shapes that include shortcut commands or shape data, such as the In-line Valve shape, right-click the shape and choose a configuration command from its shortcut menu or select a value in a shape data field.

- **Plumbing** — This stencil offers standard plumbing shapes, including Boiler, Radiator, Toilet, and Bath. Bathroom shapes include both top view and side view shapes so you can show how fixtures are connected in a plan or cutaway view.

Adding Security and Access Systems

Unfortunately, good security becomes more important every day. Whether you're designing a state-of-the-art security system for a top-secret development facility or setting up electronic access for a home or office building, the Security and Access Plan template helps prepare a diagram of security and building access features. You can start a diagram from scratch by choosing File ➪ New ➪ Maps and Floor Plans ➪ Security and Access Plan, but you can also open the following stencils (File ➪ Shapes ➪ Maps and Floor Plans ➪ Building Plan) while you work on an existing floor plan:

- **Alarm and Access Control** — Includes shapes for card readers, keypads, cameras, and other access devices

- **Initiation and Annunciation** — Includes shapes for paging and alarms

- **Video Surveillance** — Includes shapes for motion detectors, cameras, and other video equipment

Most of the shapes on the Security and Access stencils include shape data for specifying different device configurations. For example, the Mount Type property, which is associated with numerous shapes, specifies whether the device is mounted on the ceiling, on the wall, flush, hidden, or in other ways. On the drawing page, the shape label changes to reflect the type of mounting, the type of technology, and the function type.

Creating Site and Landscaping Plans

Site plans (available only in Visio Professional) come in two flavors, depending on the type of site information you want to show. Landscaping plans can be as intimate as a backyard garden for a town home — showing plants, fences, the sprinkler system, stepping stones, and recreational elements. Conversely, site plans often represent a much larger area and show landscaping, irrigation, parking, driveways, and traffic management features.

For smaller sites, an existing building plan is a perfect place to start. You can use the building plan as a background and add site details on a foreground drawing in relation to the building. For larger sites, a separate site plan is usually required simply because the larger area needs a smaller scale to fit onto the drawing page.

To create a site plan from scratch, choose File ➪ New ➪ Maps and Floor Plans ➪ Site Plan. Visio creates a new drawing using a civil engineering scale of 1" = 10'0". This scale fits a site 360 feet by 240 feet on the architectural drawing page. If your site is larger than that, you can choose predefined civil engineering scales up to 1" = 100'0" or you can set a custom scale.

If you use an existing building plan, your buildings are already on the drawing page. However, when you start a new site plan with the Site Plan template, you must draw building outlines using drawing tools such as Line, Rectangle, or Pencil.

Adding Landscaping Elements

Several stencils provide shapes for the plantings and constructed landscaping features. These shapes work equally well for commercial landscaping and home garden layouts. Drag shapes from the following stencils to add landscaping elements to the site plan:

- **Garden Accessories** — Includes shapes for fences, posts, and gates; and shapes for surfaces, such as flagstone, brick pathways, concrete, driveways, and patios.

- **Irrigation** — Includes shapes for irrigation lines, spray heads, valves, and other devices. These shapes include shape data for specifying different types of irrigation devices.

- **Planting** — Includes shapes for different types of trees, shrubs, hedges, and potted plants. Label plants with both common names and plant descriptions with the Plant Callout shape.

- **Sport Fields and Recreation** — Includes shapes for recreational equipment, such as pools, swing sets, and different types of sports fields.

Landscaping elements often end up positioned at regular intervals, whether you are planting a grid of palm trees or constructing a brick sidewalk. To create and arrange shapes at regular intervals vertically and horizontally, use the Array Shapes command.

 To learn how to apply the Array Shapes command, see Chapter 4.

Working with Roads and Parking Lots

The Site Plan template opens both the Parking and Road stencil and the Vehicles stencil. The Parking and Road stencil include shapes for roads, driveways, parking stalls and lots, sidewalk ramps, and traffic islands. By adding shapes from the Vehicles stencil to the plan, you can make sure that vehicles of different sizes can navigate the site.

Creating Roads and Parking Lots

To create roads and parking lots, follow these steps:

1. Drag guides onto the drawing page to mark reference points for the perimeter of the site, roads, and parking stalls.

2. Drag Curb and Driveway shapes onto the drawing page. To connect Curb and Driveway shapes, use the Line tool to draw lines between the shapes.

3. To add parking strips, stalls, and islands, drag shapes from the Parking and Roads stencil onto the drawing page. Glue their end points to guides on the page to simplify repositioning them.

4. Parking stall shapes and parking strip shapes glue to the end points or control points of Island shapes. Simply drag parking stall or parking strip shapes to Island shapes and drop them into place when Visio highlights their connection points with red boxes.

5. If you drag individual Parking Stall shapes onto the page and glue them to a guide, you can reposition all the Parking Stall shapes at once by dragging the guide to which they are glued.

6. Drag shapes from the Vehicles, Site Accessories, and Planting stencils, to show vehicles, parking lot components, drains, outdoor furniture, and plants. Glue shape end points to guides.

Modifying Parking and Road Shapes

Parking and Road shapes include selection and control handles for reconfiguring the shapes. Unlike mall parking at holiday time, some shapes, such as the Parking Strip shape, are extendable, and add additional parking stalls as you drag the selection handle at either end of the shape. Other shapes include control handles, such as the control handle on the Radial Strip shape, which changes the number of stalls as you drag it, as shown in Figure 27-5.

FIGURE 27-5

Use glue and control handles to position and configure parking stalls.

Drag control handle to change the number of stalls

Glue parking shapes to islands

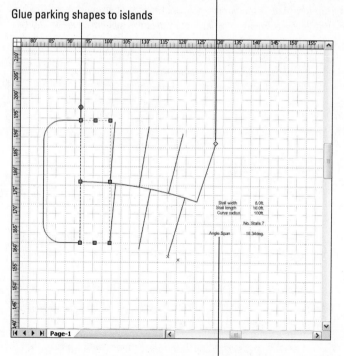

Text block showing properties of the parking shape

 To find out what a control handle does, position the pointer over the control handle until the screen tip appears.

Some shapes include shape data for further configuration. For example, the Parking Strip and Parking Stall shapes include a Stall Angle shape data field. By default, Visio parking stalls are oriented at seventy degrees. To create straight parking stalls, on the drawing page, right-click a Parking Strip or Parking Stall shape and, from the shortcut menu, choose Properties. In the Stall Angle box, type **90deg.** and click OK. The values in the Stall Width and Stall Length fields determine the size of the parking stall, which is ideal for creating parking stalls for compact cars and monster trucks alike. Shapes with curves, such as Curbs, include a Radius property that controls the radius of the curve on the shape.

Some Parking shapes include shortcut commands to reconfigure the shape. For example, with the Radial Strip shape, you can choose Stalls Inside or Stalls Outside from the shortcut menu to change whether the stalls face the inside radius or outside radius for the shape.

Drawing Directional Maps

Another type of Visio drawing that covers an even larger area than that of site plans is the *directional map*. With the Directional Map template, you can create road maps that show how to drive to your destination, or metro maps that help users navigate the transportation system in a city. These maps can include roads, transportation lines, as well as landmarks. In Visio, three-dimensional Directional Map shapes make it easy to render small geographic areas, such as villages or towns, with appealing and colorful three-dimensional shapes. By default, Visio creates maps without scale, but you can indicate an approximate scale by adding text to a Scale shape.

 The Directional Map templates are available in both Visio Standard and Visio Professional.

Creating Road and Metro Maps

To create a directional map, follow these steps:

1. Choose File ⇨ New ⇨ Maps and Floor Plans ⇨ Directional Map.

2. To add roads to the map, drag a shape, such as Road Square, Curve 2, or Railroad from the Road Shapes stencil onto the drawing page. You can resize Road shapes by dragging their end points.

3. To specify the thickness of a Road shape, right-click the shape and, from the shortcut menu, choose Thin, Standard, Thick, or Custom. If you choose Custom, drag the control handle to specify the thickness you want.

 To change the direction of a Road shape, right-click the shape and choose one of the Rotate or Flip commands. You can also click Rotate or Flip buttons on the Action toolbar or use keyboard shortcuts such as Ctrl+L and Ctrl+R to rotate a shape ninety degrees to the left or right.

4. To simplify moving multiple road segments at once, glue Road shapes to guides and then drag the guides to reposition the shapes.

5. Drag 3-Way, 4-Way, Roundabout, and Interchange shapes onto the drawing page to represent intersections. Glue end points for Road shapes to connection points on the intersection shapes.

6. Identify numbered roadways by dragging shapes, such as Interstate and State Route, onto the page near the Roadshape you want to identify. Double-click the identifier shape and type the interstate highway number, route number, or road number.

7. To draw metro or subway systems, use shapes on the Metro Shapes stencil. These shapes work similarly to Road shapes, but also include shapes to show stations and stops.

8. To indicate landmarks such as lakes, rivers, airports, malls, schools, and hospitals, drag the appropriate shapes onto the map. You can drag selection handles to resize Landmark shapes in one direction or proportionally.

9. To add transportation signs, such as one-way street signs or freeway exits, drag shapes from the Transportation Shapes stencil.

10. To indicate the location of recreational areas, drag shapes from the Recreation Shapes stencil.

Modifying Roads and Intersections

Visio Road shapes are quite simple, but the combination of straight, curved, and flexible Road shapes do a good job of representing roads for a directional map. The following are some methods for modifying roads and intersections on a map:

- **Set Default Road Thickness** — To specify the thickness for different types of roads, make sure no shapes are selected and then right-click the drawing page and choose Data ➪ Shape Data. In the Shape Data dialog box, type values for Road Width, Primary Route Width, Narrow Road/Street Width, and Metro Width.

- **3-way and 4-way intersections and interchanges** — To change the thickness of intersection roadways, right-click the shape and, from the shortcut menu, choose the thickness you want (Thin Road, Standard Road, Thick Road, or Custom). All arms of the shape change to the same thickness. To change the length of vertical or horizontal intersection arms, drag the selection handles on the sides or the top and bottom of the shape. To lengthen all the arms, drag one of the corner selection handles.

- **Flexible roads or flexible metros** — To change the thickness of intersection roadways, right-click the shape and, from the shortcut menu, choose the thickness you want. Use one or more of the following methods to modify a flexible shape:

  - **Change a curve** — To reposition points or redefine a curve, on the Drawing toolbar, click the Pencil tool. Drag a vertex or selection handle.

  - **Add a vertex** — Click the Pencil tool on the Drawing toolbar. Then Ctrl-click the shape where you want to add the vertex.

Annotating Maps

Basic Visio annotation techniques are all you need to add notes to your directional maps. The Landmark Shapes stencil includes several Text Block shapes for adding text at different font sizes. You can use the Callout shape to add text and point it to an element on the map.

Shapes on the Landmark Shapes stencil indicate scale and direction for maps. Although maps are unscaled by default, a Scale shape on the drawing page can indicate an approximate scale of the map. Select the Scale shape and type the total distance that the shape represents. To indicate direction, drag either the Direction or North shape onto the drawing page. Drag the rotation handles on either of these shapes so that north is pointing in the right direction.

Using 3-D Map Shapes

Three-dimensional directional maps are perfect for producing illustrated maps or tourist guides. These maps aren't to scale, but show buildings, roads, and landmarks in an isometric three-dimensional view. To create a three-dimensional map, you choose File ➪ New ➪ Maps and Floor Plans ➪ Directional Map 3-D and then drag and drop shapes from the Directional Map Shapes 3-D stencil onto the drawing page. You use standard Visio techniques to work with these shapes, including the following:

- Dragging selection handles to resize the shapes
- Using Ctrl+D to quickly duplicate shapes, such as road segments
- Using rotation handles to rotate shapes
- Grouping shapes so you can move them as one
- Adding text to shapes by selecting the shape and typing

Summary

Visio Professional includes templates for numerous types of architectural and engineering plans, as well as directional maps. By placing the shapes for the building shell and core on a background page, they are readily available as reference for the components for other types of plans, such as building services. Most of the shapes for architectural and engineering plans include shortcut menu commands, control handles, and shape data for modifying shape configuration. In addition, shape data is equally effective for creating component schedules and other types of reports.

Directional maps show larger areas than building or site plans. Although they are unscaled in Visio by default, they still represent real-world distances. You can drag and glue shapes together to depict roads, metros, and landmarks for road and city maps.

Chapter 28

Planning Space and Managing Facilities

S pace plans are valuable tools from the very first thoughts about building design to the ongoing management of facilities after they're built and occupied. Early on, space plans help determine the optimal arrangement of building space to meet organizational needs. It's much easier to move simple outlines around to figure out adjacencies like whether the conference room should be closer to the reception area or the executive offices. Then, when buildings are occupied, space plans help track the use of space by different departments, identify the location of resources, and put those spaces to best use and the resources in the most appropriate locations.

For basic space planning, such as arranging the rooms for your new home or office, the Office Layout template available in both Visio Standard and Visio Professional is all you need. By dragging and dropping Space and Room shapes from the Walls, Doors, and Windows stencil, you can shuffle the shapes around and resize them until you're satisfied with the layout.

However, for larger-scale space planning or facilities management (tracking and managing who and what goes where in facilities), only the Space Plan template in Visio Professional can help. This template enables you to build a model of spaces, assets, and resources, which you can use to plan the best use of space, track usage, and analyze changes and upcoming moves.

Space and facilities management plans process significant amounts of data — space assignments, asset tags, resource locations, and more. Building these plans with Space Plan template tools and incorporating Visio data features can turn your existing facilities data into a visual model of your facilities.

This chapter begins by showing you how to create space plans using the Space Plan Startup Wizard, either by importing spaces using the Import Data

Wizard or by adding shapes directly to drawings. You will then learn how to assign resources, such as people, furniture, and equipment to space plans. Finally, you will learn how to use these space plans to manage facilities.

Understanding Space Planning and Facilities Management Using Visio

Space planning typically means arranging space within a building so that groups obtain the space they need and business processes operate efficiently. High-level space planning can begin with a few hand-drawn ovals on a napkin, but eventually planning the detailed layout of spaces within a building shell involves studying organizational processes and needs, work flows, architectural requirements, and more. A more fastidious cousin to space planning, facilities management usually translates to tracking and relocating people, computers, equipment, and furniture to respond to reorganizations and office moves. Some organizations take facilities management further to include facilities maintenance. For example, by tracking when lightbulbs are changed, the maintenance department can develop plans to keep areas lit.

Visio Space Planning Tools

The tools to produce high-level space plans are available in both Visio Standard and Visio Professional. For more detailed space planning and facilities management, Visio Professional and the Space Plan template are a must. Depending on the type of planning you want to do, choose from the following Visio features:

- **Conceptual Planning** — For sketching general ideas of space arrangements, the Ink tool or the Ellipse tool can produce rough outlines of space on a drawing page. Select the Ink or Ellipse shape you've added and type the name of the area. To show spaces that should be located near each other, draw arrows between shapes.

- **Detailed conceptual planning** — When you have an existing building shell to divvy up into spaces or you know the amount of space each department needs, Visio Space shapes are convenient planning tools. Space shapes appear on the Walls, Doors, and Windows stencil and the Walls, Shell, and Structure stencil. Drag Space shapes onto the page and resize them until the value for the area displayed within the shape matches a department's allocation. Space shapes work equally well on blank drawings, existing Visio plans, or on top of imported CAD drawings.

- **Detailed space planning** — When you're planning space within a fully constructed building, you can drag Space shapes onto the page and resize them not only to meet a department allocation, but also to fit within the Wall shapes that represent constructed walls. The Auto-Fit command on a Space shape's shortcut menu fits the space to existing shapes on the floor plan.

No matter which type of space planning you want to do, Space shapes help you focus on the abstract spaces. The shape data associated with Space shapes helps track information, such as the intended use for a space (such as office or conference room), the name of the space, the department using it, its occupancy, and more. During initial space planning, the name of the space or the department might be sufficient and adding those values manually might be acceptable. As you develop more detailed plans, especially for large facilities, filling in all the shape data fields by importing facilities data is faster and less error-prone.

To help you size and lay out spaces properly, Space shapes automatically calculate and display the area they enclose. Feel free to change the color and fill pattern of Space shapes to differentiate spaces during the planning process. Later, you can convert Space shapes to Wall shapes and begin a detailed building plan, as described in Chapter 27.

Managing Facilities with Visio Features

For small facilities, management can be as simple as relocating Visio shapes on a drawing page. However, for larger facilities or more systematic tracking and management, Visio Professional offers specialized tools to track space and assets accurately, and can use external data to automatically refresh plans with data changes. Visio Professional space planning tools aid in performing the following facilities management tasks:

- Develop space plans that reflect the physical spaces you manage.

- Associate people and assets, such as equipment and furniture, with areas and space by importing data from a spreadsheet or database.

- Allocate unassigned people and assets to spaces in the plan.

- Reassign people and assets by dragging shapes from one space to another in the Space Plan model or on drawing pages.

- Automatically update the space plan when data in the spreadsheet or database changes.

Because facilities management relies so heavily on data, setting up full-blown space planning and facilities management plans requires more work than many other types of drawings. It helps to have a database administrator working with you to make sure Visio and your organization's facilities database play well together. Fortunately, the Visio Space Plan Startup Wizard simplifies these setup steps considerably. Before you can track and manage resources in a Visio plan, you must complete the following tasks, which are described later in this chapter:

- Prepare your facilities data in your external data source.

- Import your facilities data into a space plan using the Space Plan Startup Wizard or the Import Data Wizard.

- Set up your space plan to refresh when the data in your external data source changes.

- Make sure that all spaces are placed in your space plan using the Space Explorer window.

- Assign people and resources to the correct categories and spaces using the Space or Category Explorer window.

- Make sure that unassigned people and resources are assigned to spaces using the Space or Category Explorer window.

Exploring the Space Plan Template

The Space Plan template (remember, available only in Visio Professional) is one of the more powerful Visio templates with features for laying out spaces, importing data, and tracking facilities. When you create a new drawing using the Space Plan template, the template adds the Plan menu to the Visio menu bar and opens stencils associated with space planning and facilities management. The Space Plan template also launches the Space Plan Startup Wizard to help you configure the data that represents your spaces, assets, and resources. The Import Data Wizard is an additional tool for setting up data on an existing plan. For space plans, Visio opens a letter-size drawing using landscape orientation and a scale of $\frac{1}{8}$" = 1' 0".

Menus and Stencils

Don't be disappointed when the Space Plan template adds a Plan menu to the Visio menu bar. Unlike its architectural and engineering plan counterparts, the Plan menu for the Space Plan template contains more than the Convert to Walls and Set Display Options commands. This Plan menu includes the following additional entries:

- **Explorer** — This menu entry toggles the display of the Explorer window, which displays the space plan model, described in the next section.

- **Color By Values** — Choose this entry to launch the Color By Values add-on, which associates colors with space shape data values, as described in Color-Coding Space Plans.

- **Label Shapes** — This menu entry launches the Label Shapes add-on, which displays shape data values within the shapes on the drawing page.

- **Assign Category** — The shapes on a space plan fall into one of several categories: Asset, Boundary, Computer, Person, Printer, and Space. Each category comes with a set of shape data fields specifically for tracking that type of resource. The section "Assigning Categories to Resources" later in this chapter explains how to assign categories and put them to use.

- **Import Data** — This menu entry launches the Import Data Wizard for importing facilities data into shapes on a drawing page, as described in the section "Adding Spaces Using the Import Data Wizard."

- **Refresh Data** — Choose this entry to refresh the facilities data that appears on a space plan from your external data source.

The Space Plan template also opens the Resources stencil, which includes Space and Boundary shapes for delineating departments, offices, and other areas in a building. The Resources stencil also includes shapes that represent resources, such as Person, Computer, and Asset for identifying the people, computers, and other assets (such as furniture) that inhabit the spaces. The Report shapes on the Resources stencil are an easy way to create predefined facilities management reports quickly and easily.

Using the Space and Category Explorers

Because a Visio space plan is also a model of your facilities, the Space and Category Explorers provide an alternate, and sometimes preferable, way to view and modify spaces and resources. First, the Space and Category Explorers show the relationships between model components in an outline view, so you can immediately identify which resources belong to which spaces, and which spaces are contained within larger areas. In addition, the model shows all the spaces and resources for your space plan, not just the ones that appear on drawing pages.

The Space Explorer displays a hierarchy of boundaries, shapes, and resources, where boundaries represent larger areas such as departments, shapes represent offices or rooms, and resources represent people, computers, or other assets associated with spaces. For example, you can associate people and equipment to offices, and offices to departments. In the Space Explorer, you can see which resources are located in which spaces, as shown in Figure 28-1, regardless of which drawing page shows them. The Category Explorer groups elements in the space plan by category, such as Person, Computer, or Space.

When you open a new space plan, the Space and Category Explorer windows appear in one docked window, with tabs for each Explorer window. You can hide or show both of the Explorer windows by choosing Plan ➪ Explorer. To float each window individually, right-click its tab and, from the shortcut menu, choose Float Window. To float the entire window with both Explorer windows in it, drag the window title bar to a new location.

> **TIP** In Figure 28-1, the Explorer windows are docked within the Shapes window to display as much of the drawing window as possible. To dock the Explorer windows there, drag the Explorer title bar into the Shapes window and release the mouse button.

Here are methods for managing spaces and resources with the Space Explorer and the Category Explorer:

- **Locate and select spaces and resources on a drawing page** — In either Explorer window, right-click a space or resource and, from the shortcut menu, choose Show. Visio displays the drawing page on which the corresponding Space or Resource shape is located, centers the shape in the drawing window, and selects the shape on the page.

- **Move resources between drawing pages** — Instead of cutting a shape from one page and pasting it onto another, in the Space Explorer, drag a person or asset from one space to another. The corresponding shape moves on the drawing as well.

- **Enter resource information** — In either Explorer window, right-click a resource and, from the shortcut menu, choose Properties. In the Shape Data dialog box, click a shape data field and type or select a value. The value you type in the Name field appears by default in the Explorer windows and on the shape on the drawing page.

FIGURE 28-1

The Space and Category Explorers dock in the same window, showing spaces and the people and assets assigned to them.

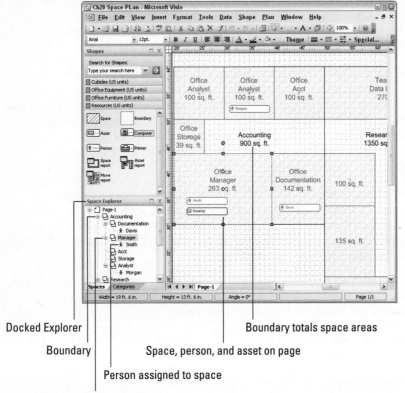

Docked Explorer

Boundary totals space areas

Boundary

Space, person, and asset on page

Person assigned to space

Space within boundary

- **Rename resources and assets** — In either Explorer window, right-click a resource or asset and, from the shortcut menu, choose Rename. Visio selects the name in the Explorer window, so you simply type the new name to replace the existing one.

- **Assign resources to spaces** — If people or assets have not yet found a home in a space in the plan, in either Explorer window, expand Unplaced Data. Drag a resource or asset from the Unplaced Data area to a Space shape on a drawing page.

NOTE   The drawing pages might not include all spaces, people, or assets for your space plan. The Unplaced Data folder in either Explorer window shows the spaces and resources that don't exist on drawing pages.

- **Delete resources** — To delete a person or asset both in the model and on the drawing page, in either Explorer window, select a resource and press Delete.

Creating Space Plans

The amount of detail in space plans depends on their purpose. Space plans for prototyping the arrangement of departments and offices can be quite spare — a name and department might be all you need to identify the spaces. Space plans used to manage the assets, personnel, and space in a facility are often loaded with detailed data and kept up to date with real-world facility changes. Because of the tools that the Space Plan template provides, you'll want to start a space plan by choosing File ⇨ New ⇨ Maps and Floor Plans ⇨ Space Plan to create your Visio drawing. After the file is open, the method you use to create your space plan depends on what you intend to do with the plan:

- **Preliminary layouts** — For prototyping, you don't need detailed building outlines or floor plans. When the Space Plan Startup Wizard opens, cancel it and start by using drawing tools on the Drawing toolbar to sketch the outline of floors. Then, drag Space shapes onto the drawing and move them around with standard Visio techniques to layout the spaces. Alternatively, you can jump right in and start dragging Space shapes onto the page without any building outline.

- **Detailed layouts** — To begin arranging spaces within the confines of a building shell, take advantage of the Space Plan Startup Wizard to display a floor plan as a backdrop. If you have a list of spaces in an Excel workbook, the wizard can creates shapes for those spaces. You can still drag Space shapes onto the drawing and move them around. The Auto Size command on Space shape shortcut menus modifies Space shape borders to fit the walls on the drawing.

- **Tracking plans** — When you want to track resources on a Visio space plan, you need accurate and detailed information. When you create a space plan with the Space Plan template, the Space Plan Startup Wizard opens, in which you can add a background floor plan and add Space shapes based on a list of room numbers and names. The shapes added by the wizard appear on the drawing page and in the Explorer windows. You still must add other spaces not in the original list, add resources, and associate them with spaces.

TIP Suppose you have a floor plan drawing file that you want to use as the basis for a space plan. The building shell and walls are there, but the Plan menu lacks the specialized space plan commands in Visio. To make all the space planning tools available on the Plan menu, choose Tools ⇨ Add-Ons ⇨ Maps and Floor Plans ⇨ Enable Space Plan.

Using the Space Plan Startup Wizard

By no means does the Space Plan Startup Wizard do everything for you. Nor is it a totally flexible and all-powerful data import tool. What it does is simplify a few of the steps in setting up a new space plan. With this wizard, you can specify a floor plan to use as a background for your space plan and create spaces based on the room numbers in an Excel workbook.

> **TIP** If you have an external data source of facilities data, the Import Data Wizard is the better tool for adding spaces to a Visio space plan. This wizard creates spaces in your model and on your drawing, but it can also import as much information about those spaces as you store in the external data source. See the section "Using Visio Space Plans to Manage Facilities" later in this chapter for instructions on using the Import Data Wizard.

Creating Space Plans Using the Wizard

To use the Space Plan Startup Wizard to create a space plan, follow these steps:

1. Choose File ➪ New ➪ Maps and Floor Plans ➪ Space Plan. The Space Plan Startup Wizard launches automatically after Visio creates a new space plan drawing.

2. On the first screen of the Space Plan Startup Wizard, select the type of image or drawing you want to use as a background for your space plan:

   - **Image** — Select this option to display a graphics file of the building, such as a JPG or GIF file.

   - **Visio Drawing** — This option enables you to select an existing Visio drawing to use as an underlying plan for your space plan.

   - **CAD Drawing** — If your building shell or floor plan resides in a CAD drawing file, select this option.

   - **None** — Select this option if you plan to draw the outline of the building with Visio tools.

3. Click Next. If you chose to use a graphics file or a drawing, a dialog box opens in which you select the graphics file or drawing file you want to use. Click Open to use that file and proceed to the next part of the wizard.

4. On the next wizard screen (the Get Room List screen), select the option for the source of spaces you want to add to your space plan. You can use an existing Excel spreadsheet, create a new spreadsheet in the wizard, or type the room numbers manually.

> **TIP** It's best to create a spreadsheet of rooms before you start the wizard, so you don't have to recreate your data in case something goes awry as the wizard runs.

5. Click Next. If you use an existing spreadsheet, which is the most dependable approach, on the next wizard screen, shown in Figure 28-2, do the following:

   a. Next to the Choose The Excel Spreadsheet File That Contains Room Information box, click Browse and open the Excel workbook that contains your room data.

   b. If the workbook includes more than one worksheet or named ranges, in the Choose The Worksheet Or Range drop-down list, choose the worksheet or range. (Visio populates the drop-down list with the worksheets and named ranges in the selected Excel workbook.)

   c. In the Select The Column That Contains Room Numbers drop-down list, choose the column that contains the room numbers.

FIGURE 28-2

You can choose an Excel spreadsheet, including the specific worksheet and column, to feed room numbers to the Space Plan Startup Wizard.

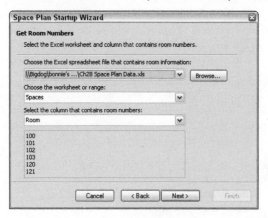

6. Click Next. On the final wizard screen, click Finish. Visio adds the spaces to the Category and Space Explorers. If Visio finds a shape on a drawing page with a matching room number, it lists the space under that drawing page in the Space Explorer hierarchy. Otherwise, it shows the spaces under the Unplaced heading.

7. For spaces that the wizard doesn't place on the drawing page, from the Space or Category Explorer window Unplaced Data folder, drag unplaced spaces onto the drawing page, making sure not to overlap Space shapes. Visio creates Space shapes for each unplaced space you drop onto the drawing. To resize a Space shape, drag any of its selection handles.

Using Shapes to Create Space Plans

If you build space plans from your facilities data, the Import Data Wizard is the easiest way to add new spaces. However, you can also add these elements by dragging shapes onto drawing pages. In the Resources stencil, the Boundary shape is intended to depict larger areas such as departments or divisions. The Space shape acts as a container for the resources assigned to a specific space or office.

CAUTION   To ensure that spaces are assigned to boundaries, and resources are assigned to spaces, add Boundary shapes to your drawing page first, followed by Space shapes, and finally Person, Asset, Computer, and Printer shapes. If you've already added these elements in a different order, correct the assignments in an Explorer window by dragging spaces to the proper boundary or resources to the proper spaces.

Manually Creating Spaces

Define spaces on a Visio space plan using any of the following methods:

■ Designate larger areas, such as departments, by dragging Boundary shapes onto the drawing page. You can drop Space shapes inside a Boundary shape to assign specific offices to a department. However, you can't assign people or assets to boundaries. The Boundary shape displays the square footage it encloses, so you can use it to show the area of multiple Space shapes.

■ From the Resources stencil, drag a Space shape onto the drawing page. Visio adds a 100-square-foot square space by default. If you drag the shape's selection handles to resize it, Visio updates the square footage that appears within the shape when you select it, as shown in Figure 28-3.

FIGURE 28-3

Space shapes automatically display the area that they enclose.

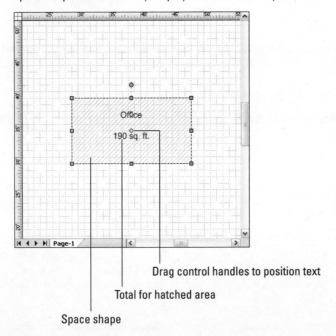

Drag control handles to position text

Total for hatched area

Space shape

> **TIP** If the floor plan contains Wall or Room shapes, you can automatically adjust a Space shape to fit an area in the floor plan. After dropping the Space shape into the area, right-click it and, from the shortcut menu, choose Auto-Fit. Visio changes the outline of the Space shape to match the perimeter defined by the Wall shapes on the floor plan and displays the new enclosed area in the Space shape text block.

■ Use drawing tools to outline spaces and then convert the drawing shapes to Space shapes. To do this, choose Plan ➪ Assign Category. In the Assign Category dialog box, select Space in the Category drop-down list and click OK. Visio converts the outlined shape into a Space shape and displays the calculated area in a text block within the shape.

CAUTION To ensure that the area on your space plan is calculated correctly, do not overlap Space shapes. When Space shapes overlap, the overall calculated space area will be greater than the area within your building.

When you intend to import facilities data from a spreadsheet or database, spaces must include a value in their Name field, so there's a way to match the space with a record in your data source. The value you specify must match a unique identifier in the spreadsheet or database. For example, if the office room number in the spreadsheet is 301, type **301** in the Space shape's Name field. To do this, follow these steps:

1. If the Shape Data window is not open, choose View ➪ Shape Data Window.

2. Select the Space shape on the drawing page.

3. In the Shape Data window, click the Name field and type the value that matches the identifier in the data source.

TIP You can also specify the intended purpose of a Boundary or Space shape in the Shape Data window by clicking the Space Use field and typing the purpose or selecting one of the predefined purposes (Office, Conference, Storage, Other) in the drop-down list.

Modifying Boundary and Space Outlines

Although the Space and Boundary shapes on the Resources stencil are 100-square-foot squares by default, you aren't stuck with rectangular areas. Boundary and Space shapes respond to basic Visio tools just like any other shapes. No matter how you indent and contort the outline or size of these shapes, Visio calculates the new total for the enclosed area and displays it within the shape. Here are methods for modifying the perimeters of Boundary and Space shapes:

■ **Resize a Space or Boundary shape** — Drag any selection handle on one of these shapes to resize it as you would any other type of shape.

■ **Reshape a Space or Boundary shape** — To change the angles of the sides or to add indents and notches to a space or boundary, right-click the shape and then, from the shortcut menu, choose Edit (or, on the Drawing toolbar, click the Pencil tool). Then choose from the following methods:

▧ Add a new vertex by Ctrl-clicking the outline of the shape.

▧ Move a vertex by positioning the pointer directly over it. When the pointer changes to a four-headed arrow and the vertex turns magenta, drag the vertex to a new location.

 To select more than one vertex to move, Shift-click each vertex and then drag to a new location.

- **Auto-size a shape** — If the Wall shapes that surround a Space shape change, you can refit the shape to the new wall configuration by right-clicking the Space shape and, from the shortcut menu, choosing Auto-Size.

Assigning Resources to Space Plans

To use Visio drawings to track and manage the people, equipment, furniture, and other assets that occupy your facilities, you must first assign those items to categories and then assign them to spaces in a space plan. The Visio Space Plan template pigeonholes spaces and resources in six categories, each with its own set of predefined shape data fields:

- **Boundary** — A location category that represents larger areas within a building, such as departments or functional uses
- **Space** — A location category that represents specific spaces such as offices or rooms and can contain other resources, such as people or assets
- **Person** — Represents the human resources assigned to work in spaces
- **Computer** — An asset category specifically for computer equipment
- **Printer** — An asset category specifically for printers
- **Asset** — A generic asset category for tracking any other kind of asset

NOTE   You can't rename the built-in categories or define your own.

Each category carries a different set of shape data fields, appropriate for the type of resource. For example, the shape data fields for the People category include Name, Title, Phone Number, E-mail Alias, Manager, and Department. The Printer category focuses on tracking information, such as Asset Number, Serial Number, Manufacturer, Product Number, and so on. By assigning spaces and resources to categories, you can view the items on a plan and produce facilities reports by category. The hierarchy in the Space Explorer starts with boundaries and then shows the spaces within boundaries, and then the people and assets within each space.

Adding Resources to Space Plans

To track resources, they must be part of your space plan model. As with spaces, you can add resources to your model manually or by importing resource data:

- **Add resources manually** — It's easy to add resources to space plans, because the Resources stencil includes a shape for each type of resource. Dragging these shapes onto a drawing page automatically assigns the shape to the correct category. Shapes on the Resources stencil also include layer assignments to a layer of the same name as the resource category, such as Space, Person, and Computer. In the Shape Data window, in the Name field, type the name or an identifier for person or asset.

> **TIP** Shapes other than those on the Resources stencil are fair game for a space plan — as long as you assign them to a category, as described in the section "Assigning Categories to Resources" later in this chapter. For example, you might want to use custom shapes to represent different types of computers, or assign furniture shapes to the Asset category.

- **Import resources** — Choose Plan ⇨ Import Data to use the Import Data Wizard to create shapes for people and assets, and link external data to the shape data. The Import Data Wizard uses the masters on the Resources stencil to create shapes on the drawing page. For instructions on how to use the Import Data Wizard, see the section "Using Visio Space Plans to Manage Facilities" later in this chapter.

Placing Unassigned Resources

The Import Data Wizard automatically assigns resources to spaces — if it can match a space identifier in a resource record in the external data source with a space identifier in the Visio space plan model. If a person or asset doesn't have a space assignment or the wizard can't make a match, it places the people or assets in the Unplaced Data folder, which you see in the Space and Category Explorer windows.

To assign an unplaced resource in the Space Explorer window, do the following:

1. Next to the Unplaced Data folder, click the plus sign to expand the folder.

2. From the Unplaced Data folder, drag an unassigned person or asset onto the icon in the Space Explorer that represents the space to which you want to assign the unplaced item. Visio adds a shape for the person or asset to the Space shape on the drawing page as well.

> **NOTE** You can also assign a person or asset by dragging the resource from the Unplaced Data folder onto a Space shape on the drawing page. Visio adds a shape to the drawing and, in the Explorer window, moves the resource from the Unplaced Data folder to the icon that represents the space containing the resource.

Assigning Categories to Resources

The shapes on the Resources stencil all look about the same — they're small rectangles with an even smaller graphic to differentiate people, computers, printers, and other assets. If you would rather use more recognizable shapes on a space plan, you can assign categories to any Visio shape. For example, you could assign categories to the shapes from the Office Equipment or Office Furniture stencils. When these shapes are assigned to categories, you get the best of both worlds — not only do the shapes on the drawing page depict the resources they represent, but they also take on the shape data fields for the category and appear in the appropriate category in the Category Explorer window.

> **NOTE** Assigning a shape to a category applies the category shape data fields to that shape. What if the shape already has shape data fields with values that you don't want to give up? The Assign Category command includes a feature for mapping existing shape data fields to the ones for the category.

To assign a category to a shape, follow these steps:

1. Select all the shapes on the drawing page that you want to assign to the same category.

2. Choose Plan ⇨ Assign Category.

3. In the Assign Category dialog box, in the Category drop-down list, select the category to which you want to assign the shapes.

4. If the shapes you selected have shape data associated with them, map these fields to the predefined space plan field by doing the following:

    a. Click Properties. The Properties dialog box opens.

    b. Under Properties, select an existing shape data field.

    c. Under Category Properties, select the predefined category property to which you want to map the original shape data field.

    d. Click Add to map the fields, as shown in Figure 28-4.

FIGURE 28-4

You can transfer data from shape data fields to category fields by mapping the two while assigning a category to a shape.

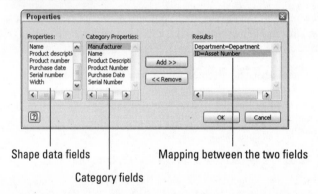

Shape data fields

Category fields

Mapping between the two fields

    e. Repeat steps b, c, and d for each field you want to map.

    f. Click OK when you're done.

NOTE Shape data fields that you don't map remain associated with shapes so you won't lose any existing data.

5. In the Assign Category dialog box, click OK.

Using Visio Space Plans to Manage Facilities

People, computers, equipment, and furniture seem to relocate continuously in response to reorganizations and office moves, so any way to simplify tracking these resources is welcome. By displaying spaces and resources on a Visio space plan, you have a graphic map to where things are within the facilities you manage. Visio can also produce facilities reports to show departments the space and resources they have or to charge groups for the space and assets they use. All this convenience requires some work up front — a Visio space plan has to have space plan or facilities data associated with the shapes. That's where the Import Data Wizard comes into play.

The Import Data Wizard can grab data from external data sources and use them to create spaces and resources on a Visio space plan. Regardless whether the data resides in Excel spreadsheets, Exchange Server Address Books, Active Directory servers, or tables in any ODBC-compliant database, the wizard can create new entities in a space plan model for the data it imports or store the entities as unplaced data in the Space and Category Explorer windows so you can place it when you're ready.

If you enthusiastically populated your space plan drawing with shapes, the wizard can import external data into the shape data fields for those shapes. Otherwise, it goes ahead and creates new shapes to hold the data. For example, the Import Data Wizard can import department assignments into the Department shape data field for the existing Space shapes on the drawing page. On the other hand, faced with a multitude of people and assets to add, you can delegate the grunt work of creating and assigning Person, Computer, Printer, and Asset shapes to the space plan Space shapes to the Import Data Wizard.

Preparing Your Space Plan Data for Use in Visio

Regardless of the quantity of space and facilities data you use with Visio space plans, importing proceeds more smoothly when the data is set up the way Visio expects. The Import Data Wizard accepts data from a variety of sources, including Excel workbooks, Microsoft Active Directory directory services, Microsoft Exchange, and any ODBC-compliant data source, such as a Microsoft Access database. Conversely, the Space Plan Startup Wizard reads data only from Excel workbooks. To properly import data into a Visio space plan, follow these guidelines to set up your data:

- **Assign unique identifiers** — In the data source, include a field or column that uniquely identifies the spaces, people, and assets you want to add to your space plan. A key field, such as employee number or asset tag number, is necessary if you want to automatically update the space plan when data in your data source changes.

- **Include room numbers to assign resources to spaces** — For people and assets assigned to specific spaces, in the data source, include a field for room numbers. Room numbers in the data source must match the room numbers for spaces in your space plan. Leave the room number field blank if a person or asset isn't assigned to a space. Use the same room number for each asset or resource assigned to a space.

> **NOTE**   Visio stacks the shapes for all these resources within the same Space shape. If you want to see all the shapes on the drawing page, drag each shape away from the shapes underneath while keeping them within the Space shape.

> **NOTE**   When a resource is assigned to a space number that doesn't exist in the space plan, Visio places the resource in the Unplaced Data folder in the Explorer windows. You can create the spaces, if necessary, and manually place the resources in the appropriate spaces.

- **Identify buildings in a data field** — For data sources that include data for more than one building, include a field that represents the building name or number. You can specify which building's data to import when you run the Import Data Wizard.

- **Include fields for other attributes** — For other data that you want to include in your space plan, set up additional fields, such as asset tag numbers, serial numbers, or maintenance dates. You can import this data into shape data associated with shapes.

Using the Import Data Wizard

The Import Data Wizard does a lot, but it doesn't take many steps to tell it what to do. To use the wizard to import facilities data and create shapes on a space plan, start by choosing Plan ➪ Import Data. Then, complete the steps described in the following sections.

> **CAUTION**   If you want the wizard to automatically assign people and assets to spaces, make sure your data source includes a field for room or space numbers. Also, verify that each Space shape on the Visio drawing includes a room or space number in its Name field. The easiest way to check for room numbers in Space shapes is to inspect the spaces in the Space Explorer window. If any space is named Space, you know that its Name property is blank.

Tell the Wizard Where to Store the Imported Data

In the first Import Data Wizard screen, you tell the wizard where to put the data you're importing. Choose one of the following two options:

- To create shapes to hold the data you import, select the Into Shapes That I Will Manually Place option. Select a sub-option to either cascade the shapes on the page so you can drag them into position or store the items as Unplaced Data so you can drag them from the Explorer directly to their position on the drawing.

- If you already have at least some shapes on the drawing, select the Into Shapes That Are Already on My Drawing option. Underneath that option, select a sub-option to add the data to the existing shape data or create new shapes on top of the existing shapes. For example, if you have Space shapes on the drawing and want to create People and Asset shapes in those spaces, select the Add As New Shapes On Top Of Existing Shapes option.

Click Next to proceed to the wizard screen for choosing the data source.

Identify the Data Source

In the What Is The Source Of The Data screen, in the Type drop-down list, select the type of data source that contains your facilities data (Excel, Active Directory, Exchange Server, or ODBC). The steps you take depend on the type of data source:

- **Excel** — Click Browse and then specify the path and file name for the file.
- **Active Directory** — Click Next to log into the server.
- **Exchange Server** — Click Next to log into the server.
- **ODBC** — Click Next to display an additional screen for selecting the data source, database file, and table within the database.

After specifying the data source, click Next to proceed to the wizard screen for specifying the data to import.

Specify the Data to Import

In the What Data Do You Want To Import screen, the way you specify the location of your data depends on the type of data source. For example, for a spreadsheet, from the Worksheet drop-down list, choose the name of the worksheet in the Excel workbook.

This screen includes two key check boxes: Include All Columns/Fields and Include All Rows/Records. If you leave these check boxes selected, you can skip to the next section, because Visio imports every record and every field in the data source. However, to choose the records and data fields to import, clear one or both of the check boxes. If you do so, the next screen that appears is your chance to tell the wizard what to import, as illustrated in Figure 28-5.

FIGURE 28-5

The Import Data Wizard can import all data in a data source or only some fields and records.

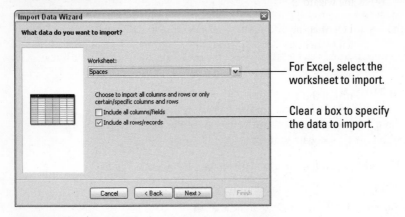

For Excel, select the worksheet to import.

Clear a box to specify the data to import.

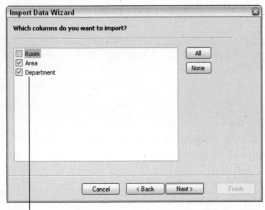

Choose the columns or records to import.

Click Next to proceed to the final section of the wizard.

Configure the Shapes with Imported Data

The final set of wizard screens that appear depend on whether you are importing data into existing shapes or adding shapes to the drawing as part of the import process. Here are the steps for both of these situations:

- **Creating shapes and importing data into them** — If, in the very first step of the wizard, you chose the Into Shapes That I Will Manually Place option, do the following to configure the new shapes the wizard adds:

  1. In the What Kind Of Shape Do You Want to Add to Your Drawing screen, Visio selects the Resource stencil by default. If the shapes you want to use are on another stencil in the Space Plan template, in the Stencil drop-down, choose the stencil or click Browse to choose from any Visio stencil.

  2. Select the thumbnail of the shape you want and click Next.

  3. To label or color-code the new shapes, specify the shape data you want to use to label or color by in the Label With and Color By drop-down list, respectively. Click Next.

  4. On the next screen, specify the field in your data source with unique IDs, such as employee ID, Name, or ID. Visio adds these values to a shape data field in the new shapes and uses those values to update data when you use the Refresh Data command. Click Next.

- **Adding data to existing shapes** — Because the shapes are already on the drawing, you don't have to specify the type of shape to add. However, you do have to identify the shapes to receive data, by doing the following:

  1. To label or color-code the new shapes, specify the shape data you want to use to label or color by in the Label With and Color By drop-down list, respectively. Click Next.

  2. On the next screen, select the type of shape into which you want to import data. For example, you can choose all shapes or specify one type of shape. Click Next.

  3. On the next screen, choose the shape data field that uniquely identifies the shapes on the drawing, for example, employee ID, Name, Room Number, or ID. Visio uses this field to match records in the data source with shapes on the drawing. When Visio finds a match, it adds the data to the shape's data fields. Click Next.

  4. On the next screen, choose the field in your data source that uniquely identifies the data records. When Visio finds a matching value in the shape data field and the data source identifier field, it adds the data to the shape's shape data fields. Click Next.

On the final wizard screen, Visio displays a message that the Import Data Wizard is completing and lists the actions the wizard took, such as the number of shapes the wizard added or the number it updated with imported data. To finish and close the wizard, click Finish.

> **NOTE** Just as you can with other shapes, you can manually add or change data in shapes by typing or selecting values in the Shape Data window. However, changing shape data does not change the corresponding data in your spreadsheet or database. Visio does not export data from your drawing back to your source data, so it's important to make data changes in the external data source.

Refreshing Space Plan Data

When you import information into a space plan, you can create a connection between the data in the data source and the shapes on your space plan. After that connection is established, it's a snap to keep your space plan current with any changes in your data source. For example, if employees move to other offices and the data source contains the new office numbers, refreshing data in the Visio drawing automatically shows the employees in their new spaces. When assets are sold or recycled and deleted from the data source, Visio removes them from the space plan as well. In addition, when the data source includes a new field such as maintenance data, refreshing the data adds that field to the shapes on drawing.

To refresh the data in a space plan, follow these steps:

1. Make sure that the data source hasn't been deleted or moved.

2. If you haven't done so already, use the Import Data Wizard and specify the field in your spreadsheet or database that contains unique identifiers.

3. Choose Plan ⇨ Refresh Data. After the Refresh Data command processes the data, click the Import Data Report link to see a log of the changes the command made. Click OK.

Labeling Facilities Shapes

Labels on shapes make it much easier to find spaces and resources on a space plan, particularly when you use the Resources stencil shapes, because they are so similar in appearance. The Plan menu in the Space Plan template includes the Label Shapes command, which runs the Label Shapes add-on. To label Space and Resource shapes, follow these steps:

1. Select the shape or shapes you want to label and then choose Plan ⇨ Label Shapes.

2. In the Label Shapes dialog box, in the Shape Type box choose `<all selected shapes>` to label the shapes you selected. If you want to label one type of shapes, select the shape name.

 To label all the shapes on the drawing, choose `<all shapes>`.

3. In the Label 1 box, select the shape data field to display as the first line of the shape's label, as shown in Figure 28-6.

4. To display additional fields in subsequent lines in the text block, select the fields for Label 2, Label 3, and Label 4.

5. Click OK. Visio adds the label text blocks to the selected shapes

FIGURE 28-6

The Label Shapes add-on can insert the values for up to four shape data fields into a text block on shapes.

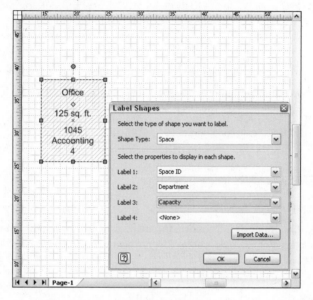

Color-Coding Space Plans

Color-coding makes it easy to spot shapes on a drawing. For example, you can color-code Space shapes by assigned departments to quickly identify which spaces belong to which departments. If the same color is splashed randomly throughout a building, an office move to co-locate all the people in that department might be in order.

Although the new data graphics in Visio 2007 include an option to color-code shapes, the Space Plan template uses the Color By Values add-on to color-code shapes. Choose Plan ➪ Color By Values to color-code shapes.

To select the shapes you want to color-code, use one of the following methods:

- To color-code all shapes or all instances of one master, click the drawing page background to make sure nothing is selected.
- To color-code specific shapes, select only those shapes.

Controlling the Display of Spaces

Space shapes include display options similar to those for Wall, Door, and Window shapes on floor plans (see Chapter 27). Right-click any Space shape and, from the shortcut menu, choose Set Display Options. In the Set Display Options dialog box, on the Spaces tab, you can perform the following tasks:

- Specify the shape data fields that appear in Labels 1 through 4 on Space shapes.
- Choose whether Space shapes auto size to Wall shapes or to wall reference lines.
- Choose the units in which Visio displays shape area.
- Specify the precision of the area calculation.

To run the Color By Values add-on, follow these steps:

1. Choose Plan ➪ Color By Values.

2. In the Color By drop-down list, select the shape data field by which you want to color-code the shapes. For example, choose Department to color spaces based on the departments assigned to them.

3. In the Shape Type list, select <all shapes>, <all selected shapes>, or the type of shape.

4. In the Range Type list, choose how to use colors to reflect values.

   - **Unique values** — This choice applies a different color to each unique value in the shape data, which is perfect for coloring shapes by department name.

   - **Discrete ranges** — Use this choice to use a different color for each range of values, such as 100 to 199, 200 to 299, and so on. This choice works well when you want to define the upper and lower limits of the ranges for each color.

   - **Continuous values** — This choice applies colors from low to high across the range of values in a field. For example, this choice would divide the range of square footage into equal ranges and apply a different color to each range.

5. In the Color field, select the colors you want to use.

   - **Unique values** — Click the color box for a value and choose the color you want to use. Repeat this step for each unique value.

   - **Discrete ranges** — Click the color box for a range and choose the color you want to use. Repeat this step for each range.

   - **Continuous values** — Click the first color to open the Color Ramp dialog box. By selecting a color in the Top Color and Bottom Color boxes, Visio creates gradations between the two colors for the range divisions it creates.

6. Visio adds a Legend shape to the drawing page showing the field used to color-code shapes and the correlation of color to values.

7. To modify the color-coding in your drawing, right-click the legend on the drawing page and from the shortcut menu, choose Edit Legend.

8. To update the color-coding to reflect changes to shape data, right-click the legend and choose Refresh Legend.

Finding and Moving Resources

The life of a facility manager is never boring with people quitting, new people being hired, and regular reorganizations thrown in for good measure. Naturally, as the locations of people and assets constantly change in a facility, so will your space plan. You can move people or assets in the Explorer window or move the corresponding shape on the drawing page. In addition, when you manage a large number of resources, you can use tools in Visio to quickly locate the resource you want.

To find resources in a space plan, use one of the following methods:

- **Using the Explorer windows** — In the Space or Category window, expand a space or category, right-click the person or asset you want to locate and, from the shortcut menu, choose Show. Visio displays the drawing page that contains the corresponding shape, selects the shape on the drawing page, and centers it in the drawing window.

- **Using the Find command** — If your space plan is crammed with so many resources that it's hard to find the one you want even in the Explorer window, choose Edit ➪ Find. In the Find dialog box, in the Find What box, type text that is associated with the shape you're looking for, such as its name. You can use text that appears on a shape, in shape data, or the shape name itself. Under Search In, select the locations you want to search and then click Find Next. The Find command finds the first shape containing the text you specified and highlights text within a shape or selects the shape if the text is found in a shape data field, shape name, or cell.

To move resources in a space plan, use one of the following methods:

- **Using the Explorer windows** — In the Space Explorer window, expand pages and space until you find the resource you want to move. Drag the icon for the resource onto a different space icon.

- **On the drawing page** — Select the shape that represents the resource you want to move and drag it into a different Space shape on the drawing page.

Finding and Installing a Printer from a Space Plan

You can use Visio space plans to locate printers. In addition, if Smart Tags are turned on, you can install that printer on computers right from the Visio plan. When others open the space plan, they can locate a printer that's convenient. To set the printer as their default printer, they position the pointer over the Printer shape, click the Printer Smart Tag button and then, from the shortcut menu, choose Set As Default Printer. To see the documents in the printer's queue, they choose Open Print Queue from the shortcut menu.

To set up a space plan to do this, follow these steps:

1. To turn Smart Tags on, choose Tools ⇨ Options, select the View tab and make sure the Smart Tags check box is selected. Click OK.

2. Drag a Printer shape from the Resources stencil onto the space plan.

3. Position the pointer over the Printer shape. When the Smart Tag appears, click Configure Printer on the Smart Tag.

4. Choose either Find a Printer in the Directory if you use an Active directory or Browse for Printer to find the printer on the network.

5. In the Connect To Printer dialog box, select a printer, click OK, and then save the space plan drawing.

Generating Facilities Reports

Similar to other management callings, facilities management produces its fair share of reports. Visio Professional includes shapes for predefined reports, such as Door and Window Schedules that are useful in facilities management. In addition, Visio includes three specialized reports for space planning and facilities management:

- **Asset Report** — For each asset in a space plan, this report shows the asset type, its name, its manufacturer, and to whom it belongs. You can use this report as part of a facilities audit to ensure that resources are located where they should be or that the proper group has responsibility for the assets under their control.

- **Move** — This report shows where people are located.

- **Space Report** — This report shows the department, room number, space use, and area for each space in a plan.

CROSS-REF   To run a report, choose Data ⇨ Reports. In the Reports dialog box, choose the report and click Run. To learn how to run or modify reports, see Chapter 10.

The easiest way to create one of these facilities management reports is to drag the corresponding report shape from the Resources stencil onto a drawing page. The shape uses the current report definition, gathers data from the shapes on the drawing, and displays the results in a shape on the page. To display other information, just modify the report definition before adding the shape to the page.

Summary

You can use Visio to plan space to varying levels of detail and then, after construction, manage the facilities and the people and assets within them. For high-level space planning, you can use Ink or the Space shapes available in both Visio Standard and Visio Professional. As you become more precise about the location and areas you're planning, the Space Plan template provides specialized shapes and tools to help you.

The Space Plan template builds a model of a space plan in addition to a drawing. You can add, modify, and delete the elements of space plan directly on the drawing page or in either the Space Explorer or Category Explorer.

Because of the amount of data that space planning and facilities management requires, the Space Plan template provides the Space Plan Startup Wizard to jumpstart creating a space plan, the Import Data Wizard to import external data into Visio shape data, and the Refresh Data command to keep data in a space plan and an external source in synch.

Chapter 29

Integrating CAD and Visio

Visio was never meant to be a substitute for a CAD program, such as AutoCAD. If you're producing hard-core architectural or engineering drawings for large or complex projects, you'll need every bit of functionality the CAD application provides — and even then, you'll probably pine for features the CAD application doesn't offer.

Where Visio shines when it comes to CAD drawings is ease of use. Suppose you've been working with an architect to remodel your house and now you want to arrange your furniture. The architect gave you the CAD drawings of the remodeled house, but you don't know the first thing about CAD. What's more, you don't own CAD software. With Visio and CAD integration, you can bring the CAD drawings into Visio and use them as a backdrop for your interior decorating.

Taking a step back from using the finished architect's drawings, Visio is also a handy review tool. When you have Visio, the architect can send the remodel plans in CAD format, and you can review them in Visio. If the palatial master bath looks like it will break your budget, you can let the architect know what you think by adding Ink shapes and text blocks to a separate layer to mark up and comment on the design.

When nothing but editing a CAD drawing will do, you can convert its contents into Visio shapes. Although this solution has its faults, including slow redraw, it's invaluable if you must edit a CAD drawing and no longer have access to the CAD program that created it. Finally, Visio can export its shapes to a few CAD formats, which is helpful when you want to prototype a design in Visio and then send it to CAD for the detail work.

This chapter describes the different methods for working with CAD drawings using Visio and when to use each one. You'll also learn how to convert CAD drawings into Visio format and vice versa.

Understanding CAD and Visio Integration

Visio-CAD integration comes in four basic flavors:

- **Importing (inserting)** — This method is, by far, the most popular way to use a CAD drawing in Visio. The imported CAD drawing is an OLE object, visible on a Visio drawing page. You can snap to its contents as you would Visio shapes. However, the drawing isn't editable. Inserted drawings offer the following benefits:

  - You can view, but not edit, inserted CAD drawings, so that you can use the CAD drawing as a reference without worrying about it being changed inadvertently.

  - You can snap Visio shapes to the geometry in an inserted CAD drawing just as you snap to other Visio shapes, rulers, grids, and guides in Visio.

  - You can crop CAD drawings to show details for part of a Visio drawing.

  - Inserted CAD drawings open and redraw more quickly than converted CAD drawings.

NOTE   CAD drawings imported into Visio reside within an ActiveX control so they remain in their original file format. In Visio, you can modify the scale of the drawing and the layers that are visible.

CAUTION   Visio works with AutoCAD DWG and DXF files saved in AutoCAD 2002 or earlier (which represents AutoCAD 2000 file formats). If you use a newer version of AutoCAD, you must save your drawings to AutoCAD version 2002 or earlier before trying in import a CAD drawing.

- **Converting CAD drawings to Visio shapes** — Typically, the reason to convert a CAD drawing to Visio is to edit a CAD drawing when you don't have access to the program that created it. Because Visio and AutoCAD take dramatically different approaches to scaled drawings, working with converted CAD drawings in Visio tends to be slow and often problematic. The process of converting large CAD drawings to Visio format voraciously devours system memory, which sends the operating system paging memory out to disk and slows performance to a crawl. In addition, some CAD drawings won't convert at all. If you must edit a CAD drawing with Visio, keep the following points in mind:

  - CAD blocks and entities are mapped to the closest Visio shapes.

  - Quite often, lines and text blocks convert to separate shapes, reducing performance and sucking up every bit of memory your computer has to offer.

  - Converted CAD drawings don't look as good as inserted CAD drawings, because accuracy and some details are lost.

CAUTION Microsoft recommends that you perform conversions in only one direction, because a CAD drawing converted to Visio loses any advanced features available only in the originating CAD program. If you subsequently reconvert the Visio drawing back into CAD, that additional information is gone.

- **Converting CAD symbol libraries to Visio masters on a stencil** — If you want to prototype plans in Visio, you can take advantage of predefined CAD objects by converting libraries of CAD blocks into Visio shapes on a custom stencil. Then, drawing in Visio is as simple as dragging and dropping your new masters onto the drawing page.

- **Exporting Visio drawings to CAD format** — If you create a sketch of a plan in Visio, you can export your Visio drawing to CAD format when it's time to get serious about drawing. During export, Visio shapes convert to CAD entities, and, if you inserted CAD drawings, they go back to CAD format as they came in. You can save Visio drawings as DWG or DXF files.

CROSS-REF Although you can export Visio drawings to the CAD format, for CAD users who want to review Visio drawings, the Visio Viewer is an easier solution for most other audiences, as described in Chapter 11.

If you're familiar with CAD programs, Visio might leave you hungry for more. Conversely, as a Visio user, you might be frustrated by the complexity of most CAD programs. However, by understanding and appreciating their differences and similarities, you'll know when to use which tool. Table 29-1 compares Visio and CAD features.

TABLE 29-1

A Comparison of Visio and CAD Features

Feature	Visio	CAD
Coordinate Systems	The Visio drawing page controls the coordinates you use, based on the drawing page size and scale you choose. For example, for a drawing page using feet for measurement units and a scale of 1:12, 1 inch on paper equals 1 foot in real-world measurements. To specify drawing size and scale, choose File ➪ Page Setup and select the Page Size and Drawing Scale tabs.	CAD programs often include two types of coordinate systems. Model space represents the true size of the objects in the model. Paper space applies a scale to the model so you can print or plot the model on paper. With this approach, you can create multiple paper spaces to display the same model space in different ways, for example, a site plan versus a building detail.

continued

TABLE 29-1	*(continued)*	
Feature	**Visio**	**CAD**
Units	For unscaled drawings, drawing units are simply the units in which you define the drawing page, such as inches. For scaled drawings, the drawing units, which you see on the ruler, are the real-world units. Measurement units are the units that Visio uses to dimension shapes, which you choose in the Page Setup dialog box on the Page Properties tab. You can select from U.S. units, metric units, publishing units, and even time-based units.	Units represent the real-world units for the model coordinate system, but aren't set to specific units such as feet or meters.
Scale	Drawing scale is the ratio of real-world measurements on the drawing page to the units on the printer paper. Every shape on a drawing page is drawn at the same scale. When you insert or convert a CAD drawing, Visio inserts the CAD drawing at the same scale as the drawing page, whether or not the CAD drawing fits on the drawing page.	In paper space, CAD creates views of your model in which each view can use a different scale. You can use one scale for an entire floor plan and another for a construction detail. Unlike Visio, you can compose a print or plot with multiple views, each using a different scale.
Layers	Visio layers categorize shapes and you can assign the same shapes to more than one layer. Visio layers don't control stacking order. You can specify attributes such as color, visibility, printability, and whether shapes are editable.	CAD layers also categorize shapes so you can control their attributes. However, unlike Visio, CAD layers can also control the order in which CAD objects appear and CAD objects can only belong to one layer.
Objects	Shapes	Blocks

Displaying CAD Drawings in Visio

Visio is ideal as a viewing tool for CAD drawings. It's easy to use and includes tools, such as markup and Ink, that even the most sophisticated CAD programs might not offer. Of all the methods for working with CAD drawings in Visio, inserting them into Visio drawings is the preferred way to go. Your CAD drawings look better, Visio responds faster; and you can still control many aspects of inserted drawings. With inserted CAD drawings, you can perform the following tasks:

- Use Visio to review CAD drawings produced by someone else, adding comments with Visio shapes, text, Ink, or markup on a separate Visio layer.

- Insert CAD drawings as backgrounds for Visio drawings. For example, if you want to use Visio to quickly prototype different office layouts, you can insert a CAD floor plan into your drawing as a reference and add Visio Furniture and Equipment shapes over it. From within Visio, you can crop or pan the CAD drawing to change the area that appears as a background, or you can hide or show layers. However, you can't edit the CAD geometry.

- Insert CAD drawings as details. For example, you can insert a CAD drawing that shows a highly detailed structural connection on a Visio drawing of an entire floor.

Reviewing CAD Drawings in Visio

If viewing a CAD drawing is your only goal and you don't plan to add Visio shapes or even markup on top of it, you don't have to insert the CAD drawing into a Visio drawing. Just open the CAD drawing directly by following these steps:

1. Choose File ➪ Open.

2. In the Files of Type box, select AutoCAD Drawing.

3. Navigate to the folder that contains the file you want to open, select the file, and click Open. Visio creates a new drawing file and one drawing page, inserts the CAD drawing onto the page, and sets the Visio measurement units and drawing scale to the units and scale of the CAD drawing.

4. To see when the CAD drawing was last updated—for example, to check that you have the latest version of the drawing—right-click the inserted CAD drawing and, from the shortcut menu, choose Data ➪ Shape Data. The Last Updated box shows the latest modification data, and the CAD File Name property shows the original file name of the CAD drawing.

5. After you open the drawing, you can crop, resize, rescale, or reposition the drawing, as well as hide or show its layers and change layer colors, as described in other sections in this chapter.

NOTE Image files embedded in DWG files don't appear when you insert the CAD drawing into Visio. However, they will appear if you convert the drawing to Visio shapes. If you have no reason to convert the CAD drawing to Visio, you can embed the image files in the Visio drawing directly using the methods described in Chapter 8.

Inserting CAD Drawings into Visio

When inserting a CAD drawing from a DWG or DXF file into Visio (available in both Standard and Professional versions), the Visio drawing assumes the last saved spatial view of the CAD drawing, either in model space or paper space. By using a CAD drawing saved in model space, you have more control over the CAD drawing after it's inserted into Visio. For example, with CAD drawings saved in model space, you can change the CAD drawing's scale in Visio. In addition, panning and resizing in Visio is faster when you use model space drawings. To insert a CAD drawing into a Visio drawing, follow these steps:

1. Open the Visio drawing file and the page that you want to contain the CAD drawing.

2. Choose Insert ➪ CAD Drawing. In the Insert AutoCAD Drawing dialog box, navigate to the folder that contains the CAD drawing, select the DWG or DXF file you want to insert, and click Open.

The CAD file name must end in DWG or DXF or it won't appear in the Insert AutoCAD Drawing dialog box.

3. In the CAD Drawing Properties dialog box, Visio automatically sets the scale to the Visio page scale and displays all the CAD layers. If you have specific requirements, choose the settings you want, as described in other sections of this chapter. The preview pane shows the size of the CAD drawing compared to the Visio drawing page, as shown in Figure 29-1. If the CAD drawing doesn't fit on the Visio drawing page, the best solutions are to either change the size of the Visio paper or change the Visio drawing scale.

FIGURE 29-1

Visio sets the CAD drawing scale to match the Visio drawing scale so CAD objects and Visio shapes appear at comparable sizes.

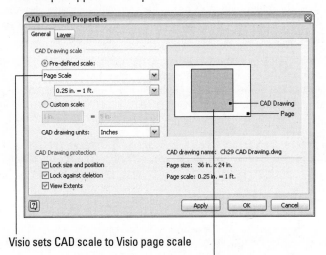

Visio sets CAD scale to Visio page scale

If the CAD drawing doesn't fit, change the Visio scale or page size

4. Click OK. The CAD drawing appears on the Visio drawing page.

Displaying CAD Layers in Visio

Although CAD drawings are typically teeming with layers, each inserted CAD drawing is assigned to only one Visio layer, named CAD Drawing. You might think that the CAD drawing layers are a lost cause, but within that inserted CAD drawing, you still can specify which CAD layers are visible, along with the color and line weight for the objects on each layer. Whether you've just inserted the CAD drawing or are working on the Visio drawing at any time afterward, right-click the CAD drawing on the Visio drawing page and, from the shortcut menu, choose CAD Drawing Object ➪ Properties. Select the Layer tab and then use one of the following methods:

Working with External File References

When you insert or convert a CAD drawing that references external files, Visio tries to open those external files as well. Visio looks for external reference files in a folder with the same path as the one used when the file was originally linked or in the same folder as the Visio drawing. Before you insert a CAD drawing—or when you receive a message that Visio can't find the externally linked files, be sure to place any external files in the same folder as your Visio drawing or create a folder structure that mirrors the original used by the CAD drawing.

- **Visibility** — To toggle the visibility of a layer, click the Visible field for the layer you want to toggle (or, select the layer and then click the Set Visibility button).

- **Color** — To specify the color of a layer, click the Color field for the layer you want to tint (or select the layer and then click Set Color). On the Standard tab, click a color cell, or select the Custom tab and specify the color you want. Click OK.

- **Line Weight** — To specify the line weight for lines on a layer, click the Line Weight field for the layer you want to modify (or select the layer and then click Set Line Weight). Type the line weight in points and then click OK.

Modifying Inserted CAD Drawings

Although you can't edit the CAD drawings you insert into Visio, you can modify them in several ways. You can change units and scale, for example to create a thumbnail image of the floor plan. If you want to show specific portions of an inserted CAD drawing, you can crop, pan, move, or modify the visibility of layers in the CAD drawing. You can drop Visio shapes on top of the CAD drawing and even position Visio shapes by snapping to the geometry of the inserted drawing.

 If you want to edit or delete individual objects in an inserted CAD drawing, convert only the layers containing those objects to Visio shapes and then make the changes you want.

Modifying Units and Scale

When you insert a CAD drawing into a Visio drawing, Visio sets the CAD drawing scale to the Visio page scale, which means that Visio shapes and CAD objects with the same dimensions automatically appear at the same size in Visio. Even when you match the drawing scales when you insert a CAD drawing, you can change the CAD drawing scale by dragging the border of the inserted drawing. If you use a scale other than the Visio page scale, the CAD drawing becomes a sizable image rather than a scaled drawing. This is helpful if you don't plan to overlay Visio shapes over the CAD drawing, for example, when you insert the CAD drawing as a thumbnail image of the overall building on a Visio drawing that lays out the office space in one wing.

NEW FEATURE   Visio 2003 inserted CAD drawings at a scale that fit them to the Visio drawing page, which immediately created inconsistencies between the size of CAD objects and the Visio shapes on the page. In Visio 2007, the Insert CAD Drawing command matches the CAD drawing size to the current Visio page scale so CAD objects and Visio shapes appear at comparable sizes.

Coordinating CAD Units with Visio Measurement Units

CAD drawings don't use pre-set drawing units. In the CAD world, a drawing unit can represent any unit — from a centimeter to an inch, or even a mile. When you insert a CAD drawing into a Visio drawing, Visio interprets CAD drawing units as Visio measurement units, which might be incorrect, especially if you insert a metric CAD drawing into a Visio drawing based on U.S. drawing units. To change the measurement unit for a CAD drawing, follow these steps:

1. To check the measurement units for the drawing page in which the CAD drawing is inserted, select the drawing page tab and then choose File ➪ Page Setup.

2. Select the Page Properties tab and check the value in the Measurement Units box. Click OK. If you prefer to change Visio units, change the value for Measurement Units here.

3. On the Visio drawing page, right-click the inserted CAD drawing and, from the shortcut menu, choose CAD Drawing Object ➪ Properties. On the General tab, in the CAD Drawing Units drop-down list, select the units that match the Visio measurement units. For example, if the Visio drawing uses feet and inches, select Feet in the CAD Drawing Units list.

4. Click OK.

Modifying Drawing Scales

CAD drawings often represent very large areas, such as the sprawling enclosure of a shopping mall. When you insert CAD drawings like this into Visio drawings, they might not fit on the Visio drawing page. You can change the Visio drawing scale and the CAD drawing scale to one that squeezes the CAD drawing onto the page. If you haven't drawn any Visio shapes yet, changing the Visio drawing page size (and possibly the scales) is another option. To change drawing scales, follow these steps:

1. If you want to change the Visio drawing scale, choose File ➪ Page Setup and select the Drawing Scale tab. For example, you can change the Visio drawing scale to fit the CAD drawing on the Visio drawing page.

2. To use a standard architectural or engineering scale, select the Pre-Defined Scale option, select the type of scale you want, and then select the scale you want to use. Click OK.

NOTE   If you want to use a custom scale, select the Custom Scale option and then enter a paper dimension in the first box and the real-world distance it represents in the second box.

3. To change the drawing scale for the inserted CAD drawing, right-click the inserted CAD drawing and, from the shortcut menu, choose CAD Drawing Object ➪ Properties.

4. Select the General tab, and select the scale using one of the following methods:

   ■ **Match CAD and Visio scale** — By default, Visio selects the Pre-defined Scale option and Page Scale in the drop-down list. This sets the CAD drawing to the same scale as the Visio drawing page scale and shows Visio shapes and CAD objects at the same size.

   ■ **Use industry-standard scale** — If you are only reviewing an inserted drawing and won't add Visio shapes on top of it, select the Pre-defined Scale option, in the drop-down list, select the type of scale you want to use, and then the specific scale.

   ■ **Define a custom scale** — Select the Custom Scale option and then enter a paper dimension in the first box and the real-world distance it represents in the second box.

5. Click Apply and then check the preview of the CAD Drawing and the Visio drawing page (shown in Figure 29-1) to make sure that the CAD drawing still fits on the Visio drawing page. If the CAD drawing doesn't fit, either change the Visio drawing size or change the scales you use.

6. When the inserted drawing is scaled the way you want, click OK.

 You can't change the scale of an inserted DWG drawing saved in paper space.

Protecting Inserted CAD Drawings

By default, Visio protects the CAD drawings you insert from deletion, resizing, and repositioning. When a CAD drawing is locked, you can't inadvertently move it, reposition it, or delete it as you work on Visio shapes overlaid on top of it. However, if you do want to reposition the CAD drawing or change it in other ways, first you must unlock the drawing from within the CAD Drawing Properties dialog box. To unlock an inserted CAD drawing, follow these steps:

1. Right-click the inserted CAD drawing and, from the shortcut menu, choose CAD Drawing Object ⇨ Properties.

2. Select the General tab and clear the Lock Size And Position check box and the Lock Against Deletion check box.

3. Click OK.

4. After you've completed the change to the CAD drawing, reopen the Cad Drawing Properties dialog box, and select the Lock Size and Position check box and the Lock Against Deletion check box.

In addition to the locking you can do with CAD Drawing Properties, you can also lock the Visio layer on which the CAD drawing is inserted. Although this technique mainly has the same effect as locking the CAD Drawing properties, it has the added advantage of preventing the drawing from moving when you pan using the Crop tool.

Finding the Invisible CAD Drawing

Placing an object far from the main drawing action is an all-too-common error in CAD. It's easy to apply a CAD command incorrectly or specify the wrong dimension. When you don't see the object where you expect it, you might just add it again. By doing this, you end up with one object where it's supposed to be, but another out in space somewhere.

If you can't see the contents of your CAD drawing in Visio, go back to the original CAD application and zoom to the drawing extents. If you see a tiny speck of color off in one corner, chances are good that the drawing has some wayward objects. If an inserted drawing balloons past the edges of the Visio drawing page, the best solution is to correct this issue in your CAD drawing. Open the drawing in the CAD program, locate and delete the objects that are misplaced, and resave the drawing. Then, insert the corrected drawing in Visio.

Positioning and Resizing Inserted CAD Drawings

Sometimes, you want to show only a portion of a CAD drawing, for example, the section of a floor plan for the offices you are laying out in Visio. Likewise, a CAD drawing smack in the middle of the drawing page isn't what you want, for example, when you offset the floor plan on the page to make room for notes on one side. What resizing and repositioning in Visio does depends on the tool you use, as described in Table 29-2. You can reposition or resize the entire CAD drawing. You can also crop the CAD drawing, reducing the CAD drawing border so that only a portion of the drawing appears. Panning within a cropped border changes the portion of the CAD drawing that you see, as shown in Figure 29-2.

TABLE 29-2

Tools for Positioning and Resizing CAD Drawings

Visio Tool	Positioning	Resizing
Pointer Tool	When the Pointer tool is active, move the entire drawing to a new location by dragging the inserted CAD drawing.	When the Pointer tool is active, change the size of the CAD drawing on the Visio drawing page (and its corresponding CAD drawing scale) by dragging a selection handle.
Crop Tool	To pan the area of the CAD drawing that appears within the CAD drawing border, on the Picture toolbar, click the Crop tool, click inside the CAD drawing border, and drag the Hand icon to a new location.	To crop the CAD drawing border so only a portion of the CAD drawing appears, right-click the CAD drawing and, from the shortcut menu, choose Crop Tool. Then drag a selection handle on the CAD drawing border until it's the size you want. Click the Pointer tool to turn off the Crop tool.
Drawing Scale	Does not apply	Change the size of a CAD drawing by specifying a different scale in the CAD Drawing Properties dialog box. This is more precise than dragging the inserted CAD drawing's selection handles.

FIGURE 29-2

Use the Crop tool to change the border of a CAD drawing.

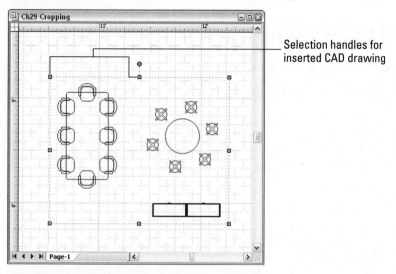

Selection handles for inserted CAD drawing

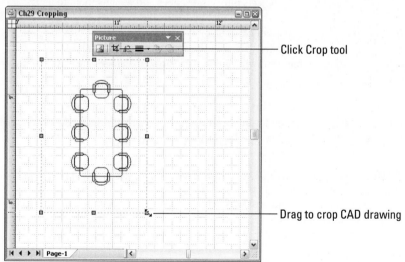

Click Crop tool

Drag to crop CAD drawing

Before you attempt to reposition or resize an inserted CAD drawing, make sure that the drawing is unlocked. To confirm that it is unlocked, right-click the drawing choose CAD Drawing Object ⇨ Properties, and then clear the Lock Size and Position check box and the Lock Against Deletion check box. To confirm that the Visio layer into which the CAD drawing is inserted is also unlocked, choose View ⇨ Layer Properties and ensure that the Lock field for the CAD Drawing layer is cleared.

 When you've completed repositioning and resizing the object, lock the drawing so that you don't accidentally move or resize it as you continue your work.

Editing CAD Drawings in Visio

Converting CAD drawings to the Visio format isn't recommended, but sometimes, it's the only solution. For example, you have a CAD drawing that you received from an architect and absolutely must make changes to it before a presentation Monday morning. After the CAD drawing is converted to Visio shapes, you can use Visio tools to modify the contents the way you want.

CAUTION If you do convert a drawing to make changes, don't convert the modified drawing back into the CAD format. This two-way conversion can reduce the quality of the drawing. Instead, ask the originator of the drawing to use the CAD program to make the same changes you did.

The Disadvantages of CAD Conversion

CAD and Visio formats differ significantly, and converting a CAD drawing into the Visio format highlights those differences. CAD drawings can contain thousands, even hundreds of thousands, of objects, each of which belongs to one of the CAD layers. When you convert a CAD drawing into the Visio format, Visio converts those CAD objects into Visio shapes and assigns them to layers using the layer names contained in the source CAD drawing. Visio doesn't always recognize all the objects that represent the same item, such as an office chair, which should convert into the same Visio shape. The result — a converted CAD drawing with thousands of different shapes, each stored separately in the Visio drawing file. This glut of unique Visio shapes raises two issues with converted CAD drawings:

- **Slow response time** — Visio must sort through thousands of shapes and perform tremendous amounts of processing for the simplest actions, such as selecting all the shapes and repositioning them. Even relatively small converted drawings generate a noticeable delay in redraws or completion of commands.

- **Large file size** — When you use Visio masters, Visio stores the definition of the master only once and uses instances of the master to show shapes on the drawing page. When you convert CAD objects into unique shapes, Visio stores each shape definition separately, greatly increasing the Visio file size. For example, a DWG file that consumes 850KB of space might require 10MB of space after it's converted into Visio shapes.

When you convert a CAD drawing, you convert the last saved view of that drawing, which might have been saved in model space or paper space. Converting a drawing saved in model space converts all the objects and text on the layers you select into Visio shapes. However, when you convert a drawing saved in paper space, Visio converts only the objects wholly contained within the paper space viewport into Visio shapes, converting anything partially contained in the viewport into lines.

Converting CAD Drawings to Visio Format

Because of the quantity of data that CAD drawings often contain, the Visio conversion tool helps you limit what you convert by specifying which CAD layers to convert. Even so, it's a good idea to eliminate unused layers, blocks, linetypes, and other types of CAD objects from the CAD drawing before you begin conversion. To convert a CAD drawing into Visio shapes, follow these steps:

1. Insert the CAD drawing into Visio as described in the section "Inserting CAD Drawings into Visio," earlier in this chapter.

2. Right-click the CAD drawing and, from the shortcut menu, choose CAD Drawing Object ➪ Convert.

3. To ensure that the conversion proceeds as fast as possible and the resulting file is a manageable size, in the Convert CAD Object dialog box, click Unselect All to ensure that no layers are selected.

4. In the Convert CAD Object dialog box, select only the check boxes for the layers you want to convert to Visio shapes.

5. To specify additional options for the conversion, click Advanced. A second Convert CAD Object dialog box opens with options for how you want to convert the CAD drawing.

6. In the second Convert CAD Object dialog box, shown in Figure 29-3, choose conversion options:

   - **Delete Selected DWG Layers** — For each layer you select for conversion, this option tells Visio to delete the layers from the original inserted CAD drawing, which means you won't see shapes duplicated between Visio and the background CAD drawing. Unconverted layers remain as part of the display-only inserted CAD drawing.

   - **Hide Selected DWG Layers** — For each layer you select for conversion, Visio hides the original CAD layers. You won't see duplication, but the original CAD objects are still there in case you need them.

   - **Delete All DWG Layers** — Visio converts the selected layers into Visio shapes and then deletes the entire inserted CAD drawing.

   - **Convert Into Visio Dimension Shapes** — This option converts CAD dimensions into Visio Dimension shapes, which means those shapes will recalculate dimensions as you edit the plan shapes.

   - **Convert Into Lines and Text** — This option converts CAD dimensions into Visio drawing elements, which won't adjust as you edit the drawing.

   - **Do Not Convert Hatch Patterns Into Visio Shapes** — Choosing this option dramatically reduces the number of shapes in the converted drawing but also means no hatching appears in the converted drawing.

   - **Convert Every Hatch Line Into A Visio Shape** — The conversion processes hatching, but also creates a separate shape for every line of hatching. A better alternative is to convert the drawing without hatching and use Visio tools to reintroduce the hatching you want.

FIGURE 29-3

Specify the layers you want to convert and how to convert their contents.

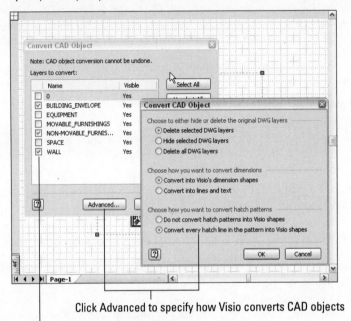

Click Advanced to specify how Visio converts CAD objects

Select the layers you want to convert

7. Click OK to close the Convert CAD Object dialog box and then click OK to convert the CAD drawing into Visio shapes. Depending on the size and detail in your CAD drawing, this process can take some time. A progress bar shows you how much of the conversion process is complete.

8. After you've converted the layers you want, be sure to lock the remainder of the inserted CAD drawing, as described earlier in this chapter.

9. Save the Visio drawing.

TIP   When blocks in a CAD drawing overlap, the resulting Visio shapes in the converted drawing overlap as well, but they might not appear in the correct order. To correct the stacking order in which converted Visio shapes appear, choose Shape, and then choose either Bring to Front, Bring Forward, Send to Back, or Send Backward.

Converting Multiple CAD Drawings

If you want to convert several CAD drawings, you can use the Convert CAD Drawings add-on. However, this add-on converts each CAD drawing into a separate Visio drawing file and converts every layer in each CAD drawing. If you need more control over the conversion process, convert each drawing separately using the Convert command. To use the Convert CAD Drawings add-on, follow these steps:

1. Copy or move all the files you want to convert into one folder.

2. Choose Tools ➪ Add-Ons ➪ Visio Extras ➪ Convert CAD Drawings.

3. In the Convert CAD Drawings dialog box, navigate to the folder that contains the drawings you want to convert, select all the files you want to convert, and then click Open.

4. After Visio converts the CAD files into Visio drawing files, save each of the converted files by choosing File ➪ Save. Although Visio opens the Save As dialog box, the Save As Type option is set to Drawing (*.vsd) for a Visio drawing file. Type the name for the new file and click Save.

Creating Stencils from CAD Libraries

When you work in Visio, you don't have to forego the symbols available to you in your CAD application. Symbol libraries are nothing more than regular DWG files that contain library objects. You can convert the symbols in those DWG files into Visio masters and store them on a stencil to drag and drop as you would any other Visio shape. When you convert a symbol library, each block becomes a Visio master, named based on the name of the original block used to create it. Block attributes are converted into Visio shape data and are stored with the master on the stencil. No matter how many symbol libraries you convert at once, Visio places the new masters on one new stencil. Block attributes become Visio shape data stored in the master on the stencil.

To convert CAD libraries to a Visio stencil, follow these steps:

1. Choose Tools ➪ Add-Ons ➪ Visio Extras ➪ Convert CAD Library.

2. In the Convert CAD Library dialog box, select all the DWG files for the libraries you want to convert and then click Open. The add-on converts each block in the selected DWG files to a master and places them on a new stencil.

3. To save the stencil, right-click its title bar and, from the shortcut menu, click Save As. In the Save As dialog box, type a name for the stencil and then click Save.

Converting Visio Drawings to CAD Format

In some circumstances, you might want to save a Visio drawing to CAD format. For example, suppose you've prototyped a plan in Visio and don't want to redraw it in your CAD application. It's easy to save Visio files as DWG or DXF files. You simply choose File ➪ Save As, in the Save as Type drop-down list, select either AutoCAD Drawing or AutoCAD Interchange, and then click Save. However, you must save each page in a multipage Visio file separately. Metafiles, such as Ink objects, inserted in Visio files are not supported when you save a Visio file as an AutoCAD drawing. Visio maps Visio entities to the most representative CAD element in the DWG or DXF file formats.

Visio drawings are saved as DWG or DXF files with fills and hatches turned off. You can turn fills and hatches back on in AutoCAD by setting the FILLMODE system variable to 1 and then using the REGEN command to regenerate the drawing.

Summary

Working with CAD drawings and Visio drawings can go several ways. When you don't need to edit CAD drawings, inserting CAD drawings as OLE objects into Visio drawings is easy and relatively quick, but you can still move and resize the CAD drawings on the Visio drawing page, and modify settings such as scale and visibility. Editing CAD drawings in Visio is possible by converting the CAD drawings to Visio shapes, but the slow performance and potential conversion problems make this a last resort. You can convert CAD symbol libraries to Visio drawings and stencils, if you want to build Visio drawings with the symbols you use in your CAD application. And, finally, converting Visio drawings to the CAD format is as easy as using Save As with one of the AutoCAD file formats.

Chapter 30

Working with Engineering Drawings

Engineering drawings can be complex, but building them doesn't have to be. The templates that Visio Professional provides for engineering disciplines include many of the shapes and symbols you need to prepare mechanical, electrical, and process engineering drawings, diagrams, and schematics. (Visio Standard doesn't include any of the engineering templates.)

What's more, you can use basic Visio techniques, such as drag and drop, shape text blocks, snap and glue, and shape data, to construct and fine-tune your engineering drawings. Although you can drag shapes from Electrical Engineering, Mechanical Engineering, and Process Engineering stencils onto a drawing page, with basic Visio tools, you can position those shapes to the precise tolerances required in parts and assembly diagrams. Connectors and glue define the relationships conveyed in process flow diagrams.

Engineering stencils include hundreds of configurable shapes that make it easy to produce the documents you want. In addition, the Process Engineering template includes tools that help you build a process engineering model. You can create components to track the elements of your model and view your model on a Visio drawing or in outline form.

This chapter shows you how to create mechanical and electrical engineering diagrams and schematics. It also describes how to build process engineering models and create process engineering diagrams. In addition, you will learn how to use components to add data to the shapes on your engineering drawings, tag and number components, and generate component lists and bills of materials.

Exploring the Visio Engineering Templates

Visio Professional includes templates for mechanical, electrical, and process engineering drawings and schematics. The Mechanical Engineering and Electrical Engineering templates include stencils of shapes that make it easy to assemble drawings. The Process Engineering templates go one better with tools to build a model and manage components. Unlike the separate engineering categories in Visio 2003, all the engineering templates in Visio 2007 are lumped into the Engineering category. Nevertheless, when you select File ➪ New or use the Getting Started pane, here's how the templates you see correspond to engineering disciplines:

- **Mechanical Engineering**
  - **Fluid Power** — Document designs for hydraulic or pneumatic controls, fluid and valve assemblies, fluid power equipment, and flow paths through systems.
  - **Part and Assembly Drawing** — Produce drawings and specifications that show the construction of parts, how to assemble equipment, or how machine parts fit together.
- **Electrical Engineering**
  - **Basic Electrical** — Produce wiring diagrams, electrical schematics, or one-line diagrams.
  - **Circuits and Logic** — Document integrated circuit designs, printed circuit boards, or digital or analog transmission paths.
  - **Industrial Control Systems** — Produce drawings of industrial control systems and power systems.
  - **Systems** — Represent components and relationships between electrical devices, particularly for large-scale systems such as utility infrastructure.
- **Process Engineering**
  - **Piping and Instrumentation Diagram** — Design and document industrial process equipment and pipelines.
  - **Process Flow Diagram** — Represent piping and distribution systems and processes.

Using Basic Visio Techniques in Engineering Drawings

In many cases, you can produce complex engineering drawings with nothing more than basic Visio tools. Consider the following Visio tools for building engineering drawings:

- **Drag and drop** — Drag shapes from Engineering stencils onto drawing pages.
- **Connectors and the Connector tool** — Drag connectors onto a page and glue the ends to Engineering shapes or use the Connector tool to draw the connector you want between shapes. Engineering shapes aren't set up to work with AutoConnect.

CROSS-REF Refer to Chapter 5 for detailed instructions on using connectors and connection points.

- **Drawing precision** — Use snap and glue settings, shape extension lines, and the Size & Position window to draw and position shapes to the tolerances you want.

CROSS-REF Refer to Chapter 4 for more information about positioning shapes precisely.

- **Shape operations** — In addition to the shapes on the Drawing Tool Shapes stencil, use Visio drawing tools and Shape Operation commands to draw unique parts.
- **Shape data** — Associate shape data with shapes and add values to shape data to configure shapes or include real-world information about the components that the shapes represent.

CROSS-REF To learn how to use drawing tools and Shape Operation commands, see Chapter 33. Shape data is discussed in Chapter 10.

- **Text features** — Label shapes by editing shape text blocks or adding callouts and other annotation shapes.
- **Background pages** — Include title blocks on background pages to identify the contents of engineering drawings.

CROSS-REF Refer to Chapter 6 for more information about using text and annotation tools. Refer to Chapter 2 for more information on background pages.

Working with Mechanical Engineering Drawings

You probably wouldn't want to rely on the Parts and Assembly Drawing template to produce complex mechanical drawings every day, but it does provide basic tools for creating detailed part specifications or showing how pieces of equipment are assembled. The Fluid Power template is intended to document hydraulic or pneumatic power systems, flow control, and fluid power schematics and assemblies. Neither template includes specialized menus or toolbars, but they open several stencils with specialized shapes — many of which include control handles or shortcut commands for configuration.

Drawing Parts and Assemblies

To manufacture mechanical parts, you need drawings that specify a part's dimensions, edges, planes, and curves. To assemble separate parts into a functional whole, you need instructions that show how the parts fit together. The Visio Parts and Assembly Drawing template includes masters that help you construct the geometric shapes found frequently on part and assembly drawings.

To create a part and assembly drawing, choose File ➪ New ➪ Engineering ➪ Part and Assembly Drawing. The US Units template creates a new ANSI B-size (17 inches by 11 inches) drawing in landscape orientation using a mechanical engineering one-quarter scale (shapes appear on the page at one-fourth of their actual size). The Metric template sets the paper to A4 landscape and the scale to 1:10. In addition, the following stencils open:

- Stencils in the Engineering stencil category:
  - **Fasteners 1** — Nuts and bolts
  - **Fasteners 2** — Rivets and washers
  - **Geometric Dimensioning and Tolerances** — Symbols used to show dimensioning origins and tolerances
  - **Springs and Bearings** — Shapes for springs and different types of bearing conditions
  - **Welding Symbols** — Standard shapes that indicate different types of welds
- Stencils in the Visio Extras stencil category:
  - **Annotations** — Annotation shapes for callouts, text blocks, north arrows, reference and section indicators, and drawing scale symbols.
  - **Drawing Tool Shapes** — Geometric shapes often used for parts and assemblies, including circle tangents, perpendicular lines, triangles, and rounded rectangles. Try these shapes before diving into Shape Operation commands for less common geometries.
  - **Dimensioning-Engineering** — Dimensioning shapes that use standard engineering dimension styles for linear and radial dimensions.
  - **Title Blocks** — Frames, tables, title blocks, and revision blocks.

Using Geometric Shortcuts

The shapes on the Drawing Tool Shapes stencil, shown in Figure 30-1, are quite fascinating. Unlike many other shapes that represent specific elements, these shapes simplify drawing geometric constructions — without resorting to basic Visio drawing tools or Shape Operation commands. The basic drawing tools provide only one way to construct rectangles and circles. The Drawing Tool Shapes stencil offers shapes for drawing circles using four different sets of data and four different types of rectangles. For example, you can create circles by specifying the circle diameter, the circle radius and one point on the circumference, or by using three points on the circumference. In addition to handles that you can drag, some shapes have shortcut menus that offer other constructions. For example, the Sector-graphical shape shortcut menu includes the Show Complementary Sector command, which switches a pie slice to the rest of the pie.

FIGURE 30-1

The shapes on the Drawing Tool Shapes stencil provide shortcuts for drawing many types of common geometric constructions.

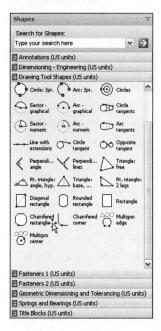

If you work on mechanical engineering drawings frequently, it's worth spending some time exploring all the shapes on the Drawing Tool Shapes stencil. Here's an introduction to a few of the shapes on the stencil:

- **Measure shapes** — In the stencil, these masters look like measuring tapes. Drag a shape, such as Measure Tool, Horizontal Measure, or Vertical Measure, onto a page and glue it to the shape from which you want to measure. As you drag the green selection handle around, the text block shows the distance from the glued point in the set measurement units.

- **Circle Tangents and Arc Tangents** — These shapes are ideal for drawing systems comprised of belts and wheels. The control handles at either end of the Arc Tangent shape adjust the radius at the corresponding end of the shape. Dragging the yellow control handle on the Circle Tangent shape changes the length of the tangent line and the position of the end points while keeping the line tangent to the circle.

NOTE To create linkages for belt systems, for example, to connect one mechanical cam to another, drag an Arc Tangents shape onto the drawing page to represent the first cam, then drag another Arc Tangents shape onto the page and glue one of its connection points to the first Arc Tangents shape.

■ **Rounded Rectangle** — Use this shape to quickly draw process storage tanks. Change the roundness of the corners by dragging the control handle.

 NOTE   For shapes specifically for tanks, open the Equipment–Vessels stencil in the Engineering ↩ Process Engineering category.

■ **Sector-graphical and Arc-graphical** — The Sector-graphical shape is a pie slice with selection handles for changing the radius, origin, and rotation of the slice, and a control handle that modifies the angle circumscribed by the slice. The Arc-graphical shape is an arc shape that works the same way as the Sector-graphical shape.

■ **Sector-numeric and Arc-numeric** — The selection handles and control handles determine the radius and origin for the sector or arc. Instead of dragging a control handle to define the angle circumscribed by the slice or arc, select the shape and type the number of degrees for the angle, even though there's no visual indicator that tells you that.

■ **Triangle shapes** — These triangle shapes take you back to those halcyon days of geometry class and Pythagoras. In addition to constructing triangles with different sets of data you can adjust the angles in the Right Triangle: Angle, Hypotenuse shape by selecting the shape and typing the angle you want.

■ **Multigon shapes** — To draw polygons with different numbers of sides, drag one of these shapes onto the drawing page, right-click the shape, and then, from the shortcut menu, choose the polygon you want — from the triangle that appears initially up to an octagon.

Creating Springs, Bearings, and Fasteners

Springs, bearings, and fasteners come in many standard shapes and sizes — too many to include one of each on the Mechanical Engineering stencils. To provide the plethora of sizes you might need, most of these shapes include shape data to configure shape dimensions. When you modify the value in a shape data field, such as the Number of Coils for a Helical Spring shape, the shape morphs accordingly. In some shapes, you can specify one dimension and let Visio adjust the other dimensions based on industry standards or you can set all the dimensions exactly the way you want. Although these shapes are locked so you can't resize them inadvertently, you can unlock and display the shape handles so you can use them to resize the shapes. The following list includes some of the commands that appear on shape shortcut menus, depending on which shape you right-click:

■ **Set Standard Sizes** — Choose this command on a shortcut menu to modify the thread diameter in the Shape Data dialog box. Visio adjusts the other dimensions for the shape to industry-standard lengths.

■ **Set Dimensions** — Modify one or more of the dimensions in the Shape Data dialog box to configure the shape the way you want.

■ **Resize with Handles** — Display the selection handles on a shape so you can resize it by dragging.

- **Hatched** — Display cross-hatching on the shape. Choosing Unhatched from the shortcut menu removes the cross-hatching.

- **Simplified** — Show a simplified version of the shape with some lines removed. When the simplified version appears, the command on the shortcut menu changes to Detailed.

- **Alternate Symbol** — Alter the appearance of the shape slightly, such as changing an X to a plus within a shape.

Creating Welding Symbols

As you probably expect, the symbols on the Welding Symbols stencil show the locations and types of welds on a drawing. Although the Visio welding shapes faithfully duplicate the appearance of industry standard welding symbols, you need to know what kind of welds are required if you want your metal pieces to hold together.

To add a weld to a drawing, follow these steps:

1. Drag one of the Arrow shapes (Arrow, Arrow with Bend, or Additional Arrow) from the Welding Symbols stencil onto the drawing page and position it so that the arrowhead points to the weld joint.

2. To specify additional weld information, right-click the arrow shape and, from the shortcut menu, choose Show All Around Circle and/or Show Tail.

3. To add symbols to specify the type of weld, follow these steps:

   a. Double-click the Arrow shape on the drawing page to open the shape's group window.

 If you can't see the drawing window while the group window is open, choose Window ⇨ Tile to display all the windows side by side.

   b. Drag symbols that represent different types of welds, such as V-groove, from the stencil onto the Arrow shape in the group window.

   c. To annotate the weld, drag Annotation shapes into the group window and edit the text. You can glue Welding Symbol shapes and Annotation shapes to the guides in the group window to keep symbols positioned correctly when you resize the Arrow shape.

4. To return to the drawing page, click the Close button in the group window.

Annotating Dimensions and Tolerances

Part and assembly drawings are typically awash with dimensions. Between the Dimensioning-Engineering stencil and the Geometric Dimensioning and Tolerancing stencil, Visio engineering templates offer shapes for adding datum points and dimensions.

Dimension shapes include control handles you can drag to define the distance to measure as well as the location of the dimension lines, as illustrated in Figure 30-2. The control handles that appear depend on the dimension shape you choose. As an example, you can add dimensions from a vertical baseline by following these steps:

1. Drag the Vertical Baseline shape onto the page near the bottom of the part you want to dimension. Drag the green end points to both position the ends of the horizontal reference lines and define the distance for the first dimension.

2. To position the text and vertical dimension line for the first dimension, drag the yellow control handle on the first dimension line.

3. Drag the yellow control handle between the green selection handles up and to the left to add another vertical dimension line.

4. After you have added the dimensions you want, you can drag the control handles at the top of each dimension up or down to change the height of the dimension. When you drag these yellow control handles to the left or right, Visio repositions the horizontal reference line.

5. To change the spacing between the vertical dimension lines, drag the yellow control handle at the bottom of the Vertical Baseline shape.

FIGURE 30-2

Drag control handles on Dimension shapes to modify the distance you want to measure and the position of the dimension lines.

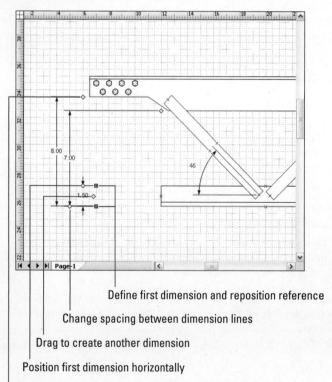

Define first dimension and reposition reference

Change spacing between dimension lines

Drag to create another dimension

Position first dimension horizontally

Change dimension and position reference line

598

Datum points and datum frames show origins and geometric characteristics. Datum points delineate positions that you can use to align shapes on different pages or drawings. To add these symbols, use one of the following methods:

- **Datum shapes** — Drag a Datum shape, such as Datum Symbol or Datum (New), from the Geometric Dimensioning and Tolerancing stencil onto the drawing page and, while the shape is selected, type the text you want in the symbol. Press Esc to complete your text entry.

- **Datum Frame** — Drag a Datum Frame shape, such as 1 Datum Frame or 2 Datum Frame, onto the drawing page, and then double-click it to open the group window. To denote geometric characteristics, drag shapes, such as Cylindricity, into the box on the left end of the Datum Frame shape. To add text to the other boxes in the shape, double-click the box and type the text you want. When you have finished editing the Datum Frame shape, click the Close button for the group window to return to the drawing window.

Constructing Fluid Power Diagrams

The Visio Fluid Power template helps you draw the pipes and equipment that make up fluid power systems. To create a fluid power drawing, choose File ⇨ New ⇨ Engineering ⇨ Fluid Power. Visio creates a new letter-size drawing in landscape orientation with no scale and opens the following stencils:

- Engineering ⇨ Mechanical Engineering stencil category:
  - **Fluid Power-Equipment** — Pumps, compressors, gauges, meters, and other types of equipment
  - **Fluid Power-Valve Assembly** — Shapes that represent valves and control equipment
  - **Fluid Power-Valves** — Different types of valves
- Visio Extras stencil category:
  - **Annotations** — Annotation shapes for callouts, text blocks, north arrows, reference and section indicators, and scale symbols
  - **Connectors** — Different types of generic connectors

Putting together fluid power diagrams requires nothing more than basic Visio techniques and a background in the design of fluid power systems. Use one or more of the following methods to construct a fluid power diagram:

- **Add shapes to drawings** — Drag shapes from Fluid Power stencils onto the drawing page.

- **Reconfigure shapes** — Right-click a shape and choose a command from the shortcut menu. For example, you can configure the Pump/motor (Simple) shape to be hydraulic or pneumatic, bidirectional or unidirectional, variable, or compensated. You can also configure the Pump/motor (Simple) shape to represent a pump, a motor, or a combination of the two.

> **NOTE** Most but not all shapes on Fluid Power stencils include configuration commands on their shortcut menus.

- **Annotate shapes** — Add text by selecting a shape and typing the text you want. After you add text, most shapes include a control handle you can drag to reposition the text if it overlaps graphics on the drawing.

- **Connect equipment** — Use the Connector tool to connect equipment, as described in Chapter 5.

> **NOTE** Some shapes on Fluid Power stencils include control handles that you can use to connect them to other shapes. To find out what a control handle does, view a screen tip by positioning the pointer over the control handle.

Working with Electrical Engineering Drawings

To create an electrical engineering drawing, choose File ➪ New ➪ Engineering and then choose the template for the type of drawing you want. Visio creates a new letter-size drawing in portrait orientation and no scale. As with mechanical engineering drawings, basic Visio techniques are the tools available for constructing electrical engineering drawings.

> **CAUTION** Electrical Engineering shapes include connection points that are not interchangeable, such as the positive or negative terminals on a battery. When you construct electrical engineering diagrams, take care to glue to specific connection points, rather than use shape-to-shape glue. When you glue connectors to Electrical Engineering shapes, make sure that Visio highlights only the connection points in red, not the entire shape.

Building Process Engineering Models

With the Visio Process Engineering template, it's easy to draw piping and instrumentation diagrams (P&IDs) and process flow diagrams (PFDs). You can drag shapes from Process Engineering Equipment stencils onto your drawing page, connect them with Pipeline shapes, and then add shapes to represent components, such as valves. Process engineering models use components to represent physical objects, such as pipelines or pieces of equipment. In addition to the graphical view of your process engineering diagram, you can use the Component Explorer and Connectivity Explorer windows to examine the hierarchy of components and connections.

Shape data holds information about process engineering components that you can use to produce reports or equipment lists. In addition, tags identify and track the components on your drawings. By default, when you drag components onto the drawing page, Visio adds tags and displays the tag text in component text blocks. You can choose to hide or show tag information to improve the readability of a drawing. Although Visio provides predefined tag formats, you can also create your own. If you construct different views that contain the same component, such as an overall plan and a detail, you can assign the same tag to multiple shapes, so you can accurately track components.

Viewing Process Engineering Diagrams in Different Ways

In addition to Visio drawing pages, the Component Explorer and Connectivity Explorer windows show components and connections in a process engineering model in outline form. The Shape Data window comes in handy when you want to examine data associated with components. The Processing Engineering templates provide the following windows for viewing and working on your process engineering models and drawings:

- **Drawing window** — Turn to this window when you want to see how process equipment is connected. Process engineering diagrams can span multiple pages or include both overview and detailed portions of a model. When you select a shape in the drawing window, Visio highlights the corresponding component in the Component Explorer window.

- **Component Explorer window** — This window categorizes the components in a model into groups, such as Equipment, Pipelines, Valve, or Instrument. The outline shows the tag number for each component as well as each shape that represents the component, for example, when a component appears on more than one page of the diagram. Component tag numbers appear in the outline.

- **Connectivity Explorer window** — This window focuses on the pipelines that connect components in your model. Pipelines appear at the top level of the hierarchy, identified by their tag numbers. Underneath each pipeline, the components connected to the pipeline are listed by tag number.

- **Shape Data window** — Whether you select a shape in one of the Explorer windows or on the drawing page, the Shape Data window is the easiest way to browse values from shape to shape. Select a shape on the drawing page or in the Component Explorer or Connectivity Explorer window and its shape data appears in the Shape Data window. Process Engineering shapes come with several properties predefined. For example, Pipeline shapes include properties for line size, material, design pressure, design temperature, and more.

Open and close Explorer windows or switch between Explorers depending on what you want to see. To open an Explorer window, choose Process Engineering ➪ Component Explorer or Process Engineering ➪ Connectivity Explorer.

 When both Explorer windows are open, you can switch between Explorers by selecting the Components or Connectivity tab at the bottom of the window.

The outline format of the Explorer windows makes it easy to see your entire model even if it spans several drawing pages. The Expand and Collapse icons filter the components you see, as shown in Figure 30-3. What's more, you can perform any tasks in the Explorer windows that you can do on a drawing page — create new components, rearrange them, and rename them. And the Explorer windows are particularly helpful when you want to find a component on a complex drawing — right-click a component in an Explorer window and, from the shortcut menu, choose Select Shapes. Visio zooms into the shape on the drawing.

FIGURE 30-3

The Explorer windows not only perform all the tasks you use with shapes on drawing pages. They also make it easy to find shapes on large and complex drawings.

Work with components using any of the methods described in Table 30-1.

TABLE 30-1

What You Can Do in the Explorer Windows

Task	Description	Explorer
Expand or collapse hierarchy levels	Click the plus icon to expand the outline to show the next lower level in the hierarchy. Click the minus icon to hide the lower level.	Both
Select components	Double-click a component or right-click a component and then, from the shortcut menu, choose Select Shapes. Visio selects the shape that represents the component on the drawing page.	Both
Rename components	Right-click a component in an Explorer window and, from the shortcut menu, choose Rename. Type the new name and press Esc when you're done. Visio renames the component in the Explorer windows and the drawing.	Both
Create components	Create a new component by right-clicking the category to which you want to add a component and then, from the shortcut menu, choose New Component.	Component Explorer only
Associate shapes with other components	To change the component to which a shape belongs, in the Components Explorer window, drag a shape from one component to another component in the same category.	Component Explorer only

Updating Visio 2000 Process Engineering Projects

Visio process engineering drawings changed drastically between Visio 2000 and Visio 2002. Projects contained all the drawings and documents for a model, including PFDs, P&IDs, and other files. Projects could contain multiple drawings, each in its own Visio drawing file, so the Project Explorer helped you navigate a Visio 2000 model.

To take advantage of all the features in the current Visio Process Engineering solution, you must update models built in Visio 2000. To do this, migrate a Visio 2000 Process Engineering project to Visio 2002 and then open it in Visio Professional 2007. The migration process converts each drawing in a Visio 2000 project to a separate Visio 2002 process engineering drawing, but doesn't affect the Visio 2000 files in any way. In moving to Visio 2007, you can include several process engineering drawings in the same Visio drawing file by adding each drawing as a separate page.

To migrate a Visio 2000 Process Engineering project to Visio 2007, follow these steps:

1. To save the migrated files in a new folder, create a destination folder using Windows Explorer.

2. Open the Visio 2000 Process Engineering project (.vsd) in Visio 2002. When Visio prompts you, click Enable Macros to begin the migration. When Visio prompts you to migrate all drawings, click Yes to convert all the drawings in the Visio 2000 project to Visio 2002 drawings.

3. In the Browse for Folder dialog box, select the destination folder you created in step 1 and then select a file name. Visio begins to migrate the files into the new format. The migration process might take some time, particularly if you are migrating a large Visio 2000 project. To avoid tying up your computer, you can start the migration and let it run on its own.

4. After the migration is complete, click OK. Visio presents a summary of the names and paths of the migrated files. Review the summary and then open the migrated files in Visio 2007.

Creating P&ID and PFD Drawings

Creating process engineering drawings isn't much different than creating any other type of engineering drawing, although Visio provides a few additional tools to simplify some tasks, such as adding valves to pipelines. The steps in this section provide an overview of the basic sequence for creating a process engineering drawing. Following sections provide more detailed instructions about building specific portions of a process engineering drawing.

To create a process engineering drawing, choose File ➪ New ➪ Engineering and then choose the template for the type of drawing you want. Visio creates a new ANSI B-size (17 inches by 11 inches) drawing in landscape orientation and with no scale. In fact, whether you want to create a PID or a PFD, you can choose either template, because they both open the same stencils from the Process Engineering category:

- Equipment-General
- Equipment-Heat Exchangers
- Equipment-Pumps

- Equipment-Vessels
- Instruments
- Pipelines
- Process Annotations
- Valves and Fittings

To assemble a process engineering drawing, follow these basic steps:

1. Add major equipment first by dragging shapes from the Equipment-Vessels, Equipment-Pumps, Equipment-Heat Exchangers, and Equipment-General stencils onto your drawing. As you drop them onto the page, Visio adds tags that identify each piece of equipment as a component.

2. Connect the equipment with Pipeline shapes. The easiest way to connect Equipment shapes using Pipeline shapes is to click the Connector tool on the Standard toolbar, select the Pipeline master you want to use in the Pipelines stencil, and then, on the drawing page, drag the mouse pointer from one equipment shape to another.

 To change to a different type of pipeline, select another Pipeline master in the Pipelines stencil and then continue to drag between Equipment shapes on the drawing page.

 To modify the direction of a pipeline or the type of pipeline, change the line style of the Pipeline shape, as described in the next section.

3. Insert valves into pipelines by dragging Valve shapes from the Valves and Fittings stencil onto Pipeline shapes on the drawing page. When a red square appears on the Pipeline shape and the Valve shape rotates to the orientation of the Pipeline shape, release the mouse button to glue the two together. Visio automatically splits the Pipeline shape into two pieces, both of which are glued to the Valve shape.

4. Attach monitoring instruments to equipment and pipelines by dragging Instrument shapes from the Instruments stencil onto the drawing page near the shapes for the equipment that the instruments monitor. If the Instrument shape includes a control handle, you can drag it to glue the Instrument shape to the shape for the component it monitors.

5. Annotate the drawing by dragging shapes from the Process Annotations stencil.

6. Add data to components. Choose View ⇨ Shape Data Window. Select a shape, click a shape data field, and then type or select a value.

Building Pipelines

Pipelines are components that connect equipment, such as vessels or centrifuges. However, when you connect Pipeline shapes to one another or add shapes for other equipment components to them, Visio splits the Pipeline shapes into separate shapes. Although a pipeline might comprise several separate shapes on the drawing page, Visio keeps them all associated with the same component and sharing the same tag and shape data values.

Specifying How Pipeline Shapes Behave

To control how pipelines behave when you add valves or connect other pipelines, choose Process Engineering ➪ Diagram Options and then specify the following options:

- To split Pipeline shapes when you drop Valve and Fitting shapes onto them, select the Split Pipelines Around Components check box.

- To split Pipeline shapes when you connect other Pipeline shapes to them, select the Split Pipelines When Branches Are Created check box. In the Use This Shape At Pipeline Branches drop-down list, you can choose the shape that Visio inserts to designate branches, such as Junction.

> **NOTE** If Pipeline shapes don't split despite turning on the above settings, be sure that you are gluing to shape geometry. Choose Tools ➪ Snap & Glue and check the Shape Geometry check box under the Glue To heading.

- To repair Pipeline shapes when you delete components or other Pipeline shapes, select the Repair Split Pipelines check box.

- To instruct Visio to automatically number the components you add, select the Number Components When They Are Added To The Drawing check box.

Adding Components to Pipelines

When you drop a Valve shape onto a Pipeline shape, Visio splits the Pipeline shape into two pieces, with the ends of the two segments glued to the component you inserted, as shown in Figure 30-4. The two shapes still belong to the same component and share the same tag and properties as the original Pipeline shape.

When you connect one pipeline to another, the original Pipeline shape also splits into two pieces. Visio adds a Junction shape at the point where the three Pipeline shapes intersect (the added pipeline and the pipeline that's been split in two). Although you don't see Junction shapes on the drawing page, they do appear in the Connectivity Explorer window.

> **TIP** If you want to display Junction shapes — for example, to validate your drawing — choose Process Engineering ➪ Diagram Options, check the Split Pipelines When Branches Are Created check box, and then select Junction in the drop-down list.

To add a valve or fitting to a pipeline, follow these steps:

1. Drag a Valve or Fitting shape from the Valves and Fittings stencil and position it on top of a Pipeline shape.

2. When Visio displays a red square and rotates the valve or fitting to the pipeline, release the mouse button. Visio splits the Pipeline shape in two, gluing the two Pipeline shapes to either end of the valve or fitting shape.

FIGURE 30-4

When pipelines split, the shapes still belong to the same component.

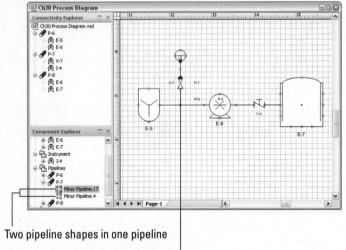

Two pipeline shapes in one pipeline

Pipeline shape splits when valve is added

NOTE When you delete a valve or other component that splits a pipeline into two pieces, the pipeline heals into a single shape. However, if you delete a component between two different pipeline components, the pipelines remain separate components. If pipelines that belong to the same component don't repair themselves when you delete a component, choose Process Engineering ➪ Diagram Options and check the Repair Split Pipelines check box.

Continuing Pipelines on Other Pages

In some circumstances, you might want to continue the same pipeline on different pages, for example, a long pipeline that connects hundreds of equipment components. The Process Annotations stencil includes shapes for indicating that a pipeline continues on another page. Despite their names, these shapes are identical to Off-Page Reference shapes. To learn how to use these shapes to continue pipelines, follow the instructions for Off-Page Reference shapes in Chapter 6.

Working with Components

In a process engineering model, components represent individual real-world objects such as valves, pumps, or pipelines. In Visio process engineering drawings, components are made up of one or more Visio shapes. For example, a main pipeline with several intersecting branches requires a separate Visio shape for each segment, but to you, it's still one component. Each component includes properties, such as pressure, temperature, or material, which apply to all the shapes in it.

You can categorize components in a process engineering model so that it's easier to track and report on different types of components. Visio includes several categories corresponding to the Process Engineering stencils: Equipment, Instrument, Pipelines, and Valve. You can create your own categories as well.

CAUTION Only shapes from the Process Engineering stencils appear automatically in the Component and Connectivity Explorer windows. If you add shapes from other stencils or draw your own, they won't appear as components. However, you can convert these shapes so that they will work with Visio Process Engineering features by following the steps outlined in the section "Converting Shapes and Symbols into Components."

Associating Shapes with Components

Visio automatically creates components when you add shapes from Process Engineering stencils to drawings. Each shape you place on the drawing page receives a tag number that identifies the component to which the shape belongs. However, you can associate more than one shape with the same component or transfer ownership of a shape from one component to another.

Use one of the following methods to associate a shape with a component:

- In the Component Explorer window, drag an entry from one component to another component in the same category.

- On the drawing page, select a shape and type the tag for the component to which you want the shape to belong. Visio doesn't give you any indication that typing will change the component tag, but as soon as you begin typing, the shape text box appears with the current tag number. When you click away from the text box, you'll see the item switch to its new component in the Component Explorer window.

NOTE If you want to remove a component association completely, you must delete the shape on the drawing page.

Working with Component Data

To accurately model and engineer processes, your drawings must include information about the components in the process, such as the operating temperature range for a piece of equipment or the design pressure of a pipeline. Visio uses shape data to store process engineering data as well as to configure shapes — for example, displaying different versions of the shape depending on the type of instrument you choose.

Process Engineering shapes come with several shape data fields by default, as shown in Table 30-2. Visio uses shape data sets to associate groups of fields with each category of component. You can use these fields and shape data sets or modify them to suit your organization's needs. Although you can add data to components individually in the Shape Data window, Visio's Database wizards can import and link data directly from engineering databases.

CROSS-REF To learn how to create your own shape data fields and shape data sets and apply them to Visio shapes, see Chapter 33.

 To learn how to add values to shape data fields as well as how to import data from databases or link Visio shapes to database records, see Chapter 10.

TABLE 30-2

Default Custom Properties for Component Categories

Equipment	Instruments	Pipelines	Valves and Fittings
Description	Description	Description	Description
Material	Connection Size	Line Size	Line Size
Manufacturer	Service	Schedule	Valve Class
Model	Manufacturer	Material	Manufacturer
	Model	Design Pressure	Model
	Instrument Type	Design Temperature	
	Local/remote		

Displaying Component Data

Visio displays the component tag in the text block of each Process Engineering shape by default, but you can show or hide component tags depending on whether you want to unclutter a busy diagram or make information readily available. To hide or show the tag for a shape, right-click the shape on the drawing page and, from the shortcut menu, choose Hide Tag or Show Tag (the command toggles between the two).

In addition, you can display other component data on the drawing page. For example, Callout shapes from the Process Annotations stencil can show key design attributes on the drawing, such as temperature or pressure.

To display component data using Callout shapes, follow these steps:

1. Drag one of the Callout shapes from the Process Annotations stencil onto the drawing page near the shape whose properties you want to display.

2. Drag the control handle from the Callout shape to any point on the shape that you want to annotate.

3. In the Configure Callout dialog box, select the check boxes for the shape data fields you want to display in the Callout shape. With the Callout shape linked to the shape data, changes to those fields appear in the Callout shape automatically. If you select more than one field, be sure to specify a separator, so you can distinguish individual values in the Callout shape.

If you want to change the order in which fields appear in the Callout shape, select a field and then click Move Up or Move Down to reposition it in the order.

4. To show the property name in addition to the value, check the Show Property Name check box.

5. Click OK. The properties appear in the Callout shape.

To change the properties that appear or change whether the property name is shown after you add the Callout shape to the drawing page, right-click the Callout shape and, from the shortcut menu, choose Configure Callout. Choose Show Leader to draw a leader from the Callout shape to its associated shape.

Tagging and Numbering Components

Visio identifies the components in your process engineering model with a tag. In Process Engineering shapes, the tag appears in the shape's text block. By default, Visio formats tags as `<tag name>-<tag counter>`, for example, P-10. By default, the tag name is the first letter of the component category and the tag counter is a number that increments by one every time you add a component from that category to the drawing. If you want to number components in a specific way, you can define your own custom tag format.

 If you don't want Visio to automatically number components as you add them, choose Process Engineering ⇨ Diagram Options and uncheck the Number Components When They Are Added to the Drawing check box.

Applying Tag Formats to Shapes

The easiest way to change the tag format for all instances of a shape is to change the tag format for a master on a stencil. To use a different tag format for shapes already on a drawing, follow these steps:

1. Select the shape or shapes you want to change on the drawing page and then choose Process Engineering ⇨ Apply Tag Format.

2. In the Apply Tag Format dialog box, in the Tag Format drop-down list, select the tag format you want, select the Apply To Shapes Selected In Drawing option, and click OK.

To change the tag format for masters in a stencil, follow these steps:

1. Choose Process Engineering ⇨ Apply Tag Format.

2. In the Apply Tag Format dialog box, select the tag format you want in the Tag Format drop-down list.

3. Select the Apply To Shapes In A Stencil option and then click Choose Shapes.

4. In the Choose Shapes dialog box, the Document drop-down list includes all the open stencils. Select the stencil you want to modify.

 If you want to choose masters on the Document stencil, select the drawing name in the Document drop-down list.

5. Select the check boxes for the masters whose tag formats you want to change and click OK. In the Apply Tag Format dialog box, click OK to apply the new tag format to the selected masters.

6. Right-click the stencil title bar in the Shapes window and, from the shortcut menu, choose Save As to save the modified stencil as a custom stencil.

Defining Tag Formats

Although Visio includes a default tag format for each category of components in the Process Engineering templates, you can modify these formats to fit your organization's standards or create formats of your own. Tag formats can include text, punctuation, the values of shape data, and numeric sequences, and can span more than one line.

To create a new tag format, follow these steps:

1. Choose Process Engineering ⇨ Edit Tag Formats and click Add.

2. Type the name for the new tag format in the Name box.

3. Use one of the following options to specify the basis for the new format:

   - **Create a New Format** — Choose this option to create a brand-new format based on the default tag format, `<tag format name>-[Counter]`.

   - **Create from an Existing Format** — Choose this option to use an existing format as the basis for the new tag format. Select the drawing or stencil that contains the format you want to use in the Document drop-down list. Then, from the Format drop-down list, select the tag format you want to use.

4. Click OK. Visio adds the new name to the Tag Format list.

To edit a format, follow these steps:

1. In the Edit Tag Formats dialog box, select the tag format you want to modify and click Modify. Visio opens the Tag Format Properties dialog box and selects the text in the Tag Expression box.

 The Sample Tag Value box shows a preview of the current tag expression.

2. To insert text in the tag expression, position the insertion point in the tag expression and type the text you want.

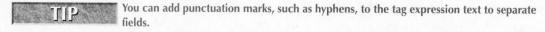 You can add punctuation marks, such as hyphens, to the tag expression text to separate fields.

3. To replace text in the tag expression, select the text and type the new text you want.

4. To create a multi-line tag, position the insertion point where you want to start a new line in the tag expression and press Enter.

5. To add a shape data field to the tag expression, position the insertion point where you want to insert the property, select a field in the Available Shape Data Properties list, and then click Insert Property.

6. To insert a sequential counter to the tag expression, position the insertion point where you want to insert the counter and then click Insert Counter.

> **TIP** You can specify the number of digits that the counter occupies by selecting an entry in the Format drop-down list. Visio adds leading zeroes to the counter. For values larger than the number of digits specified, Visio simply adds more digits to the counter.

7. When you have completed the tag expression, click OK. The tag expression appears in the Expression column of the Edit Tag Formats dialog box.

> **NOTE** You can also rename or delete tag formats in the Edit Tag Formats dialog box by clicking Rename or Delete.

Renumbering Components

As you add components to your model, Visio numbers tags using a numeric sequence. For example, the tag for the first piece of equipment you add is E-1, followed by E-2, and so on. However, as you work on a model, you might want to clean up the tag sequence. For example, if you delete components or reassign shapes from one component to another, you can end up with sequence numbers that are no longer used. You can renumber components to reuse those numbers, specifying the starting value you want to use and the increment between each tag.

To renumber the components in a model, follow these steps:

1. Choose Process Engineering ⇨ Renumber Components.

2. To specify which components you want to renumber, click one of the following options:

   - **Document** — Renumbers all the components in the current drawing file

   - **Page** — Renumbers all the components on the current drawing page

   - **Selection** — Renumbers the selected components

3. Clear the check boxes for any tag format you don't want to renumber. By default, all the tag formats are checked.

4. To specify how to renumber components that use a tag format, select a tag format in the Include Tag Formats list. Type the starting value in the Starting Value box and type the increment between numbers in the Interval box. Repeat these steps for each tag format.

5. Click OK to renumber the components.

> **NOTE** After Visio renumbers the components, the tags for new components begin where the last renumbered components left off. In addition, new tags use the settings from the renumbering you applied. For example, if you renumber pipelines starting at 100 using an interval of 2 and the pipelines in the model are tagged from 100 to 128, the next pipeline you add will start at 130.

Generating Component Lists and Bills of Material

You can generate reports about components in your model from the values in shape data. Visio includes a predefined report for each category of components, which lists specific information about each component. You can use these reports as provided or use Visio's report features to define your own reports. Even if you don't add values to shape data, you can still run these built-in reports to see a list of components by tag number, because Visio adds tag numbers automatically.

Visio provides the following predefined reports:

- **Equipment List** — Includes tag number, description, manufacturer, material, and model
- **Instrument List** — Includes tag number, description, connection size, service, manufacturer, and model
- **Pipeline List** — Includes tag number, description, line size, schedule, design pressure, and design temperature
- **Valve List** — Includes tag number, description, line size, valve class, manufacturer, and model
- **Inventory** — Shows the number of shapes on the page grouped by shape name

To run one of these reports, choose Data ⇨ Reports, select it in the Report list, and then click Run.

CROSS-REF To learn more about creating, modifying, and running reports, see Chapter 10.

Converting Shapes and Symbols into Components

In order to work with Visio Process Engineering features, shapes must belong to a component category and have a tag format assigned to them. Without these, you won't see the shapes in the Component Explorer or Connectivity window and they won't function as other Process Engineering shapes do. However, you can convert shapes or objects from other sources into Process Engineering shapes using the Shape Conversion command. You can convert the following elements:

- Shapes you draw with Visio drawing tools
- Existing shapes on a drawing page
- Shapes from stencils other than the Process Engineering stencils
- Symbols created in AutoCAD

NOTE Process Engineering shapes can lose their attributes if you perform some actions — for example, ungrouping a grouped Process Engineering shape or applying Shape Operation commands to them. When this happens, you can use the Shape Conversion command to reassign a category and tag format.

To transform shapes or symbols into Process Engineering shapes, follow these steps:

1. If you want to convert shapes on the drawing page, select the shapes you want to convert.

2. Choose Process Engineering ➪ Shape Conversion. Then, under the Source heading, choose one of the following options:

   ▪ **Selected Shapes** — Converts the shapes you selected on the drawing page.

   ▪ **Shapes in a Visio Stencil** — Converts masters on a Visio stencil. Click Choose Shapes, select the stencil in the Document list, check the check boxes for the masters you want to convert, and click OK.

> **TIP** To convert masters on the current drawing's Document stencil, in the Choose Shapes dialog box, select the drawing name in the Document list. By doing this, you can convert all the shapes in the current drawing file.

   ▪ **Symbols in a CAD File** — Converts symbols in a CAD file. Click Browse and then locate and select the CAD file containing the symbols you want to convert. To set the drawing scale in Visio, enter a positive value for the number of Visio measurement units that equals one CAD unit and select the units you want to use in the Units drop-down list.

3. Select or type the name of a category in the Category box to assign it to the converted shapes.

> **NOTE** If you type a category name that doesn't exist, Visio creates a new category for you.

4. In the Tag Format list, select a tag format to assign it to the converted shapes.

> **CROSS-REF** To apply shape data sets to the converted shapes, see Chapter 33.

5. Click OK to convert the shapes. If you converted CAD symbols, Visio creates a new stencil that contains the shapes you converted. To save the stencil, right-click the stencil title bar and choose Save from the shortcut menu.

Summary

Visio provides templates for mechanical, electrical, and process engineering drawings. You can use basic Visio techniques such as drag and drop to perform much of the work for creating drawings. Visio uses shape data not only to add engineering information to shapes but also to configure shapes to show different varieties of equipment.

The Process Engineering templates help you build a model and a drawing. Visio uses components with identifying tags to track and report on the objects in a process engineering model. The Component and Connectivity Explorer windows present your model as an outline. You can create, delete, rename, and move components around on the drawing page or in these windows.

Part VI

Customizing Stencils, Templates, and Shapes

Chapter 31

Creating and Customizing Templates

Visio comes with dozens of built-in templates for many different kinds of drawings and diagrams. Visio Standard includes about two dozen templates, whereas Visio Professional offers more than 60. If the built-in templates don't have what you need, additional Visio templates are available for download from plenty of web sites, some of which are listed in Chapter 40. Even with all these templates at your disposal, unique requirements can lead you to customizing a built-in template or creating your own.

If you're reading this chapter, no doubt you already appreciate the power of Visio templates. By customizing templates to your specific needs, you can benefit even more — for example, an environment set up to simplify the construction of a unique type of drawing, settings that conform to your organization's standards, or perhaps your preferred combination of settings, toolbars, and windows. Just by creating a drawing based on a customized template, anyone in your organization can obtain the right paper size, scale, and other page settings; a set of stencils, whether built-in or customized; any number of standard drawing pages, including backgrounds; pre-populated shapes, such as your company logo; and even the Visio windows you typically use docked just where you want them.

There's no need to build a custom template from scratch, if a built-in Visio template offers most of what you need. You can copy a built-in template and customize it as you see fit. Another shortcut to a customized template is to create a template by saving an existing drawing. In this chapter, you'll learn how to create templates and add stencils and styles to the templates you create.

Reasons to Customize Templates

Developing customized templates makes sense for a number of reasons. Even if many of the settings in a built-in template are what you want, you can create a customized template to perfectly match your requirements. Customized templates are great time-savers if you:

- Use specific paper sizes, scales, and other page settings for different types of drawings
- Create drawings using unusual page sizes
- Draw plans using unusual drawing scales
- Use backgrounds with standard sets of reference shapes
- Include the same shapes on all your drawings, such as company logos or title blocks
- Apply special formatting, such as color themes, to different types of drawings
- Use stencils other than the ones that a built-in Visio template opens
- Use different stencils to produce the same type of drawing for different departments
- Prefer specific positions for Visio windows, such as the Shape Data window

TIP Even when an existing drawing is similar to a new drawing you want to create, it's better to start from a template. When you open a template, Visio automatically uses a copy of the template for the new drawing file. If you open an existing drawing file and forget to use the Save As command, you overwrite your original drawing instead of creating a new one. In addition, you must remove or revise the content of an existing drawing.

Customizing Templates

Whether you want to tweak one of the built-in Visio templates or create a wildly unique template from an existing drawing, creating a customized template is easy. In short, you set up a drawing with the content, settings, and stencils you want, and then save it as a template or XML template.

Creating and Saving Customized Templates

Existing Visio drawings, built-in Visio templates, and even other customized templates are all potential fodder for a new customized template. If you plan to modify a built-in template, it's better to create a new template based on the built-in template. In that way, you can use your customized template, while still keeping a copy of the original if you need it.

A template assembles practically everything about a drawing file into a convenient reusable package. Make sure you set up a file exactly the way you want before you save it as a template, because every new drawing based on the template will inherit the same settings. Of course, you can modify the template to correct omissions, but you also must make those modifications in any drawings you created from the template in the meantime. Here are some of the items that a template stores:

- Page settings

- Print settings

- Snap and glue options

- Drawing pages with any existing shapes

- Layers

- Styles

- Color palettes

- Macros

- Window sizes and positions

The next section discusses the types of customizations you can make in a new template, but the following are the bare bones steps for creating a template:

1. Open an existing Visio drawing (VSD file) or create a new drawing based on the template you want to customize.

2. Set up the file with the content and settings that you want, as described in the next section.

3. Choose File ➪ Save As.

4. Navigate to the folder in which you want to save the template.

5. In the Save as Type list, select either Template or XML Template.

6. In the File Name box, type the name for your template.

7. Click Save. Visio saves the contents and settings to a VST file if you chose a template, or to a VTX file if you chose an XML template.

TIP If you want to create a template that others can use but not edit, click the arrow next to the Save button, and then choose Read Only. You can create a new drawing using the template. However, if you try to open the template itself, Visio displays an error message.

NOTE XML templates contain the same information as regular templates, so you can save a drawing file as either type of template without worrying about losing information.

Setting Up a File for Use as a Template

Templates store as many or as few settings as you want, and can contain predefined foreground and background pages with or without shapes. Make your changes in any order — as long as you make all the changes before you save the file as a new template. Use any of the following methods to customize a template:

- **Open stencils** — To define the set of stencils that the template opens by default, simply open the stencils that you want and close any stencils that you don't want. Whatever stencils appear in the Shapes window when you save the template are the stencils that will appear in the Shapes window when you use the template to create a new drawing.

CAUTION   If you make changes to a stencil as you are setting up your template, be sure to save the stencil file as well as the template file. Otherwise, your stencil changes will be lost even though Visio opens the stencil as part of the template.

- **Pre-populate pages** — To create standard pages that you want in every drawing file based on the template, insert the number of pages you want and specify the page settings you want for each page.

TIP   A background page in a template is the easiest way to ensure that your company logo, standard title block, and other stock information appear for every new drawing based on the template. To do this, create the background page, add the shapes you want to it, and assign it to foreground pages before you save the template.

- **Pre-define layers** — If you want drawings to use a standard set of layers, create the layers you want for each page in the drawing.

- **Pre-populate shapes** — To begin drawings with a standard set of shapes, add the shapes you want to a drawing page (and layer, if you use them).

- **Pre-define print settings** — Choose File ⇨ Page Setup, select the Print Setup tab, and specify the print settings you want, such as the size of the printer paper.

- **Pre-define page settings** — Choose File ⇨ Page Setup, select other Page Setup tabs, and specify the settings you want, such as page orientation, drawing scale, line jump settings, and shadow settings.

- **Specify snap and glue options** — Choose Tools ⇨ Snap & Glue and select the options you want.

- **Customize a color palette** — If you use specific colors to present information, Visio Color Themes won't work. Instead, you can define a color palette for the template by choosing Tools ⇨ Color Palette and selecting the color palette or colors that you want.

CAUTION   When you customize both the stencils and colors in a template, make sure that the styles and colors for the template and stencil files are compatible. The style and color settings for the template file override the settings in the stencil file, which can lead to strange results when they conflict.

- **Make styles available** — Custom styles belong to the drawing that's open when the styles are created. To make a custom style available for future drawings, modify or create the styles you want before saving the template.

- **Set up Visio windows** — To open and position windows automatically when you create a new drawing, open the windows you want, such as the Shape Data Window or the Pan & Zoom Window, and position them where you want.

- **Include macros** — Create any macros that you want to use.

NOTE   You can also specify where stencils appear in the Visio window and whether they are docked or floating. Position the stencils where you want them to appear when Visio opens them.

Displaying Thumbnail Images of Templates

If you like the thumbnail images of templates that Visio displays in the Getting Started window (see Chapter 2), you can mimic that behavior be creating previews of the customized templates you create. To create a preview image that illustrates the type of drawing the template produces, do the following:

1. Open the template file and produce a sample drawing by adding some shapes that represent typical contents, for example, flowchart shapes for a customized business diagram template.

2. Choose File ⇨ Properties. In the Properties dialog box, select the Save Preview check box.

3. Save the template. If you want to be able to modify the template in the future and keep the sample drawing, choose File ⇨ Save As to save a second copy of the template, in which you keep the shapes for the sample drawing.

4. Reopen the original template file. (Don't create a new drawing from the template.)

5. Choose View ⇨ Drawing Explorer and right-click the root of the hierarchy, which shows the file path and name for the template.

6. From the shortcut menu, choose Protect Document and in the dialog box that opens, select the Preview check box. This saves the current preview image regardless of the changes you make to the template file.

7. If you want the template to open a blank drawing, delete the shapes in the template and save it.

When you open the Browse Templates dialog box to create a drawing from a template, you'll see the preview if your folders are set to display thumbnails, as shown in Figure 31-1.

Accessing Customized Templates

By default, Visio 2007 installs its built-in templates in `C:\Program Files\Microsoft Office12\1033\`. However, it's a good idea to keep your customized templates separate from the built-in templates. Not only do you protect your templates from being removed during a software upgrade, but it's far more likely that you'll remember to back up your customized files with your other data. To store and retrieve your customized templates, follow these steps:

1. To keep your templates organized, save them in a folder dedicated to customized templates, such as `My Documents\My Visio Templates`.

TIP   Take a hint from the Visio categories and create subfolders in your main folder, such as `My Documents\My Visio Templates\Scaled Drawings`. If you don't store your templates in a subdirectory, they show up in a category called (Other).

2. To specify the file path for your templates, choose Tools ⇨ Options and select the Advanced tab.

3. Click the File Paths button and, in the Templates box, type the file path that points to your templates. If a file path already exists in the box, type a semicolon as a delimiter and then type your file path. You can save templates on your own computer or on a network device.

4. Click OK to close the Advanced dialog box. Click OK again to close the Options dialog box.

5. To create a drawing using one of your templates, choose File ➪ New ➪ New Drawing From Template. In the Browse Templates dialog box, navigate to the folder that contains the template you want to use, as illustrated in Figure 31-1, and double-click the template name. Visio creates a new drawing file based on the template.

FIGURE 31-1

Double-click a template to create a new drawing file based on the template.

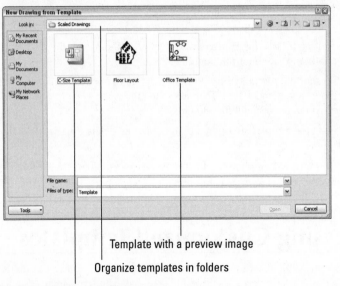

Template with a preview image

Organize templates in folders

Template without a preview image

Summary

Whenever you create several drawings that use the same settings or configuration, a template that incorporates those settings and configurations can save you time and effort. Every time you want to create a drawing with those settings, you simply create a new drawing based on that template. Visio automatically sets up the new drawing with all the settings, configurations, and content you added to the template. You can specify a variety of settings with templates, including page settings, style definitions, color palettes and schemes, drawing pages, and common shapes, such as company logos.

Chapter 32

Creating and Customizing Stencils

O ne way that Visio templates simplify your diagramming is by opening stencils that include shapes — *shape masters*, to be precise —
that you can drag onto a drawing page to build diagrams quickly. If
you want different masters, you simply open the stencils that contain them.
Sometimes, the masters on built-in stencils simply aren't enough. Or, you
might use so many different stencils that the Shapes window barely has room
to show any shapes. Fortunately, you can build your own custom stencils —
to put the masters you use all the time in one convenient stencil or to store
custom shapes that you create.

Custom stencils reside within your My Shapes folder, by default. Opening
them is as easy as opening any built-in stencil. Maybe easier. You choose
File ⇨ Shapes ⇨ My Shapes and navigate to the stencil you want.

In this chapter, you learn how to create custom stencils for the shapes you
use frequently or custom masters you create from scratch. You'll also learn
how to add masters to the Favorites stencil. Finally, you will learn how to
modify the shape information that appears in stencils, and how masters are
arranged.

IN THIS CHAPTER

Creating stencils

**Using Visio tools to store your
favorite shapes**

Saving stencils

Adding shapes to stencils

**Modifying the master
information that appears in
stencils**

**Changing spacing and colors in
stencils**

Rearranging shapes in stencils

> **NOTE** Microsoft has copyrighted the masters on the stencils that Visio
> installs. You can copy and modify masters to suit your require-
> ments and distribute drawings that contain copyrighted shapes. However, you
> can't sell or distribute original or modified Visio masters.

Creating and Saving Stencils

Whether you start with existing stencils or want to create new stencils from scratch, you are free to create, edit, and save as many customized stencils as you want. You can choose from several methods for creating new stencils from stencils that already exist. When you go to the trouble of creating your own stencils, saving them as read-only protects them from editing by others. Besides the formal process of creating and saving custom stencils, Visio has a few shortcuts for quickly storing your favorite shapes.

CROSS-REF The Visio Extras stencil category is chock-full of useful stencils, such as Annotations, Dimensioning, and Icon Sets. To open one of these stencils, choose File ⇨ Shapes ⇨ Visio Extras and then choose the stencil you want. See Chapter 42 for a list of the stencils in the Visio Extras category.

Creating a Stencil from an Existing Stencil

The fastest and easiest way to produce a custom stencil is by starting with an existing stencil that contains most of the shapes you want. If the stencil you want to copy is already open, all you do is right-click the stencil's title bar and, from the shortcut menu, choose Save As. In the Save As dialog box, navigate to the folder in which you store stencils, type the name for the new stencil, and click Save.

Creating a new stencil from a stencil that isn't open requires only a few more steps:

1. Open or create a drawing so that the Shapes window opens.

2. Choose File ⇨ Shapes ⇨ Open Stencil. In the Open Stencil dialog box, Visio displays the contents of the My Shapes folder by default.

3. In the Open Stencil dialog box, navigate to the folder that contains the stencil you want to copy and select the stencil name. To create a stencil from one of your custom stencils, navigate to the folder that contains the stencil, by default `My Documents\My Shapes` or a subfolder within My Shapes. To start with a built-in Visio stencil, navigate to `C:\Program Files\Microsoft Office\Office12\1033` and select one of the stencil names.

4. Click the arrow on the Open button and choose Open As Copy from the drop-down menu. Visio opens a copy of the stencil in the Shapes window using a default name, such as Stencil1. The icon in the stencil title bar includes an asterisk, indicating that the stencil is editable, as shown in Figure 32-1.

5. Add, remove, or rearrange masters on the stencil, as described in later sections in this chapter.

6. To save the new stencil, right-click the stencil title bar and, from the shortcut menu, choose Save.

7. In the Save As dialog box, navigate to the folder in which you want to store the stencil, typically My Shapes or a subfolder within My Shapes. Type a name for your stencil and then click Save.

FIGURE 32-1

A stencil open for editing has an asterisk in the title bar.

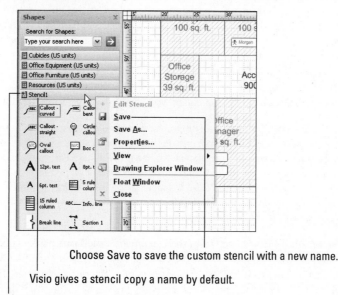

Choose Save to save the custom stencil with a new name.

Visio gives a stencil copy a name by default.

Asterisk indicates stencil is editable.

Creating Stencils from Shape Search Results

If you track down shapes you want to use by typing keywords in the Search for Shapes box in the Shapes window, Visio creates a stencil for you. The program names the stencil using the keywords you searched for and adds the shapes it finds to that search results stencil. When you find the shapes you want, you can avoid searching for them again by saving the entire search results stencil or just the individual shapes you want as a custom stencil. If the shapes were stored online, saving them to stencils stored on your hard drive saves you from going online to access those shapes in the future.

To save the search results stencil as a custom stencil, right-click the title bar of the search results stencil and, from the shortcut menu, choose Save. The Save As dialog box opens to the My Shapes folder. In the File Name text box, type a name for the custom stencil. If you want, navigate to a subfolder to categorize the new stencil, and then click Save.

CROSS-REF See the section "Adding Shapes from Other Stencils" later in this chapter to learn how to save individual shapes in the search results stencil to another stencil.

Creating Stencils from Scratch

To begin with a new, blank stencil, choose File ➪ Shapes ➪ New Stencil. When you have finished adding masters to the stencil, right-click the stencil's title bar and, from the shortcut menu, choose Save. In the Save As dialog box, type a name for the stencil and click Save.

Creating a Custom Stencil from a Document Stencil

If you use only a few shapes from each stencil that a Visio template opens by default, you can save switching between stencils by creating a custom stencil that contains only the shapes you use for the current type of drawing. With the Document stencil, this is easier than you would expect. Each drawing you create contains a Document stencil, which stores a master for each shape you add to the drawing. Whether you create shapes with drawing tools or drag shapes from stencils or shape search results, Visio places a master for each unique shape on the Document stencil. At any point, the Document stencil contains the shapes you added to the drawing—and only those shapes. Create a new stencil based on the Document stencil and you have a one-stop-shop stencil for your drawing.

NOTE   The Document stencil is a component of a Visio drawing file, not a separate file like other stencils. It comes in handy for seeing which shapes you've used in a drawing or to modify every instance of a master across all pages in a drawing file. For example, to make the Manager boxes in an organization chart smaller, you can modify the Manager master on the Document stencil and every instance of the Manager shape in your drawing file resizes to match that master. Because the Document stencil is part of a drawing file, it's easy to copy shapes unique to a drawing by adding them to the Document stencil.

Unfortunately, creating a new stencil from the Document stencil isn't particularly intuitive. In fact, it's a little unnerving, but the results are worth the anxiety. First, you open the Document stencil and edit it so that it contains the masters that you want in the order you want. Then, to reduce the size of the file, you delete all the shapes and drawing pages within the Visio file and save the drawing file as a stencil file.

CAUTION   Because you delete all the drawing file contents except the Document stencil, you must take extra care to use the Save As command. Otherwise, you overwrite your drawing file with empty drawing pages and could lose important information.

To create a new stencil from the Document stencil, follow these steps:

1. To prevent overwriting your drawing file, make a copy of the Visio drawing file that contains the Document stencil you want to work with.

2. To display the drawing file's Document stencil, choose File ⇨ Shapes ⇨ Show Document Stencil.

3. Modify the masters in the Document stencil as needed. For example, edit the master properties or rename the masters. If you want the masters in a different order, drag them into the sequence you want.

4. Delete all the shapes on all the pages in the drawing. The fastest way to delete all the shapes on a page is to press Ctrl+A and then press Delete.

CAUTION   If disk space is scarce, it's very important that you delete the shapes on the drawing pages before saving the file as a stencil. If you don't, the shapes on the pages remain in the file and take up disk space even though they aren't visible.

5. To save the Visio drawing file as a new stencil, choose File ⇨ Save As. In the Save as Type list, choose Stencil. In the Save As dialog box, type the name for the stencil and click Save.

Quickly Storing Your Favorite Shapes

As with other Microsoft programs, in Visio, the concept of Favorites is alive and well. The Favorites stencil is a readily accessible place to store shapes you can't live without. Visio creates the Favorites stencil automatically during installation and stores it in My Documents\My Shapes. If you build up an overabundance of favorites shapes, you can move shapes from the Favorites stencil to additional custom stencils.

To store and retrieve favorite shapes, use one of the following methods:

- **Store a favorite shape** — In the Shapes window, right-click the shape and, from the shortcut menu, choose Add to My Shapes ➪ Favorites.

- **Open your Favorites stencil** — To access all the masters in your Favorites stencil, choose File ➪ Shapes ➪ My Shapes ➪ Favorites.

- **Open other custom stencils** — If you created other custom stencils in your My Shapes folder, choose File ➪ Shapes ➪ My Shapes and then choose the custom stencil you want.

Saving Stencils

As with other types of Visio files, saving stencils spans several methods. The Save command itself pops up in several places in the Visio environment. But, you can also choose where you want to store your stencils, for example, in subfolders that categorize the type of stencil. Likewise, you can increase consistency on drawings by specifying stencils as read-only so others can't edit the masters.

> **TIP** When you use custom stencils regularly, it's worthwhile to tell Visio where to look for your custom stencil files. When you choose Open Stencil or New Stencil, the dialog boxes automatically open to the folder you specify. To define the file path for stencils, choose Tools ➪ Options. In the Options dialog box, select the Advanced tab. Click File Paths. In the File Paths dialog box, in the Stencils box, type the file path to the folder in which you store your custom stencils.

To save a stencil, use one of the following methods:

- **Save a docked stencil** — To save a stencil docked in the Shapes window or elsewhere in the Visio window, right-click the stencil title bar and, from the shortcut menu, choose Save.

- **Save a floating stencil** — Click the icon in the stencil title bar and then choose Save.

- **Copy a stencil** — To save a stencil as a new stencil, right-click the stencil title bar and then choose Save As.

- **Prevent others from editing your stencil** — To prevent others from editing a stencil, right-click the stencil title bar and then choose Save As. In the Save As dialog box, set the Read Only option by clicking the Save arrow and then choosing Read Only.

Read-Only Stencils

By default, Visio opens all stencils as read-only, but enables you to edit your custom stencils when you use the Edit Stencil command. This behavior protects the contents of a stencil from inadvertent modifications. If you want to edit built-in Visio stencils, you must first save them as copies.

However, when **you** save a stencil as read-only, Visio sets the Windows Read-Only flag on the stencil file. If you try to use Edit Stencil on this type of Read-Only file, Visio displays a message that the stencil can't be edited at this time. To edit a stencil whose Read-only flag is set — perhaps to update the masters to new company standards — you must reset the Read-Only flag on the file. To do this, use Windows Explorer to locate the file. In the Windows Explorer window, right-click the file and, from the shortcut menu, choose Properties. In the Properties dialog box, select the General tab, clear the Read-Only check box, and click OK.

Adding Shapes to Stencils

In most cases, custom stencils exist because you want quick access to the shapes you use frequently. Any shape on a drawing page can become a shape master, whether you create it with Visio drawing tools or modify an existing shape. All you have to do is add the shape to an editable stencil. So, the first thing you do is make a stencil editable by performing the following steps:

1. Open the stencil you want to edit with any of the following methods:

   ▪ **Create a new stencil** — Choose File ➪ Shapes ➪ New Stencil.

   ▪ **Open your Favorites stencil** — Choose File ➪ Shapes ➪ My Shapes ➪ Favorites.

   ▪ **Open another custom stencil** — Choose File ➪ Shapes ➪ My Shapes, choose the custom stencil you want, and click Open. If you store custom stencils in a folder other than My Shapes, choose File ➪ Shapes ➪ Open Stencil. In the Open Stencil dialog box, navigate to the folder, choose the stencil, and click Open.

2. To make the stencil editable, right-click the stencil title bar and, from the shortcut menu, choose Edit Stencil. The icon in the stencil title bar changes to include an asterisk, indicating that the stencil is editable, as shown in Figure 32-1.

Adding Shapes from Drawing Pages

To add a shape from a drawing page to a stencil, follow these steps:

1. Make sure the stencil is open and editable.

2. On the drawing page, select the shape you want to add to the stencil and then use one of the following methods:

   ▪ **Copy the shape** — Hold the Ctrl key as you drag the shape to the stencil.

   ▪ **Move the shape** — Drag the shape from the drawing page to the stencil.

When you add a shape to a stencil, it becomes a master. In the stencil, the master appears as an icon with a label, Master.*x*, where *x* is the number that is one greater than the largest ShapeSheet number for that stencil.

3. To give a new master a more descriptive name, in the stencil, right-click the master icon and, from the shortcut menu, choose Rename Master. Visio selects the icon label, so you simply type the new name and press Enter when you're finished.

4. When you have finished adding shapes to the stencil, save the stencil file. See the "Saving Stencils" section in this chapter for more information about the different ways you can save a stencil.

NOTE To remove a master in a stencil, select the master and then press Delete.

Adding Shapes from Other Stencils

The Add to My Shapes command appears on the shortcut menu when you right-click a master in a stencil. You can use it to quickly copy a master to another stencil by doing the following.

1. Open both the stencil containing the master you want to copy (the source) and the stencil to which you want to copy it (the destination).

NOTE Even if you don't open the destination stencil, Visio copies the shape, but doesn't open the stencil so that you can confirm that the copy worked.

2. In the source stencil, right-click the master, choose Add to My Shapes, and then choose one of the following options:

   ▪ **Favorites** — Adds the master to your Favorites stencil.

   ▪ **A Custom Stencil** — Adds the master to the custom stencil you choose from the shortcut menu. Visio adds the master to the stencil without opening the stencil and saves the stencil automatically.

   ▪ **Add to New Stencil** — Creates a new stencil and then adds the master to it.

   ▪ **Add to Existing Stencil** — Opens the Open stencil dialog box so you can choose the stencil to which you want to copy the master.

Copying and Pasting Masters Between Stencils

Copy and Paste commands are another way to copy masters to other stencils. To do so, right-click the master you want to copy and, from the shortcut menu, choose Copy or press Ctrl+C. Then right-click the stencil to which you want to copy the master and, from the shortcut menu, choose Paste or press Ctrl+V. One of three things happens:

- If you try to paste a master into a built-in stencil, Visio displays a message that you can't edit the stencil and provides instructions for adding the master to a custom stencil.

- If the destination stencil is editable, Visio copies the master.

- If the destination stencil is not editable, Visio asks you whether you want to open the stencil for editing so you can complete the paste operation.

Modifying Stencil Appearance

Depending on the size of your screen or the amount of space your drawing window consumes, more compact stencil windows can be desirable. Displaying only icons for masters minimizes the space needed for a stencil window, but makes it harder to identify the correct shapes. For more information about masters, you can display icons, master names, and even details about each master. Displaying icons and names is a good compromise and is the setting that Visio chooses by default. The settings for stencil appearance go further. You can specify how much text appears in the master labels, rearrange the shapes in the stencil, or use different background colors to make the stencils easier to read. Positioning the pointer over a master displays a screen tip about how to use the shape.

Displaying Master Information

To specify the amount of information that Visio displays for each master in a stencil, right-click the Shapes window title bar and choose one of the following options:

- **Icons and Names** — Displays icons and master names, which is the default setting and a compromise between identifying masters and minimizing the screen real estate for a stencil window.

- **Icons Only** — Displays only icons, which minimizes screen area, as illustrated in Figure 32-2.

- **Names Only** — Displays only names, which uses less space in the window but requires more familiarity with the masters.

- **Icons and Details** — Displays icons, names, and a brief description of the master, which consumes the most screen area but also provides brief tips on using the masters, as illustrated in Figure 32-2.

NOTE You can also specify the information that Visio displays by right-clicking the title bar of a stencil, choosing View and then selecting the option you want. Although you right-click a stencil's title bar, Visio changes the information that appears in every open stencil.

FIGURE 32-2

Change the information displayed in a stencil window to modify the amount of screen area the stencil requires.

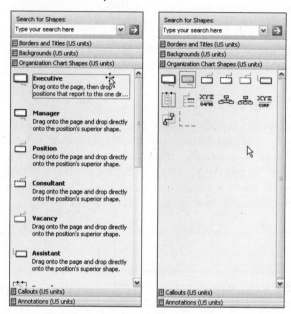

Changing Spacing and Color Settings

If you recognize shapes visually, limiting the text displayed with each master reduces the space required for a stencil window. Changing the amount of text displayed in labels also increases or decreases the spacing between master icons. In addition, you can change the background color for stencils to make them easier to read. Use any of the following methods to change the appearance of stencils:

- **Characters per line** — Specifying more characters per line increases the horizontal space between masters, in addition to providing more room for the master name. To specify how many characters of text appear in each line of a master label, choose Tools ➪ Options and select the View tab. Underneath the Stencil Spacing label in the Characters Per Line box, type the number of characters you want.

- **Lines per master** — Increasing the number of lines increases the vertical space between masters, in addition to providing more room for the master name. To specify the number of lines of text for each master label, choose Tools ➪ Options and select the View tab. Underneath the Stencil Spacing label in the Lines Per Master box, type the number of lines you want.

- **Background color** — To change the background color for stencils, choose Tools ⇨ Options and select the Advanced tab. Click Color Settings and specify any of the following colors:
  - **Text Color** — Specify the color for master labels.
  - **Background Color 1** — Specify the background color for a stencil.
  - **Background Color 2** — If your monitor is set to 32-bit color, the stencil background grades smoothly from the first background color to the second.

Rearranging Shapes in Stencils

As long as you go to the trouble of creating custom stencils, you might as well order the masters in a way that makes sense to you or puts the more commonly used masters at the top of the stencil. Rearranging shapes in a stencil is as simple as dragging the masters to different positions. Of course, you must first open the stencil you want to rearrange and make it editable (right-click the stencil title bar and, from the shortcut menu, choose Edit Stencil). Be sure to save the stencil after you have moved the masters into the positions you want.

Summary

Custom stencils are handy for a variety of reasons. You can use them to consolidate your favorite built-in Visio shapes onto a smaller number of stencils. They are a must for storing your custom shapes. Visio provides several easy ways to create custom stencils. In addition to creating blank stencils, existing stencils, search results stencils, and Document stencils are all viable candidates for the foundation for new custom stencils. After creating a stencil, you can add, remove, or rearrange masters. You can also modify the information that appears in stencils to provide more detail or minimize the space the stencil requires.

Chapter 33

Creating and Customizing Shapes

Whether you use shapes only from the built-in Visio stencils or craft highly specialized shapes unique to your organization, you can modify shapes in many ways — change their appearance on a drawing, make them behave differently, assign special behaviors to them, or edit the data they contain. If you plan to use a customized shape frequently or on more than one drawing, creating a master of the customized shape enables you to reuse that shape by dragging it from a stencil onto any drawing page.

For customized shapes, Visio drawing tools and Shape Operation commands help build just the shape you want, whether it is an open path or an enclosed area — a shape that acts like a line or a box with two dimensions. You can draw geometry from scratch with Visio's drawing tools, but it's often easier to build customized shapes by starting with existing shapes or objects from other applications as the basis for customized masters. What's more, if every instance of a shape on a drawing carries the same flaw, the fastest and easiest way to correct them is by making the changes you want on the master that spawned all those instances.

This chapter explains the difference between shapes and masters as well as the features that make shapes smart. You will learn how to create and modify masters and how to customize the graphic elements that comprise shapes and the behaviors that shapes exhibit. Finally, you will learn how to create and apply shape data to shapes as well as how to create shape data sets to quickly apply several properties to shapes and masters.

Understanding Shapes and Shape Data

If you do nothing more than drag built-in shapes from an existing stencil onto a drawing page, you might not care one whit about what makes shapes tick. Chances are you'll stumble across something that the built-in Visio shapes don't do or the shape you want is frustratingly absent from every stencil you open. There's no need to despair. The customizability of Visio shapes is limited mainly by the effort you're willing to expend. By understanding the different types of shapes, and their components, behaviors, and properties, you can build shapes that do exactly what you want. Conversely, if a shape doesn't behave as you would expect, this knowledge makes it easier to troubleshoot the problem by analyzing the shape's configuration.

What Makes Shapes Smart?

Built-in Visio shapes are preprogrammed to perform their tasks. Modular furniture shapes snap together much like they do in real life. Electrical components connect to represent the wiring for a building. Organization charts show information about employees, their positions in the organization, and their reporting relationships to other employees. Helpful behaviors such as these stem from features that anyone can exploit — not just the Microsoft employees who build the masters on Visio stencils.

Basic Visio concepts have a lot to do with what makes shapes smart. Dragging pre-defined shapes onto a drawing and connecting them with glue is pretty smart to begin with. So is displaying data within shape boundaries or modifying behavior and appearance with a couple of clicks. Regardless which shape tricks amaze you with their usefulness, the following are the foundation for SmartShapes:

- **ShapeSheets** — No matter what a shape does or how it looks, you can find the fields that control every aspect of that shape in its ShapeSheet. You can change shape behaviors and properties by modifying the values in fields in a ShapeSheet. And if you want to make a shape do even more, the ShapeSheet is the place to add custom fields or define customized formulas.

CROSS-REF   Although you can modify almost everything about a shape on a drawing page, in the master drawing window, or by setting shape behavior options, Chapter 34 provides an introduction to obtaining more advanced behaviors with ShapeSheets.

- **Shape behavior options** — Although ShapeSheets contain the fields that control shape behavior, the options in the Behavior dialog box are the shortcut to modifying key aspects of shape selection, editing, positioning, and connections. For example, when you select a two-dimensional shape, you see selection handles, control handles, and the alignment box; whereas selecting a one-dimensional shape shows only selection and control handles. See the section "Customizing Shape Behavior" later in this chapter to learn more.

- **Shape data** — If data is an important part of a diagram, such as the department and employee associated with an office on a facilities plan, you can store additional information in shape data associated with shapes. You can display shape data values in shape text, use them in calculations, or present them in reports. See the section "Customizing Shape Data" later in this chapter as well as Chapter 10 to learn more.

CROSS-REF To learn about linking shape data directly to databases, see Chapter 10.

Understanding Shapes and Masters

Inheritance can be a powerful influence, whether it comes from a rich aunt or a Visio master. In Visio, a master is the original specification for a shape — what it looks like, what it does, and what data it contains. A Visio master is a lot like the plate that Uncle Sam uses to produce money. It's an original form which you can use to churn out as many copies as you want, with each copy initially possessing the same appearance, behavior, and properties as the original. By using masters to create the shapes on your drawings, you accomplish the following:

- Configure a shape once and reuse it again and again
- Ensure consistency on all your drawings
- Share shapes with other Visio users
- Set up standard shapes for your entire organization
- Save file space, because instances of masters take up less space

You can construct drawings by dragging masters that have the appearance, behaviors, and data you want onto the drawing page. As soon as a master hits the drawing page, it becomes an *instance* of the master, known more familiarly as a shape. Initially, Visio adds a copy of the master to the drawing's Document Stencil and places an instance on the drawing page. The instance is linked to the master on the Document Stencil so that changes made to that master propagate to every instance in the active drawing file. The instance contains only information that is different from the master.

For many shapes, the only thing that differs from instance to instance is the location of the shape — which is defined by the *x*- and *y*-coordinates in the PinX and PinY fields in the ShapeSheet. However, if you make other changes to the shape, such as thickening the line style of a shape border, Visio has to keep track of whatever information differs from the master definition.

You can edit the instances of masters in any way you want, but not without the occasional consequence. First, as described in the sidebar "Masters, Shapes, and Memory," every modification you make to instances increases the size of your Visio drawing file and infinitesimally reduces the performance of Visio. In addition, some modifications sever the link between an instance and its shape master, which immediately eliminates all the benefits that masters provide. For example, if you ungroup a shape created from a grouped master, you'll see the message "This action will sever the object's link to its master." What the warning is trying to say is that the instance won't inherit changes made to the master on the Document Stencil and Visio has to store the instance as a separate entity in the drawing file.

Masters, Shapes, and Memory

Even if you've been content so far with editing shapes, you might want to consider creating customized masters. Each change to a shape on the page translates into more ShapeSheet cells with values not inherited from the master shape and, therefore, stored with the shape. Although each change brings only a small increase, these custom values require more memory and disk space for the drawing file. In turn, extra bytes increase the time the Visio takes to display your drawings.

Even a simple shape can require more than 10,000 bytes of memory and formatting can multiply that number several times over. To the contrary, an instance often requires as little as a third of the bytes of the original master. If a drawing contains hundreds of copies of a master, the difference in memory consumption and file size diverges rapidly.

You can conserve memory and disk space, while improving display performance, by considering the following techniques:

- If you use drawing tools a lot, look for repeating elements and create masters for them.

- If you customize the instances on the page, see whether you can create a few customized masters for similar instances.

- For repeating sets of formatting, define Visio styles and apply them to instances. (You must turn on Developer mode to work with styles by choosing Tools ➪ Options and then on the Advanced tab, checking the Run In Developer mode.)

- Use Visio Operations to combine shapes instead of creating an excessive number of shape groups.

Understanding Shape Geometry

Shape geometry isn't that important if you simply drag built-in shapes onto a drawing and use them as is. However, as soon as you begin to modify existing shapes or create your own, understanding shape building blocks can demystify your work. By learning the features and benefits of different shape elements, you can choose the tools that produce shapes that look and behave the way you want. Learning shape terminology comes in handy when you have to resort to reading topics in Visio help or asking a Visio expert for assistance.

Exploring Shape Components

From a simple line to a graphically complex representation of a computer, shapes are composed of line segments, arcs, and the occasional text block. Even the simplest shapes come with a few basic Visio components. Visio has two basic types of shapes, 1-D and 2-D:

- 1-D shapes are basically lines or connected series of line segments. Because 1-D shapes can have width, such as the fancy arrow shapes that come with Visio, you can identify 1-D shapes when you select them, because they have start points and end points.

- 2-D shapes truly have both length and width, which you can identify by the eight selection handles that appear when you select a 2-D shape.

By assembling line segments and arcs to create shapes, you can access additional points to help you position, resize, and connect shapes. When you select a shape, Visio marks these points, called *handles,* so you know where to drag on a shape to make the changes you want. Here are the handles and what they do:

- **Start points and end points for 1-D shapes** — Each line segment and arc has a start point and an end point, which appear when you select the 1-D shape. The start point is a green square with an X inside it, whereas the end point is subtly different — a green square with a plus symbol inside it.

- **Selection handles** — When you select a 2-D shape, red or green boxes appear at each corner and in the middle of each side of the shape's alignment box. You can drag the green squares to resize the shape. A double-headed arrow appears when you position the pointer over a green square to indicate the direction of change. When you position the pointer within the alignment box, a four-headed arrow appears indicating that you can reposition the entire shape.

- **Vertices** — When you create a connected string of line segments and arcs, a diamond-shaped vertex appears at each intersection when the Line, Arc, FreeForm or Pencil Tool is selected. You can change the length of lines and arcs or reposition them by dragging their end points or vertices.

- **Rotation handles** — For 2-D shapes, you can drag the green circle of the rotation handle to rotate a shape.

- **Control handles** — Yellow diamonds that appear on some shapes are control handles that represent special controls for modifying the shape's appearance. For example, you can use a control handle to change the swing on a door or to change the width of all the bars in a bar graph.

- **Connection points** — Blue Xs mark locations where you can glue connectors or other shapes.

- **Eccentricity points** — When you select a shape with the Pencil tool, small green circles appear, which you can drag to change the curve or symmetry of a line or arc, or to adjust the angle and magnitude of the eccentricity of an arc.

Understanding 1-D and 2-D Behavior

Visio shapes are either 1-D or 2-D, but the distinction has nothing to do with the way the shape looks. The difference between 1-D and 2-D shapes is completely in the way they behave. One distinguishing characteristic of 1-D shapes is that you can both rotate and change the length of a 1-D shape in one step by moving one of its end points.

One-dimensional shapes don't have to look like thin lines. A Fancy Arrow or Road shape can transform into a mild-mannered 1-D shape with a quick behavior option switch, as described in the section "Switching Between 1-D and 2-D Behavior" later in this chapter. Because 1-D shapes include start points and end points, they're great for indicating direction, for example, with arrowheads at the end points to differentiate the source and destination of a data flow.

With 1-D shapes, you can change the relative proportion of the shape, as shown in Figure 33-1. For example, you can increase the length of a Fancy Arrow shape by dragging an end point, but the width of the arrowhead and tail remain the same. However, when you drag the selection handles on the sides of a Fancy Arrow shape, the proportions of the arrowhead change to fit the new width, even though the length remains the same. Even if you want to, you can't change the length and width of a 1-D shape at the same time, because 1-D shapes don't have selection handles at their corners.

FIGURE 33-1

1-D shapes include start points and end points to change their lengths and two selection handles to adjust their widths.

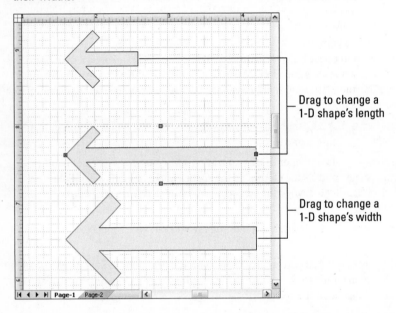

Two-dimensional shapes come packed with up to eight selection handles for resizing. The selection handles on the sides of 2-D shapes change their height or width independently, whereas dragging the selection handles at shape corners resize the shape proportionately, which is the behavior that differentiates 2-D shapes from their 1-D siblings.

NOTE   The distinction between 1-D and 2-D shapes is merely a behavioral difference that you set by choosing an Interaction Style option in the Behavior dialog box. For example, by default, a line segment is a 1-D shape with the characteristics you would expect: start and end points that you can drag to lengthen or shorten the line or to change its orientation. To transform a single line segment into a 2-D shape, all you have to do is right-click the 1-D shape, choose Format ➪ Behavior, and select the Box (2-dimensional) option on the Behavior tab. When you do so, the shape sports selection handles at the midpoint of the line to drag the line into a two-dimensional rectangle.

In addition to 1-D and 2-D shapes, it's important to recognize the difference between *open shapes* and *closed shapes*. You create an open shape when you draw a series of connected line segments that zigzag across the page. Because open shapes don't enclose an area, you can format the lines and text, but you can't apply fill formatting. Conversely, a closed shape is born when you connect a path back to its starting point. Visio indicates that a shape is closed by applying a default fill format to it. However, you can apply a different fill format if you choose.

Creating and Editing Masters

Although Visio provides hundreds of built-in shapes, with thousands more available from other sources, there are always some situations that demand custom masters. For example, if you design and sell unusually-shaped furniture, you can create Visio masters for each of your designs so you can help clients lay out their furniture. How you create your masters depends on whether there's something you can copy and where that copy resides. Starting with existing masters or shapes is by far the fastest path to custom masters, but you generate masters using converted CAD objects or data in a database. For wildly unique shapes, Visio drawing tools might be the best solution. If your needs change, you can modify the appearance, behaviors, and properties of your custom masters, and apply those changes to the instances on your drawings.

NOTE To create and edit masters, you must first open a drawing so that the Shapes window opens. Otherwise, you won't be able to access commands to make stencils editable or edit masters.

Creating Masters

Although the Visio drawing tools are easy to use, there's nothing more satisfying than getting a jump start with work you've already done. Create your own masters with the easiest of the following methods:

- **Copy an existing master** — Save an existing master on a stencil as a new custom master and then modify it to suit your needs.

NOTE If you customize the built-in (and copyrighted) masters, take care not to infringe on the copyright. For example, if you want to create shapes to sell, create the shapes from scratch.

- **Use an existing shape** — Drag an existing shape from a drawing page to an editable stencil and, if necessary, modify it.

- **Create a master from an object from another application** — Convert an object from another application, such as Microsoft PowerPoint, AutoCAD, or clipart into a Visio master.

- **Build a master from database information** — Generate new masters by combining data from a database table with existing masters in stencils.

- **Build a master with drawing tools** — Create a master and add graphics to it with the Visio drawing tools and Shape Operation commands.

Opening a Stencil for Editing

In order to create masters, you first need a stencil in which to store them. Microsoft won't let you modify their built-in stencils, so you must either create your own stencil or save a built-in one as a custom stencil. For example, to use a built-in stencil as a starter for a custom stencil, in the Shapes window, right-click the title bar and choose Save As.

When you try to add a master to a custom stencil that isn't editable, Visio asks if you want to edit the stencil. You can click Yes and get on with creating your masters. The more proactive approach is to get your custom stencil open and editable before you start. To make a stencil editable, follow these steps:

1. Open a drawing file so that the Shapes window opens. In reality, you can open any old drawing file, but if you have a template that opens the custom stencil you want to edit, you might as well open that one.

2. If necessary, open the stencil in which you want to save a master. If you store the stencil within the Visio My Shapes folder, the easiest way to open the stencil is to choose File ➪ Shapes ➪ My Shapes and then choose the stencil.

> **TIP**   If the stencil you want is somewhere else, choose File ➪ Shapes ➪ Open Stencil. In the Open stencil dialog box, navigate to the folder that contains the stencil and double-click the stencil file name.

3. When the stencil is visible in the Shapes window, right-click its title bar and, from the shortcut menu, choose Edit Stencil. Visio indicates that the stencil is editable by adding a red asterisk to the stencil icon in the title bar.

Copying Existing Masters to Create New Ones

Copying an existing master is a quick way to customize a master and make it your own. To copy a master to another stencil, follow these steps:

1. To display a stencil, in the Shapes window, click the title bar of the stencil that contains the master you want to copy.

2. Right-click the master and, from the shortcut menu, choose Copy.

3. In the Shapes window, click the title bar of the editable stencil that is the new home for the master.

4. Right-click within the stencil, and, from the shortcut menu, choose Paste.

5. To modify the master, right-click it and, from the shortcut menu, choose Edit Master. See the section "Editing Masters" later in this chapter for detailed instructions.

6. To save the stencil with the modified master on it, right-click the stencil's title bar and, from the shortcut menu, choose Save Stencil.

Saving Existing Shapes as Masters

If you customize a shape on one of your drawings and want to use it on other drawings, you can save it as a master on a stencil. To save a shape on a drawing page as a master, follow these steps:

1. With the Shapes window open, make editable the stencil in which you want to save a master.

2. To copy a shape from the drawing page into the stencil, press and hold the Ctrl key while dragging the shape into the stencil. (To move the shape from the drawing page to the stencil, simply drag the shape.) In the stencil, Visio adds an icon for the master with a generic name such as Master.*x*, where *x* is a number, as shown in Figure 33-2.

FIGURE 33-2

When you copy a shape into a stencil, Visio assigns a name by default.

3. Click the icon label. When the master name is highlighted, type the name you want for the master and press Enter.

4. To modify the master, right-click it and, from the shortcut menu, choose Edit Master. See the section "Editing Masters" later in this chapter for detailed instructions.

5. Save the stencil with the new master by right-clicking the stencil's title bar and, from the shortcut menu, choosing Save Stencil.

Creating Masters from Database Information

You can also construct new masters by combining existing masters with data within a database. For example, if you have a master for an office cubicle and store data about the cubicle's color and components in a table in a database, you can create masters for each cubicle configuration in the database. When you drag a master to a drawing page, the instance shows the data for the associated record in the database. To create masters in this way, use the Database Wizard to create a stencil containing a master for each record in a database table.

 To learn how to use the Database wizard, see Chapter 10.

Drawing Masters from Scratch

For masters that don't have any comparable cousins to start with, you can create a master from scratch by following these steps:

1. With the Shapes window open, make editable the stencil in which you want to save a master.

2. Right-click inside the stencil window and, from the shortcut menu, choose New Master.

3. In the New Master dialog box, type the name for the master. You can also include a prompt to give people hints about the purpose of the master. See the section "Modifying Master Properties" in this chapter to learn more. Click OK when you are done. In the stencil, Visio adds a blank master icon with the master name as a label.

NOTE   If you want to locate your master using the Search for Shapes command, add keywords separated by commas to the Keywords property in the Master Properties dialog box.

4. Right-click the new master and choose Edit Master ⇨ Edit Master Shape.

5. In the master drawing window, create the graphics for the master. You can use any of the following methods:

   ■ **Draw components** — Use Visio drawing tools or Shape Operation commands to create the master graphics that you want.

   ■ **Use an existing shape or master** — Drag a shape from a drawing page or drag a master from a stencil into the master drawing window.

   ■ **Paste an object from another application** — Copy an object from another application such as AutoCAD and paste the object in the master drawing window.

CROSS-REF   To learn how to convert CAD objects into Visio shapes on a customer stencil, see Chapter 29.

6. When you've drawn the master, choose File ⇨ Close to close the master drawing window. When Visio prompts you to update the master, click Yes.

7. To save your changes, right-click the stencil title bar and, from the shortcut menu, choose Save Stencil.

Editing Masters

Whether you've copied a built-in Visio master as a model or you want to adjust a master to meet changing requirements, modifications to masters come in three forms:

- Altering the graphics of the master, for example, to add new standard equipment to office cubicle masters.

- Changing the master icon, perhaps to better represent what the master does if its name doesn't appear in the stencil window.

- Modifying the master's properties, for example, to change the master name or add new keywords to search for the master.

You make many of these edits when you first create masters, but you can revise them at any time you want. To modify an existing master, make the stencil that contains it editable. Then, you can perform one or more of the following actions:

- **Rename a master** — Right-click the master you want to rename, and, from the shortcut menu, choose Rename Master. When the master name is highlighted, edit the name in the master label, and press Enter.

- **Modify master graphics** — The next section in this chapter explains how to edit the drawing elements for a master, including lines, arcs, text, and connection points.

- **Modify master properties** — See the section "Modifying Master Properties" later in this chapter to learn how to specify properties for a master, such as Name, Icon Size, and Keyword.

- **Change the icon image** — You can revise the icon that represents the master in a stencil. See the section "Modifying Master Icons" later in this chapter to learn how.

- **Add or edit shape data** — You can associate shape data with a master. See the section "Customizing Shape Data" in this chapter to learn more.

- **Delete a master** — Right-click the master you want to delete and then, from the shortcut menu, choose Delete Master.

 When you finish modifying masters on an editable stencil, be sure to save your changes by saving the stencil.

Editing Master Graphics

One way to modify the appearance of a master is by editing a shape on a drawing and then saving the shape as a master. However, you can also edit a master's drawing elements in the master drawing window. Working in the master drawing window is much like working on a Visio drawing. You can use Visio drawing tools to edit existing graphics or add other shapes and objects to the master by dragging them into the window. However, unlike working in a Visio drawing, Visio reminds you to update the master shape when you close the master drawing window.

To edit a master in the master drawing window, first open any Visio drawing file so that the Shapes window opens, and then follow these steps:

1. With the Shapes window open, make editable the stencil in which you want to save a master.

2. Right-click the master you want to edit and choose Edit Master ⇨ Edit Master Shape. Visio opens the master drawing window with the master in the middle of the master drawing area, as shown in Figure 33-3.

FIGURE 33-3

The master drawing window works like a regular drawing window except that the title bar shows the stencil name and the master you are editing.

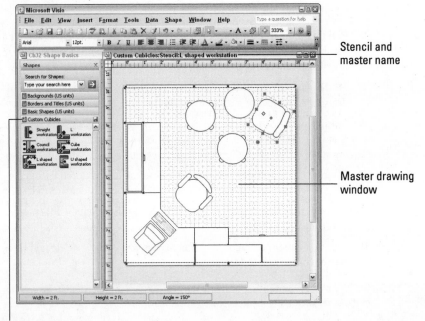

Stencil and master name

Master drawing window

Editable stencil

3. Make any of the following changes you want to the master's graphics:

   ■ **Edit drawing elements** — Use Visio drawing techniques to modify the lines, closed shapes, and text in a master.

   ■ **Format drawing elements** — Use Visio formatting commands to change colors, line styles, text styles, and fills.

   ■ **Modify connection points** — See the section "Controlling Glue with Connection Points" later in this chapter to learn how to add or modify connection points on shapes and masters.

4. When you have finished editing the master graphics, click the Close button in the master drawing window. When Visio prompts you to update the master, click Yes.

5. If you want to continue editing masters, click Cancel when Visio prompts you to save changes to the stencil containing the master you edited. Repeat steps 2 through 4 to edit another master.

6. When you finish modifying masters, save the changes you made by right-clicking the stencil's title bar and choosing Save from the shortcut menu.

NOTE The Document stencil in each Visio drawing file is the repository for a copy of each master used on the drawing. When you modify a master on a custom stencil, you won't see those changes on the drawing automatically, because the drawing file references the copy of the master on the Document stencil. If you want the drawing file to use the master that you modified on a custom stencil, copy the modified master to the Document stencil.

Modifying Master Properties

In addition to the graphic elements of a master, you might want to change some of its other characteristics, such as its name, the prompt that appears when you point to a shape in a stencil, the icon size, or the keywords that hunt it down in a search for shapes. To modify a master's properties, right-click a master in an editable stencil and choose Edit Master ➪ Master Properties. In the Master Properties dialog box, modify the properties you want and then click OK to close the dialog box when you're done. To save the changes to the master, be sure to save the stencil. Here are the properties in the Master Properties dialog box and what they do:

- **Name** — On their own, icons often don't say enough about what a master is for. You can increase the information you provide about a master with the label, which appears under the master icon in a stencil and can be up to 31 characters. Visio might truncate the name you see depending on the number of characters per line and the number of lines per master you specify on the View tab of the Options dialog box.

- **Prompt** — To provide your audience with even more information about a master and even hints about how to use it, create a prompt. When you point to a master in a stencil, the prompt appears in a balloon.

- **Icon Size** — The size of the icon for a master in a stencil can be Normal, Wide, Tall, or Double. By default, icons are Normal, which represents 32 × 32 pixels.

- **Align Master Name** — This is one of those options that someone somewhere might find helpful. Selecting the Left, Center, or Right option aligns the master name to the left, center, or right of the master icon in the stencil — if you set the stencil to show names below icons and the master name is longer than the icon is wide.

- **Keywords** — The Search for Shapes command uses the words in this property to locate the master. When you enter more than one keyword, separate the keywords with commas. This field is available only when the master drawing page contains at least one shape.

- **Match Master by Name on Drop** — Selecting this check box tells Visio to look for a master by the same name in the Document stencil and use it whenever you drag this master from the custom stencil onto the drawing page. This behavior is desirable when the master on the stencil is generic and likely to receive formatting on the drawing page. If you clear this box, Visio copies the master from the shape stencil, so that you always use the original version of the master.

- **Generate Icon Automatically from Shape Data** — For most masters, you can leave this check box cleared, because you usually edit the master icon to make it simple enough for a 32 × 32 pixel icon. However, if you want the icon to be a true depiction of the shape and you want Visio to update the master icon every time you edit the master graphics, check this check box.

Modifying Master Icons

Visio uses a miniature version of the shape as the icon that appears in the stencil. For all but the simplest masters, this results in an unreadable icon. Editing a master icon is a bit different from editing master graphics. Modifying icons proceeds pixel by pixel by pressing a mouse button and dragging the mouse over the pixels you want to change. Visio applies the color that you assigned to the mouse button. For larger changes, you can move or delete groups of pixels. The changes you make to a master icon apply only to the icon and don't affect the appearance of the master in any way.

Edit a master icon by following these steps:

1. In an editable stencil, right-click the master you want to edit and choose Edit Master ⇨ Edit Icon Image. Visio opens the icon editing window, displays the icon in the window, and displays two toolbars containing icon editing tools, as shown in Figure 33-4.

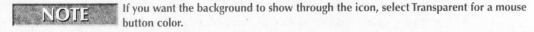 **CAUTION** If you make changes to an icon manually, be sure to clear the Generate Icon Automatically from Shape Data check box in the Master Properties dialog box. Otherwise, Visio will replace your hand-crafted icon with the revised master you edited in the master drawing window.

2. While the icon editing window is open, you can use the following methods to edit the master icon:

   - **Select colors** — To work as if you hold paint brushes with different colors in each hand, assign colors to the left and right mouse buttons by clicking the Left Button Color or Right Button Color box and then selecting the color you want to apply with that mouse button. Drag or click with a mouse button to apply its assigned color.

   NOTE If you want the background to show through the icon, select Transparent for a mouse button color.

   - **Change single pixel colors** — On the Icon Tools toolbar, click the Pencil tool and then, in the icon editing window, with the mouse button assigned to the color you want, click the pixel you want to color.

- **Change the color of an area** — On the Icon Tools toolbar, click the Bucket tool and then, with the mouse button assigned the color you want, click a pixel within the area you want to change. Visio changes the color of all contiguous pixels of the same color as the pixel you clicked.

- **Move pixels** — On the Icon Tools toolbar, click the Lasso Tool or Selection Net Tool and drag around the area you want to move. When Visio displays a selection box around the pixels, drag them to a new location.

- **Delete pixels** — On the Icon Tools toolbar, click the Lasso Tool or Selection Net Tool and drag around the area you want to delete. When Visio displays a selection box around the pixels, press Delete.

 NOTE When you move or delete pixels in a master icon, the stencil background color appears in the area from which you move or delete pixels.

FIGURE 33-4

The icon editing window provides two specialized toolbars for editing icons and automatically previews your edits in the stencil window.

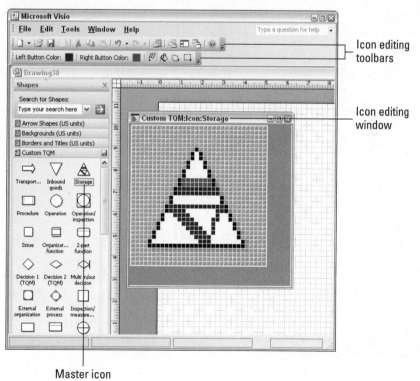

Master icon

3. When the master icon is a masterpiece of pixel art, choose File ⇨ Close to close the icon editing window. Be sure to save your changes by right-clicking the stencil's title bar and, from the shortcut menu, choosing Save.

> **NOTE** If an errant twitch sends your mousing hand across the window, you can undo changes as long as the icon editing window is still open. Simply choose Edit ⇨ Undo. However, once you close the icon editing window, the Undo command is powerless to reverse the changes you made. You must edit the icon and use the icon editing tools to return the icon to its original appearance.

Creating Shortcuts to Masters

Suppose you find one custom master incredibly useful, for example, your smiley face version of an off-page reference that you want to add to every stencil you customize. If there's any chance that you will modify that master in the future, creating shortcuts to that original master means you can modify the master on one stencil and propagate the changes automatically to the other stencils.

> **CAUTION** The shortcuts you create to masters work as long as the master stays on the original stencil. If you copy the stencil with the shortcut to another machine, you must also copy the stencil that contains the original master as well.

To create a master shortcut, follow these steps:

1. Open the stencil that contains the master you want to copy and the stencil to which you want to add the shortcut. If the destination stencil is not editable, right-click its title bar and, from the shortcut menu, choose Edit Stencil.

2. In the stencil that contains the master, right-click the master and, from the shortcut menu, choose Copy.

3. Click the destination stencil to display it, right-click the destination stencil window and, from the shortcut menu, choose Paste Shortcut.

4. To save your changes, right-click the editable stencil's title bar and, from the shortcut menu, choose Save Stencil. When you edit the master in any of the stencils, Visio changes the master in every stencil linked with a shortcut.

Drawing Shapes and Masters

No matter how many stencils and masters Visio provides, sooner or later you'll want to create a new shape or master or make changes to existing ones. Yet, even with the most specialized shapes, there's usually an existing shape that can give you a head start. For example, you might create or revise shapes to do the following:

- Create masters for the standard cubicle configuration your organization uses for office layouts.
- Create new masters by modifying existing Organization Chart masters to emphasize each level of management with a different shape.

- Modify the line and fill styles for shapes to show status.

- Draw a shape directly on a drawing when you plan to use it only once.

The Visio drawing tools and Shape Operation commands are the basic tools for drawing graphics from scratch. The Line tool and the Pencil tool are perfect for drawing straight lines. For curves, you can choose the Pencil tool, Arc tool, or Freeform tool, depending on how baroque the curve. If you change your mind, the Pencil tool can convert arcs to lines and vice versa. Visio drawing aids and shape extension lines help you define the geometry you want. For example, you can use drawing aids or shape extension lines to draw lines at 45 degrees, create a line perpendicular to another, or draw a line tangent to a circle.

CROSS-REF For additional instructions on drawing lines, curves, and closed shapes, see Chapter 2.

Drawing Line Segments and Paths

Drawing a single line in Visio is easy. On the Drawing Tool toolbar, click the Line tool and then drag between two points on the drawing page. To produce a connected path or a closed shape, which acts like a single entity, you have to draw a series of connected line segments and that requires just a tad more finesse. To create connected line segments, follow these steps:

1. Select the Line tool or the Pencil tool.

2. Drag the pointer from the starting point to the end point to define the first line segment and release the mouse button without moving the mouse away from the end point.

3. With the pointer still positioned over the second point, drag from point to point to create the next line segment.

TIP If you draw a line segment incorrectly or in the wrong place, don't stop to correct the segment. Simply continue to create line segments until the path is complete. You can then use the Pencil tool to edit the line segments in the path.

4. Repeat step 3 until you have drawn all the line segments you want. To close the shape, make sure that the last point in the drawing path overlaps the very first point in the path.

CROSS-REF If you want to create construction lines to align shapes, don't use the Visio drawing tools. Visio guide lines, which you drag from a ruler onto a page, perform the same function and they don't print by default. To learn how to create guidelines, see Chapter 4.

Creating Closed Shapes

The Rectangle and Ellipse drawing tools automatically create closed shapes — on the Drawing toolbar, click the tool and then drag between two points on the drawing page. Visio creates the closed shape and applies a default fill style so there's no question that the shape is closed.

If you use the Line, Pencil, Arc, or Freeform tool, you must be sure to close the drawing path if you want to apply a fill pattern or color to the shape. To close a drawing path, make sure the last point is the same position as the first point on the path, as shown in Figure 33-5.

> **NOTE** If the drawing grid is visible behind your shape after you've completed the entire path, the shape is not closed. See the section "Creating and Editing Shapes with the Pencil Tool" later in this chapter to learn how to close an open shape.

FIGURE 33-5

Overlap the first and last points to close a shape and Visio applies a fill pattern to the shape.

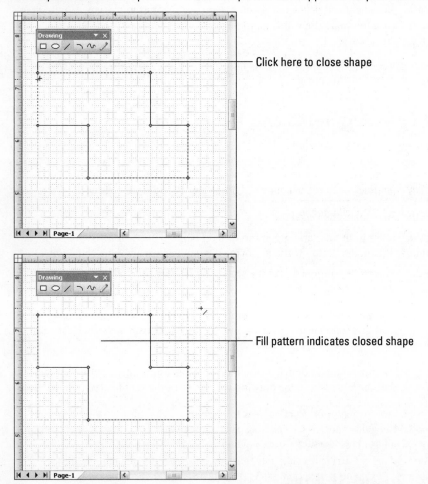

Click here to close shape

Fill pattern indicates closed shape

Drawing Graphics Precisely

When you create your own shapes, the accurate positions of lines and curves are often crucial. For example, for a precision machine part, a hundredth of an inch off and the chain won't fit the sprocket. Visio drawing aids and shape extensions help you create a variety of precise geometric constructions, such as tangent lines.

CROSS-REF To specify precise coordinates for the end points of lines, open the Size & Position window and type coordinates, as described in Chapter 4.

Drawing aids are temporary dotted lines that show you where to click to draw squares, circles, and angled or perpendicular lines. Shape extensions work similarly to drawing aids, but snap to additional geometry, as shown in Figure 33-6. Drawing aids and shape extensions work in conjunction with the Visio drawing tools, such as Line, Pencil, and Rectangle, to help you perform the following tasks:

- **Draw circles and squares** — When you use the Ellipse or Rectangle tools, drawing aids show you where to click to create a circle or square.

- **Draw lines** — When you use the Line tool or Pencil tool, drawing aids appear when the line you are constructing approaches an increment of 45 degrees. Depending on the shape extensions you enable, shape extensions show you where to click to draw lines such as tangents to a curve or perpendicular to other lines.

- **Edit lines** — When you edit a line segment, drawing aids extend at 45-degree increments as well as at the line's original angle.

FIGURE 33-6

Shape extensions facilitate the creation of precise geometry.

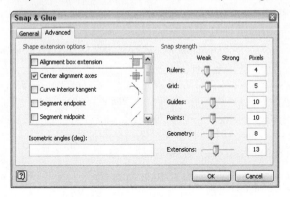

To display drawing aids, choose Tools ➪ Snap & Glue and, in the Currently Active column, select the Drawing Aids check box. To enable snapping to shape extensions, in the Snap To column, select the Shape Extensions check box, then select the Advanced tab, and select the check box for each shape extension option you want to enable. Table 33-1 describes what the different shape extensions do.

NOTE You can adjust the attraction that shape extensions exert on the mouse pointer by dragging the Extensions slider to the left or right. As you drag the slider, the proximity of the pointer, in pixels, required to activate snapping appears in the Pixels box.

TABLE 33-1

Shape Extension Options

Option	Function
Alignment Box Extension	Extends lines from a shape's alignment box, so you can snap to a position aligned with the edges of a shape.
Center Alignment Axes	Extends a line from the center of a shape's alignment box, so you can snap to a position aligned with the center of a shape.
Curve Interior Tangent	Displays an extension line tangent to a curve. To draw a tangent to a curve, drag the pointer from the edge of the curve until the tangent extension line appears, and then drag the pointer along the extension line until the extension line turns red. As you continue to move the pointer, Visio shows the tangent line from the current pointer position to the curve.
Segment Endpoint	Highlights and snaps to the end point of a line segment or arc.
Segment Midpoint	Highlights and snaps to the midpoint of a line segment or arc.
Linear Extension	Displays an extension of a line beyond its end point, so you can extend the current line.
Curved Extension	Extends an arc to show where to click to create an ellipse. For splines, this shape extension extends the curve from the nearest end point.
Endpoint Perpendicular	Displays a line perpendicular to the nearest end point of a line or arc, so you can create a line perpendicular to an existing line or arc.
Midpoint Perpendicular	Displays a perpendicular line from the midpoint of a line or arc, so you can create a perpendicular line that bisects a line or arc.
Horizontal Line at Endpoint	Displays a horizontal line from the end point of a line or arc, so you can create a horizontal line starting at the end of an existing line.
Vertical Line at Endpoint	Displays a vertical line from the end point of a line or arc, so you can create a vertical line starting at the end of an existing line.
Ellipse Center Point	Highlights and snaps to the center of an ellipse.
Isometric Angles	Displays extension lines at isometric angles to simplify the construction of isometric diagrams.

Creating and Editing Shapes with the Pencil Tool

The Pencil tool not only creates new lines and arcs, but edits existing ones in ingenious ways. Correcting crooked lines, repositioning vertices and end points, switching between straight lines and circular arcs, or closing open shapes, the Pencil tool can do it all. On the Drawing Tool toolbar, click the Pencil tool and then use one of the following methods to create or edit lines and arcs:

- **Draw a straight line** — Click a point on the drawing page and then drag the pointer relatively straight in any direction. Visio indicates that it is in Line mode by changing the pointer to crosshairs, with an angled line below and to the right.

- **Draw an arc** — Click a point on the drawing page and then sweep the pointer in a curve. Visio indicates that it is in Arc mode by changing the pointer to crosshairs, with an arc below and to the right. By moving the pointer, you can adjust the radius of the arc as well as the angle it circumscribes.

- **Switch between Line and Arc mode** — Move the pointer back to the starting point of the current segment. When the pointer changes to crosshairs only, drag or sweep the pointer to switch to Line mode or Arc mode.

- **Correct a crooked line** — If you created line segments that aren't orthogonal, straighten them by selecting the shape to display its vertices and then dragging an end point or vertex to a new location.

- **Change an arc into a straight line** — Select the shape to display its handles. Drag the green eccentricity handle at the midpoint of an arc until the line is straight.

- **Change a straight line into an arc** — Select the shape to display its handles. Drag the green eccentricity handle at the midpoint of a line until the arc is the radius you want.

- **Close an open shape** — Select the shape to display its vertices. You can close a shape using one of the following methods:

  - Drag a vertex to the starting point of the first segment in the shape.

  - Draw a line from the last vertex in the existing path to the first.

Transforming Shapes into New Ones

Shape Operations are no more than a heaping handful of commands, but, combined in the right ways, they can produce just about any shape you can imagine. You can assemble simple shapes into more complex shapes; deconstruct shapes into more elementary components, or tweak shape geometry into exactly what you want. Some shapes, particularly those with holes or cutouts, are feasible only by fusing several shapes into one. Complex shapes that you could construct with repeated use of the Pencil and Group commands are often easier to build with Shape Operation commands, and the resulting Visio files are more compact. If you've worked with CAD programs, you're probably familiar with many of these functions. If not, the only way to truly master them is to experiment to find the magical combination that works.

It sounds simple. To use Shape Operation commands, select the shapes you want to transform, choose Shape ➪ Operations, and then choose the command you want. Unfortunately, to produce the results you seek typically requires selecting shapes in the correct order as well as combining Shape Operation commands in the right way. Consider it a chance to show off your creativity and ingenuity.

NOTE For Shape Operation commands, the order in which you select shapes is important for several reasons. For some commands, such as Subtract, Visio modifies the first shape selected using additional selected shapes, so the results vary depending on which shape you select first. In addition, Visio applies the formatting from the first shape you select to the resulting shape.

Updating the Alignment Box

A shape's alignment box is the boundary that you use to align it to the drawing grid or other shapes. For most shapes, the alignment box exactly matches the outside boundaries of the graphic elements it contains. Some shapes come with alignment boxes that are larger or smaller than the graphics, usually to make the final diagram look the way it should.

However, when other shapes don't land where you expect when you snap them to a shape, the shape's alignment box could be out of whack. To reset an alignment box to match the boundaries of a shape or group, select the shape or group and then choose Shape ➪ Operations ➪ Update Alignment Box.

Assembling and Disassembling 2-D Shapes

Some Shape Operation commands transform several shapes into one, while others explode shapes into smaller pieces. For some constructions, you can achieve the same results using different Shape Operation commands. For other results, there's only one solution. If one command doesn't produce the results you want, try another command or experiment with a combination of them.

CAUTION   The 2-D Shape Operation commands delete your original shapes, so it's a good idea to make a copy of your shapes before you start. If a command doesn't produce the results you expect, you can also reverse the changes by immediately choosing Edit ➪ Undo.

Creating a Union from Shapes

The Union command is aptly named because it produces one new shape that envelops the area of all the original shapes. For example, if you created Visio shapes for each country in Europe, you could create a shape for all of Europe by uniting all the country shapes, in Visio not politics.

When you apply the Union command to overlapping shapes, it creates a new shape that snakes around the outermost edges of the original shapes and then deletes the original shapes. For example, you can produce a single shape that looks like a mountain range from a collection of triangles. To create results similar to those shown in Figure 33-7, first select the shape whose format you want to apply to the resulting shape, Shift-click the other shapes you want to merge, and then choose Shape ➪ Operations ➪ Union.

FIGURE 33-7

The Union command creates one shape that encloses the area of several.

Select additional shapes

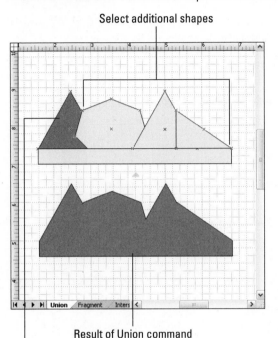

Result of Union command

Select shape with formatting for result

> **TIP** You might consider turning to the Union command to create a single shape from several shapes that don't overlap. However, by creating a group out of several shapes with the Group command, you gain control over how the shapes in the group behave, such as whether they resize when you resize the group or simply move.

Combining Shapes

For shapes with holes and cutouts, such as the frame of a window, the Combine command is the only game in town. It makes the overlapping areas of the selected shapes transparent so they look like holes, as demonstrated in Figure 33-8. The formatting of the first shape you select determines the formatting of the result. The command also deletes the original shapes when it's done. To combine shapes, select the shape with the format you want for the resulting shape, Shift-click the other shapes you want to process, and then choose Shape ➪ Operations ➪ Combine.

FIGURE 33-8

Create shapes with holes using the Combine command.

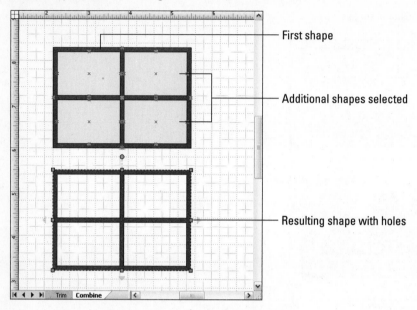

First shape

Additional shapes selected

Resulting shape with holes

Breaking Shapes into Smaller Pieces

Sometimes, the easiest way to get an odd-looking shape or a many-sided polygon is to carve it out of other shapes. The Fragment command breaks overlapping shapes into smaller pieces. Any areas that overlap become new 2-D shapes and the remaining areas that don't overlap also become 2-D shapes, as shown in Figure 33-9. In addition, you can draw lines through a 2-D shape to indicate where you want to break it. For example, by drawing lines through the center of a circle, you can break it into pie-shaped slices. If you don't need some of the shapes resulting from the fragmenting, delete them.

To fragment shapes, select the shape with the format you want for the resulting shape, Shift-click the other shapes you want to process and any lines you want to use as breaks, and then choose Shape ⇨ Operations ⇨ Fragment.

Removing Areas That Don't Overlap

The Intersect command produces one shape that includes *only* the overlapping areas of the shapes you select. For example, you can intersect two ellipses to produce an irregularly shaped island for a traffic intersection, as shown in Figure 33-10. The Intersect command deletes the original shapes and formats the resulting shape like the first shape you select. To intersect shapes, select the shape with the format you want for the resulting shape, Shift-click the other shapes you want to process, and then choose Shape ⇨ Operations ⇨ Intersect.

FIGURE 33-9

The Fragment command breaks shapes into smaller pieces.

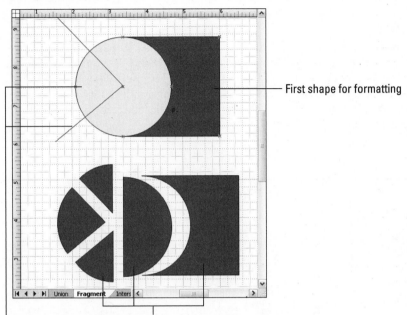

First shape for formatting

Results of Fragment command, separated for emphasis

Additional shapes for fragmenting

FIGURE 33-10

The Intersect command strips away any areas that don't overlap.

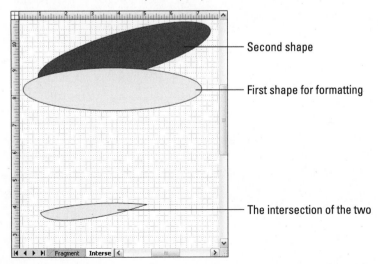

Second shape

First shape for formatting

The intersection of the two

CAUTION Because shapes that don't overlap don't have any intersecting areas, applying the Intersect command to non-overlapping shapes or individual shapes is a circuitous method for deleting the original shapes.

Creating Cutouts

The Subtract command acts a bit like a cookie cutter — it creates a new shape that looks like the first shape you select minus the overlapping areas from additional shapes. The order in which you select shapes is particularly important for the Subtract command because it affects both the resulting shape and the formatting that Visio applies, as illustrated in Figure 33-11. To create a cutout, select the shape from which you want to cut out areas, Shift-click the other shapes you want to use as cutouts, and choose Shape ⇨ Operations ⇨ Subtract.

FIGURE 33-11

The Subtract command removes parts of shapes that overlap the first shape you select.

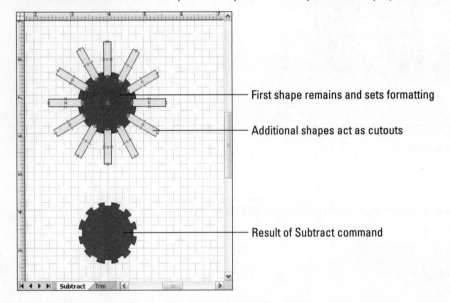

First shape remains and sets formatting

Additional shapes act as cutouts

Result of Subtract command

Manipulating Lines

The Shape Operation commands don't apply only to 2-D shapes. There are a few specifically for creating and manipulating 1-D shapes. For example, you can produce a set of parallel lines with the Offset command or use the Join command to create 2-D shapes from lines and arcs.

Joining Lines into Paths

The Join command and the Combine command have a lot in common. Just as Combine produces a single 2-D shape from several separate shapes, the Join command turns the 1-D lines and arcs you select into a single 2-D shape. For example, you can use the Join command to transform individual line segments into a single path. Although the Join command produces a 2-D shape, it applies only text and line formatting, not fill formatting, to the resulting shape. To join several lines and arcs, select the 1-D shape with the format you want to apply to the resulting shape, Shift-click the other shapes you want to join, and then choose Shape ⇨ Operations ⇨ Join.

Breaking Shapes into Lines and Arcs

The Trim command breaks shapes into separate lines and arcs. You can use lines to denote break points for the Trim command just as you can for the Fragment command. Visio converts shapes and lines into separate snippets wherever the lines intersect. For example, suppose you draw a long free-form curve — lovely except for the last few inches at one end. You can draw a line through the curve where you want to split it and then use the Trim command to break it into separate pieces. Select the shape with the format you want to apply to the resulting shape, Shift-click the other shapes you want to trim, and then choose Shape ⇨ Operations ⇨ Trim.

Creating Parallel Lines and Curves

The Offset command gets a lot of use creating parallel lines, grids, cross-hatching, concentric circles, or other repeating patterns. When you use the Offset command, it creates a set of parallel lines or curves at the distance you specify on *both* sides of the original shape, as shown in Figure 33-12. Unlike other shape operations, Offset does not delete the original shape. For example, if you use Offset on a single line, you end up with three equally spaced lines, the original in the middle and the two new ones on either side. Don't be scared off by the lines created on both sides. Sometimes, it's easier to use Offset and delete a few extraneous lines than to create graphics any other way. To create parallel lines or curves, select the shape or line you want to copy for the offsets and choose Shape ⇨ Operations ⇨ Offset. In the Offset dialog box, type the offset distance and click OK. Visio creates the additional lines on either side of the original.

FIGURE 33-12

Offset creates new lines or shapes on either side of the original shape.

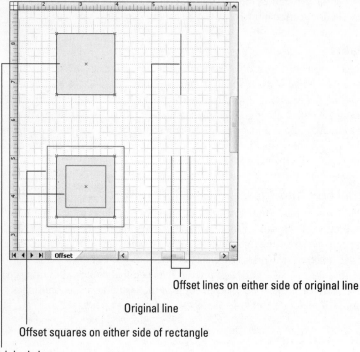

Offset lines on either side of original line

Original line

Offset squares on either side of rectangle

Original shape

Reversing Line Ends

Sometimes, you draw a line in the wrong direction, which is especially obvious with lines that have arrowheads or other line ends. If the line direction is important, as in data flow diagrams, you can correct your error without recreating the line. To switch the start and end points of a line, select the line and then choose Shape ➪ Operations ➪ Reverse Ends.

Creating Curves from Lines

The Fit Curve command transforms lines with multiple segments into curves. Unfortunately, this command can produce very different results depending on the settings and error tolerance you choose. The only solution is trial and error. If you don't like the results you obtain, press Ctrl+Z to undo those changes and try different settings. To create a free-form curve from a connected series of line segments, select the path that you want to convert to a curve and choose Shape ➪ Operations ➪ Fit Curve.

You can specify how Visio transforms a path into a combination of lines, arcs, and splines by setting the following curve parameters:

- **Periodic Splines** — Select this check box to create a seamless spline from a closed and smooth shape. For example, applying Fit Curve to a rectangle using periodic splines creates an ellipse.

- **Circular Arcs** — Select this check box to replace the vertices in a path with either circular arcs or line segments.

- **Cusps And Bumps** — Select this check box when you want to preserve sharp angles in the original shape, as shown in the shape at the bottom right of Figure 33-13. With the Cusps and Bumps check box unchecked, fitting a spline smoothes out the angles, as illustrated by the shape at the top right of Figure 33-13.

FIGURE 33-13

Selecting the Cusps And Bumps check box keeps the sharp angles and other features of a shape while adding curves.

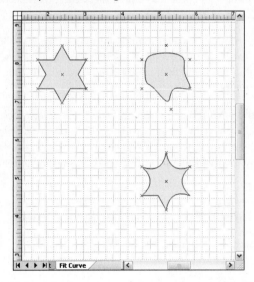

- **Error Tolerance** — To produce a simpler curve with fewer points, type a larger value in the Error Tolerance box. To create a curve that exactly fits the vertices of the selected path, use an error tolerance of zero.

Controlling Glue with Connection Points

Glue is one of the things that makes Visio drawings so easy to work with. When you glue shapes and connectors, you can move them around but they remain steadfastly connected to one another. If you've ever wondered what identifies the sticky spots on shapes and exactly how you can glue connectors and shapes to them, now you know. They're called connection points. Built-in shapes come with connection points in place, but you can add, move, and delete connection points to change the spots to which you can connect other shapes.

Connection points come in different types, which determine how shapes connect at those points. For example, you can use connection point types to ensure that the wiring on an electrical plan connects to electrical outlets the way it should in the real world.

Adding, Moving, and Deleting Connection Points

Although you can't modify built-in shapes on Microsoft stencils, you can add, move, or delete connection points on any shape on a drawing page. If you want to modify the connection points on a built-in Visio master, create a custom copy of the master and then edit the connection points the way you want.

Working with connection points requires the Connection Point tool, which lurks behind the more frequently used Connector tool. To activate the Connection Point tool, on the Standard toolbar, click the arrow next to the Connector tool and then choose Connection Point Tool. The basic techniques for working with connection points are as follows:

- **Add a connection point** — With the Connection Point tool activated, select the shape to which you want to add a connection point. Ctrl-click the selected shape at the position where you want the new connection point. As soon as you press Ctrl, Visio changes the pointer to blue crosshairs so that it's easier to position the connection point.

NOTE   Make sure that the shape is selected before trying to add a connection point. Visio creates connection points only for the selected shape, regardless where you click on the drawing page. And if no shape is selected, Ctrl-clicking won't do anything but waste your time.

- **Delete a connection point** — With the Connection Point tool activated, select a connection point on a shape. When the connection point turns magenta, indicating it is selected, press Delete.
- **Move a connection point** — With the Connection Point tool activated, select a connection point on a shape. When the connection point turns magenta, indicating it is selected, drag it to a new location.

Working with Types of Connection Points

For some drawing types, such as electrical plans or piping and instrumentation diagrams, shapes glue to each other instead of to linear connectors. In the systems that these diagrams represent, components connect in specific ways. For example, it's important that you connect the wiring to an electrical outlet the correct way or sparks will fly. Visio provides different types of connection points so you can control how shapes glue together.

Understanding Types of Connection Points

Visio provides three types of connection points to model the way components connect in the real world. Built-in Visio shapes already include the types of connection points needed to glue them properly. However, if you're creating your own shapes and solutions, it's important to understand how the three types work so that you can choose the right ones:

- **Inward connection point** — This type is the default connection point used almost exclusively on drawings in which connectors connect shapes, such as organization charts. Inward connection points connect to pretty much anything except other Inward connection points — they connect to end points of 1-D shapes (such as connectors), Outward connection points, and Inward & Outward connection points. Visio indicates Inward connection points with a blue X.

- **Outward connection point** — This type connects to Inward connection points, Inward & Outward connection points, shape geometry, and 1-D endpoints. For example, the Work Peninsula shape on the Cubicles stencil includes an Outward connection point. You can glue a Work Peninsula shape to the connection points on Work Surface or Corner Surface shapes, to extend those work areas. But you can't glue two Work Peninsula shapes together, because in real life, the resulting cubicle arrangement wouldn't have a leg to stand on. Visio indicates Outward connection points with a blue square.

- **Inward & Outward connection point** — This type connects to all types of connection points. For example, modular furniture on the Office Furniture stencil include Inward & Outward connection points so you can glue the components together in any order. The indicator for Inward & Outward connection points appropriately looks like a combination of the blue X of an Inward connection point and the blue square of an Outward connection point.

If shapes don't glue the way you would expect, you might be trying to glue the wrong connection points together. Refer to Table 33-2 for a quick reference of valid connection points.

TABLE 33-2

Valid Connections for Connection Points

	Inward Connection Points	Outward Connection Points	Inward and Outward Connection Points	Shape Geometry Points	1-D Shape End Points
Inward connection point	No	Yes	Yes	No	Yes
Outward connection point	Yes	No	Yes	Yes	Yes
Inward and Outward connection point	Yes	Yes	Yes	Yes	Yes

Changing the Type of Connection Point

By default, Visio adds Inward connection points when you use the Connection Point tool. However, you can modify the type of connection point so your shape glues to the points you want. You can change connection points on shapes on drawings or on masters in custom stencils. If you want to change the type of connection point for a built-in shape, create a custom version of the shape and then change the connection point. To change the type of connection points, follow these steps:

1. On the Standard toolbar, activate the Connection Point tool.

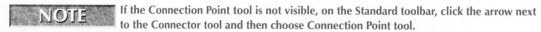 If the Connection Point tool is not visible, on the Standard toolbar, click the arrow next to the Connector tool and then choose Connection Point tool.

2. Right-click the connection point you want to modify. When it turns magenta and the shortcut menu appears, choose the type of connection point you want.

Customizing Shape Behavior

In addition to a shape's visible components and connection points, shapes come with dozens of settings that control how they behave. Whether a shape prints, how the shape resizes, what action launches when you double-click the shape, and how the shape interacts with other shapes when you place it on a drawing page are all behaviors that you can specify with shape behaviors. Whether you want to direct the behavior of shapes already on drawings or specify the default behaviors for masters in custom stencils, the Behavior dialog box (right-click a shape and, from the shortcut menu, choose Format ⇨ Behavior) is the first place to look.

CROSS-REF If the behavior you want to specify isn't available in the Behavior dialog box, you can tweak additional behaviors by modifying cells and formulas in a shape's ShapeSheet, as described in Chapter 34.

Specifying What Happens When You Double-Click a Shape

By default, when you double-click a shape, Visio opens the shape's text block so you can edit its text. That's great for most shapes, but completely useless for shapes that don't use text blocks as labels or obtain their text values from a linked database. You can pick from several other actions to run when you double-click a shape depending on what action you perform most often. For example, for groups, opening the group in a new window makes it easy to modify the group. In addition, to launch more complex actions, you can tell Visio to run a macro when you double-click a shape. To specify the action that occurs when you double-click a shape, follow these steps:

1. Right-click the shape you want to customize and, from the shortcut menu, choose Format ⇨ Behavior.

2. In the Behavior dialog box, select the Double-Click tab and choose one of the Double-Click options, described in Table 33-3.

3. Click OK. Don't forget to confirm that the behavior is what you want by double-clicking the shape.

TABLE 33-3

Double-Click Behaviors for Shapes

Double-Click Option	Resulting Behavior
Perform Default Action	Performs the default double-click action defined for the shape, such as opening the shape text block for editing.
Perform No Action	Does nothing when you double-click the shape.
Edit Shape's Text	Opens the shape's text block so you can edit its text.
Open Group in New Window	When you select a group, this option opens the group in the group editing window so that you can modify the group. This option is not available if you select a shape.
Open Shape's ShapeSheet	Opens the ShapeSheet for the shape.
Custom	This option performs the custom behavior defined in the shape's ShapeSheet in the EventDblClick cell (in the Events section). The option is always dimmed, because it sets itself when the EventDblClick cell contains a double-click formula.
Display Help	This option displays a help topic for the shape, which is ideal if you are building custom shapes to distribute to others and want to open a custom-built Help file. To specify a help topic, enter either *FILENAME!keyword* or *FILENAME!#number*, where *FILENAME* is the name of a .hlp or .chm Windows help file, *keyword* is a term associated with the help topic, and *number* is an ID referenced in the MAP section of the help project file.
OLE Verb	When you select a linked or embedded object, this option activates an OLE command, such as Open. This option is not available when you select a Visio shape.
Run Macro	Runs the macro or add-on that you select in the drop-down list. The list box includes your custom macros as well as built-in Visio tools, such as Database Update.
Go to Page	Jumps to the page number specified in the drop-down list. To open the page in a new window, check the Open in New Window check box.

NOTE   If you define a custom double-click formula for a shape in its ShapeSheet, you can overwrite the custom behavior by choosing a double-click behavior option other than Custom. However, if the custom formula is protected with the Guard function, Visio ignores the double-click option you choose and continues to apply the custom formula instead.

Specifying Placement Behavior

The Placement tab in the Behavior dialog box specifies how a 2-D shape reacts when you use the Layout and Routing tools. For drawings in which connections are key components, such as flowcharts and database models, built-in shapes come preconfigured to behave the way you would expect. However, you can specify whether Visio lays out and routes around a shape, and, if it does, what happens when you place the shape on a drawing page.

 Because Placement options apply only to 2-D shapes, they are not available in the Behavior dialog box when you select a 1-D shape.

To specify Placement behaviors for a shape, right-click a shape and, from the shortcut menu, choose Format ⇨ Behavior. Select the Placement tab and choose the Placement options you want, which are described in Table 33-4. Click OK.

TABLE 33-4

Placement Options for Shapes

Placement Option	Description
Placement Behavior	The default choice, Let Visio Decide, gives Visio the reins for deciding how to lay out a shape when you glue a Dynamic connector to it. You can also tell Visio to always lay out and route around a shape or to ignore the shape during layout.
Do Not Move During Placement	Prevents Visio from moving the shape during automatic layout. This option is helpful if you've carefully positioned a shape for reasons that Visio knows nothing about and you want the shape to remain where it is, no matter what.
Allow Other Shape to Be Placed on Top	Allows Visio to place other shapes on top of the shape during automatic layout. To make sure that every shape on a page is visible, clear this check box.
Move Shapes Away on Drop	Specifies whether the shape moves other shapes out of the way when you position it on the page. You can choose to keep other shapes where they are, always move other shapes away, or use the option specified in the Page Setup dialog box. This option overrides the Move Other Shapes Away on Drop option on the Layout and Routing tab in the Page Setup dialog box.
Do Not Allow Other Shapes to Move This Shape Away on Drop	Specifies whether the shape remains where it is when other shapes are dropped onto the page, regardless of which Move Other Shapes Away on Drop option you choose.
Route Through Horizontally	Specifies whether connectors can route through the shape horizontally. Clear this check box if you want Visio to route around the shape.
Route Through Vertically	Specifies whether connectors can route through the shape vertically. Uncheck this check box if you want Visio to route around the shape.

Modifying Other Shape Behaviors

The Behavior tab in the Behavior dialog box is the catch-all for other popular shape behaviors. For example, you can specify the following:

- Whether a shape acts like a line or a 2-D shape
- Whether connectors split when you drop shapes on them
- Which elements are highlighted when you select a shape
- How you can resize a shape
- How groups behave

Preventing Shapes from Printing

If you add reference shapes to a drawing, such as construction lines, you want to view them while you work on the drawing but you don't want them to print and ruin those award-winning architectural drawings. On the Behavior tab, under the Miscellaneous heading, select the Non-printing Shape check box and click OK.

Switching Between 1-D and 2-D Behavior

No matter what a shape looks like, you can make it behave like a 1-D line or a 2-D shape. For example, if you define a line as a 2-D shape, you drag its selection handles to turn it into a rectangle. Conversely, you can create a shape with 2-D graphics, such as the Fancy Arrow shape on the Basic Shapes stencil, and make it act like a line. You can change the thickness of the Fancy Arrow shape by dragging its side selection points, but you can glue only to its start and end points.

To specify whether a shape behaves like a 1-D or 2-D shape, on the Behavior tab, under Interaction Style, select the Line (1-dimensional) or Box (2-dimensional) option, and then click OK.

CAUTION Don't be cavalier about switching a shape's behavior from 1-D to 2-D or vice versa, because the switch changes the contents of the shape's ShapeSheet and can produce unexpected results. For example, it can break formulas you define in the ShapeSheet or glued connections between the shape and other shapes.

Showing That a Shape Is Selected

No doubt, you're used to seeing the alignment box, selection handles (shape handles), and control handles when you select a shape. By default, Visio displays all three, but you can choose to highlight different shape elements. On the Behavior tab, under the Selection Highlighting heading, check one or more of the following check boxes:

- **Show Shape Handles** — Select this check box to display the shape's selection handles. Clearing this check box is one way to prevent resizing a shape. Without shape handles, you can move the shape but can't resize it.
- **Show Control Handles** — Typically, you want to keep this check box selected so you can see the shape's control handles and easily access the features they provide.

> **CAUTION** If you clear the Show Control Handles check box, you can activate the editing features that the control handles provide, but, like a game of hide and seek, you must know where the control handles are without seeing them.

- **Show Alignment Box** — Select this check box to see the alignment box for the shape.

> **CAUTION** If you clear all the Selection Highlighting options, Visio provides no visual indication that the shape is selected and that is asking for trouble. Chances are you will make changes to the wrong shapes.

Controlling the Splitting of Connectors

If you work on a data flow diagram and realize that you've missed a process, the easiest solution is to drop a shape onto the connector where you want to insert the process, have Visio split the connector, and glue the two connector pieces to the new shape. You can specify exactly that behavior with the following options under the Connector Splitting heading in the Behavior dialog box:

- **Connector Can Be Split By Shapes** — When you select a connector, you can select this check box so that Visio splits the connector in two when you drop a shape (that is configured to split connectors) onto it.

- **Shape Can Split Connectors** — When you select a shape, you can select this check box so that Visio splits a connector (that is configured to split) when you drop the shape onto it.

> **CROSS-REF** For these splitting options to work, you must also select the Enable Connector Splitting option on the Layout and Routing tab of the Page Setup dialog box and on the General tab of the Options dialog box, as described in Chapter 5.

Controlling Group-related Behaviors

Although groups are comprised of shapes, the groups are entities with properties and behaviors of their own. For example, you can control how shapes are added to groups, how to select groups, and how to display group text and data. For shapes that belong to a group, you can control how the shapes resize when you resize the group.

On the Behavior tab, specify one or more of the following behaviors:

- **Resizing shapes in groups** — In a predefined office cubicle, you don't want work surfaces or partitions to resize, even if you enlarge the cubicle area, because the furniture shapes represent a scaled version of the furniture's actual size. Conversely, if you create a group of cells for a tabular form, you probably want the cells to change size as you increase or decrease the overall size of the table. To control how a shape resizes when it belongs to a group, under the Resize Behavior heading, choose one of the following options:

  - **Scale with Group** — Select this option to have the shape resize proportionally when you resize the group. This is helpful when the shapes don't represent real-world dimensions.

  - **Reposition Only** — Select this option to have the shape remain the same size when you resize the group. For example, if you create a group to represent a kitchen, you can set the appliances to Reposition Only so you can rearrange them within the kitchen, but not change their dimensions.

- **Use Group's Setting** — Select this option to have the shape resized based on the Resize behavior defined for the group.

- **Adding shapes to groups** — You can always add shapes to groups by selecting the shape and group and choosing Shape ➪ Grouping ➪ Group. However, with behavior options, you can use a drag and drop shortcut for adding shapes to groups. To control the methods you can use to add shapes to groups, use the following options:

  - **Add Shape to Groups on Drop** — Under the Miscellaneous heading, select this check box to add a shape to a group (that accepts dropped shapes) when you drop the shape on top of the group on the drawing page. To ensure that a shape remains separate no matter what option is set for a group, clear this check box.

  - **Accept Dropped Shapes** — Under the Group Behavior heading, select this check box if you want a group to absorb a dropped shape (whose Add Shape To Groups On Drop check box is selected). For self-contained groups, such as factory-configured equipment, clear this check box to prevent shapes from joining the group.

- **Selecting shapes and groups** — In some groups, such as office workstations, the group is the important element. You're more likely to select the group to arrange your office layout. For other groups, such as a title block, you usually want to select the shapes within the group to add the text to identify your drawing. To specify whether clicking selects the group or a shape within the group, choose one of the following options in the Selection drop-down list:

  - **Group Only** — Select this option to prevent the selection of shapes within a group. By doing so, you can edit the shapes within the group only when you ungroup them.

  - **Group First** — Select this option to select the group the first time you click and then subselect the shape within the group upon subsequent clicks.

  - **Members First** — Select this option to select the shape within a group the first time you click and then select the group with the next click.

- **Snapping in groups** — If you want to snap and glue to shapes in a group, such as connecting cables to the Equipment shapes within a Cubicle group, under the Group Behavior heading, select the Snap to Member Shapes check box.

- **Displaying and editing text** — To specify the display and editing of group text, choose one of the following:

  - **Edit Text of Group** — Under the Group Behavior heading, select this check box to add text for the group. For example, you can use the group text box to show the catalog number for a standard office workstation.

  - **Hide** — In the Group Data drop-down list, select this option to hide the group text box.

  - **Behind Member Shapes** — In the Group Data drop-down list, select this option to display the group text behind the shape text.

  - **In Front of Member Shapes** — In the Group Data drop-down list, select this option to display the group text in front of the shape text.

Adding Screen Tips to Shapes

ScreenTips are a great way to provide help or additional information about shapes. They appear when you position the pointer over shapes and remain out of view otherwise. Use one of the following methods to add or edit ScreenTips:

- **Add a ScreenTip** — Select the shape to which you want to add a ScreenTip and then choose Insert ➪ Shape ScreenTip. Type the text for the ScreenTip and click OK.

- **Edit a ScreenTip** — Select the shape and choose Insert ➪ Edit Shape ScreenTip. Edit the text the way you want and then click OK.

- **Delete a ScreenTip** — Select the shape and choose Insert ➪ Edit Shape ScreenTip. Delete the text in the Shape ScreenTip dialog box and click OK.

Customizing Shape Data

In Chapter 10, you learned about all sorts of ways to represent data in Visio drawings. Whether that information comes from external databases or is stored directly in shapes, shape data is the data repository in Visio. Using shape data, you can search for shapes, add or review shape data values as you work, annotate shapes, or produce reports. Built-in masters often have shape data properties already assigned. Because Visio is built to satisfy a broad audience, chances are good that your organization might track more or fewer fields than you find on the built-in shapes. You can add or modify the shape data properties assigned to shapes.

CROSS-REF With formulas in a ShapeSheet and values from shape data, you can even configure the appearance of a shape. For example, the Pie Chart shape includes a Slices shape data property. By typing a number in that field, you create or modify a Pie Chart shape to contain that number of slices. To learn how to configure a shape using shape data and the ShapeSheet, see Chapter 34.

In many situations, all the masters and shapes for a specific type of drawing use the same set of fields. In Visio, shape data sets contain one or more shape data properties that you can apply all a once. For example, many of the shapes in the Business Process stencils include Duration, Cost, and Resources properties so you can track process statistics. In the Shape Set window, you can create, rename, or delete shape sets and assign shape data to those sets. You can also apply the properties in shape sets to all the shapes you want at one time.

Creating and Editing Shape Data

Regardless what you plan to do with shape data, the first step is to create shape data properties and associate them with your masters or shapes. Sure, you can add shape data to individual shapes on a drawing, but it's much more efficient to apply shape data properties to masters. Whenever you drag a master onto the drawing page, the instance you create has the shape data properties already assigned.

> **NOTE** When you work with shape data for masters, the stencil in which you make your changes determines the scope of your changes. If you want every instance you create in the future to incorporate your changes, edit the master in the Visio stencil. However, if you only want to change the instances on your current drawing, edit the master in the Document stencil for that drawing.

Whether you create fields individually or define them through shape data sets, the real work occurs in the Define Shape Data dialog box, shown in Figure 33-14. The easiest way to open the Define Shape Data dialog box is from within the Shape Data Window: Right-click within the window and, from the shortcut menu, choose Define Shape Data.

FIGURE 33-14

The Define Shape Data dialog box includes settings to define a new field and list the fields already assigned to the selected shape or master.

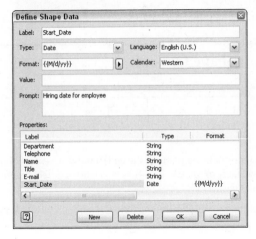

There are a couple of other ways to open the Define Shape Data dialog box, depending on the shape that is selected:

- **Shape with existing shape data** — If the selected shape has existing fields of shape data, right-click the shape and, from the shortcut menu, choose Data ➪ Shape Data. In the Shape Data dialog box, click Define to open the Define Shape Data dialog box, which lists the fields currently assigned to the selected shape.

- **Shape without existing shape data** — If you right-click a shape and choose Data ➪ Shape Data, you might see a message that states that the shape has no shape data and asks if you want to define data. Click Yes to open the Define Shape Data dialog box.

- **Master in a stencil** — To add shape data to a master in the Document stencil or any editable stencil, in the stencil window, right-click the master and choose Edit Master ⇨ Edit Master Shape. In the master drawing window, right-click the master in the master drawing window and, from the shortcut menu, choose Data ⇨ Shape Data. In the Shape Data dialog box, click Define. The Define Shape Data dialog box lists the fields currently assigned to the master.

- **Drawing page** — Click any blank area of the page to ensure that nothing is selected. Right-click the blank area and, from the shortcut menu, choose Data ⇨ Shape Data. Depending on whether the page has shape data defined, either click Yes to create data or click Define.

With the Define Shape Data dialog box open, you have access to the settings for the fields you might add to shapes, as well as the fields that already apply to the selected shape. Table 33-5 lists the settings you can specify to define shape data properties.

TABLE 33-5

Shape Options

Option	Description
Label	The label is the name of the shape data property, which appears in the Shape Data window to the left of the cell in which you enter the field value. This field consists of alphanumeric characters and underscore characters.
Name	The Name field represents the name of the cell in the ShapeSheet and is visible only if you run in Developer mode (choose Tools ⇨ Options, select the Advanced tab, and select Run In Developer Mode check box).
Type	The data type for the property, including String, Number, Fixed List, Variable List, Boolean, Currency, Date, and Duration. The most flexible choice is String.
Language	Specifies the language to correctly format the date and time when you create a Date property. For example, English (U.S.) uses mm/dd/yy, whereas English (U.K.) uses dd/mm/yy.
Calendar	Specifies whether to use the Arabic Hijri (Islamic), Hebrew Lunar (Jewish), Saka Era (Hindu), or Western (Gregorian) calendar to convert a date entered in a Date field.
Format	The format for the data type. The options available depend on the Type and Calendar options selected. You can select from lists of predefined formats when you define data types such as String, Number, Fixed List, Variable List, Currency, Boolean, Date, and Duration. To specify fixed lists or variable lists, type each item in the list separated by semicolons. For example, you can create a color list by entering **red;white;blue**. If you create a fixed list, you can only select one of the entries on the list. With variable lists, you can enter another value, such as **green**.
Value	Specifies the initial value for the property. For existing properties, this box shows the current value. Omit this value if you want the property to be blank initially.

Option	Description
Prompt	Specifies text that appears when you select the property in the Shape Data dialog box or, in the Shape Data window, pause the pointer over the shape label. You can use the prompt to describe the field or provide instructions on its use.
Properties	Displays the existing shape data properties for the selected shape. When you select a property in the list and modify it, Visio applies the changes you make to its definition immediately.
New	Creates a new shape data property.
Delete	Deletes the shape data property selected in the Properties list.

NOTE In Developer mode, you can also do the following: specify the order in which shape data appear in the Shape data dialog box and Shape data window, hide the property, or prompt users to enter shape information when they create, duplicate, or copy shapes.

To define, edit, or delete shape data for a shape or master, select the shape or master and open the Define Shape Data dialog box, as described earlier in this section. Then, choose one of the following tasks:

■ **Create a property** — If the shape has no shape data assigned, the Label field is filled in automatically with Property1. Simply begin specifying the property settings. If the shape already has at least one shape data property, click New and then specify the settings.

TIP In the Define Shape data dialog box, you can't copy properties from one shape to another. To reuse shape data, add them to shape sets and apply those sets to your shapes and masters.

■ **Edit a property** — In the Properties list, select the property you want to edit and then make the changes you want to the settings.

■ **Delete a property** — In the Properties list, select the property you want to delete and click Delete.

CAUTION If a shape has predefined shape data, deleting them might affect the shape's behavior.

After you complete your work with that shape's shape data properties, click OK to close the dialog box. If you worked with properties for a shape, you're finished. If you added, edited, or deleted shape data for a master, click the Close button in the master drawing window. When Visio prompts to update the master, click Yes and then save the stencil.

NOTE If you edit a master in the Document stencil, Visio will update the master as well as all the shapes in the drawing file based on that master.

Using Shape Sets

Most of the time, you want to apply the same set of shape data properties to multiple shapes or masters. Shape data sets are an easy way to work with a group of shape data properties and apply them to multiple shapes. After you create and name a shape set, you can add the properties you want to it or modify the definitions of its properties. When you have configured the shape set the way you want, you can apply the collection of properties in the set to the selected shapes on a drawing page, the Document stencil, or any open editable stencil.

Shape sets also simplify adding shape data properties to shapes that weren't created by dragging masters from stencils. For example, built-in Workstation shapes on the Cubicles stencil include a dozen different shape data properties. You can create a shape data set from the properties associated with a built-in Workstation shape. If you import additional Workstation shapes or create you own, you can easily associate those shape data properties to your new shapes by applying the shape set to your selected Workstation shapes or masters.

Creating Shape Sets

You can create shape sets from scratch, but it's often easier to create them from the properties associated with a shape, or to expand on a shape set that already exists. For example, you can create a new shape data set based on the properties assigned to TQM shapes: Cost, Duration, and Resources. You might add Work and Department properties and then apply the new shape data set to all the masters in your custom TQM Shapes stencil.

To create a new shape set, follow these steps:

1. If you intend to create a shape data set based on the properties associated with a shape or master, select the shape on a drawing page or a master in an editable stencil.

2. Right-click the Shape Data Window and, from the shortcut menu, choose Shape Data Sets.

3. In the Shape Data Sets window, click Add to open the Add Shape Data Set dialog box, shown in Figure 33-15.

4. In the Name box, type a name for the set of properties. For example, you might name the set after the type of drawing to which it applies.

5. Select an option to specify whether you want to define properties from scratch or start with properties from another set. The easiest approach is to choose the Create A New Set From The Shape Selected In Visio option. If your set applies to shape on a drawing or stencil, choose Create A New Set From An Existing Set and then in the drop-down list, choose the set.

6. Click OK to save the new shape data set and store it in the current drawing or stencil. To add properties to a set, modify properties in a set, or apply a set to shapes, see the following sections.

FIGURE 33-15

From the Shape Data Sets window, you can add or edit shape data sets.

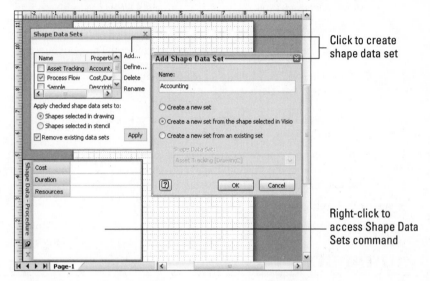

Click to create
shape data set

Right-click to
access Shape Data
Sets command

Editing Shape Sets

Within the Shape Data Sets window, shown in Figure 33-15, you can add or modify the properties in a shape data set, as well as rename or delete existing shape data sets. To make changes to shape data sets, use one of the following methods:

- **Rename a shape data set** — In the Shape Data Sets window, select the set you want to rename in the list and click Rename. Type the new name and press Enter.

- **Delete a shape data set** — In the Shape Data Sets window, select the set you want to delete in the list and click Delete.

- **Modify the properties in a shape data set** — In the Shape Data Sets window, select the set whose properties you want to modify and click Define. In the Define Shape Data dialog box, create new properties, modify existing property definitions, or delete properties, as described in the section "Customizing Shape Data" earlier in this chapter. Click OK when you're done.

Applying Shape Sets to Shapes and Masters

You can apply shape sets to different shapes and masters depending on where you save the shape data sets. When you create a shape data set in a drawing, the set is saved in the drawing file and you can apply the set to shapes on the drawing pages or to masters in the Document stencil. However, if you create a shape data set in a stencil, you can apply the shape data set only to the masters in the stencil.

To apply shape data sets to shapes or masters, follow these steps:

1. Select the shape(s) or master(s) to which you want to apply the shape data set using one of these methods:

   ▪ **Select shapes** — On a drawing page, select the shapes you want.

   ▪ **Select masters** — In an editable stencil, select the master you want.

 To select additional shapes or masters, Shift-click the other shapes or masters you want to add to the selection.

2. In the Shape Data Sets window, select the check boxes for all of the shape data sets that you want to apply.

3. Choose an option to either apply the set to selected shapes or selected masters. If you want to remove the existing shape sets from the selected shapes or masters, make sure the Remove Existing Property Sets check box is selected.

4. Click Apply.

Summary

You can customize shapes on your drawings in numerous ways or create brand-new shapes using Visio drawing tools and Shape Operation commands. To reuse customized shapes, save them as masters on custom stencils so you can drag and drop them onto drawing pages just like built-in shapes. In this chapter, you learned the concepts that make shapes and masters so smart. You also learned how to perform the following actions:

- Create masters
- Modify master graphics and properties
- Modify master icons
- Use drawing tools and Shape Operation commands to create or edit shapes and masters
- Configure connection points to control how shapes glue together
- Specify shape behavior, including what shapes do when you double-click them, how layout and routing lays them out, and whether shapes act like 1-D or 2-D shapes
- Create and apply shape data to shapes and masters

In the next chapter, you learn how to further customize your shapes by modifying properties and creating formulas in ShapeSheets.

Chapter 34

Customizing Shapes Using ShapeSheets

Throughout this book, you'll run across tips and techniques that talk about making changes in a shape's ShapeSheet when there's no way to make a shape do what you want from a drawing page. In reality, shapes are collections of fields that control every aspect of shape appearance and behavior. The shapes you see on a drawing page are graphical representations of the Visio shapes. Visio commands and dialog boxes are simply convenient interfaces to these fields from the graphical representations on a drawing page. Alternatively, the *ShapeSheet* in Visio provides access to these same fields in a spreadsheet-like view. The values and formulas in the fields you see in the ShapeSheet specify every line, arc, and text block of a shape; how the shape looks, and how it behaves.

So, a shape on a drawing page and its ShapeSheet are two views of the same Visio shape. Whether you modify a shape on a drawing page or in a ShapeSheet, Visio changes values in the underlying shape definition.

Of course, not every shape requires a value for every field. For example, if a shape doesn't use connection points, the ShapeSheet doesn't waste space by filling in fields with connection point attributes. Even so, Visio needs a significant number of fields to define everything about shape appearance and behavior. If a field doesn't contain a value, Visio fills in the field with a default value or the value from the shape's master.

The ShapeSheet is organized in tabular sections so it's easier to target the section you want to modify and locate the field you want to change. For example, the Shape Transform section specifies attributes for the location, size, and rotation for a shape, whereas the Fill Format section specifies attributes for filled areas.

This chapter acts as an introduction to ShapeSheets. It explains the different sections of the ShapeSheet and describes how to modify the value in a field. The chapter also shows you how to create a simple ShapeSheet formula.

 **CROSS-REF** To learn how to perform more advanced tasks with ShapeSheets, consider reading the book *Visio 2003 Developer's Survival Pack* by Graham Wideman, Trafford Publishing, 2003.

Viewing ShapeSheets

Because a ShapeSheet is another view of a shape, it opens in a separate window, as shown in Figure 34-1. For the casual customizer, you can open every ShapeSheet in the same ShapeSheet window. For more serious customizing, opening a new window for each ShapeSheet might be more to your liking. ShapeSheets have settings for specifying which sections of the ShapeSheet appear in the ShapeSheet window, which is handy when you want to focus on a few key sections without slogging through the ones you don't care about. If you want to customize other aspects of a shape, you can insert sections that aren't currently used but applicable to the type of shape.

FIGURE 34-1

A shape on a drawing page and the ShapeSheet are different views of the same Visio shape.

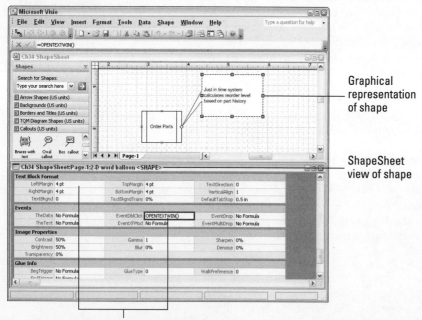

Opening ShapeSheets

Every item on a Visio drawing page has a ShapeSheet associated with it—from shapes and masters to guides and pages themselves. To open a ShapeSheet, select the item and then choose Window ➪ Show ShapeSheet. In the ShapeSheet window, the title bar identifies the Visio drawing file, the drawing page, and the item represented in the window. For example, the title bar Ch34 ShapeSheet: Page-1:2-D word balloon <SHAPE> tells you that the drawing file is Ch 34 ShapeSheet; the page is Page-1; and the selected item is a shape using the 2-D word balloon master.

To open a ShapeSheet, use one of the following methods:

- **Shape, group, guide, guidepoint, or OLE object**—Click the element to select it, and then choose Window ➪ Show ShapeSheet.

- **Shape in a group**—Subselect the shape by clicking once to select the group, clicking again to select the shape, and then choosing Window ➪ Show ShapeSheet to open the ShapeSheet for the shape.

- **Master**—Open the master's stencil for editing by right-clicking the stencil's title bar and, from the shortcut menu, choosing Edit Stencil. Right-click the master and, from the shortcut menu, choose Edit Master. Select the master in the master drawing window and choose Window ➪ Show ShapeSheet. For a grouped shape, you can select the shape by right-clicking the shape in the Master Explorer.

- **Page**—To show a page ShapeSheet, click an empty area of the drawing page to make sure no shapes are selected and then choose Window ➪ Show ShapeSheet.

TIP If you access ShapeSheets frequently, you can display the Show ShapeSheet command automatically on every shape's shortcut menu (displayed when you right-click the shape). Choose Tools ➪ Options, select the Advanced tab, and select the Run in Developer Mode check box.

Depending on your enthusiasm for ShapeSheets, you can open one ShapeSheet window and swap the shapes you work on, or you can open separate windows for each shape. Opening separate ShapeSheet windows is ideal when you want to compare values for more than one shape. If screen real estate is at a premium, using one ShapeSheet window consumes less screen area. To specify how you want to manage ShapeSheet windows, follow these steps:

1. Choose Tools ➪ Options and select the Advanced tab.

2. To use only one ShapeSheet window, select the Open Each ShapeSheet In The Same Window check box. Leaving this check box cleared opens a new window for each ShapeSheet you view.

Viewing and Adding ShapeSheet Sections

When you consider all the things you can do with Visio shapes, it should come as no surprise that Visio ShapeSheets contain hundreds of fields. Visio attempts to limit information overload by grouping fields into sections. Even so, there are 30 different sections that a shape might use, depending on its configuration and features. For example, one-dimensional shapes include the 1-D Endpoints section, whereas only shapes with control points use the Controls section. Visio automatically adds the sections required to construct a shape to its ShapeSheet. As you work, you can show or hide those sections or add others. Table 34-1 shows some of the more commonly used ShapeSheet sections.

TABLE 34-1

Commonly Used ShapeSheet Sections

Section	Purpose
1-D Endpoints	Specifies the coordinates of each end point on a 1-D shape.
Shape Transform	Specifies the dimensions of the shape, the coordinates of its position on the drawing page, rotation, pin position, and more.
User-defined Cells	Includes custom values or formulas, such as keywords used for searching shapes and scaling formulas. For example, formulas and attributes for employee pictures in Organization Chart shapes are stored in this section.
Shape Data	Includes the labels and values for the shape data properties associated with a shape (formerly known as custom properties).
Geometry	Each path in a shape has its own Geometry section. Each Geometry section contains coordinates for each vertex in a path and specifies attributes such as whether shapes can be filled.
Protection	Specifies lock settings set with Format ⇨ Protection. This section also includes a few locking options available only in the ShapeSheet, such as LockGroup.
Miscellaneous	Includes fields for controlling various shape behaviors, some of which are available by choosing Format ⇨ Behavior, such as with control handles, selection handles, and the alignment box appear when the shape is selected.
Group Properties	Specifies group behaviors, such as whether a group accepts a dropped shape.
Line Format	Includes settings for lines and line ends.
Fill Format	Includes settings for fill formats.
Paragraph	Specifies paragraph formatting as defined on the Paragraph tab after choosing Format ⇨ Text.
Text Block Format	Specifies text block formatting as defined on the Text Block tab after choosing Format ⇨ Text.
Text Transform	Specifies size and position of a shape's text block.
Events	Defines behavior in response to specific events, such as double-clicking or dropping the shape onto the drawing page.
Shape Layout	Specifies settings for layout features as defined on the Layout and Routing tab of the Page Setup dialog box, as well as Placement Behavior settings.

To hide or show the sections in a ShapeSheet, right-click within the ShapeSheet window (not the title bar) and, from the shortcut menu, choose View Sections. In the View Sections dialog box, select a Section check box to display the section; clear a Section check box to hide it, as demonstrated in Figure 34-2. To quickly modify the sections that appear, click All or None and then select and clear sections you want or don't want.

FIGURE 34-2

Select or clear Section check boxes to hide or show ShapeSheet sections.

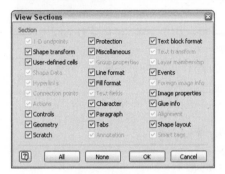

In the View Sections dialog box, the check boxes and labels for sections not currently in use are dimmed. To insert a section that isn't in use, right-click the ShapeSheet window and, from the shortcut menu, choose Insert Section. The Insert Section dialog box opens, showing the sections you can insert. Sections that don't apply to the selected shape or other item are dimmed. Select the check box for each section you want to insert and then click OK.

Printing ShapeSheets

When a ShapeSheet window is active, you might notice that the File menu does not offer a Print command. That's no great loss, because the rows and rows of fields in a ShapeSheet would consume paper faster than a dog eating steak off a kitchen counter. The Visio 2007 SDK (software development kit) includes the Print ShapeSheet Tool program, which simplifies transferring ShapeSheet information to other programs, files, or printers. Perhaps the best way to view and print ShapeSheet information is to copy it to the Clipboard and then paste it into Excel. You can use Excel features to sort the ShapeSheet information or search for what you want.

To download the SDK, navigate to the Visio page of the MSDN web site (`http://msdn .microsoft.com/office/program/visio/`) and look for the link to the Visio 2007 SDK. Download and install the SDK on your computer and then follow these steps to use the Print ShapeSheet Tool:

1. Select a drawing window (not a ShapeSheet window) and choose Tools ⇨ Add-ons ⇨ SDK ⇨ Tools ⇨ Print ShapeSheet to launch the Print ShapeSheet tool.

2. In the Print ShapeSheet dialog box, select the ShapeSheet type (Document, Styles, Page, All Shapes).

3. In the Send To drop-down list, choose whether you want to send the ShapeSheets to the Clipboard, a file, or a printer. Because you can't select the ShapeSheet for only one shape, it's a good idea to transfer the data to the Clipboard. Then, you can paste the data into another program, such as Excel, which gives you total control over what you print.

4. Select the check boxes for the ShapeSheet sections you want to print.

5. Click OK.

6. If you chose to send the data to the Clipboard, open Excel.

7. In an empty worksheet, choose Paste or press Ctrl+V. The ShapeSheet information pastes into the first three columns of the worksheet.

8. Using Excel techniques, select the rows you want to print and then use the Excel Print function to print the ShapeSheet information.

Exploring ShapeSheet Sections

A Visio ShapeSheet is a powerful tool that controls every aspect of a shape. Much like with other powerful tools (chain saws come to mind), using one without knowing what you're doing can be disastrous. With 30 different ShapeSheet sections and numerous fields within each section, this chapter can't come close to covering every possible ShapeSheet field. Instead, it highlights fields that are particularly useful and provides examples of their use.

CROSS-REF   For a complete reference to the purpose of each section and field, peruse the Microsoft Office Visio ShapeSheet Reference. The quickest way to find out what a ShapeSheet field does is to select the cell that contains the field value and press F1. The Visio Help window opens right to the ShapeSheet Reference topic for that field.

To really study ShapeSheet fields — or when you have trouble sleeping at night — open the entire Microsoft Office Visio ShapeSheet Reference. Choose Help ➪ Microsoft Office Visio Help. In the bar near the top of the Help window, click the drop-down arrow next to Search. From the drop-down menu, choose Visio ShapeSheet Help. Click ShapeSheet Reference.

TIP   You can also learn a lot about ShapeSheet cells and formulas by looking at the settings for some of the built-in Visio shapes. Select a built-in shape that you have dragged onto the drawing page and open its ShapeSheet.

Here are some examples of ShapeSheet cells and what you might do with them:

- **Shape Transform** — Controls the position and orientation of a shape:
    - **Width** — Use a formula to set a shape's width based on the size of the picture it contains, as in the User.Width cell for an Organization Chart shape.

■ **Height** — Link this cell to a database field to define a shape's height with a dimension stored in a parts database.

CROSS-REF To learn about linking ShapeSheet cells to fields in external data sources, see Chapter 10.

■ **LocPinX and LocPinY** — Use formulas to position the center of rotation of a shape based on the shape's width and height. By default, the pin is located at the center of a shape, but you can position it wherever you want.

■ **Angle** — Use a formula to constrain a shape's rotation to multiples of ninety degrees, as the Balloon Vertical Callout shape does.

■ **Geometry** — Each path in a shape has its own Geometry section, which contains coordinates for each vertex in the path, and attributes that control the path's appearance:

■ **MoveTo** — Use a reference to the ShapeSheet fields for control handle coordinates to start a line at the location of a control handle, similar to the behavior of Dimensioning shapes.

■ **LineTo** — Use a reference to other fields in the Geometry section to intersect one line with another, also illustrated by Dimensioning shapes.

■ **EllipticalArcTo** — If you draw several lines and arcs with the Pencil tool and accidentally create an arc when you wanted a straight line, you can correct the mistake in the ShapeSheet. In the Geometry section for the Pencil tool path, it's easy to spot arcs even if they look like straight lines on the screen, as shown in Figure 34-3. Click an EllipticalArcTo cell to highlight the vertex on the drawing page. When the correct vertex is highlighted, right-click the EllipticalArcTo cell and, from the shortcut menu, choose Change Row Type. For arcs, the row type is set to EllipticalArcTo. Select LineTo and click OK to draw a straight line instead.

■ **Infinite Line** — You can convert a line drawn on the drawing page to an infinite line, which acts like a true construction line. In the Geometry section, right-click the LineTo cell for the line and then, from the shortcut menu, choose Change Row Type. Select Infinite Line to create a construction line.

■ **Geometry.NoFill** — By default, Visio changes the NoFill value from TRUE to FALSE when you close a path, which applies a fill format to the now-closed shape. If you use Shape Operation commands to create compound shapes, you can set the NoFill fields for each Geometry section to TRUE or FALSE, depending on whether you want that geometry to accept fill formatting. For example, if you transform a checkerboard of squares into one shape, you can set alternating squares to NoFill = TRUE. When you apply fill formatting to the shape, half the squares apply the fill format.

FIGURE 34-3

Change arcs to straight lines and vice versa within a ShapeSheet.

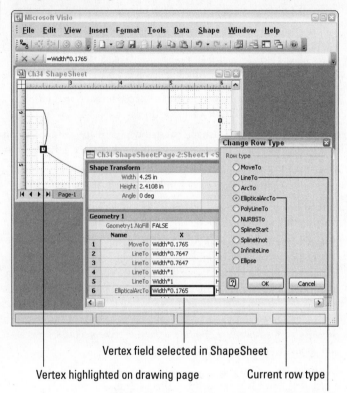

Vertex field selected in ShapeSheet

Vertex highlighted on drawing page

Current row type

Select LineTo to convert segment to straight line

- **Miscellaneous** — This section sets various attributes, including handle selection and visibility. Many of these attributes can be modified by choosing options on the Behavior tab in the Behavior dialog box:

  - **NoObjHandles** — To ensure that a shape is not resized, hide its selection handles by setting this field to TRUE.

  - **Comment** — Use logical functions in formulas along with field references to shape data to display comments that vary based on shape data values.

  - **NonPrinting** — Use logical functions and field references so that a shape that includes an infinite line doesn't print.

  - **NoAlignBox** — To prevent someone from snapping to a shape's alignment box, hide the alignment box by setting NoAlignBox to TRUE.

  - **DropOnPageScale** — Use formulas to specify scaling factors based on the scale for the drawing page.

- **Fill Format** — This section specifies fill formatting and shadows:
  - **FillForegnd** — Link this field to a database field — for example, to show a Chair shape in the color specified by a catalog database.
  - **FillBkgnd** — Use the RGB function to specify the exact color you want, such as FillBkgnd = RGB(250,150,100) to create a sepia tone for a pattern background.
- **Group Properties** — This section specifies behaviors and other attributes for groups, not the individual shapes within them:
  - **DontMoveChildren** — If you can't move a shape that you subselect within a group, in the group's ShapeSheet, change the value in this field to FALSE.
  - **IsTextEditTarget** — Use formulas and field references to make the group's text block editable when the grouped shapes already include text.
- **Protection** — This section includes fields for locking different aspects of shapes (some of these can only be modified in the ShapeSheet):
  - **LockCustProp** — Locks shape data properties so they can't be edited.
  - **LockCrop** — Locks an OLE object to prevent it from being cropped with the Crop tool.
  - **LockGroup** — Locks a group so it can't be ungrouped. If you drop a built-in master onto a drawing page and can't move a shape within it, change this field value to 1 in the group's ShapeSheet to unlock the group.
  - **LockCalcWH** — Locks a shape's selection rectangle so it doesn't change when vertices are edited.
  - **LockVtxEdit** — Locks shape vertices so you can't edit them with drawing tools.
- **Actions** — This section defines menu items on a shape's shortcut menu or a shape's SmartTags.
  - **Action** — Type the formula to execute when a user chooses the command from the shortcut or SmartTag menu, as demonstrated in Figure 34-4.
  - **Menu** — Specify the name of a menu item on a shortcut or SmartTag menu.
  - **Checked** — Indicates whether an item is selected.
  - **Disabled** — Indicates whether an item is dimmed.
- **Events** — This section specifics the actions to take when events occur:
  - **EventDblClick** — Specify a formula to execute when a user double-clicks a shape, such as editing the shape text block, as most shapes do by default.
  - **EventDrop** — Specify a formula to execute when a user drops a shape onto the drawing page.

FIGURE 34-4

Add shortcut commands in the ShapeSheet.

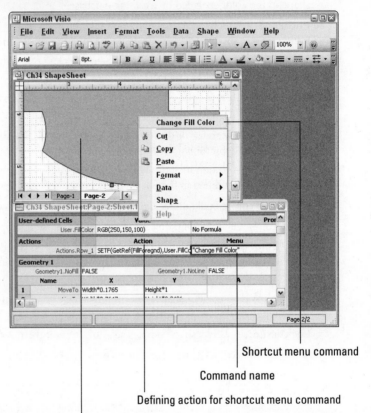

Shortcut menu command

Command name

Defining action for shortcut menu command

Result of choosing shortcut menu command

- **Shape data** — This section defines the shape data associated with a shape, as well as attributes of each property, such as language or sort order:
  - **Ask** — If you want Visio to query users to enter shape data when they drop a shape on a drawing page, set this field to TRUE.
  - **SortKey** — Influences the order in which items are listed in the Shape Data window and dialog box.
  - **Type** — Specifies the data type for the property.
- **User-defined cells** — You can specify formulas that you can reference from other cells. You can name the fields in this section so that references to them are meaningful. The fields in the Scratch section can also contain user-defined formulas, but you can't name Scratch fields.

- **Prompt** — Specify the prompt for the User-defined field.
- **Value** — Specify a formula. For example, Visio Organization Shapes include User-defined fields to specify employee photo attributes. Built-in scaled objects often include an anti-scaling User-defined field.

Writing ShapeSheet Formulas

To Visio, anything you type in a ShapeSheet field is a formula, whether it's a numeric value, a reference to another field, or a full-blown formula with functions and operators. You can write formulas for ShapeSheet fields to define shape behavior. For example, you can specify that the height of a shape is always half its width by writing a formula in the Height field in the Shape Transform section. Visio evaluates formulas and shows the results in units appropriate for the field. You can display field contents in a ShapeSheet window as either formulas or values.

Visio creates many formulas by default when you create a shape. For example, Visio positions a shape's pin at the shape's center by default. If you look in a ShapeSheet, you'll see the following formulas:

```
LocPinX = Width*0.5
LocPinY = Height*0.5
```

To conserve file space and simplify the propagation of changes from masters to instances, Visio instances inherit formulas from masters whenever possible. Shape instances inherit formulas from their masters on the Document stencil and inherit formatting from style definitions stored with the drawing.

If you want to customize a shape on a drawing page, you can write local formulas in any field in that shape's ShapeSheet. A local formula replaces the formula inherited from the master and is shown in blue text in the ShapeSheet window. If you change the formula for that field in the master, the local formula in the instance's field prevents the shape from inheriting the master's formula. To restore the value from the master, just delete the contents of the cell.

Learning More About Programming Shapes

Programming shapes and automating Visio provides enough material to produce a second volume to this book. However, if you want to learn more about how to use ShapeSheets to program shapes, you can find resources on the Microsoft Developer Network web site at http://msdn.microsoft.com/visio. This site includes articles about Visio shape programming, Visio developers' documentation, as well as links to books, training resources, and much more. To find specific information, enter keywords in the Search For box on the site.

Visio automatically updates some fields when you change a shape. When you apply a style to a shape, Visio deletes local formulas in the related fields unless you select the Preserve Local Formatting check box in the Style dialog box. If you want to prevent local formatting formulas from being overwritten, you can use the GUARD function.

Exploring the Elements of Formulas

Visio formulas are similar to Excel formulas. They always start with an equals sign, although Visio inserts the equals sign automatically. A formula can comprise any of the following elements:

- Numbers
- Coordinates
- Boolean values
- Operators
- Functions
- Strings
- Cell references
- Units of measure

NOTE   The ShapeSheet Reference explains how each element works and how to include it in formulas. Choose Help ➪ Microsoft Office Visio Help. In the bar near the top of the Help window, click the drop-down arrow next to Search. From the drop-down menu, choose Visio ShapeSheet Help. Click ShapeSheet Reference.

Creating Formulas

If you're familiar with creating formulas in an Excel spreadsheet, you'll find that ShapeSheet formulas work much the same way. Select a field cell and type a formula; or double-click a field cell to display the insertion point so that you can edit the formula it contains. However, if you are developing a long formula, it's easiest to select the cell and then edit the formula in the formula bar.

To create or edit a formula in the formula bar, follow these steps:

1. In the ShapeSheet window, click a field cell to select it and display the formula in the formula bar.

2. Type the formula, for example to calculate the position of a shape's pin, as shown here.

   `LocPinX = GUARD(Width*0.5)`

3. To include a reference to another cell in the formula, type the name of the cell, such as Width in the LocPinX formula.

4. To include a function, type the function name and then type its parameters in parentheses, as in the GUARD function in the LocPinX formula. To choose the function you want from a list, position the insertion point where you want to insert the function in the formula and then choose Insert ➪ Function. Select the function you want and click OK.

> **TIP** The GUARD function is used frequently to prevent a formula from being modified by actions performed on the drawing page, such as moving or resizing. Fields often affected by moving and resizing are Width, Height, PinX, and PinY.

5. To accept the formula, to the left of the formula bar, click the Accept button; or press Enter.

6. If the formula contains an error, Visio displays an error message box. Click OK in the message box. Visio highlights the area, if not the exact location, of the error in the formula.

7. After you correct the error, click Accept or press Enter.

> **TIP** If you want to add shapes to a scaled drawing but don't want the shapes to scale, you can define a formula in the ShapeSheet to prevent the shape from scaling. You can read an article about scaling formulas at `http://msdn.microsoft.com/library/default.asp?url=/library/en-us/devref/HTML/DVS_12_Scaled_Shapes_and_Measured_Drawings_487.asp`. Type **Formula Scale Visio** in the Search For box on the MSDN Web site to find other topics about controlling scaling with the ShapeSheet.

Summary

ShapeSheets and shapes on drawing pages are just different views of the same Visio shape. Although many aspects of Visio shapes are editable using commands and shortcut menus on the drawing page, some shape settings are available only on a shape's ShapeSheet. Opening each ShapeSheet in a separate window helps you compare values between ShapeSheets, but opening each ShapeSheet in the same window saves screen space.

You can also define formulas in a shape's ShapeSheet to specify how a shape appears or behaves based on shape settings, values in other ShapeSheet fields, or even values in an external database. Formulas in ShapeSheets are powerful, as demonstrated by the behaviors exhibited by built-in Visio shapes. Studying the ShapeSheets for built-in shapes is a good way to learn how to use formulas to produce specific behaviors. The Visio ShapeSheet Reference is a comprehensive document about using ShapeSheet fields and formulas.

Chapter 35

Formatting with Styles

With the introduction of the Themes feature in Visio 2007, the everyday Visio user won't have as much use for styles as in the past. So many built-in shapes are designed to work with Themes that attractive color-coordinated diagrams are yours with only a few clicks. Although some shapes don't work with Themes, they often have pre-defined formatting applied.

Styles still have a purpose in Visio 2007, although the primary audience is people who build or customize shapes for others to use. Let's say you're building a customized set of masters and you want them formatted the same way. Or you are fine-tuning a diagram and want to apply the same formatting to several shapes to indicate their behind-schedule status. Instead of applying one format command after the other, applying a style is faster and produces more consistent results.

Styles can apply as much or as little formatting as you want — from something as simple as a line weight to the gamut of line, text, and fill formatting. It's easy to apply a collection of format settings to shapes by applying the same style to them. In addition, changing the formatting for shapes formatted with styles is nothing more than redefining the style. Although Visio templates offer sets of styles designed to work together, you can customize styles and define new line, fill, and line end patterns to fit your needs.

This chapter shows you how to work with existing styles as well as create and edit styles. You'll also learn how to define line patterns, such as dots and dashes, fill patterns for hatch patterns, and line end patterns for the marks at the ends of lines.

Understanding Styles

If you've used styles in Microsoft Word, you'll find the concepts for Visio styles to be similar. In Visio, styles handle formatting for more than text — a style is a named collection of format settings, including line, text, and fill formats. When you apply a style to a shape, the formatting specified in the style sets the formatting for the lines, text, and fill for that shape. Because styles take care of applying multiple format settings, they increase the consistency of formatting and reduce the time you spend making your diagrams look the way you want.

NEW FEATURE The big difference with styles in Visio 2007 is that they tend to hide. Because Themes offer easy-to-apply and attractive formatting options, styles are more useful to shape developers. Therefore, in Visio 2007, the commands for defining and applying styles are available only if you run Visio in developer mode. To do this, choose Tools ➪ Options. In the Options dialog box, select the Advanced tab. Select the Run In Developer Mode check box and then click OK. The Style and Define Styles commands appear on the Format menu.

CROSS-REF For information about applying styles to shapes, see Chapter 7.

For shapes with special requirements, you can override the style formatting by applying shape-specific formatting. For example, you could change the fill color for shapes that represent vacant positions in an organization chart that have offer letters in the mail. This formatting that you apply manually with Format Text, Format Line, or Format Fill commands is known as *local formatting* because Visio stores the formatting locally in the shape's ShapeSheet. When you manually format shapes and then apply a style to those shapes, by default, the styles overwrite the local formatting. If you want Visio to keep your manual formatting in place no matter what, in the Format Style dialog box, select the Preserve Local Formatting check box.

NOTE Visio does not apply Theme colors to shapes with local formatting. However, if you remove the local formatting, they automatically conform to the current Theme. The easiest way to remove local formatting is to select a shape that uses the current Theme, on the Standard toolbar, click the Format Painter icon, and then click the shape with local formatting.

You don't have to apply every type of formatting a style has to offer. For example, you might want to apply a style's line formatting, while keeping the text formatting you've applied manually. You can also define styles so they don't override your local formatting. For the ultimate in formatting protection, you can protect your shapes in several ways to prevent their formatting from changing. If all else fails, you can start over by restoring the default styles for a shape.

NOTE To reset styles for shapes dragged from a stencil, select Use Master's Format in a style list. Shapes that you draw directly on the page don't have masters, so there are no master's formats to apply.

Assigning Default Styles to Drawings

Every Visio template includes a few basic predefined styles as well as any styles specific to the shapes associated with the template. For example, every template includes Guide, No Style, None, Normal, and Text Only styles. The Guide style is what makes guide lines appear as dotted, blue lines, by default. The None style, as you would expect, formats elements by removing lines and fills and applying a basic text formatting to text.

If you create a template by saving a Visio drawing file, the styles in the drawing file are available automatically for every new drawing file you create using the template. In addition, you can specify the default styles Visio applies when you draw shapes with drawing tools. First, make sure you are using Visio in developer mode. Then, click an empty area on the drawing page to ensure that no shapes are selected. Choose Format ⇨ Style, select the styles you want to use as defaults in the Text Style, Line Style, and Fill Style drop-down lists, and then click OK.

To apply styles to your shapes, use one or more of the following methods:

- **Apply specific types of formatting or all formatting** — Right-click a shape and, from the shortcut menu, choose Format ⇨ Style. In the Style dialog box, in the Text Style, Line Style, and Fill Style drop-down lists, choose the styles you want.

TIP If the formatting you apply isn't what you want, press Ctrl+Z to undo the formatting.

- **Preserve local formatting** — If you want to apply styles to shapes without resetting any local formatting that you've applied, choose Format ⇨ Style, choose the styles you want to apply, select the Preserve Local Formatting check box, and click OK.

- **Format locked shapes** — Shapes can be locked in two different ways, but you can remove these locks to format the shapes in question. When you see an error message about shape protection preventing the execution of your command, the shape is locked against formatting. To remove this formatting lock, right-click the shape, from the shortcut menu, choose Format ⇨ Protection, and then clear the Format check box. If a shape doesn't accept your formatting and no error message appears, the formatting settings in the ShapeSheet are guarded, which prevents you from changing the value in the formatting field. To remove the guard, select the shape and choose Window ⇨ Show ShapeSheet. In the ShapeSheet, find the cell for the formatting you want to change and remove the GUARD function in that cell.

CAUTION Typically, shapes include the GUARD function for formatting locks when the shapes' formatting is essential to the proper behavior or appearance of those shapes. Although you can remove formatting locks and guards, be aware that shapes might behave differently or change their appearance in unexpected ways.

- **Restoring default styles for shapes** — If you want to remove local formatting so shapes revert to using style formatting, choose Format ⇨ Style. For each type of style, select Use Master's Format from the drop-down list.

Creating and Editing Styles

If the formatting in existing styles leaves you cold, go ahead and edit the styles to use the line, text, and fill formatting you want. If you frequently create your own custom shapes, you might want to create your own styles to accompany them. Style names are perfect for identifying which styles go with which shapes, for example, the Flow Normal style used by shapes on flowcharts.

As with other Visio elements, renaming styles, deleting styles, and copying styles between Visio drawings are all tools in the style editing toolbox. No matter how you want to create or edit your styles, the Define Styles dialog box has all the commands you need. You can create or modify multiple styles in one editing session. By using the Apply button in the Define Styles dialog box, you can apply the styles you've created or edited to selected shapes.

Creating Styles

Like styles in Microsoft Word, Visio styles can be based on other styles, inheriting formatting from the parent style. For example, consider a set of styles that apply the same color fills and fonts, but use progressively thicker line weights. You can create a base style with the colors, fill formatting, text formatting, and line patterns you want. Then, you can base each style on the original, changing only the line weight. For unique styles, creating a style from scratch might be the easiest approach. In this situation, base the style on the No Style style that is built into every Visio file by default.

Creating Styles Based on Shape Formatting

The best way to create styles that do what you want is to first format a shape with the settings you want. When the shape looks good (at least to you), you can save those settings as a style. In addition to creating a style quickly, this approach provides a preview of the formatting before you save the style. To create a style based on a shape's formatting, follow these steps:

1. Select a shape that includes most, if not all, of the formatting you want. To make any additional formatting changes, choose Format ➪ Text, Format ➪ Line, or Format ➪ Fill and select additional format options.

2. When the shape is formatted the way you want, choose Format ➪ Define Styles.

3. In the Define Styles dialog box, in the Name box, select <New Style> at the top of the drop-down list. Visio fills in the settings to match those of the selected shape.

4. Type the name you want for the style in the Name box and click Add to save the style to the drawing file. You can continue to define or edit other styles or click Close to end the style session.

Creating Styles from Scratch

To create a style from scratch, follow these steps:

1. Click an empty area on the drawing page to ensure that no shapes are selected and then choose Format ➪ Define Styles.

2. In the Name box, type the name you want for the new style.

3. In the Based On list, select No Style or an existing style that has similar formatting to what you want.

4. Under the Includes heading, select the check box for each type of formatting you want the style to apply. For example, to define a style for connectors, which don't use fill, you can select the Text and Line check boxes, leaving the Fill check box cleared.

 Selecting the Includes check boxes also determines which style lists display the style.

5. To edit the formatting settings for the style, in the Change area, click the buttons to open the dialog box for that type of formatting. For example, if you click Line, the Line dialog box opens, as shown in Figure 35-1. Specify the formatting you want and click OK to close the formatting dialog box. Repeat this step to edit formatting options for each type of formatting you want the style to apply.

FIGURE 35-1

Specify the formatting options you want for a style within the Define Styles dialog box.

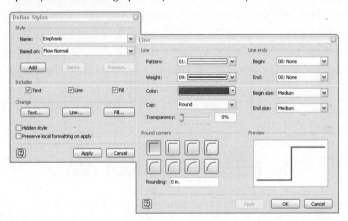

TIP You can prevent a style from appearing in the style lists — for example, when styles are designed for specific masters and you don't want those styles used for anything else. To hide a style so it doesn't appear on the style lists, select the Hidden Style check box before you save the style. Hidden styles still appear in the Name list in the Define Styles dialog box and in the Drawing Explorer window.

6. To save the style in the current drawing file, click Add.

7. Use one of the following methods to close the dialog box:

   ▪ If shapes are selected and you want to apply the style to those shapes, click Apply and then click Close.

   ▪ If you don't want to apply the style to the selected shapes, click Close.

   ▪ If no shapes are selected, click OK.

Editing Styles

Style definitions are not set in stone — you can change them any time you want. The best thing about editing styles is that any shapes in the drawing file formatted with that style reflect the changes you make.

You can also delete and rename styles in the Define Styles dialog box. For example, if you want to create a template, it's a good idea to clean up the file by removing any unused styles before you save it. When you copy styles between drawing files, the styles you copy can't use the same names as styles in the destination file. You can rename the styles in the source drawing file so that the styles copy correctly.

To delete a style, select the name of the style you want to delete and then click Delete. To rename a style, select the name of the style you want to rename and then click Rename. In the Rename Style dialog box, type the new name and click OK.

To edit the formatting that a style applies, follow these steps:

1. Choose Format ➪ Define Styles.
2. In the Name box, select the style you want to edit in the drop-down list.
3. For each type of formatting you want to edit, in the Change area, click the corresponding button. In the formatting dialog box that opens, make the changes you want and click OK.
4. Under the Style heading, click the Change button to update the style definition and all shapes formatted with the style.
5. Repeat steps 2 through 4 to edit another style. Click Close when you're finished.

Copying Styles Between Files

If you intend to use the styles you create in many drawing files, the easiest approach is to save the drawing file containing the styles as a template (see Chapter 31). When you create new drawing files based on that template, they automatically contain those styles. However, you can also copy styles from one drawing file to another by copying shapes formatted with those styles. When you paste shapes into a destination file, Visio copies any styles associated with the shapes. The styles remain in the destination file even if you delete the pasted shapes.

The easiest way to copy several styles is to create a new drawing page and paste all the shapes with the styles you want to copy onto that drawing page. After the styles are copied, you can delete the drawing page to remove the shapes all at once.

If the destination drawing file already includes styles with the same names as the ones you want to paste, Visio doesn't copy the styles from the source. Instead, it applies the style with the same name to the pasted shapes. To copy styles in this situation, you must rename the styles in the source before you copy the shapes to which they are applied.

Creating Fill, Line, and Line End Patterns

Visio comes with numerous built-in patterns for creating dotted or dashed lines, adding marks at the ends of lines, or defining hatch patterns to fill in areas. You can apply these patterns to lines and shapes directly or associate them with styles. Of course, perpetual tinkerers can create new patterns, for example, a fill pattern that looks just like the tiles being installed in a bathroom.

Whether you want to create a line, line end, or fill pattern, you start by creating a *pattern shape*, which represents one copy of the pattern. Pattern shapes can be Visio shapes and fills or bitmaps you import from another application. A pattern shape can work as a line pattern, a line end, or a fill pattern, as demonstrated in Figure 35-2.

FIGURE 35-2

You can use a pattern shape for line patterns, for fill patterns, or as line ends.

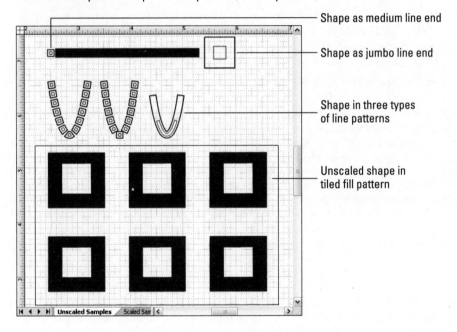

Shape as medium line end

Shape as jumbo line end

Shape in three types of line patterns

Unscaled shape in tiled fill pattern

Pattern shapes can behave in different ways depending on the circumstances, such as with straight or curved lines, or when closed shapes are resized. When you define patterns, you specify how you want the pattern shapes to behave as well as whether the pattern should conform to the drawing scale or stay at a constant size.

When a custom pattern is complete, you can apply it to shapes as you would with any built-in pattern. The Fill and Line dialog boxes list custom fill and line patterns at the bottom of the Pattern drop-down list. Custom line ends appear in the Line dialog box at the bottom of the Begin and End drop-down lists.

 When you create custom patterns, you can't use text, metafiles such as Visio Ink shapes, or gradient fills, rendered as solid fills.

Designing Pattern Shapes

Upfront design is the key to success with patterns. Decisions you make while drawing a pattern shape can lead to significantly different results in patterns. Visio uses the following attributes of a pattern shape to determine the position and appearance of the shape in patterns:

- **Overall pattern shape** — The pattern shape is one copy of the design that you want to repeat, such as a single tile in a tile floor. If you want to use multiple shapes as a pattern shape, either group them (select them and press Ctrl+G) or choose Shape ➪ Operations ➪ Combine to make them one shape.

- **Pin position** — Visio aligns a pattern shape in a pattern by the shape's pin or center of rotation, so the pin position is crucial. If the pin is in the center of the pattern shape, a line pattern using the shape centers the shape on the line. If the pin is at the top of the shape, the pattern shape is offset below the line in the line pattern.

- **Alignment box** — When you tile a pattern shape in a pattern, Visio positions the alignment boxes for the pattern shape side by side. You can put this to good use creating the appearance of spaces between pattern shapes by making the alignment box larger than the graphics of the pattern shape. If you make the alignment box smaller than the graphics, the shapes overlap in the pattern.

TIP To change the alignment box of a shape, use the Rectangle tool to draw a rectangle at the size of the alignment box you want and located where you want the alignment box to be. Select both the rectangle and the pattern shape and press Shift+Ctrl+G to group them. With the group selected, choose Edit ➪ Open Group to open the group window. In the group window, delete the rectangle and close the group window. The result is the pattern shape with an alignment box of the size and position of the deleted rectangle.

- **Color** — After you apply a pattern to a shape, you can change the color of any black areas in the pattern by changing the color of the line or fill. Areas created in any other color remain that color. When you design a pattern shape, be sure to draw elements in black if you want them to change color. If you want areas to remain black, use a color that looks black, but really isn't, such as (0,0,1).

- **Line weight** — If you draw your own dash patterns, draw the line segments with lines of zero weight. The line segments inherit the line weight of the shape to which the pattern is applied.

NOTE When you use a bitmap for a pattern shape, transparent areas on the bitmap aren't transparent in the pattern. When you use a Visio shape as a pattern shape, the background shows through.

How Pattern Shapes Behave in Line Patterns

It's important to take into account how a pattern shape will accommodate straight and curved lines and lines of varying widths. Visio line patterns use behavior options to determine how Visio applies a pattern shape under different circumstances, as shown in Figure 35-3.

FIGURE 35-3

Visio can arrange pattern shapes in several ways to create different line patterns.

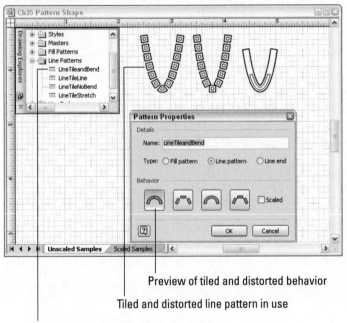

Preview of tiled and distorted behavior

Tiled and distorted line pattern in use

Line pattern in Drawing Explorer window

When you define a line pattern, as described in the section "Creating New Custom Patterns" later in this chapter, choose from the following behavior options:

■ **Tile and bend** — In the Pattern Properties dialog box, this behavior is the first icon in the Behavior section. Visio tiles the pattern along the path of the line. If the path is not straight, Visio also distorts the pattern to conform to the path. For example, this behavior is perfect for a line pattern that represents the painted lines for a highway. When you apply that style to a curved line, you want each of the painted lines (the pattern shapes) to bend with the curves in the road, and you want the dashes for the passing lane to stay the same length.

- **Tile without distortion** — Visio tiles the pattern along the path of the line, but doesn't distort the pattern to account for curves. For example, engineering plans often indicate pipelines with lines interspersed with letters that identify the type of pipeline. In this case, you want the line to bend with the curves, but you don't want the letters (the pattern shape) to distort.

- **Stretch** — Visio stretches a single copy of the pattern over the length of the line. You might use this behavior to create an arrow whose head and tail stretch to fit the length of the line.

- **Tile over line** — Visio tiles the pattern without distortion but retains the original line formatting as well. You might use this behavior to draw a line pattern that looks like barbed wire, with a solid line that follows the curve but undistorted shapes at regular intervals along it.

 Line patterns can have no more than 1,000 instances of the pattern shape along a line.

How Pattern Shapes Perform as Line Ends

Line end patterns are simply pattern shapes attached to the ends of lines, so options for how they can behave are fewer than for line and fill patterns. You can add pattern shapes to the ends of lines in two ways:

- **Oriented with the line** — Visio rotates the pattern shape so it is aligned with the line, such as arrowheads, which are typically angled to match the line.

- **Always upright** — The pattern shape is always upright with respect to the drawing page no matter what the orientation of the line is. If you use a symbol with letters in it, use this behavior to ensure that the letters are always right side up.

When you use a pattern shape as a line end, Visio positions the pin of the pattern shape at the vertex or end point of a line. Since the pin is typically in the middle of a shape, the pattern shape would extend beyond the end of the line, as shown in the top line in Figure 35-4. To simulate the behavior of the built-in Visio arrowheads, whose graphics end exactly at the end point of the line, move the pin to the side of the shape that attaches to the line and orient the shape with the line, as illustrated in the bottom line in Figure 35-4.

TIP If a line end implies direction, such as the arrowheads on connectors on an organization chart, draw your pattern shape so it points to the right. By doing this, Visio points the line end away from the line whether you apply it to the start point or end point of a line.

FIGURE 35-4

The pin position of the pattern shape determines where the graphics appear in relation to the end point of a line.

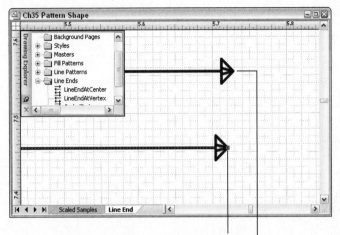

Pattern shape with pin on the side ends at line end point

Pattern shape with pin at center extends beyond end point of line

Scaling Line Ends

Scaled and unscaled line ends appear at different sizes when you apply them to lines. Unscaled line ends appear at the same height as the weight of the line, which is mostly invisible for very thin lines. If your new line end doesn't show up when you apply it to a line, you can try two different methods to make it larger. First, increase the size of the line end in the pattern drawing page. For example, in the Drawing Explorer, right-click the line end and, from the shortcut menu, choose Edit Pattern Shape. In the pattern shape drawing window, drag selection handles until the shape fills the drawing page and then close the window to update the pattern shape. The other approach is to change the weight of the line on the drawing page, such as increasing it to .25 inches.

Conversely, when you create a line pattern that includes a shape that represents a real-world object, you can specify a scaled pattern shape. When you apply a scaled line end to a line, the line end resizes to match the scale of the drawing. In this situation, changing the weight of the line or selecting a size in the Begin Size or End Size drop-down list has no effect on the line end.

Designing Fill Patterns

Fill patterns are for filling in closed Visio shapes. Although fill patterns are often solid colors, you can create hatching or more interesting fill patterns with Visio shapes or bitmaps. If you've ever used a photograph as a Windows background, the options for applying a shape or bitmap to a fill pattern should be familiar. To apply a shape or image as fill, use one of the following three techniques, illustrated in Figure 35-5:

- **Tile** — Visio copies the shape or image over a grid, like tiles on a kitchen floor.
- **Center** — Visio positions one copy of the shape or image at the center of the closed area.
- **Stretch** — Visio stretches a single copy of the shape or image to fill the closed area.

FIGURE 35-5

You can apply images to fill patterns in three ways.

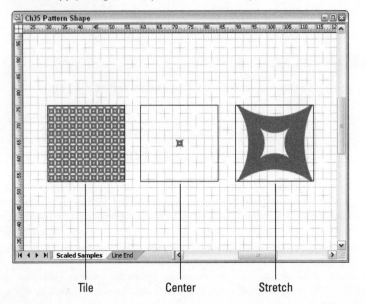

Tile   Center   Stretch

NOTE Fill patterns are limited to 40,000 instances of a pattern shape or a grid of 200 by 200 shapes.

Fill Patterns for Scaled Drawings

For best results, design separate fill patterns for scaled and unscaled drawings. Unscaled fill patterns appear at the size you created them. If the fill pattern uses a shape that is 1 inch square, the fill pattern uses a 1-inch square no matter what scale the drawing uses. When you apply an unscaled fill pattern to a scaled drawing, the pattern might look as if it's a solid black fill.

 Scale and pattern size don't affect fill pattern when you use the option that stretches the image to fill the area.

Transparent Versus Solid Backgrounds

Fill patterns that use shapes show the area between the repeated shapes as transparent. For example, if you tile a circle over an area, the pattern looks like colored polka dots over a transparent background. Here's how it works. When you create a pattern shape, any areas that you fill with the color black appear in the fill color that you choose. In the circle example, the filled circle would take on the fill color. However, the shape's alignment box is a square, so there's no color between the borders of the alignment box and the outside of the circle. That area looks transparent when you use the pattern shape in a fill pattern.

If you want to create an opaque background for a fill pattern, simply add a colored square the size of the alignment box behind the circle. If you use a Theme color or an indexed color, the color of the background changes with the Theme. To apply a color that doesn't change, set the color with the RGB function, for example, RBG(254,254,254) for white.

Creating New Custom Patterns

When the pattern shapes exist, the steps to create any type of custom pattern are similar. Creating patterns always takes place in the Drawing Explorer window. However, where you save the pattern depends on where you open the Drawing Explorer window. Choose one of the following methods:

- **In current drawing** — To save the pattern in the current drawing file, choose View ⇨ Drawing Explorer Window.

- **In an existing stencil** — To add the pattern to a stencil, choose File ⇨ Shapes ⇨ My Shapes and then choose the stencil in which you want to save the pattern. Right-click the stencil's title bar and choose Edit Stencil. Right-click the title bar a second time and choose Drawing Explorer Window.

- **In a new stencil** — Choose File ⇨ Shapes ⇨ New Stencil. Right-click the title bar for the new stencil and choose Drawing Explorer Window.

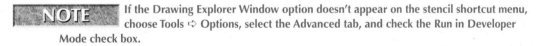 If the Drawing Explorer Window option doesn't appear on the stencil shortcut menu, choose Tools ⇨ Options, select the Advanced tab, and check the Run in Developer Mode check box.

Creating a New Pattern

To create a line pattern, line end pattern, or fill pattern, follow these steps:

1. In the Drawing Explorer window, right-click the folder for the type of pattern you want to create (Fill Patterns, Line Patterns, or Line Ends). From the shortcut menu, choose New Pattern.

2. In the New Pattern dialog box, in the Name box, type the name for the pattern.

3. By default, Visio selects the Type option based on which Drawing Explorer folder you right-clicked. If you change your mind about the type, select the option for the type of pattern you want to create (Fill Pattern, Line Pattern, or Line End).

4. In the New Pattern dialog box, in the Behavior section, click the icon for the behavior you want the pattern to exhibit. The icons that appear depend on the type of pattern you are creating (as described in the "Creating Fill, Line, and Line End Patterns" section earlier in this chapter).

5. To specify a pattern that scales as the drawing scale changes, select the Scaled check box.

6. Click OK. Visio adds the new pattern to the appropriate folder in the Drawing Explorer window.

7. In the Drawing Explorer window, right-click the new pattern and choose Edit Pattern.

8. In the drawing page for the pattern, draw the pattern shape you want or copy a shape from another drawing page and paste it onto the pattern drawing page. You can also insert a bitmap onto the drawing page. For fill patterns, be sure to cover the entire drawing page with the pattern shape so that the pattern fills in areas completely.

9. Click the Close button at the top-right corner of the pattern shape drawing window and click Yes to update the pattern.

Sizing Line Ends

If you have added line ends, follow these steps to make the line ends appear at the size you want:

1. Draw or select a line on the drawing page and choose Format ⇨ Line.

2. Click the arrow next to the Begin or End list and scroll down until you see the line end pattern you want. Select the line end in the drop-down list.

3. Click Apply.

4. If the line ends don't appear at the size you want, click the arrow in the Weight box and choose Custom.

5. Type a heavier line weight, such as .25 in. Check the appearance of the line end in the preview window. Repeat this step until the line end is the size you want.

6. Click OK. To simplify the application of the pattern, save the current settings as a line style.

Summary

If the formatting that Visio 2007 Themes provide isn't enough for you, styles are the next best thing. They apply a collection of format settings for lines, text, and fill, which saves time and increases the consistency of formatting on shapes. Although Visio includes quite a few built-in styles, you can define your own, either based on the formatting applied to a shape or by specifying options within the Define Styles dialog box.

In addition to styles, you can design patterns for lines, fills, or line ends. Line, line end, and fill patterns all use pattern shapes to build their repeating patterns. Designing pattern shapes requires some care, as the pin position, alignment box, color, and line weights within a pattern shape can produce drastically different patterns. When you create patterns, you can specify how Visio applies the pattern shape to create the pattern and whether the pattern scales with the drawing page scale.

Chapter 36

Customizing Toolbars and Menus

L ike the user interface for most applications, Visio menus, toolbars, and keyboard shortcuts provide easy access to commonly used commands. By default, Visio tries to help by displaying the commands you use most frequently and hiding the ones you don't. If you don't care for hand-holding, you can show full menus at all times.

Although many people never venture beyond the standard menus and toolbars that Visio offers, it's actually easy to set up menus and toolbars that display only the commands you want in the order you prefer. If the keyboard is your tool of choice, you can assign keyboard shortcuts so you can execute commands without switching between keyboard and mouse.

This chapter shows you how to customize menus and toolbars to include the commands you use frequently. You will learn how to modify button images on toolbars and specify menu and toolbar options. In addition, you'll learn how to create keyboard shortcuts to increase your productivity.

Customizing Toolbars and Menus

Working with drawings, you don't want to give up screen space — particularly to display commands that you don't use. By customizing existing toolbars and menus or creating your own, you can minimize the space that toolbars and menus consume while still accessing the commands you want.

> **NOTE** It's easy to customize the Visio interface, but the changes you make to menus and toolbars appear for every drawing you open. If you don't want to change Visio's built-in toolbars and menus or you want specialized menus for specific tasks, you can create new menus and toolbars that contain only the commands you want.

Using Personalized Menus

By default, Visio personalizes toolbars and menus by tracking the commands you use and restricting the commands that initially appear to those. At first, menus and toolbars contain commands popular with most users. As you work, Visio adds the commands you choose to the menu and hides the commands you rarely pick. To display the full menu of commands at any time, click the chevron at the bottom of the menu.

If you would rather see full menus all the time, right-click any menu and, from the shortcut menu, choose Customize. In the Customize dialog box, select the Options tab and select the Always Show Full Menus check box. If you like personalized menus but your recent work has skewed the commands that appear, reset personalized menus to Visio's default selections. In the Customize dialog box, on the Options tab, click Reset Menu and Toolbar Usage Data.

Customizing Toolbars

Toolbars are such handy creatures. They're small but they pack in a lot of commands. And you can move them to any spot that's not in your way. You can add, remove, and rearrange buttons on toolbars and specify the appearance of buttons. To open a toolbar so you can customize it, follow these steps:

1. If the toolbar you want to customize is not visible, choose View ⇨ Toolbars and then choose the toolbar you want.

2. Choose Tools ⇨ Customize.

3. In the Customize dialog box, select the Toolbars tab.

TIP   **One way to reduce the area that toolbars take up is by displaying the Standard and Formatting toolbars on one line. Right-click any toolbar and then choose Customize on the menu. Select the Options tab and clear the Show Standard And Formatting Toolbars On Two Rows check box.**

If a command you want doesn't appear when these toolbars share a row, on the right end of the toolbar, click the Toolbar Options arrow and choose the toolbar button you want.

To modify the contents or order of a toolbar, choose one of the following methods:

- **Add a button** — To add a button to a toolbar, follow these steps:

  1. In the Customize dialog box, select the Commands tab. In the dialog box, Visio displays categories of commands on the left side and the commands within the selected category on the right side.

  2. Choose the category that contains the command you want to add. (At first, this might require trial and error.)

3. Drag the command you want from the Commands list to the toolbar. When the I-beam (which resembles a standard cursor but looks more like a bolded, uppercase I) is where you want to add the command, release the mouse button.

■ **Remove a button** — To remove a button from a toolbar, open the Customize dialog box, right-click the button you want to remove on the toolbar, and, from the shortcut menu, choose Delete.

 If the Customize dialog box isn't open, you can remove a button from a toolbar by pressing and holding the Alt key as you drag the button you want to remove off the toolbar. When an X appears below and to the right of the button, release both the Alt key and the mouse button.

■ **Rearrange buttons** — To move a button to another position on a toolbar, open the Customize dialog box and drag the button on the toolbar until the I-beam is positioned where you want to place the command, as illustrated in Figure 36-1, and then release the mouse button.

FIGURE 36-1

Use the I-beam to position buttons on a toolbar.

You can hide or display buttons without opening the Customize dialog box. To do this, at the end of the toolbar, click the Toolbar Options button, choose Add or Remove Buttons, and choose the name of the toolbar you want to customize. Visio displays a list of all the buttons associated with the toolbar. Check a command to display it or uncheck a command to hide it.

 Although you can hide or display toolbar buttons using Toolbar Options, you can add buttons to or remove buttons from toolbars only when the Customize dialog box is open.

Adding Menus to Toolbars

To add a menu to a toolbar, open the Customize dialog box and select the Commands tab. To add a built-in menu to a toolbar, select the Built-In Menus category and drag the menu you want from the Commands list onto the toolbar.

To add a custom menu to a toolbar, select New Menu in the Categories box and drag New Menu from the Commands list to the toolbar. Right-click the new menu on the toolbar, on the shortcut menu, type a name in the Name box, and press Enter. You can add commands or menus to the new menu by dragging items from the Customize dialog box to the pull-down area below the new menu name. To modify a menu entry, right-click it.

Customizing Menus

Just like toolbars, menus can contain commands or other menus. Although menus use more space than toolbars, that room provides space for descriptions of commands or submenus. In addition to adding, removing, or rearranging commands and submenus on a menu, you can change the appearance of menu buttons and even specify whether the menu displays buttons or text, as shown in Figure 36-2.

FIGURE 36-2

The appearance of commands on menus is in your control.

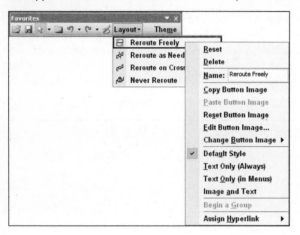

To customize a menu, open the Customize dialog box by choosing Tools ⇨ Customize and then using one of the following methods:

- **Add a command** — Select the Commands tab in the Customize dialog box, choose the category for the command you want to add, and drag the command from the Commands list to the position you want on the menu.

- **Add a menu** — Select the Commands tab in the Customize dialog box and choose the New Menu or Built-In Menus category. To create a new menu, drag the New Menu command from the Commands list to the desired position on the menu. To add a built-in menu to another menu, drag the menu from the Commands list to the menu.

- **Remove a command from a pull-down menu** — To remove an entry from a menu, with the Customize dialog box open, navigate to the command you want to remove, right-click it, and then, from the shortcut menu, choose Delete.

- **Rearrange commands** — To move a command or submenu to another position on a menu, with the Customize dialog box open, drag the command in the menu until the I-beam is positioned where you want to place the command and then release the mouse button

■ **Remove a menu from the menu bar** — With the Customize dialog box open, right-click the menu you want to remove from the menu bar and choose Delete from the shortcut menu.

NOTE If you want to reset menus you have customized to the commands Visio provides by default, open the Customize dialog box. Select the Toolbars tab and click Reset. When Visio prompts you, click OK to reset your menus.

Creating Toolbars and Menus

When you constantly use a small number of commands from several built-in toolbars or menus, a custom toolbar or menu will soon be your favorite.

To create a new toolbar, open the Customize dialog box and then follow these steps:

1. Select the Toolbars tab and click New.

2. In the New Toolbar dialog box, type the name of the toolbar and click OK. Visio adds the toolbar name to the Toolbars list and displays the empty toolbar on the screen.

3. Follow the instructions in the section "Customizing Toolbars" earlier in this chapter to add commands to the toolbar.

You can create new menus in the process of customizing a toolbar or another menu. To create a new menu on a toolbar or menu, select the Commands tab in the Customize dialog box, choose the New Menu category, and drag the New Menu command from the Commands list to the position you want on the menu. Name the menu and drag the commands you want onto the new menu.

NOTE When you create a new menu, it is available only on the toolbar or menu on which you place it. Visio doesn't maintain a list of user-defined menus that you can use to add customized menus to other toolbars or menus.

Modifying Toolbars and Menus

You can rename toolbars and menus, group commands on them, modify button images, or specify whether Visio displays a button, text, or both. As with customizing menus and toolbars, the trick is to open the Customize dialog box.

Renaming Toolbars and Menus

You can rename both user-defined and built-in menus, but you can only rename user-defined toolbars. Visio uses the menu name as the text that appears in a menu or toolbar, so it's best to keep menu names short and descriptive. To rename toolbars and menus, open the Customize dialog box and then use one of the following methods:

- **Rename a user-defined toolbar** — Select the Toolbars tab, choose the toolbar you want to rename, and click Rename. In the Rename Toolbar dialog box, type the new name and click OK.

- **Rename a menu** — Right-click the menu, type the new name in the Name box on the shortcut menu, and press Enter.

 To create a keyboard shortcut to open a menu, when you type the menu name in the Name box, type an ampersand (&) in front of the letter you want to use as the shortcut.

Deleting Toolbars and Menus

You can only delete custom toolbars that you create. To delete a user-defined toolbar, open the Customize dialog box, select the Toolbars tab, choose the user-defined toolbar you want to delete, and, from the shortcut menu, choose Delete. When prompted, click OK to confirm the deletion.

Although you can't delete built-in toolbars, you can reset them to their original configuration. When you select a built-in toolbar in the Customize dialog box, the Delete button is dimmed. To reset a built-in toolbar, select the Toolbars tab, select the toolbar in the Toolbars list, and then click Reset. When prompted, click OK.

 Resetting a toolbar removes any custom buttons you created. To save custom buttons, copy them to another toolbar before you reset the current one.

Grouping Commands and Menus

To group buttons on a toolbar or menu, right-click the button you want as the first in the group and, from the shortcut menu, choose Begin Group. To remove a separator, drag the two buttons or commands on either side of the separator closer together.

Changing the Width of a Drop-Down List

To change the width of a drop-down list in a toolbar, follow these steps:

1. Open the Customize dialog box and select the drop-down list in the toolbar whose width you want to change.

2. Position the pointer over either end of the drop-down list and drag the pointer until the drop-down list is the width you want.

Modifying the Appearance of an Entry

When you right-click a menu or toolbar entry while the Customize dialog box is open, a shortcut menu appears with commands for changing the appearance of the entry. The commands on this shortcut menu include the following:

- **Reset** — Restores the original button, associated command, and settings for the button on a built-in toolbar.

- **Delete** — Removes the button.

- **Name** — Displays a box in which you can type the ToolTip for the button.

- **Copy Button Image** — Copies the selected button image to the Clipboard so you can paste it to another button.

- **Paste Button Image** — Pastes the image on the Clipboard to the selected button. You can copy graphics or images from other applications or buttons.

- **Reset Button Image** — Restores the default button image.

- **Edit Button Image** — Opens the Button Editor dialog box so that you can edit the image or create your own.

- **Change Button Image** — Displays a selection of images that you can choose as the new button image. Unless the command is generic, you're not likely to find an image that conveys the function of the button.

- **Default Style** — Displays only a button image on a toolbar and the button image and text on a menu.

- **Text Only (Always)** — Displays only text in both toolbars and menus.

- **Text Only (In Menus)** — Displays a button image on a toolbar and only text on menus.

- **Image and Text** — Displays a button image and text in toolbars and menus.

- **Begin a Group** — Adds a divider to the toolbar or menu.

- **Assign Macro** — Opens the Customize Tool dialog box so you can designate a macro for the button.

Modifying Button Images

You can modify the button images on a toolbar. For example, if you have different Print buttons for each printer you use, you can modify button images to indicate the associated printer.

To modify a button image, follow these steps:

1. Open the Customize dialog box, right-click the button you want to edit, and then click Edit Button Image. The Button Editor dialog box appears.

2. To modify the image, choose a color in the Colors section of the dialog box. To erase colored boxes in the image, click the Erase box.

3. In the Picture section, click individual cells or drag the mouse pointer over cells to change their color. You can see what the image looks like in the Preview section of the dialog box.

4. To move the image within the Picture section, click a directional arrow in the Move section of the dialog box. When the image fills the Picture section in one or more directions, Visio dims the appropriate directional arrows.

Creating Keyboard Shortcuts

If you prefer the keyboard over the mouse, you can assign keyboard shortcuts to any command on any menu. To use a keyboard shortcut, press and hold the Alt key, press the shortcut letter for the menu, and then press the shortcut letter for the command. For example, to save a file, you can press Alt+F+S, which first opens the File menu and then selects the Save command.

To create a keyboard shortcut, follow these steps:

1. Open the Customize dialog box and right-click the menu or command for which you want to define a keyboard shortcut.

2. In the Name box on the shortcut menu, type an ampersand (&) before the letter you want to use as the shortcut. In the menu, Visio underlines the shortcut letter for the command.

> **NOTE**   Use a different letter for each keyboard shortcut. If you choose a letter that is already in use by another menu entry, you might have to press the letter more than once to select the command you want.

Sharing Customized Toolbars and Menus

When you customize toolbars and menus, they are associated with your Visio application, not a drawing file. To share a customized toolbar with someone else, you must copy it to a drawing file and send that file to your colleague. To share a customized toolbar or menu, follow these steps:

1. Choose View ➪ Toolbars ➪ Customize.

2. In the Customize dialog box, select the Toolbars tab and then click Attach.

3. In the Custom Toolbars list, select a toolbar that you want to share and click Copy. Click OK after all the toolbars you want to share appear in the Toolbars in Drawing list.

4. Save the drawing file with the toolbars by pressing Ctrl+S.

Summary

Customizing built-in toolbars and menus is a great way to gain more screen area for drawings and perhaps increase your productivity. Whether you customize menus and toolbars to include the commands you want or create new ones to consolidate your favorites, you can add, remove, or rearrange entries. In addition, you can specify whether Visio displays a button image, a text description of the entry, or both. You can also customize or create button images to better represent the commands they represent.

Chapter 37

Automating Visio

Visio is more than software for creating diagrams by dragging shapes onto a drawing page. Almost the entire Visio user interface and object model is open to interrogation and modification. You can write code to create or update Visio drawings based on information stored in a database; update a database as changes are made to a Visio drawing; or manipulate the shapes on a page. In every corner of capabilities of Visio, you can extend or utilize those features through automation.

If you're not a programmer, you might think that any kind of Visio automation is beyond your reach, but you would be wrong. Even if you use Visio only for creating simple diagrams, the Visio macro recorder is a time-saving tool that saves the steps you perform so you can easily repeat them in the future. You can save the macro as is or modify it using Visual Basic for Applications (VBA) to make it more flexible.

If you are a programmer, or can at least find your way around a programming language such as Visual Basic, you can write code to develop more sophisticated automated solutions. Visio documents, windows, drawing pages, shapes, and ShapeSheet cells are all accessible through code. Extending capabilities of Visio is achieved with the help of Component Object Model (COM) add-ins, similar to the add-ons that are packaged with Visio, such as the Number Shapes add-on. Your custom-built code can launch from menus, from shape shortcut menus, or in response to events.

Another level of automation is to integrate Visio into an external application, such as Microsoft Excel. The Visio ActiveX control provides complete access to the Visio object model and user interface. For example, you could develop an application that helps an engineer design a sound system for a customer and, when the design is complete, scan the resulting diagram to produce a parts list, an estimate, and a purchase order.

This book is primarily for people who draw diagrams with Visio, so this chapter sticks to an overview of the development features that Visio offers. If you want to delve more deeply into automation, this chapter also identifies resources for learning more both within Visio Help and online.

Development Features Added in Visio 2003

In case, you're upgrading from Visio 2000 to Visio 2007, you'll find several significant new features for developers that first appeared in Visio 2003. The biggies are the macro recorder and the Visio Drawing control. The new additions in Visio 2007 include Automation support for features introduced in Visio 2007, such as Data Graphics and Themes. Here are some of the new features that debuted in Visio 2003:

- **Macro Recorder** — See the section "Working with Macros" to learn more about recording the actions you perform within the Visio application.

- **Microsoft Office Visio Drawing Control** — Available with both Visio Standard and Visio Professional, this ActiveX control provides complete access to the Visio object model and user interface so you can fully integrate Visio into an external application.

- **SmartTags** — Add SmartTags to shapes to display drop-down menus and make shape actions and settings easier to find.

- **Publishing add-ons** — Instead of specifying file paths for Visio add-ons, you can publish add-ons using a Microsoft Windows Installer package to take advantage of Microsoft Office System application features such as language switching, installation on demand, and repair.

- **Formula Tracing Window** — This window helps identify interdependencies between ShapeSheet cells by showing cells that use or reference a ShapeSheet cell.

- **Keyboard and mouse events** — Visio 2003 introduced new events to handle mouse and keyboard events, such as mouse movements, mouse clicks, and keyboard actions.

- **ShapeStudio** — Available with the Visio Software Developers Kit (SDK), this add-on provides a development environment for creating Visio shapes.

- **Primary interop assemblies (PIAs)** — Access the Visio object model from applications that use the common language runtime (CLR) 1.1.

- **XML Web Service support** — Integrate XML Web Service into diagrams by selecting Web Service references from a dialog box.

Working with Macros

At their simplest, macros offer us mere mortals an easy way to automate tasks we perform frequently in Visio. For example, if you like to display different sets of toolbars for different types of drawings, you can record the steps you perform to show a specific set of toolbars. Then, when you work on one type of drawing, you can run the macro to set up your toolbar environment for that drawing type.

The macro recorder can also act as an educational tool if you're still groping your way around VBA. When you record steps in Visio, the macro recorder generates the basic code for your actions in a macro. You can edit the macro to see what the code looks like for the steps you performed. In addition, you can add prompts for input or write code for additional actions. Of course, for experienced programmers, the macro recorder opens the VBA development environment so you can show off and write macros from scratch.

To record a macro, follow these steps:

1. Choose Tools ➪ Macro ➪ Record New Macro.

2. In the Record Macro dialog box, in the Macro Name box, type a brief but meaningful name for the macro.

3. In the Store Macro In drop-down list, select the location in which you want to store the macro. The default setting, Active Document, stores the macro in the current drawing file, which is what you usually want. You can also choose to store the macro in any open stencil.

4. To assign a keyboard shortcut to launch the macro, in the Ctrl+ box, type a letter.

5. In the description box, type text that explains what the macro does. This helps others find the macro they want and often helps jog your memory about why you wrote the macro in the first place.

6. Click OK to start recording. The Stop Recording toolbar appears, although the toolbar is so short you can't read the title.

7. Execute the commands and actions you want to automate. If you want to pause recording to perform some other task, on the Stop Recording toolbar, click Pause Recording. Click Pause Recording again to resume recording.

8. When you're done with the steps in your procedure, on the Stop Recording toolbar, click Stop; or choose Tools ➪ Macro ➪ Stop Recording.

9. To confirm that your macro has been recorded, choose Tools ➪ Macro ➪ Macros. In the Macros dialog box, in the list box, you should see the name of the macro you recorded.

To edit a macro using VBA, follow these steps:

1. Choose Tools ➪ Macro ➪ Macros.

2. In the Macros dialog box, in the list box, select the macro you want to edit and click Edit. The Microsoft Visual Basic window opens and within it, the Project Explorer, the Properties window, and the Code window, as shown in Figure 37-1.

> **TIP**   If the macro you want to edit doesn't appear in the list box, you might have stored it in a different location. In the Macros In drop-down list, select the location in which the macro is stored.

FIGURE 37-1

The Visual Basic Editor window provides tools for writing code or editing macros you record.

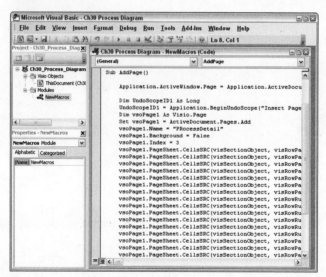

3. Make the changes you want to the VBA code.

4. When you're finished editing, choose File ➪ Close And Return To Visio.

The following methods are the most common ways to run macros:

■ Choose Tools ➪ Macro ➪ Macros. Select the macro and click Run.

■ Double-click a shape whose double-click behavior is set to run a macro.

To set a shape's double-click behavior to run a macro, select the shape and choose Format ⇨ Behavior. Select the Double-Click tab and select the Run Macro option. In the Run Macro drop-down list, select the macro name and click OK.

■ Right-click a shape and choose a menu item that runs a macro. To add a shortcut command to run a macro, add the menu item in the Actions section of the ShapeSheet window as described in Chapter 34.

Writing Add-Ins to Automate Visio

Visio takes direction from programs written in VBA, Microsoft Visual Basic, C++, or any programming language that supports Automation, a proprietary Microsoft development environment. You can use COM add-ins the same way you do in other Microsoft Office System applications. In fact, if written appropriately, COM add-ins can work in more than one application.

With Automation, programs control Visio elements by accessing and using the Visio object model — an interface to Visio objects (drawings, windows, shapes, and so on) and their properties, methods, and events. The Visio type library defines the objects, properties, methods, events, and constants that Visio exposes to Automation clients. Automation components include the following:

■ **Objects** — Elements within the Visio application, such as documents, drawing pages, windows, shapes, and ShapeSheet cells containing formulas.

■ **Properties** — Attributes that determine the appearance or behavior of objects, much like the shape data associated with shapes.

■ **Methods** — Actions that an object provides. For example, applying the Delete method to the Page object deletes a Page object and can renumber the remaining pages.

■ **Events** — Occurrences that trigger the execution of code or programs.

To use the Visio type library, a development environment must reference it. VBA projects in a Visio document automatically reference the Visio type library. If you use other development environments, you must use the appropriate commands or steps to reference the library. When you use the Visio type library, you can use an *object browser,* such as the one in VBA, to view descriptions of objects supplied by an Automation server. The object browser displays the syntax of a Visio property, method, or event, and might include code you can paste into your program. In addition, by using the type library, the development environment you use can bind your code to Automation server code at compile time, which can result in faster program execution. For example, you can use objects such as Visio.Shape instead of the generic Object.

If you write your add-ins using VBA, you can access the Visual Basic Editor from within Visio. To open the Visual Basic Editor window, choose Tools ⇨ Macro ⇨ Visual Basic Editor.

Using the Visio Drawing Control

The Microsoft Office Visio Drawing Control is an ActiveX control that integrates the Visio drawing surface into applications you develop. It provides full access to the Visio object model and user interface so you can build applications that take advantage of Visio's drawing features instead of writing your own drawing tools. For example, you might develop an application for finding someone's office on a large campus. The application asks for information such as the person's name and department and then uses the Visio Drawing Control to display a drawing that shows the location of the office.

 The Visio Drawing control is installed when you install Visio, even if you choose the Minimal Install.

You can embed the Visio Drawing control in ActiveX control containers, such as the ones provided by Microsoft Visual Studio .NET 2005. However, the Visio Drawing control isn't available when you develop a solution in VBA within Visio. You can insert more than one instance of the Visio Drawing control in an application, but each instance displays only one drawing window and one Visio drawing file.

By default, the Visio Drawing control opens a blank Visio drawing, but you can specify that the control open an existing Visio document, either at design time or at run time. Unlike Visio, the Drawing control doesn't display the Visio Getting Started screen or a docked Shapes window on startup. However, if you load an existing drawing that already displays a docked Shapes window, the window appears in the Visio Drawing control window. You can also display the Shapes window in a blank drawing by using the `Document.OpenStencilWindow` method from the Visio object model. Other methods are available for displaying menus and toolbars.

 Although the Visio Drawing control does not expose the ShapeSheet in the user interface, Automation methods and properties provide access ShapeSheet cells.

Learning More About Automating Visio

This chapter acts only as an introduction to programming features available in Visio. Regardless which programming language you use or whether you develop macros with Visio VBA or the Visio Active X control, you can find dozens of books devoted to writing code in each language or environment. You can use the following Visio Help and Microsoft's Web resources to obtain a great deal of material about developing solutions with Visio:

- **Visio Help** — Choose Help ➪ Developer Reference to access the Visio Developer Help. This Help site includes links to the Visio Automation Object Model Reference and the Visual Basic for Applications Language Reference. You can read about the Visio object model, Visio programming concepts, and obtain detailed information about objects, interfaces, methods, properties, events, and enumerations. To access the Visio ShapeSheet Reference, in the Visio Help window, click the down arrow on the Search button and then choose Visio ShapeSheet Help.

- **VBA Help** — When you're working in the VBA window, press F1 or choose Help ⇨ Microsoft Visual Basic Help to read about generic Visual Basic topics and obtain assistance with VBA editing tools.

- **MSDN** — The URL for the Visio Developer Portal is `http://msdn.microsoft.com/office/program/visio/`. On this section of the MSDN site, you can access all kinds of Microsoft documentation for Visio development as well as technical articles about development topics. The site also includes links to Visio blog topics that discuss Visio development.

- **Books** — The classic book for Visio development is *Developing Visio Solutions*, AKA DVS. The current version of this book is available from Microsoft Press, titled *Developing Microsoft Visio Solutions*. You can also obtain it online at the MSDN web site, `http://msdn.microsoft.com/library/default.asp?url=/library/en-us/devref/HTML/DVS_01_Introduction__561.asp`. For the true fanatic, read Graham Wideman's *Visio 2003 Developer's Survival Pack* (Trafford Publishing).

- **MVPS.org** — The MVPS web site is a great online resource (`http://mvp.support.microsoft.com/`) for many of Microsoft's programs and is maintained by Microsoft MVPs. For Visio information, navigate to `http://visio.mvps.org`. This web site includes headings for VBA Information, ShapeSheet development, stencil downloads, and Visio History.

- **Blogs** — Bill Morein writes a blog that often dips into development topics (`http://blogs.msdn.com/wmorein`). There is also a blog for the Visio team (`http://blogs.msdn.com/visio/`). The Visio MVPs have their own Visio blog with a smattering of development topics as well as other interesting information (`http://msmvps.com/blogs/visio`). Chris Roth, one of the original shape developers, provides new and enhanced shapes through his blog, `http://visguy.com/`.

Summary

For non-programmers or less experienced developers, Visio offers the macro recorder, which transforms the actions you perform in Visio into VBA code. For more advanced automation assignments, you can use any programming language that supports Automation, such as Visual Basic or Visual C++, to write COM add-ins with access to the entire Visio object model. In addition, the new Visio Drawing control is an ActiveX control that enables you to include the Visio interface and Visio functionality in external applications you develop.

Part VII

Quick Reference

Chapter 38

Installing Visio 2007

I f you have installed other Microsoft products on your computer, the Visio installation should be quite familiar. The Visio Setup program and the Microsoft Windows Installer use procedures similar to those for other Microsoft Office applications. Whether you are running Windows XP or Windows Vista, the process of installing programs thankfully remains the same.

This chapter describes how to install Visio 2007 from a CD or use other methods designed for deploying it in large organizations. You'll learn how to install multiple versions of Visio on your computer, which is helpful if you use some of the features discontinued in previous versions. Although the installation procedure includes default options, this chapter will show you how to choose where to install Visio as well as which components you want to install.

To verify that you are installing a legal copy of the software, Microsoft requires that you activate your copy after installation. If you don't, you can run Visio only a few times and can use only some of Visio's features during those sessions.

Exploring Visio Installation Methods

If you are installing Visio on a computer that is not attached to a network, it's easy to install Visio 2007 from the Visio 2007 CD. However, Microsoft Office 2007 offers additional installation methods when you want to deploy programs throughout an organization and allow users to easily maintain the software on their computers.

 To learn more about the advanced deployment features of Microsoft Office, review the Office Review Kit on Microsoft's web site at www.microsoft.com/office/ork.

Using a Local Installation Source

When you install Microsoft Office 2007 applications from a CD or from a compressed CD image on the network, the Setup program copies any required installation files to a hidden folder on the local computer. Windows Installer uses this local installation source to install Office. Keeping the installation files on a local or network computer enables you to do the following:

- **Easily repair, reinstall, or update your Office programs** — When you keep the installation files on your computer or another computer on the network, you can repair, reinstall, or update the programs without tracking down your installation CD.

- **Add features on demand** — If your disk space is at a premium, you can install features when you need them without accessing the installation source on the network or CD.

- **Install smaller updates** — As long as the installation files are on your computer, system administrators can distribute smaller client patches for you to apply.

Deploying with an Administrative Installation Point

If you're a Visio user, you aren't likely to take on installing from an administrative installation point. However, if you work in an organization with hundreds of computers, your IT department might use one. In that case, it's nice to understand why they use this approach for installation, even if it sometimes makes it harder for you to install the software you want.

With administrative installation points, administrators customize Microsoft Office client installations and deploy them to users throughout an organization. The administrative installation point resides on a network server from which you can run Setup. When you use an administrative installation point, updates and maintenance originate from the administrative installation point on the network. Although the Setup program can't create a local installation source for local updates and repairs, an administrative installation point provides the following benefits:

- **Centralized management** — The administrative installation point stores one set of Office files located in a central location. By patching a single administrative image, you can update all installations from that image.

- **Standardized installations** — You can create standard Office configurations suited to the needs of groups of users.

- **Flexible installation options** — You can specify whether features are installed on first use or run over the network. You can also use other deployment tools, such as Microsoft Systems Management Server to install software.

Using a Compressed CD Image

If you want the benefits of an administrative installation point but also want a local installation source, a compressed CD image for installing Office applications is an alternative. This approach can come in handy even for individuals, particularly when you have multiple computers connected to a home or small office network. For example, you can use a compressed CD image to simplify installation to your desktop and laptop computers.

To create a compressed CD image, you simply copy the compressed files from the Visio CD or Office CD to a network share. Then, instead of inserting a CD into a CD drive, you install from the file on a hard disk. With a compressed CD image, you can perform the following actions:

- **Create multiple configurations** — You can modify the Setup.ini file to customize Office and create multiple configurations from the same compressed CD image.

- **Set features for install on demand** — You can set features to be installed on demand, but you can't run Office applications over the network.

- **Create chained packages** — You can attach additional packages to the Office installation to install standalone products such as Microsoft Office Publisher or Microsoft Office Project.

- **Use software deployment tools** — You can use deployment tools such as Microsoft Systems Management Server to install Office on users' computers.

Installing Visio on Your Computer

If you want to install Visio from a CD onto your computer, the Visio Setup program walks you through the steps for the installation. To start your Visio installation, follow these steps:

1. Close any programs that are running — including your virus protection software — and insert the Visio 2007 CD into your computer. If AutoRun is turned on, the installation wizard launches and you can begin installation.

2. If the Setup program doesn't start automatically, do the following:

    a. Click Start to open the Windows Start menu.

    b. Choose All Programs ⇨ Control Panel.

    c. In the Control Panel window, double-click Add or Remove Programs.

    d. Click Add New Programs and then click CD or Floppy.

    e. In the Install Program from Floppy Disk or CD-Rom dialog box, click Next.

    f. In the Open box, type **D:\Setup** (or replace D with the letter that represents your CD-ROM drive). If you store the compressed installation file on a hard drive, type the path and name of the file.

g. Click Finish.

h. In the Open File dialog box, you can ignore the dire security message and simply click Run to launch the installation program. You might have to watch a progress bar for a few minutes while the program extracts the files it needs. Grab a cup of coffee and make sure you have your product key ready.

3. On the Enter Your Product Key screen, type the 25-character product key from your CD-ROM. Although the product key appears in capital letters and numbers, you can type lowercase letters. If the product key is valid, a green check mark appears to the right of the box.

4. Click Continue.

5. On the Read the Microsoft Software License Terms screen, read the terms of the agreement either for amusement or caution, but read them nonetheless. To continue with the installation, select the I Accept The Terms Of This Agreement check box and then click Continue.

 If you want to run multiple versions of Visio, be sure to choose Customize in step 6. See the section "Installing When a Previous Version Is Present."

6. On the next screen are two buttons: Install Now and Customize. To install the most typical components, click Install Now. Another progress bar appears and you can refill that cup of coffee. On the other hand, if you want to choose the features to install or set components to install on first use, click Customize. The steps for customizing the installation are described in the section Customizing Your Installation.

7. When the installation is complete, click Register for Online Services to obtain access to features at Microsoft Office Online. Otherwise, click Close to close the dialog box and get on with using Visio.

8. Because Microsoft releases updates to its programs from time to time, it's a good idea to check for and install any updates to Visio. In addition, be sure to apply the latest updates to printer drivers. The easiest way to update features is to activate and run Visio and then choose Help ➪ Check For Updates.

Installing When a Previous Version Is Present

By default, the Visio Setup program removes previous installations of Visio. If it finds previous versions of Visio, it displays them in the Previous Versions screen. To keep those Visio installations, be sure to click Customize for the type of installation in step 6 in the previous section. In the Previous Versions screen, select the Keep All Previous Versions option and continue with the installation.

TIP If you plan to work with multiple versions and you don't keep the installation source files on your computer, make sure that your Visio CD for each version you want to use is handy. You will need your CD every time the Auto-repair program runs.

NOTE When you use multiple versions of Visio on the same computer, Visio might display a dialog box when you start Visio. Wait while the Visio Auto-repair process sets up the files properly for the version you want to run and then work with Visio as you would normally.

Customizing Your Installation

When you install Visio, you can specify the type of installation you want and where you want to install the program files on your computer. When you click Install Now, the Setup program automatically installs the most commonly used components. However, you can also choose to install every feature, only required features, or only the features you want. With Customize, you can specify exactly the features you want to install and when; where to install the software, which is helpful if your C: drive is getting perilously full; or type your name, initials, and organization.

In the installation procedure, click Customize and then perform one or more of the following customization tasks:

- **Installation Options** — By default the dialog box selects the Installation Options tab. On this tab, you can control which features are installed and when. To customize the installation of a feature, click the + in front of a category and then click the icon to the left of the feature you want, as shown in Figure 38-1. From the drop-down menu, choose one of the following entries:

  - **Run From My Computer** — Installs the selected feature on your computer.

  - **Installed on First Use** — This choice seems innocuous but can be inconvenient in practice. It does not install the feature during the current installation. Later, when you try to use the feature for the first time, Visio prompts you to install it, at which time you have to find your Visio CD or access the installation files on your computer or the network. And of course, you have to stop what you are doing to install the feature.

  - **Not Available** — Does not install the selected feature.

- **File Location** — To install Visio in a location other than the default path, `C:\Program Files\Microsoft Office\`, select the File Location tab and then click the Browse button. Navigate to the installation path you want, and click OK.

- **User Information** — Select the User Information tab and then type your name, initials, and organization in the boxes. These values identify you as the person who makes changes in shared Office files.

NOTE Visio provides two versions of each template and add-on: one version using U.S. units and the other using metric units. By default, the Setup program selects the version to install based on the locale and configuration on your computer. If you want to include both versions, you must perform a custom install and specify that you want to install the following: Solutions (US units), Solutions (Metric Units), Add-ons (US units), and Add-ons (Metric units).

Installation Option Icons

When you customize your installation, the following icons indicate how features will be installed:

- A white box with a drive icon indicates that features and options will be installed as you've specified.
- A gray box with a drive icon means some options for a feature won't be installed.
- A white box with a red X means the features and its options won't be installed.
- A white box with a drive icon and the number one means the feature will be installed the first time you use it.

FIGURE 38-1

Choose features to install and also how to run them.

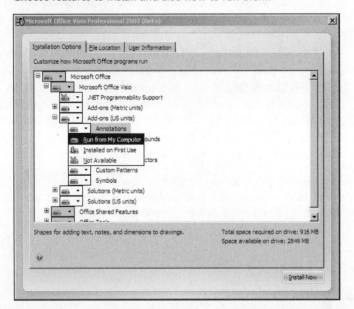

Activating a New Visio Installation

Activation is a technique that Microsoft uses to ensure that you've installed a legal copy of a Microsoft product. If you run Visio without activating it, the product runs in Reduced Functionality mode, which enables you to evaluate Visio features, but limits your ability to perform meaningful work. In addition, after you run Visio several times without activating, you won't be able to run the program again until you reinstall and activate it.

When you activate Visio, Microsoft requires that you indicate your country or region, not personal information. The Microsoft Office Visio Activation Wizard creates a unique hardware identifier that identifies only the configuration of your computer solely for the purpose of activating Visio. You don't have to reactivate your copy of Visio for minor upgrades, although you might have to reactivate if you completely rebuild your computer. During activation, you can also register your copy of Visio. The Activation Wizard starts automatically when you run Visio for the first time. However, you can also start the wizard by choosing Help ⇨ Activate Product.

NOTE Although registration requires personal information such as your name and contact information, Microsoft employs a privacy policy and uses security mechanisms to protect your personal information and privacy.

NOTE If you have trouble activating Visio, your Internet connection might be disconnected. To activate Microsoft Office Visio by telephone, start the Microsoft Office Visio Activation Wizard and follow the instructions.

If you do not activate Microsoft Office Visio after you install it, Visio starts in Reduced Functionality mode. You can open, close, and print existing drawings in Reduced Functionality mode, but you can't perform the following tasks:

- Create new files.
- Save changes to existing files.
- Display anchored windows, such as Pan & Zoom, Size & Position, and Custom Properties.
- Display built-in or custom stencils or drag shapes from stencils.
- Cut, copy, or paste content. This includes the Paste Special command and placing content on the Clipboard.
- Access or assign shape properties or custom properties.
- Import or export data.
- Use Microsoft Visual Basic for Applications (VBA) and the Visio object model.
- Enable existing macros or create new macros.
- Access the ShapeSheet window.

Maintaining and Repairing Visio

After your initial installation of Visio, you can uninstall Visio, add or remove Visio-specific features, or reinstall features that aren't working properly. The Add or Remove Programs tool within the Windows Control Panel is the place to go to do all these things. To access the Add or Remove Programs window, click Start to open the Windows Start menu and then choose All Programs ⇨ Control Panel. In the Control Panel window, double-click Add or Remove Programs.

 To uninstall Visio completely, in the Add or Remove Programs window, in the Currently Installed Programs list, click the Remove button for Microsoft Office Visio.

Adding and Removing Components

To add or remove features, in the Add or Remove Programs window in the Currently Installed Programs list, click the Change button for Microsoft Office Visio. The Setup wizard appears. Adding or removing specific Visio features is a lot like a customized installation. In the Setup wizard, select the Add Or Remove Features option and click Continue. The next screen that appears looks exactly like the one in Figure 38-1. To change the installed features, follow the instructions in the section Customizing Your Installation earlier in this chapter. After you have made all your selections, click Continue to initiate the reconfiguration. When the installation is complete, click Close.

Repairing Your Visio Installation

If Visio begins to act strangely — really strangely — repairing the installation is often the best solution. This approach corrects small problems with the installation, such as missing files and registry settings.

 If repairing the installation does not fix the problem, try reinstalling Visio.

To repair your Visio installation, in the Add or Remove Programs window in the Currently Installed Programs list, click the Change button for Microsoft Office Visio. The Setup wizard appears. In the Setup wizard, select the Repair option and click Continue. While you watch the progress bar inexorably move from left to right, Visio analyzes your installation and repairs any problems it finds.

Summary

Whether you want to install Visio on one computer, a few, or hundreds, there's a method that makes installation easy. When you install Visio on your own computer, you can keep a copy of the installation on your hard disk so you can easily repair or reconfigure your installation without having to find your Visio CD. The Visio Setup program can also add or remove Visio features, repair your Visio installation, or reinstall or uninstall the software.

In addition to the installation process, you must activate your copy of the software to use all its features. Activation doesn't require any personal information; it is required solely to verify that you are installing a legal copy of the software.

Chapter 39

Visio Help Resources

So many Visio features are intuitive; you could go for months without needing any help at all. Even so, the Visio shortcuts and special features make it worthwhile learning about what Visio has to offer. And once in a while, even an easy-to-use program like Visio can have you scrambling for a second opinion.

This chapter starts with explaining how you can make the most of the help that is available in Visio. You'll also learn about additional online resources including web sites and Visio blogs. Finally, you'll discover how to use message boards to obtain answers from Visio experts as well as your diagramming peers.

Finding Microsoft Visio Help

When you access Visio help by any means, Visio takes you to the Microsoft online help by default. Online help provides the most up-to-date information, but timeliness might take a back seat to raw speed, if you have a slow Internet connection. If you are offline or use your dial-up connection sparingly, Visio can search the help topics stored on your computer until you tell it to go online.

Accessing Help Quickly

If you need a hint about a shape or a toolbar command, ScreenTips are faster than help topics. To display ScreenTips, use one of the following methods:

- For a shape ScreenTip, pause the pointer over a master in a stencil. To obtain more information about the master, click More in the ScreenTip dialog box.
- For a toolbar ScreenTip, pause the pointer over the toolbar icon.

Viewing Visio Help Topics

Only the sleep-deprived are likely to read help topics for a diversion. When you get stuck and don't know what to do, you look for help about the issue at hand. Visio Help is often the first place you look. It's easy to access, but it's typically helpful only when you need to know how to do something. If you want to learn what a feature is for or need to know the limitations of a tool, read the other sections in this chapter to learn where to find more in depth answers.

Here are several ways to obtain Visio help and the benefits of each:

- **Help in dialog boxes** — The Help icon that appears in most dialog boxes opens Visio Help directly to information about the features, options, and buttons in that dialog box. In many cases, your reaction might border on "Please tell me something I don't know." However, if you want to know what things do before you click them, these Help topics sometimes clarify the result of selecting an option or checking a check box.

 If the Help topic explanation doesn't shed any light on the matter, remember you can experiment and choose Edit ➪ Undo (or press Ctrl+Z) if you don't like the results.

- **The Type A Question box** — At the top-right corner of the main Visio window sits a text box for typing questions, shown in Figure 39-1. Whether you type keywords or a fully formed question, pressing Enter opens Visio Help to search results related to the text you typed.
- **The Help menu** — To open the Visio Help window to an index of help topics, so you can browse through topics, choose Help ➪ Microsoft Office Visio Help, shown in Figure 39-1 or, from the Standard toolbar, click the Help icon.

FIGURE 39-1

Ask a question or browse through Visio help for answers.

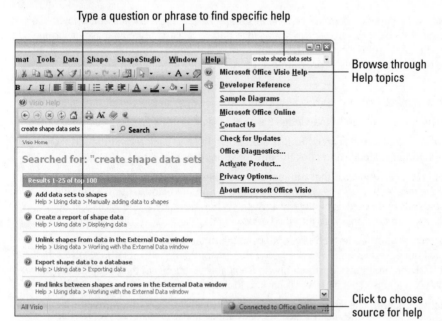

Type a question or phrase to find specific help

Browse through Help topics

Click to choose source for help

Telling Visio Where to Go for Help

Most of the time, online help is the best option. The help is up-to-date and you can expand your search for relevant information to other online areas. However, when you're crammed into economy class on an airplane or without Internet access for other reasons, you can tell Visio to display the help topics installed on your computer. To change your source for help, click the Connected to Office Online box in the bottom right corner of the Visio Help window, shown in Figure 39-1. On the Connection Status drop-down menu, a check mark appears to the left of Show Content From Office Online by default. To switch to offline help, click Show Content Only From This Computer.

Searching the Microsoft Knowledge Base

If Visio doesn't behave as you expect or your question stumps Visio Help, the Microsoft Knowledge Base is a good place to look for answers. The Knowledge Base is a searchable database of support articles that are particularly helpful for troubleshooting error messages, odd program behavior, or results that you suspect are the results of program defects. To search the Microsoft Knowledge Base, follow these steps:

1. In your browser, navigate to `http://support.microsoft.com`.

2. In the menu bar, click Search Knowledge Base.

3. Scroll to the bottom of the Search Product drop-down list and click More Products. Scroll down the Product List page and click the link for your version of Visio.

4. In the For box, type keywords that summarize your question or issue. To increase the relevance of the results, type words you would expect to find in an answer to your problem. Don't use words that could apply to any answer, such as *how*, *why*, and *is*. To find help on errors, copy an error message into the For box.

5. To further clarify your request, choose options in the other boxes, such as Using, which specifies whether to search for all your keywords, any of them, an exact phrase, or more sophisticated combinations constructed with Boolean operations. In the Modified drop-down list, you can also specify how far back to look.

6. In the Include section, check the check boxes for the different types of information to search, such as How To Articles, Troubleshooting, Office Online Content, MSDN Articles, and TechNet Articles.

7. Click Search to find results.

Microsoft Office Online

When you want more than a small serving of help, Microsoft Office Online offers a smorgasbord of help topics, online training courses, downloadable templates, clip art, articles, demos, discussion groups, and more. Microsoft Office Online (in Visio, choose Help ➪ Microsoft Office Online) got a face lift to go along with Microsoft Office 2007. Unlike its predecessor, the new Microsoft Office Online web site has a tabbed navigation bar for accessing the different categories of features it provides. However, when you're looking for features specific to Visio, the best approach is to click the Products tab and then, in the left navigation bar, click Visio to open the Microsoft Office Visio web page, shown in Figure 39-2.

FIGURE 39-2

The Microsoft Office Visio web page provides access to Visio help, templates, articles, discussion groups, and more.

Here are features that the Microsoft Office Visio web page offers:

- **Search** — When you click the Visio link, the search section at the top of the page is tuned to Visio. The label says Microsoft Office Visio and the default text in the search box is Search Visio 2007. Type keywords in the box and click Search to find any kind of content related to your search terms. If you want to change the version of Visio for the search or the scope of the web pages, click the down arrow on the Search button and choose the version (such as Visio 2003) or the web area (such as All Office Online or All Microsoft.com).

- **Help and How-to** — In the left navigation bar, under Help and How-to, click one the Help links to browse the help for a specific version of Visio (as far back as Visio 2000). This help is the same online help that you find when you choose Help ➪ Microsoft Office Visio Help within Visio.

- **Templates** — Clicking the Templates for Visio link takes you to a web page with a broader selection of templates than in the past. Besides templates in different categories, such as calendars, you can click links in the left navigation bar to peruse templates submitted by other Visio uses. Or, if you're the proud parent of a Visio template, click Submit a Template to post your creation.

- **Technical Resources** — If you're expanding into advanced customizations, in the left navigation bar, under Technical Resources, click Developer Center to navigate to the Visio Developer Portal on the MSDN web site. You can download the Visio SDK or code samples, or read about integrating Visio with other applications using the Visio ActiveX Control.

NEW FEATURE The new Microsoft Office Online enables you to limit what you see on the Help and How-to web page to help for the Office programs you have installed. Microsoft Office Online can detect the programs on your computer and build the list. To build a customized list of programs, click the Help and How-to tab. On the Help and How-to web page, in the Browse Help and How-to by Product area, click Edit List and select the check boxes for the programs you use.

- **Work Essentials** — In the left navigation bar, under Additional Resources, click Work Essentials to read articles specifically for your occupation. To increase your Visio productivity, obtain products from other companies or find a solution provider to help you customize Visio.

- **Discussion Groups** — In the left navigation bar, under Additional Resources, click Discussion Groups to open the Office Discussion Groups home page to topics within the Visio General Questions category. You can post questions and obtain answers from other users and Visio MVPs. As with any discussion group, you stand a better chance of receiving a helpful answer if you clearly state your question and provide background to help readers understand what you're trying to do or the problem you're experiencing.

- **Training** — The Getting Started Tutorial that used to appear on the Visio Help menu in earlier versions is no longer available. However, online Visio training courses cover the basic concepts to courses about using specific templates, such as organization charts. Click the Help and How-to tab and, under Help Resources, click Training. On the Training page, in the Search box, type Visio and click Search to see all the Visio training courses available.

Visio Blogs

Blogs (Web logs) have become a popular way to disseminate information, for individuals and corporations alike. Part of their appeal is the informality of the posts, regardless whether an individual publishes the blog or it's sponsored by Microsoft itself. Visio boasts several blogs. Although some of them are definitely developer material, a few discuss topics that every Visio user can put to use. Here are the blogs you might want to keep tabs on or add to your RSS (Real Simple Syndication) feeds.

- **Visio Insights** — This blog (`http://blogs.msdn.com/visio/`) is authored by members of the Microsoft Visio team. For people who love to understand what makes a program tick, this blog launched in May 2006 with a series of posts that explain in detail the features that make Visio so easy and powerful. It also holds a distinguished seat on the Microsoft Office Visio web page on Microsoft Office Online. To the right of the search section, click the Visio Team Blog link to open the Visio Insights web page.

- **Visio 12 - Eric Rockey** — Eric Rockey is the Lead Program Manager for the Visio team, so his blog (`http://blogs.msdn.com/eric_rockey/`) focuses on the new features in Visio 2007 (aka Visio 12).

- **Visio - The Blog** — The Microsoft Most Valuable Professional web site (`http://msmvps.com`) sponsors a Visio blog (`http://msmvps.com/blogs/visio/`). This blog runs the gamut of topics, mainly because the authors scan the internet for Visio tidbits and write about what they uncover. Topics include discussions of Visio background pages versus headers and footers, custom line patterns, and using the Macro Recorder.

- **Bill Morein's WebLog** — This blog (`http://blogs.msdn.com/wmorein/`) by a Visio Program manager focuses on Visio data and how to present it, whether you use built-in tools, such as Data Graphics, or delve into development.

- **Chris Castillo's WebLog** — Chris's blog (`http://blogs.msdn.com/chcast/`) focuses on developing solutions using the Visio development tools.

Summary

Help for Visio comes in many forms. Within Visio, you can open Help topics by clicking the Help icon wherever it appears in the program. The Ask a Question box at the upper right of the Visio page opens Help topics related to the text you type. Or, you can choose entries on the Visio Help menu to open Visio Help, Microsoft Office Online, or the Developer Reference. Once you navigate to Microsoft Office Online, there's a whole world of features to help you use Visio more effectively.

Chapter 40

Additional Resources for Templates and Stencils

Visio comes with dozens of templates and thousands of shapes, and yet, you might yearn for other templates and stencils that better fit your requirements. Microsoft Office Online has become a rich source of templates and shapes, from both Microsoft and Visio users around the globe.

Even with all of the Microsoft Visio resources, for highly specialized diagrams, you often need to look elsewhere, for example, when you want a template for diagramming with Juniper Networks equipment or setting up a devilishly difficult dog agility course.

Many companies provide Visio templates and stencils that work with their products. For example, you can obtain Visio stencils for some equipment when you purchase it from the manufacturer. Other companies sell third-party templates for a variety of applications. This chapter shows you how to find the templates and stencils that Visio offers, as well as templates, stencils, and other solutions from third-party resources and web sites.

IN THIS CHAPTER

Finding all-purpose stencils in Visio Extras

Downloading templates from Microsoft Office Online

Finding third-party templates and stencils

Exploring the Visio Extras Stencils

No matter what type of drawing you're creating, the shapes you crave might be as close as the stencils in the Visio Extras category. Many of the most useful stencils are found in this catch-all category. To open one of these stencils, choose File ➪ Shapes ➪ Visio Extras and choose the stencil you want. Visio Extras stencils include shapes for annotations, backgrounds, borders and titles, callouts, connectors, patterns, dimensioning, symbols, title blocks, and more.

CROSS-REF For a complete list of Visio Extras stencils, see Chapter 41.

Starting with Sample Diagrams

The Diagram Gallery, which offered examples of diagrams in almost every Visio template category, is one of the casualties in the 2007 edition of Visio. On the Help menu, instead, is the Sample Diagrams command, which opens the Getting Started window and selects the Samples template category.

It's disappointing that Visio comes with only a few samples, but these samples aren't intended to show you how to create different types of diagrams. Each sample diagram puts several new Visio 2007 features through their paces. For example, the IT Asset Management sample includes a data link to external data. The Sales Summary sample illustrates the power of the new PivotDiagram template.

When you select one of the sample thumbnails, the right side of the window shows a larger image of the sample diagram and two buttons:

- **Open Diagram** — As you would expect, clicking this button opens the sample as a Visio diagram, so you can see how that type of diagram works. For example, the IT Asset Management sample includes a link to data and uses the networking template.

- **Open Data** — Clicking this button opens the sample data in an Excel workbook, which is helpful if you want to learn how to import data from another application.

> **TIP** The package of 20 sample diagrams for Visio 2003 is still available at Microsoft Office Online. These samples give you an example of how to put together several different types of Visio diagrams, and you can use them with Visio 2007. To download this file, navigate to `http://office.microsoft.com`. Click the Downloads tab. In the Downloads box, type "Visio 2003 sample" and click Search. Then, click the Visio 2003 Sample: 20 Sample Diagrams link that appears.

Downloading Templates from Microsoft Office Online

With the release of Microsoft Office 2007, Microsoft Office Online (`http://office.microsoft.com`) has expanded the templates it offers. In addition to the templates that Microsoft or its partners create and publish on Microsoft Office Online, you can now obtain templates from your peers, which are called community-submitted templates. For example, you can download templates for kitchen layouts, crime scene documentation, or project timelines. One template from the community is a Visio template for a gift certificate for babysitting.

> **NOTE** In addition to the demise of the Diagram Gallery, Visio 2007 no longer offers other avenues for finding templates. In Visio 2003, the New Drawing Task Pane included links for searching for templates on your computer, on Microsoft Office Online, or other web sites. The New Drawing Task Pane is not available in Visio 2007, so neither are these search methods.

There are several ways to track down templates at Microsoft Office Online, and the best one to use depends on how narrow your requirements are. To obtain templates from Microsoft Office Online, use one of the following methods:

- **Search** — If you're working on a project or assignment and want a template for a specific function, you can look for templates specifically for that purpose. Suppose you want to create a school calendar quickly. On Microsoft Office Online, in the top navigation bar, click Templates and then, in the Templates search box, type *school calendar* and click Search. Initially, the site returns school calendar templates created in Visio, PowerPoint, Word, Excel, Publisher, and OneNote — and in many cases, you can choose the version of the program that you use as well from Office 97 through Office 2007. If you want templates for one product, in the left navigation bar, in the Filter By Product drop-down list, choose the product you want, such as Visio.

- **Visio templates only** — When you are looking specifically for Visio templates, in the top navigation bar, click Products and then in the left navigation bar, click Visio. On the Microsoft Office Visio web page, in the left navigation bar, click Templates for Visio. On the Template Categories page, you can click links for different categories of templates, as shown in Figure 40-1. If a category has subcategories, you can click the top-level category or any of the subcategories. To locate all the templates in a category, click More.

- **Visio templates from users** — You can submit your own templates to Microsoft Office Online or download templates that others have submitted. These templates are often highly specialized, so this area is worth searching if you don't find what you want within the Microsoft offerings. On the Templates page, under Community Templates, click Recently Submitted.

When you locate a template you want to use, follow these steps to download it:

1. In the template list, click the icon for the template or the name of the template. The download page shows a large image of the diagram, the download size, the version of Visio required to use the template, the number of downloads, and the average rating the template received.

2. Click Download Now. Visio opens a new drawing based on the downloaded template, ready for you to modify and save.

> **NOTE** The proud authors of templates can submit them to Microsoft Office Online to share with the world. In the left navigation bar, click Submit A Template. You must sign in. Then, you simply choose the file for your template and submit it. Templates must be created in Office 2007 and be smaller than 2 MB.

FIGURE 40-1

The Templates Web page offers several methods for finding templates to download.

Click any link to find templates for a category,
subcategory, or additional templates in that category.

Check out community templates submitted by Visio users.

Choose a product to find templates for only that product.

Finding Third-Party Visio Templates and Stencils

One of the benefits of Visio shapes is that anyone can create them. Many equipment vendors offer Visio templates and stencils with shapes that represent the equipment they sell. For example, you can build accurate equipment layouts and network diagrams that show exactly where cables connect by downloading and using Visio templates from companies such as Dell, Cisco, and Hewlett-Packard. These templates include shapes that look just like their real-world components and contain connection points where network cables plug into the devices in real life.

In addition, third-party vendors and many service companies sell Visio templates for a variety of applications. If you are looking for a particular kind of template, enter keywords, such as *Visio*, *template*, and *networking* in your favorite search engine.

CROSS-REF The Visio Search for Shapes feature, which sits right in the Shapes window, can look for shapes online. See Chapter 4 for detailed instructions for using Search for Shapes.

The following are a few online resources for finding, downloading, and purchasing templates and stencils:

- **Visio templates assembled by Visio MVPs** — visio.mvps.org includes links to web sites that offer stencils and templates for download or purchase. On the home page, click Visio Download Sites (or navigate to http://visio.mvps.org//3rdparty.html). The web page lists categories in alphabetical order, from Aerobatics to Woodworking, with links to sites under each category heading.

NOTE The Visio MVP web site offers far more than links to stencils and templates.

- **Visiocafe.com** — This web site offers collections of free stencils from equipment manufacturers and individuals alike. For example, you can download stencils from HP, IBM, Sun, and EMC, or opt for Bruce Pullig's templates for Sun equipment.

- **Shapesource.com** — This site, hosted by Visimation, sells stencils for numerous applications such as manufacturing, firefighting, restaurant management, and biology.

Summary

Microsoft online resources offer lots of additional templates and shapes. In addition, thousands of third-party vendors publish templates and stencils for a variety of specialized applications, some for free and others for a fee. Individuals and companies also offer templates and stencils, which you can find simply by searching the Internet.

Chapter 41

Keyboard Shortcuts

Keyboard shortcuts increase your productivity and reduce fatigue. With keyboard shortcuts burned into your brain cells, your hands remain rooted to the keyboard, saving the time it takes to switch between keyboard and mouse or to move the pointer to choose a command. The time savings really add up if you use keyboard shortcuts for frequently used tasks, such as zooming in and out or saving files. Switching between keyboard shortcuts and using the mouse gives your fingers and muscles a break during long work sessions.

Finding the spare brain cells in which to store the inscrutable key combinations is another story. Fortunately, Microsoft programs share many keyboard shortcuts; for instance, those for using the Help window or moving around in windows and dialog boxes. This chapter explains how to use keyboard shortcuts and identifies some of the more useful keyboard shortcuts built into Visio.

Using Keyboard Shortcuts

Many commands and tasks have their own keyboard shortcuts, whether it's a function key, such as F1 to open the Help window, or some combination of Ctrl; Shift, and letters of the alphabet, such as Ctrl+S to save. When a command has an associated keyboard shortcut, any menu entry for that command shows the keyboard shortcut after the command name. In Visio, choose the File menu and you see that the keyboard shortcut for the Open command is Ctrl+O.

NOTE   Keyboard shortcuts in Help topics, menus, and dialog boxes refer to the U.S. keyboard layout. If the language for the keyboard layout you are using and the one you chose for Microsoft Office Visio or Visio Help are different, the keyboard shortcuts might not be the same.

Another keyboard method for issuing commands on menus and submenus is holding the Alt key while you press the letter underlined in a menu entry. For example, on the Visio menu bar, the F in File is underlined, so you can open the File menu by pressing Alt+F. While the File menu is open and you are still holding the Alt key, you can save your file by pressing S.

NOTE   You can't use Alt and a key for a second-level menu item unless the top-level menu is open. For example, Alt+S works only after you press Alt+F to open the File menu.

Handy Keyboard Shortcuts

Visio provides keyboard shortcuts for almost every command and menu entry, so dedicated keyboardists might forego mousing for all but the most obscure tasks. However, most people can't remember all those combinations of Ctrl, Shift, and Alt. You can still increase your productivity simply by memorizing the keyboard shortcuts for the commands you use most frequently. This section includes a sampling of the more useful keyboard shortcuts.

For a thorough list of keyboard shortcuts, in the Search box in the upper-right corner of the Visio main window, type **keyboard shortcuts**. In the Visio Help window, the first topic you see should be Keyboard Shortcuts for Microsoft Office Visio.

File Shortcuts

Although file shortcuts aren't unique to Visio, they are some of the commands you use most often.

Ctrl+N	Open a new drawing (File ⇨ New ⇨ New Drawing).
Ctrl+O	Launch the Open dialog box (File ⇨ Open).
Ctrl+S	Save the active drawing (File ⇨ Save).
Ctrl+P	Open the Print dialog box (File ⇨ Print).
Ctrl+F4	Close the active drawing window (File ⇨ Close).

Action Shortcuts

You can quickly undo, redo, or repeat your actions in any Microsoft Office program with the following shortcuts.

Ctrl+Z	Undo the last action you performed (Edit ⇨ Undo).
Ctrl+Y	Redo the action undone by the Undo command (Edit ⇨ Redo).
F4	Repeat the previous action. For example, if you just pasted a shape onto the page, you can paste another copy by pressing F4.

Zoom Shortcuts

Zoom around your drawing with the following keyboard shortcuts.

Ctrl+Shift+Left-click	Zoom in.
Ctrl+Shift+Right-click	Zoom out.
Ctrl+Shift and drag with left mouse button	Zoom into the area you drag over.
Ctrl+W	Zoom to display the whole page.
Ctrl+Shift and drag with right mouse button	Pan window to new position.

Window and Viewing Shortcuts

It's easy to access different views or display different areas of a drawing with the following keyboard shortcuts.

Ctrl+PageDown	Display the next page in the drawing file.
Ctrl+PageUp	Display the previous page in the drawing file.
Ctrl+Tab	Make the next open drawing file the active drawing.
Ctrl+F10	Maximize the active drawing window.
Ctrl+F5	Restore the active drawing window size after maximizing it.
F6	Cycle focus through all open stencils, anchored windows, the task pane, and the drawing window.
Print Screen	Copy a picture of the screen to the Clipboard.
Alt+Print Screen	Copy a picture of the selected window to the Clipboard.
Ctrl+F1	Toggle the task pane (View ⇨ Task Pane).

Editing and Formatting Shortcuts

Editing and formatting go much more quickly if you memorize and use the following keyboard shortcuts.

Ctrl+X	Cut the selection from the active drawing and place it on the Clipboard (Edit ⇨ Cut).
Ctrl+C	Copy the selection to the Clipboard (Edit ⇨ Copy).
Ctrl+V	Paste the contents of the Clipboard to the active drawing page (Edit ⇨ Paste).
Del (Delete key)	Delete the selection (Edit ⇨ Clear).
Ctrl+A	Select all the shapes on the active page (Edit ⇨ Select All).
Ctrl+D	Create another copy of the selected shapes onto the active drawing (Edit ⇨ Duplicate). **Note:** You can copy a shape or text to the active drawing page by selecting it and then pressing Ctrl+D. You can make additional copies by pressing Ctrl+D again. To copy selected shapes or text to another drawing page or another application, use Ctrl+C to copy the selection to the Clipboard. After switching to the other page or application, press Ctrl+V to paste the contents of the Clipboard.
Ctrl+F	Open the Find dialog box (Edit ⇨ Find).
Ctrl+B	Toggle bold on or off.
Ctrl+I	Toggle italic on or off.
Ctrl+U	Toggle underline on or off.

Shape Shortcuts

As you work on drawings, use the following shortcuts to work with and edit shapes.

Ctrl+Drag with the left mouse button	Duplicate shape.
Shift+Left Arrow, Shift+Right Arrow, Shift+Down Arrow, Shift+Up Arrow	Nudge shape to the left, right, down, or up, respectively. Each nudge moves the shape 1/10th of the distance between grid lines.
Ctrl+L	Rotate the selected shape to the left (Shape ⇨ Rotate or Flip ⇨ Rotate Left).
Ctrl+R	Rotate the selected shape to the right (Shape ⇨ Rotate or Flip ⇨ Rotate Right).
Ctrl+Shift+F	Bring the selected shape to the front in shape order.
Ctrl+Shift+B	Send the selected shape to the back in shape order.
Ctrl+H	Flip the selected shape horizontally (Shape ⇨ Rotate or Flip ⇨ Flip Horizontal).
Ctrl+J	Flip the selected shape vertically (Shape ⇨ Rotate or Flip ⇨ Flip Vertical).

F2	Toggle between Text Edit and Shape Selection mode on a selected shape.
Ctrl+1	Select the Pointer tool.
Ctrl+2	Select the Text tool.
Ctrl+G	Group the selected shapes (Shape ⇨ Grouping ⇨ Group).
Tab	Change focus from shape to shape on the drawing page. Visio displays a dotted box around the shape with focus.
Shift+Tab	Change focus from current shape to previous shape on the drawing page.
Enter	Select a shape that has focus. **Note:** To select multiple shapes, hold down Shift while you press Tab to cycle focus to another shape. When the shape you want has focus, press Enter to add that shape to the selection. Repeat for each shape you want to select.
Escape	Clear the selection of or focus on a shape.

Edit Box Shortcuts

An edit box is a field in a dialog box in which you type an entry, such as the name of a new page. The following keyboard shortcuts help you move and select edit box contents.

Home	Move to the beginning of the entry.
End	Move to the end of the entry.
Left Arrow	Move one character to the left.
Right Arrow	Move one character to the right.
Shift+Home	Select everything from the insertion point to the beginning of the entry.
Shift+End	Select everything from the insertion point to the end of the entry.

Summary

Keyboard shortcuts save time and, paradoxically, reduce the finger and arm fatigue that arises from marathon Visio sessions. Visio provides keyboard shortcuts for almost every command it offers, but you can increase your productivity by memorizing just a small subset of the available shortcuts. Even if you don't know the function key or keyboard sequence for a command, you can select it from a menu or submenu by using the Alt key with the letters underlined in menu entries.

Chapter 42

Template and Stencil Reference

Visio comes with dozens of templates perfect for getting a jumpstart on diagrams for work or play. Visio Standard 2007 provides templates for common diagramming tasks, such as building block diagrams, documenting business processes, or creating charts and graphs. You can also create a few basic architecture, engineering, and information technology diagrams, such as office space layouts or basic network diagrams.

For the whole enchilada of Visio specialized diagrams, you want Visio Professional 2007. The Professional version includes most common building plan diagrams, including electrical, HVAC, and mechanical plans. For information technology, you can create diagrams to document databases, networks, software, and web sites using many of the popular modeling methodologies.

Visio automatically opens stencils associated with templates. However, any stencil is fair game if you want to add other shapes to a drawing. This chapter identifies the templates in each Visio category, the stencils these templates use, and the chapter of the book that describes how to use them. In addition, this chapter provides a list of the stencils available in the Visio Extras stencil category.

Templates for Office Productivity

Visio templates for office productivity run from the mundane block diagram to specialized diagrams for business process improvement methodologies such as Total Quality Management. Table 42-1 shows the templates for office productivity and business process improvement.

TABLE 42-1

Office Productivity Templates and Stencils

Category	Template	Stencils	Chapter
General	Basic Diagram	Backgrounds, Borders and Titles, Basic Shapes	12
	Block Diagram	Backgrounds, Borders and Titles, Blocks Raised, Blocks	12
	Block Diagram with Perspective	Backgrounds, Borders and Titles, Blocks with Perspective	12
General, Flowchart, Business	Basic Flowchart	Arrow Shapes, Backgrounds, Basic Flowchart Shapes Borders and Titles	15
Flowchart	IDEF0 Diagram	IDEF0 Diagram Shapes	15
	SDL Diagram	Backgrounds, Borders and Titles, SDL Diagram Shapes	15
Business	Audit Diagram	Arrow Shapes, Audit Diagram Shapes, Backgrounds, Borders and Titles	16
	Brainstorming Diagram	Brainstorming Shapes, Backgrounds, Borders and Titles, Legend Shapes	18
	Cause and Effect Diagram	Arrow Shapes, Backgrounds, Borders and Titles, Cause and Effect Diagram Shapes	16
	Charts and Graphs	Backgrounds, Borders and Titles, Charting Shapes	13
	Cross Functional Flowchart	Arrow Shapes, Basic Flowchart Shapes, Cross Functional Flowchart Shapes (horizontal/vertical)	15
	Data Flow Diagram (Professional only)	Arrow Shapes, Backgrounds, Borders and Titles, Data Flow Diagram Shapes	15
	EPC Diagram	Arrow Shapes, Backgrounds, Borders and Titles, Callouts, EPC Diagram Shapes	16
	Fault Tree Analysis Diagram	Arrow Shapes, Backgrounds, Borders and Titles, Fault Tree Analysis Shapes	16
	ITIL Diagram (Visio Professional only)	Backgrounds, Basic Flowchart Shapes, Borders and Titles, Computers and Monitors, Cross Functional Flowchart Shapes, ITIL Shapes, Servers	16
	Marketing Charts and Diagrams	Backgrounds, Borders and Titles, Charting Shapes, Marketing Shapes, Marketing Diagrams	13
	Organization Chart	Backgrounds, Borders and Titles, Organization Chart Shapes	14
	PivotDiagram (Visio Professional only)	PivotDiagram Shapes	19
	TQM Diagram	Arrow Shapes, Backgrounds, Borders and Titles, TQM Diagram Shapes	16

Category	Template	Stencils	Chapter
	Value Stream Map (Visio Professional only)	Value Stream Map Shapes	16
	Work Flow Diagram	Arrow Shapes, Backgrounds, Borders and Titles, Department, Work Flow Objects, Work Flow Steps	16
Project Schedule	Calendar	Calendar Shapes	17
	Gantt Chart	Backgrounds, Borders and Titles, Gantt Chart Shapes	17
	PERT Chart	Backgrounds, Borders and Titles, PERT Chart Shapes	17
	Timeline	Backgrounds, Borders and Titles, Timeline Shapes	17

Templates for Information Technology

Visio information technology templates help you document, design, and model your IT infrastructure and services. Whether you are working on software, hardware, or networks, you can find Visio templates to assist you. Table 42-2 shows the templates for information technology. Visio Standard 2007 provides only the Basic Network Diagram.

TABLE 42-2

Information Technology Templates and Stencils

Category	Template	Stencils	Chapter
Network	Basic Network Diagram	Backgrounds, Borders and Titles, Computers and Monitors, Network and Peripherals	24
	Active Directory (Professional only)	Active Directory Objects, Active Directory Sites and Services, Exchange Objects	24
	Detailed Network Diagram (Professional only)	Annotations, Borders and Titles, Callouts, Computers and Monitors, Network and Peripherals, Network Locations, Network Symbols, Servers	24
	LDAP Directory Diagram (Professional only)	LDAP Objects	24
	Rack Diagram (Professional only)	Annotations, Callouts, Free-standing Rack Equipment, Network Room Elements, Rack-Mounted Equipment	24

continued

TABLE 42-2	(continued)		
Category	**Template**	**Stencils**	**Chapter**
Network, Software and Database	Conceptual Web Site (Professional only)	Backgrounds, Borders and Titles, Callouts, Web Site Map Shapes, Conceptual Web Site Shapes	23
	Web Site Map (Professional only)	Web Site Map Shapes	23
Software and Database	COM and OLE (Professional only)	COM and OLE	22
	Data Flow Model Diagram (Professional only)	Gane-Sarson	22
	Database Model Diagram (Professional only)	Entity Relationship, Object Relational	20
	Enterprise Application (Professional only)	Enterprise Application	22
	Express-G (Professional only)	Express-G	20
	Jackson (Professional only)	Jackson	22
	ORM Diagram (Professional only)	ORM Diagram	20
	Program Structure (Professional only)	Memory Structure, Language Level Shapes	22
	ROOM (Professional only)	ROOM	22
	UML Model Diagram (Professional only)	UML Activity, UML Collaboration, UML Component, UML Deployment, UML Sequence, UML Statechart, UML Static Structure, UML Use Case	21
	Windows XP User Interface (Professional only)	Common Controls, Icons, Toolbars and Menus, Wizards, Windows and Dialogs	22

Templates for Architecture and Engineering

Visio Professional 2007 covers all the bases with templates for architecture and engineering plans of all kinds. You can build plans with Visio shapes or convert the contents of existing CAD drawings into Visio shapes. Table 42-3 shows the architecture and engineering templates. A few stencils for architecture and engineering don't open automatically for any templates. You can access these stencils by choosing File ⇨ Shapes, choosing the category you want, and then the stencil you want. Visio Standard 2007 provides only the Office Layout, Directional Map, and Directional Map 3-D templates.

TABLE 42-3

Visio Templates and Stencils

Category	Template	Stencils	Chapter(s)
Engineering	Basic Electrical (Professional only)	Fundamental Items, Qualifying Symbols, Semiconductors and Electron Tubes, Switches and Relays, Transmission Paths	30
	Circuits and Logic (Professional only)	Analog and Digital Logic, Integrated Circuit Components, Terminals and Connectors, Transmission Paths	30
	Fluid Power Diagram (Professional only)	Annotations, Fluid Power–Equipment, Fluid Power–Valve Assembly, Fluid Power–Valves, Connectors	30
	Industrial Control Systems (Professional only)	Fundamental Items, Rotating Equip and Mech Functions, Switches and Relays, Terminals and Connectors, Transformers and Windings, Transmission Paths	30
	Part and Assembly Diagram (Professional only)	Annotations, Dimensioning–Engineering, Drawing Tool Shapes, Fasteners 1, Fasteners 2, Geometric Dimensioning and Tolerancing, Springs and Bearings, Title Blocks, Welding Symbols	30
	Piping and Instrumentation Diagram (Professional only)	Equipment–General, Equipment–Heat Exchangers, Equipment–Pumps, Equipment–Vessels,Instruments, Pipelines, Process Annotations, Valves and Fittings	30
	Process Flow Diagram (Professional only)	Equipment–General, Equipment–Heat Exchangers, Equipment–Pumps, Equipment–Vessels, Instruments, Pipelines, Process Annotations, Valves and Fittings	30
	Systems (Professional only)	Composite Assemblies, Maintenance Symbols, Maps and Charts, Switches and Relays, Telecom Switch and Peripheral Equip, Terminals and Connectors, Transformers and Windings, Transmission Paths, VHF-UHF-SHF	30
Maps and Floor Plans	Directional Map	Landmark Shapes, Metro Shapes, Recreation Shapes, Transportation Shapes, Road Shapes	27
	Directional Map 3-D	Directional Map Shapes 3-D	27

continued

TABLE 42-3 *(continued)*

Category	Template	Stencils	Chapter(s)
	Electric and Telecom Plan (Professional only)	Annotations, Drawing Tool Shapes, Electrical and Telecom, Walls, Shell and Structure	27
	Floor Plan (Professional only)	Annotations, Building Core, Dimensioning-Architectural, Drawing Tool Shapes, Electrical and Telecom, Points of Interest, Walls, Shell and Structure	26, 27
	Home Plan (Professional only)	Annotations, Appliances, Bath and Kitchen Plan, Building Core, Cabinets, Dimensioning-Architectural, Drawing Tool Shapes, Electrical and Telecom, Furniture, Garden Accessories, Walls, Shell and Structure	26, 27
	HVAC Control Logic Diagram Plan (Professional only)	Annotations, HVAC Controls, HVAC Controls Equipment	27
	HVAC Plan (Professional only)	Annotations, Building Core, Drawing Tool Shapes, HVAC Ductwork, HVAC Equipment, Registers Grills and Diffusers, Walls, Shell and Structure	27
	Office Layout	Cubicles, Office Accessories, Office Equipment, Office Furniture, Walls, Doors, and Windows	26, 27
	Plant Layout Plan (Professional only)	Annotations, Building Core, Dimensioning-Architectural, Drawing Tool Shapes, Electrical and Telecom, Shop Floor-Machines and Equipment, Shop Floor-Storage and Distribution, Vehicles, Walls, Shell and Structure, Warehouse-Shipping and Receiving	26
	Plumbing and Piping Plan (Professional only)	Annotations, Drawing Tool Shapes, Pipes and Valves–Pipes 1, Pipes and Valves–Pipes2, Pipes and Valves–Valves 1, Pipes and Valves–Valves 2, Plumbing, Walls, Shell and Structure	27
	Reflected Ceiling Plan (Professional only)	Annotations, Building Core, Drawing Tool Shapes, Electrical and Telecom, Registers Grills and Diffusers, Walls, Shell and Structure	27
	Security and Access Plan (Professional only)	Annotations, Alarm and Access Control, Initiation and Annunciation, Video Surveillance, Walls, Shell and Structure	27
	Site Plan (Professional only)	Annotations, Dimensioning-Architectural, Drawing Tool Shapes, Garden Accessories, Irrigation, Parking and Roads, Planting, Points of Interest, Site Accessories, Sport Fields and Recreation, Vehicles	27
	Space Plan	Cubicles, Office Equipment, Office Furniture, Resources	28

Visio Extras Stencils

Several handy stencils are grouped within the Visio Extras category. Many templates open the following stencils, but you can use them to annotate and embellish any type of drawing:

- **Annotations** — Callouts, text boxes, north arrows, drawing scale symbols, and reference shapes
- **Backgrounds** — Graphics and patterns to spiff up any diagram
- **Borders and Titles** — Drawing borders and title block shapes
- **Callouts** — All kinds of callouts shapes
- **Connectors** — All the connectors strewn through other stencils all gathered in one convenient place
- **Custom Line Patterns** — Open this stencil to add its custom line patterns to the Pattern list in the Line dialog box
- **Custom Patterns–Scaled** — Open this stencil to add its custom scaled fill patterns to the Fill list in the Fill dialog box
- **Custom Patterns–Unscaled** — Open this stencil to add its custom unscaled fill patterns to the Fill list in the Fill dialog box
- **Dimensioning–Architectural** — Dimensioning shapes typically used for architectural plans
- **Dimensioning–Engineering** — Dimensioning shapes typically used for engineering plans
- **Drawing Tool Shapes** — A quizzical collection of shapes for drawing specialized constructions, such as tangents to arcs
- **Embellishments** — Shapes that show aesthetic details, such as Greek borders and types of tile
- **Icon Sets** — Shapes for status icons, such as trend arrows, traffic signals, and ratings
- **Symbols** — Shapes for informational signs, such as No Smoking or Telephone
- **Title Blocks** — Shapes for individual components of title blocks, such as Drawing By and Scale

Summary

Visio Standard 2007 provides templates for basic diagramming tasks, but Visio Professional is required for most specialized diagrams. Visio automatically opens stencils when you use a template to create a drawing, but you can open other stencils if you need different shapes.

Index

NUMERICS

E

W

Office heaven.

Get the first and last word on Microsoft® Office 2007 with our comprehensive Bibles and expert authors. These are the books you need to succeed!

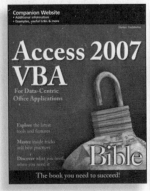
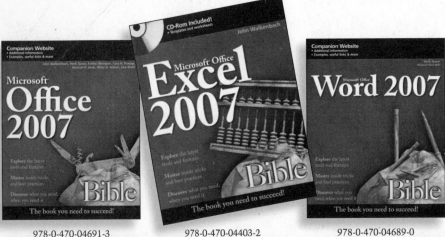
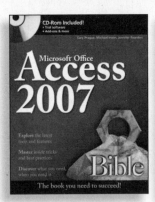
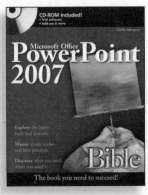
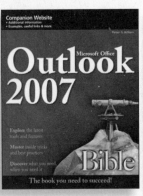